Inventing the Cloud Century

Marcus Oppitz • Peter Tomsu

Inventing the Cloud Century

How Cloudiness Keeps Changing Our Life, Economy and Technology

Springer

Marcus Oppitz
Klosterneuburg
Austria

Peter Tomsu
Leitzersdorf
Austria

ISBN 978-3-319-87018-2 ISBN 978-3-319-61161-7 (eBook)
DOI 10.1007/978-3-319-61161-7

Softcover reprint of the hardcover 1st edition 2017

Printed on acid-free paper

This Springer imprint is published by Springer Nature
The registered company is Springer International Publishing AG
The registered company address is: Gewerbestrasse 11, 6330 Cham, Switzerland

To our wives, Irmgard and Tanja for their patience
And to our families

Preface

The idea for this book was formed in the spring of 2015. We were working for Cisco at that time, one of the large players in network technology. Like every other company in the market, Cisco was on its way to embrace the new opportunities generated by cloud computing and the Internet of Things. Fascinated with bringing together the concept of cloud services and new network architectures to create new business models, we started to work on a model involving different types of ownerships to create a more precise definition of what cloud-based services could offer. The result was a first initial publication; a short summary is part of the chapter entitled "Cloud Computing." Soon we discovered that we had touched the tip of an iceberg. Cloud computing and cloud services seemed to be nothing more than the momentary status of an evolution that was started long ago and that was on its way to change economy, technology, and society in an accelerating and dramatic way.

Both of us had started our careers as engineers in the mid-1970s at the University of Technology Vienna at the time when computing and computer sciences began its journey toward a key technology for businesses. Our working environment was dictated by mainframe computers, by punch cards, and—if you were lucky—by very simple green-screen terminals. After university, we went on different paths in our professional careers. Marcus started to work in the software business building own companies and start-ups. Peter concentrated a great part of his professional life on the development and deployment of new networking technologies and cloud architectures. When we met again, 40 years later, everything had changed completely. Computers went into the background; they became a kind of commodity in your shirt pocket. Networks, the Internet, the Web, and Web-based services had become the driving power for computer science, business, and society. Smart environments using cognitive computing and the Internet of Things had started to disrupt many businesses and industry segments. Digitalization had become a prerequisite for all kinds of organizations or corporations, requiring the acceptance of new technologies but also creating a demand for change and transition of business models. The social and political impact of social media pulled communities into the global village and created many new challenges for politics and media.

Within those 40 years, we had been part of a huge transition starting with the first PCs and networks in the 1970s and moving to the expansion of the Internet, to the revolution triggered by the Web, and to the concept of cloud computing and cloud business today.

Those changes and transitions gained speed over the last decades and seem to point to a future that would be influenced by the economy of cloud-based services. Exploring the path of this evolution and trying projections into the future became a fascinating idea for both of us. There are Terabytes of literature about technological developments, social and political impacts, and the rapidly changing economy. What we had in mind is the interlock of these three dimensions to explore the making of today's cloud ecosystems as witnessed by followers of older service ecosystems that were based on networks. We also wanted to describe the move of services to the cloud and the long-term trend that is still progressing at high velocity. Successful technology is always accompanied by compelling business models and ecosystems including private, public, and federal organizations. Our target was to explore the evolution of service ecosystems, describe their similarities and differences, and analyze the way they created and changed industries. Based on the status of cloud computing and related technologies like virtualization, Internet of Things, fog computing, big data, and analytics, we tried to provide an outlook into the possibilities of future technologies, the future of the Internet, and the possible impacts on business and society moving to the cloud century.

This book is our result.

We address readers like engineers, historians, or economists who are interested in an interdisciplinary view on the history, status, and future projection of the Internet, the Web, and cloud computing. We aimed to connect the technical view with the economic history and the social effects of service ecosystems based on networks. We have tried to follow a storytelling approach, moving along the lines of historical evolution. While sometimes drilling down into technical details, this is not a technical textbook.

Vienna
2017

Marcus Oppitz
Peter Tomsu

Contents

Abbreviations

10Base2	10 Mbps Baseband 200 meter
10Base5	10 Mbps Baseband 500 meter
10BaseT	10 Mbps Baseband Twisted Pair
3G	Third generation of wireless mobile telecommunications technology
3GPP	3rd Generation Partnership Project
4G	Fourth generation of mobile telecommunications standard
5G	Fifth generation of wireless mobile telecommunications technology
AAL	ATM Adaptation Layer
ACE	Automatic Computing Engine
ACI	Application Centric Infrastructure
ADSL	Asymmetric Digital Subscriber Line
ALE	Address Lifetime Expectations
ALU	Arithmetic Logical Unit
AMD	Advanced Micro Devices
ANSI	American National Standards Institute
AOE	ATA over Ethernet
API	Application Program Interface
APIC	Application Centric Infrastructure Controller
ARP	Address Resolution Protocol
ARPA	Advanced Research Projects Agency
ARPA IPTO	Advanced Research Projects Agency Information Processing Techniques Office
ARPANET	Advanced Research Projects Agency Network
AS	Autonomous System
ASCII	American Standard Code for Information Interchange
ASIC	Application Specific Integrated Circuit
ATA	Advanced Technology Attachment

ATM	Asynchronous Transfer Mode
ATM	Automated Teller Machine
AWS	Amazon Web Services
BCF	Big Cloud Fabric
BGP	Border Gateway Protocol
B-ISDN	Broadband Integrated Services Digital Network
BLE	Bluetooth Low Energy
BNC	Bayonett Neill Concelman
BUS	Broadcaste and Unknown Server
CAD	Computer Aided Design
CAF	C++ Actor Framework
CBR	Committed Bit Rate
CCITT	Comité Consultatif International Téléphonique et Télégraphique
CD	Compact Disk
CDMI	Cloud Data Management Interface
CDPI	Control to Data Plane Interface
CEP	Complex Event Processing
CHS	Cylinders Heads and Sectors
CIDR	Classless Inter Domain Routing
CISC	Complex Instruction Set Computing
CLI	Command Line Interface
COBOL	Common Business Oriented Language
COM	Component Object Model
CORBA	Common Object Request Broker Architecture
COTS	Commercial Off The Shelf
CPS	Cyber Physical System
CPU	Central Processing Unit
CRC	Cyclic Redundancy Check
CRM	Customer Relationship Management
CRT	Cathode Ray Tube
CRUD	Create, Read, Update, Delete
CSMA/CD	Carrier Sense Multiple Access with Collision Detection
CSNET	Computer Science Network
DARPA	Defense Advanced Research Projects Agency
DBMS	Data Base Management System
DC	Data Center
DCOM	Distributed Component Object Model
DDS	Data Distribution Service
DEC	Digital Equipment Corporation
DHCP	Dynamic Host Configuration Protocol
DHCPv6	Dynamic Host Configuration Protocol Version 6
DIX	Digital Intel Xerox
DNS	Domain Name System
DOD	Department of Defense

DoS	Denial of Service
DQDB	Distributed Queue Dual Bus
DRAM	Dynamic Random Access Memory
DSL	Digital Subscriber Loop
DVD	Digital Versatile Disc
EBCDIC	Extended Binary Coded Decimal Interchange Code
EC-GSM-IoT	Extended Coverage GSM for IoT
EGPRS	Enhanced General Packet Radio Service
ELAN	Emulated LAN
ERP	Enterprise Resource Planning
ESCON	Enterprise System Connection
ESG	Enterprise Study Group
ESX	Elastic Sky X
FC	Fiber Channel
FCOE	Fiber Channel Over Ethernet
FDDI	Fiber Distributed Data Interface
FIB	Forwarding Information Base
FICON	Fiber Connection
FLOPS	Floating Point Operations Per Second
FPGA	Field Programmable Gate Array
FTP	File Transfer Protocol
FTTH	Fiber To The Home
GC&CS	Government Code and Cypher School
GE	General Electric
GFC	Generic Flow Control
GFLOPS	Giga Floating Point Operations per Second
GIG	Global Information Grid
GMO	Genetecally Modified Organism
GMR	Giant Magneto Resistive
GNSS	Global Navigation Satellite System
GNU	GNU's not Unixe
GPL	General Public License
GPRS	General Packet Radio Service
GPS	Global Positioning System
GRE	Generic Route Encapsulation
GSM	Global System for Mobile Communications
GSMA	Global System Mobile Association
GUI	Graphical User Interface
HA	High Availability
HC	Hop Count
HCC	Homebrew Computer Club
HDFS	Hadoop Distributed File System
HEC	Header Error Correction
HMI	Human Machine Interface

HPC	High Performance Computing
HRMS	Human Resource Management System
HSM	Hierarchical Storage Management
HSPA	High Speed Packet Access
HTTP	Hyper Text Transfer Protocol
IaaS	Infrastructure as a Service
IBM	International Business Machines
IC	Incubation Committee
ICMPv6	Internet Control Message Protocol Version 6
IDC	International Data Corporation
IEEE	Institute of Electrical and Electronics Engineers
IETF	Internet Engineering Task Force
IGP	Interior Gateway Protocol
IIoT	Industrial Internet of Things
Intel VT	Intel Virtualization Technology
IoE	Internet of Everything
IoT	Internet of Things
IOTC	Internet of Things Consortium
IoTSF	IoT Security Foundation
IoTWF	Internet of Things World Forum
IP	Internet Protocol
IPng	IP Next Generation
IPU	Instruction Processing Unit
IPv4	Internet Protocol Version 4
IPv6	IP Version 6
iscsi	Internet Small Computer System Interface
ISDN	Integrated Services Digital Network
ISIS	Intermediate System to Intermediate System
ISM	Industrial Scientific and Medical
ISO	International Standards Organization
ISP	Internet Service Provider
IT	Information Technology
ITS	Intelligent Transportation System
ITU	International Telecommunication Union
ITU-T	Telecommunication Standardization Sector of the International Telecommunications Union
IXP	Internet Exchange Point
KVM	Kernel Virtual Machine
LAN	Local Area Network
LANE	LAN Emulation
LBA	Logical Block Addressing
LCD	Liquid Crystal Display
LDAP	Lightweight Directory Access Protocol
LEC	LAN Emulation Client

LECS	LAN Emulation Client Server
LED	Light Emitting Diode
LES	LAN Emulation Server
LPWA	Low-Power Wide Area
LTE	Long Term Evolution
LTE-M	Long Term Evolution for Machines
LTE-MTC	LTE optimized for advanced Machine Type Communications
LUN	Logical Unit Numbers
M2M	Machine to Machine
MAC	Media Access Control
MAN	Metropolitan Area Network
MAP	Manufacturing Automation Protocol
MAU	Medium Access Unit
MHS	Message Handling System
MI6	Military Intelligence, Department 6
MIPS	Millions Instructions Per Second
MIT	Massachusetts Institute of Technology
MMU	Memory Management Unit
MP3	MPEG-1 and/or MPEG-2 Audio Layer III
MPLS	Multi Protocol Label Switching
MQTT	Message Queue Telemetry Transport
MRAM	Magnetic Random Access Memory
MTU	Maximum Transmission Unit
NAP	Network Access Point
NAS	Network Attached Storage
NASA	National Aeronautic And Space Administration
NAT	Network Address Translation
NBI	North Bound Interface
NB-IoT	Narrow Band IoT
NCP	Network Control Program
NCR	National Cash Register
NDP	Neighbor Discovery Protocol
NFC	Near Field Communication
NFV	Network Function Virtualization
NIC	Network Interface Card
NLRI	Network Layer Reachability Information
NNI	Network Network Interface
NoSQL	Non Relational Structured Query Language
NSAP	Network Service Access Point
NSCI	National Strategic Computing Initiative
NSF	National Science Foundation
NSFNet	National Science Foundation Network
NSX	VMware NSX Network Virtualization
NVP	Network Virtualization Platform

OAM	Operations Administration Maintenance
OCP	Open Compute Project
ODL	OpenDaylight
OEM	Original Equipment Manufacturer
OMG	Object Management Group
ONF	Open Networking Foundation
ONIE	Open Network Install Environment
ONL	Open Network Linux
OPC	Open Platform Communication
OPEX	Operating Expense
OS	Operating System
OSI	Open Systems Interconnection
OSS	Operational Support System
OT	Operations Technology
OVF	Open Virtualization Format
P2P	Peer to Peer
PAC	Programmable Automation Controller
PC	Personal Computer
PCM	Phase Change Memory
PDH	Plesiochronous Digital Hierarchy
PDU	Protocol Data Unit
PFE	Packet Forwarding Engine
PGP	Pretty Good Privacy
PLC	Programmable Logic Controller
PLS	Physical Layer Signalling
PMA	Physical Medium Attachment
PNNI	Private Network to Network Interface
PSTN	Public Switched Telephone Network
PT	Payload Type
PVC	Permanent Virtual Circuit
QoS	Quality of Service
QuAIL	Quantum Artificial Intelligence Laboratory of NASA
RAID	Redundant Array of Independent Disks
RAM	Random Access Memory
RCA	Radio Corporation of America
RDBMS	Relational Data Base Management System
REST	Representational State Transfer
RFC	Request for Comments
RFID	Radio Frequency IDentification
RIB	Routing Information Base
RISC	Reduced Instruction Set Computing
ROM	Read Only Memory
RSA	Rivest, Shamir and Adleman Encryption
SAN	Storage Area Network

SAP	Systems, Applications, Products
SCADA	Supervisory Control and Data Acquisition
SCV	Smart Connected Vehicle
SD	Secure Digital
SDDC	Software Defined Data Center
SDH	Synchronous Digital Hierarchy
SDK	Software Development Kit
SDN	Software Defined Networking
SDS	Software Defined Storage
SIMD	Single Instruction Multiple Data
SLA	Service Level Agreement
SLAAC	Stateless Address Auto Configuration
SMTP	Simple Mail Transfer Protocol
SNA	Systems Network Architecture
SNIA	Storage Networking Industry Association
SNMP	Simple Network Management Protocol
SOAP	Simple Object Access Protocol
SONET	Synchronous Optical Network
SQL	Structured Query Langage
SSH	Secure Shell
SSL	Secure Sockets Layer
STL	Smart Traffic Light
STP	Spanning Tree Protocol
STS	Supranet Transaction Server
STSL	Smart Traffic Light System
STT-RAM	Spin Transfer Torque Random Access Memory
SVC	Switched Virtual Circuit
TCP	Transmission Control Protocol
TCP/IP	Transmission Control Protocol/Internet Protocol
TEPS	Traversed Edges Per Second
TOR	Top Of Rack
TRILL	Transparent Interconnection of Lots of Links
TSN	Time Sensitive Networking
TTL	Time To Live
UCLA	University of California Los Angeles
UDP	User Datagram Protocol
UML	User Mode Linux
UNI	User Network Interface
UNIX	Family of multitasking, muktiuser computer operating systems
USENIX	The Advanced Computing Systems Association
UTF-8	UCS (Universal Character Set) Transformation Format
UUCP	Unix-to-Unix-Protocol
VBR	Variable Bit Rate

VC	Virtual Circuit
VCI	Virtual Circuit Identifier
VDS	vSphere Distributed Switch
VDSL	Very High Bit Rate Digital Subscriber Line
VLAN	Virtual LAN
VM	Virtual Machine
VMM	Virtual Machine Monitor
VNI	Visual Networking Index
VP	Virtual Path
VPI	Virtual Path Identifier
VPN	Virtual Private Network
VSS	vSphere Standard Switch
W3C	World Wide Web Consortium
WAN	Wide Area Network
WiFi	Trademark of the WiFi Alliance for wireless local area networking
WLAN	Wireless LAN
WPAN	Wireless Personal Area Network
WPS	Word Processing System
WSAN	Wireless Sensor and Actuator Network
XEN	Linux Foundation Collaboration Projects
Xerox PARC	Xerox Palo Alto Research Center
XML	Extended Markup Language

Introducing Cloudiness

Our daily life is largely determined by using services and consuming products delivered by companies and organizations far beyond our horizon. The food we eat is produced in remote countries and delivered to our local food store via a complex logistic system, the electricity we use to light our homes and power our tablets is generated by public or private enterprises and delivered via a continent-wide distribution network. Furthermore, the information, media and data we access are created by thousands of different media companies around the world. Most of the goods and services we receive and consume on a daily basis are delivered by technical and economic structures we neither fully realize nor are able to influence or control.

Not long, maybe 250 years ago, everybody knew where, how and who was producing the food, clothing, furniture they were consuming or who built the house they were living in. Within a relatively short time span of a little more than two centuries, technical and economic structures were created which produce goods and provide services to a large and globally distributed society. Triggered by technical progress and industrial revolutions, production was shifted from local, small units to nationwide or global enterprises. The production of goods and the delivery of services moved from the local environment to a faraway structure owned and operated by anonymous organizations.

The general term used today for these economic and technical structures is "cloud" and the number of these different clouds is constantly growing, which made us choose the title "Inventing the Cloud Century," to describe this phenomenon on a general basis. Cloudiness is an increasing phenomenon. In this book, we will focus on several fundamental findings and, as an introduction, we highlight these fundamental assumptions as our starting points throughout the rest of this chapter.

The number of services we consume as users and participants of service ecosystems like water supply, public traffic, electrical power supply, telecommunication, radio and television has been constantly increasing over time. More than two thousand years ago, the first public water pipelines were constructed, but it was not

M. Oppitz, P. Tomsu, *Inventing the Cloud Century*,
DOI 10.1007/978-3-319-61161-7_1

until the ninteenth century that service ecosystems based on network structures became available for the public and brought to a commercial success. Drilling down into those technical and economic success stories shows that the creation of service ecosystems is a major trend. What we see today as "The Cloud" and specifically cloud computing is the progression of a technical, economic and social evolution. This evolution started with the foundation of the first complex structured communities thousands of years ago, and ushered in a much broader effect with the first industrial revolution at the end of the eighteenth century.

In Search for a Better Life

The last 250 years of mankind's history were marked through a dramatic growth in population, life expectancy and economic power. This change was triggered by political, social and technical evolutions. It is a fact that the world population increased by the factor of 10 between 1700 and 2000. Life expectancy more than doubled and the global GDP increased by the factor of 100 within the last 300 years (see Fig. 1).

One major driver behind that development was the fact that service ecosystems took over the responsibility for production, operation and delivery of vital goods and services. This went together with taking over ownership from the consumer. It is a fact that life expectation doubled in North America and Europe since the beginning of the twentieth century while poverty and hunger, though still existing, were reduced dramatically (see Fig. 2).

At the same time the level and number of services that can be consumed today is as high as ever and still growing. Taking life expectancy poverty and hunger as parameters for measuring the quality of life, it seems that service ecosystems are contributing to that trend significantly. Political and social changes in the eighteenth century accompanied the movement towards a more scientific and physical understanding of the world and are related to the precepts of the first industrial revolution. Today we describe these evolutionary steps as industrial revolutions triggered by major technical inventions and leading to a global increase of productivity. The ninteenth century was influenced by the usage of steam for machines and industrial mass production. At the end of the century, electricity was first used for communication and power. The twentieth century became the electronic century, using electronic tubes and later semiconductor technology for communication, mass media and automated industrial production (Chandler, 2005). With the development of digital computers in the 1940s and 1950s, the foundation for a new technical, economic and social change was laid (see Fig. 3).

In the last decade of the twentieth century, the Internet and the World Wide Web started to conquer the consumer market and began to change complete industries by creating new business models, reshaping value chains within companies and forming new types of collaboration between corporations, providers and consumers. It seems that the twenty first century will become the cloud century—

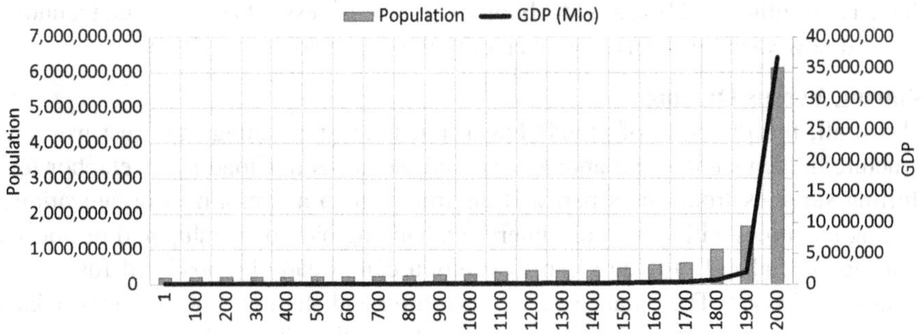

Fig. 1 World population and GDP growth in the last 2000 years

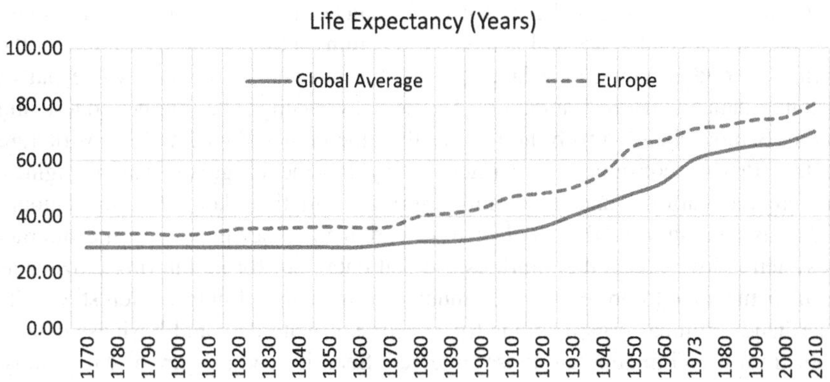

Fig. 2 Life expectancy growth since 1770

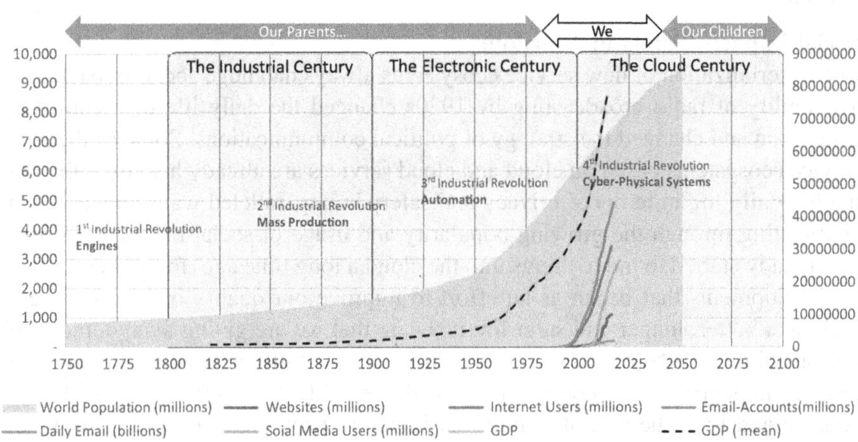

Fig. 3 Population, GDP growth and industrial revolutions

shaped and influenced by cloud technology, new businesses based on that technology and new social patterns empowered by the expansion of the Web.

Sharing Versus Owning

Above all, the principle of clouds has a much richer meaning than just using a structure hidden somewhere above or behind the horizon. Cloud is a metaphor for shifting services from one's personal environment to a centralized organization, sharing resources with other consumers and handing over ownership and responsibility to specialized providers. This evolution can already be observed for many hundreds of years. The balance between owning and sharing tools, devices, locations, and even knowledge is changing towards the direction of sharing. Surrendering responsibility and ownership to a central provider of a service seems to be relevant to survive in modern times.

In tandem, many new challenges arrived. Data became more and more important over the past several years and is already becoming a kind of new currency in the world of cloud services, as a large part of the business is creating value out of customer data and using that data for focused advertising. The Internet advertising market is more than $ 150 billion larger than the cloud service and infrastructure market. Private consumers in particular pay for the usage of search engines, e-commerce platforms and collaboration services with their data. This huge amount of data is also an attractive pool of information for other players in the Internet ecosystem. Governmental authorities use that data pool for monitoring citizens as per their mandate to protect the community in the name of public interest. On the other hand, criminal organizations try to access private data and business data to profit from it and manipulate those involved. Security and privacy of data amidst this connected world is one of the major challenges. Finding the right balance between freedom of information exchange and protection against misuse is difficult to achieve.

Social Impacts, Safety or Freedom

The materialization of new service ecosystems always had huge social impact. The availability of radio broadcasting by 1920s changed the daily life of a complete generation and changed the strategy of political communications. Some of the new service ecosystems based on cloud and cloud services are already heavily influencing our daily life in terms of privacy and safety in unparalleled ways and are even culminating through the growing popularity and usage of social media platforms. We already started to move things into the cloud a long time ago, following a chain of developments that began as an effort to improve our quality of life. We were seeking a better, longer and safer life unaware that we are giving away a piece of responsibility, maybe even freedom. Pushing responsibility into a cloud and giving away ownership to somebody else may, on the one side, free us from responsibility, concern and limit the risk of failure, but it also restricts freedom. Freedom is not absolute and static, it is balance between freedom and safety and it is quite impossible for a society to achieve both at the same time. Having more freedom means living in a society with fewer laws, regulations and executive forces, but always at the risk of lower safety. On the other side a society perfectly organized and kept safe through rules, regulations and executive power to enforce these rules and laws will give its members a high level of safety, but will also limit everybody in their personal decisions. The perfect, safe world for an individual would be prison.

Public, Private or Federal

Many of the new service ecosystems went through a wave-like evolution between private enterprises delivering the service and public or federal organization taking over the responsibility. Service ecosystems like the telephone were a nexus of private entrepreneurs in US, but started as a federal organization in many European countries. The question "Who is the owner?" of a service ecosystems and the underlying network is legendary and, in many cases of service ecosystems, an epic story. We find these movements and effects repeated when looking at information technology, computers and cloud services today. Initially, the first wide area networks were owned by federal organizations, some of them military. Today the Internet has no single owner, it's infrastructure (cables, connections, date centers) is owned by a nexus of private enterprises (the Internet providers) delivering their services to the consumers.

Future Projections

New information and cloud technologies have the potential to rapidly generated new applications and create new branches of applied technologies like the Internet of Things, data analytics or cognitive computing. In addition, this happens in a very quick and sometimes unpredictable way. From a ten-thousand-foot point of view the Information Technology nexus created during the last three decades starts to act like an artificial intelligence presence of its own. The possibilities of how this nexus will change our life and the global economy in the future are numerous. The global availability of information processing services together with other types of service ecosystems will lead to different types of fusions between information, data, power supply, public and private transportation and many other today unknown and unexpected services. It is a given that capable networks and data processing will be core elements of these future ecosystems. It is also clear that we are just at the beginning of even more transitions of economic structures and that we will face much more impact than today on our social life. These developments will gain speed and may lead to more obscurity in how economy and technology influences or supports our quality of life. As we already have experienced the level of understanding economic structures and the dependency on technology and cloudy processes becomes a vital part to further guarantee quality of life. As an extreme scenario, this could lead to the evolution of artificial intelligence, which would completely change not only societies, but also the self-awareness and the role of mankind.

Service Ecosystems: The Five Magic Elements

The basic paradigm behind cloud computing and services is the stepwise creation of service ecosystems based on network structures. These ecosystems offer private users and enterprises the possibility to delegate their demand for services to a public or private provider. What we experience today as cloud computing or cloud services is the current climax of an evolution that started centuries ago. It was triggered by the idea of specialization and division of labor, mainly following the demand for lower costs and higher quality (see Table 1).

What we call cloud today fits with other types of services and how they are delivered to consumers and end-users. Cloud services today are a successor of service ecosystems created and operated for centuries. These types of ecosystems consist of five major elements: a basic, sometimes new technology, a central network structure, a successful business model, a set of accepted technical and legal standards, and a class of standardized end devices. All these elements form a specific type of ecosystem, which has in all cases the purpose of serving a large community of users, providing products or services in a more economic, reliable way and with a higher quality than before (see Fig. 4).

From a bird's eye view the paradigm of what we call cloud services today is based on those five elements. These elements can be found in each service ecosystem. The figure below shows the example of a railway service ecosystem, where the railways are the network, connecting the train stations acting as end systems, using specific provider technologies like different types of locomotives, are based upon standards like voltage or rail width and make use of business models like schedules and tickets for the consumers of the railway service (see Fig. 5).

Network as Binding Element

All types of service ecosystems are based on a type of network. The network is the binding element between providers and consumers of services. It is the basic prerequisite for building and running service distribution structures. This is true for networks supplying water or energy, networks connecting people via telegraph

Table 1 Samples for service ecosystems based on networks

2011	Cloud computing and cloud services
1992	World Wide Web
1980	Internet
1960	Wireless mobile communication
1930	Television
1920	Wireless radio broadcasting
1880	Power and light using electricity
1876	Telephone
1844	Electric telegraph
1832	Railway
1812	Light using gas
1791	Optical telegraph
25 BC	Public transport using horse power
312 BC	Water distribution

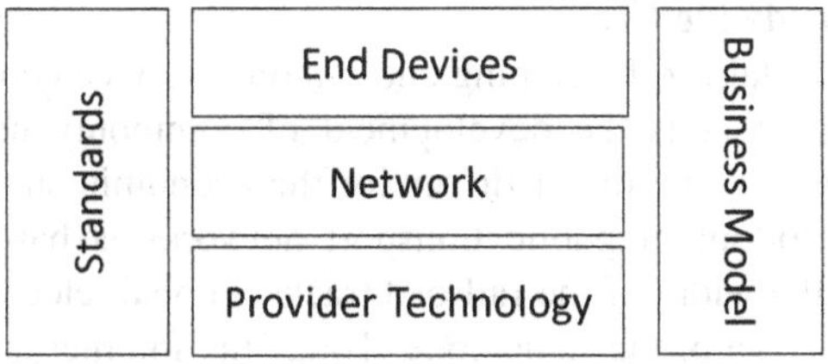

Fig. 4 The five magic elements of a service ecosystem

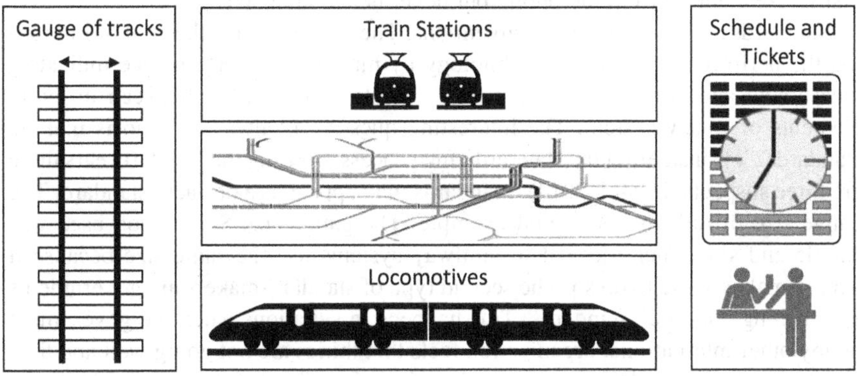

Fig. 5 Railway as a service ecosystems

or phone, networks broadcasting radio or television signals and finally the networks providing information services like those owned by companies, offered by network service providers or the Internet.

New Technologies as Basement

Networks are always based on new technology like the ability to build water pipes, the invention of transportation methods for electricity, the development of railroads or the invention of data communication over wires and, recently, over wireless media. For most of these network technologies it is also true that pure technology alone would not have led to success. Throughout this book, we will look in detail into the networking technologies which form the foundation of the modern communication infrastructure by showing from where these technologies evolved, where we stand today and give an outlook what might happen in the future.

End-Devices for Everybody

The ability of distributing services or products via a network is worthless if there are no standardized and low cost end devices available. Consuming water from a public network is much easier if you can buy water taps in the shop around the corner, lighting our homes demands mass, cheap production of light bulbs, and radio broadcasting became successful by the availability of low cost receivers for everyone. Today the same is true for information services, which are delivered at low cost to a worldwide consumer market where individuals use either personal computers or smart devices.

Standards for Worldwide Use

Another element is vital for developing and building service ecosystems based on network technology, that is the development of commonly used and accepted standards. Standards are the major driver for the economic success of networks. Usage and global rollout of public transport networks is based on a common standard, which is the width of the railroad tracks. Global telegraph networks had been based on the usage of the same code. Today the Internet is based on a set of standards like TCP, IP, HTTP and HTML. The development of these standards is always the second step after the basic technology breakthroughs. Its history is, in many cases, unknown to the public but nevertheless an epic story.

Standards are the followers to inventions. They usually have the goal of improving the performance of a new technology by making key performance indicators measurable and, additionally, they try to create compatibility between different elements of a new system. The interesting question is always: "Who is making standards?". Considering the standardizing processes during the last 200 years there are three suspects (Russel, 2014). Standards may appear as a de-facto standard. The gauge size of railways is a good example. The gauge size Stephenson used was simple and since then most of the railway systems use the same size because it simply makes sense to do so. The second type of standard-makers are governments establishing "de-jure" standards. This happens in situations when the government or any other international organization feels that things are not going well and they should use their legislative power to interfere. Samples for de-jure-standards are regulations for tariff or pricing, security regulations or also regulations referring to the privacy of data (or their use by the government). Lastly, there are a lot of standards, which evolved out of fruitful discussions between engineers and representatives of enterprises. These types of standards are called consensus standards. Many of the technical standards, including most of the standards used with the Internet and the WWW are consensus standards. The sources of these types of standards are mostly associations or unions established by the industry or interested engineers. Well known examples are the ITU (International Telecommunication Union) founded in 1865 by 20 Telegraph companies or the W3C World Wide Web Consortium founded in 1994 by Tim Berners-Lee.

Business Models for Success

Technology and standards for service ecosystems are worthless if there are no business models to create a successful market of providers and consumers. Business models have the simple purpose of creating a win-win situation between those who provide the technology or service and the consumers. Those models may be very different. For example, the first distribution networks for water had a simple business model. In ancient Rome, the senate simple wanted to reduce the risk of illness for the citizens. Fresh water was supplied to the public at no costs, improving health and—of course—the satisfaction with the government consolidating the power of the ruling class. It was not until the ninteenth century that more economically driven business models did evolve. The public railroad networks and later the first telegraph networks are the first examples of entrepreneurs investing in building ecosystems and selling their services to the public (see Fig. 6).

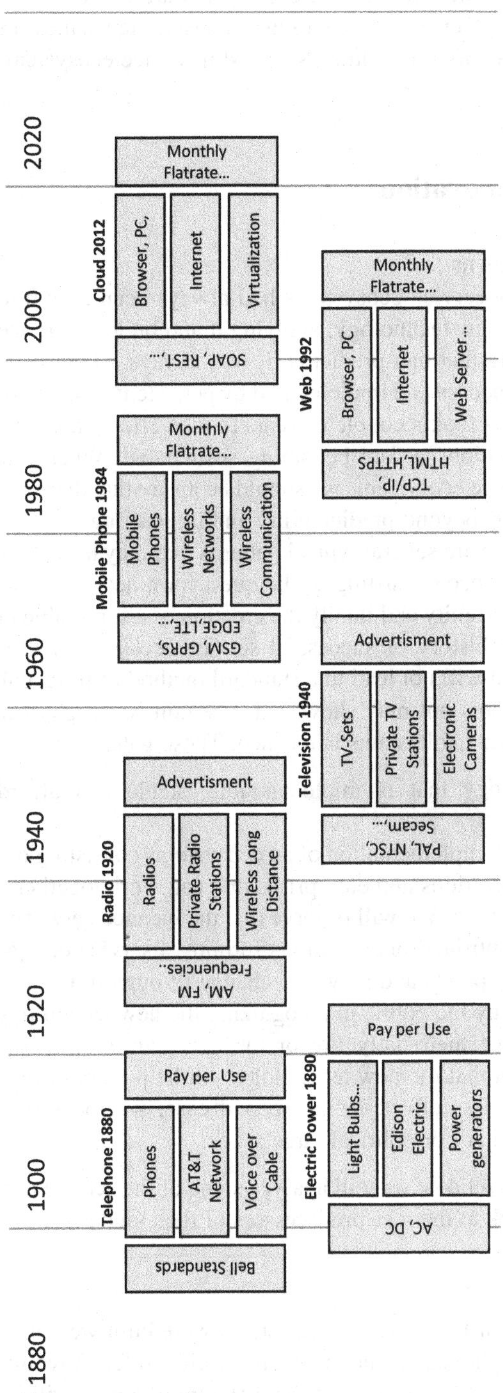

Fig. 6 Service business ecosystems

Today business models for service ecosystems are complex constructions that usually undergo fast changes. The creativity in designing business models is one of the major success factors for building successful service ecosystems.

Creation and Innovation

Timelines and Patterns
The creation of new service ecosystems had always been an unforeseen event in history. Inventing a new technology, applying it and building an operable business ecosystem is nothing that can predicted, it was always a kind of magic moment, triggered by genius people and implemented by persistent business people. Some of these magic moments took a couple of years to take effect and some of them went through several iterations before becoming sustainable. When talking about the creation of new service ecosystems we should be aware that these events are "black swans" (Taleb, 2007) beyond predictability and forecasting.

Nevertheless, there are several typical patterns that they have in common. Those patterns form a sequence consisting of the quest for a new technology, the acceptance by a large community and finally the creation of a survivable business model. Thus, analyzing the history of successful service ecosystem and following their track of development will not lead to a standard method of producing new innovative service ecosystems but may show that a certain set of gates must be passed through before a sustainable status is reached. These gates are:

(1) A new technology that is implementable, stable and affordable in mass production.
(2) The creation and implementation of an ecosystems consisting of private, public or federal organizations and enterprises that can work together in creating and operating the service. We will observe that the balance between private, public and federally controlled organizations forming a service ecosystem may have different starting points and may also change through time.
(3) The acceptance by the public in recognizing the new technology as something that will improve their daily life or the acceptance by a specific group in convincing them that the new technology will help them to reach their targets quicker or with less effort. In the second case, we focus on economic and (sometimes) military or political targets.

In the following sections, we will analyze some of the service ecosystems based on networks and look at them as predecessors of the cloud services we experience today (see Fig. 7).

Players and Roles
Following the story of how service ecosystems were built we will always find the same characters on the stage playing their specific role. Travelling through the centuries and decades the scenery will be different, the play will be performed in

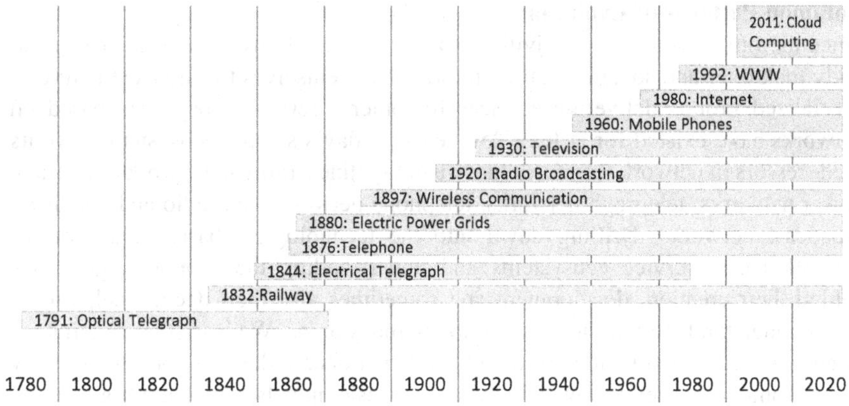

Fig. 7 Service ecosystems based on networks between 1780 and 2020

different theatres around the world and of course we will meet different actors playing or—because its real life—introducing their characters into that specific drama. Analyzing the different players involved in the development and building of service ecosystems is vital to understand the evolution of the technical and economic structures. And, as in every good play, there is always a magician and a hero at the start.

In the beginning, there are the inventors, or people who are simply fascinated by designing and creating something that was never done before. Their target is not primarily to build large economic structures, rather they want to make the world a little bit better. We will meet these inventors as scientists, engineers, technicians or amateurs building astounding things like steam engines, electric telegraphs, computing machines or global networks. Some of them had been working in the ivory tower of science unknown to the public, others were talented craftsmen, who found a technical solution for a specific task and built the first devices. Many of them had not been financial successful, and handed over their invention unintentionally or voluntary to others who had the talent to create a business out of a great idea. These are the entrepreneurs, the heroes of the economy, vital in every society as those who make things happen through the availability of new products and services for everybody. Entrepreneurs can build companies and enterprises sometimes at their own risk, sometimes at the risk of others—the investors. In some very lucky cases an inventor and an entrepreneur join forces to get investors on board by convincing them that economic success is around the corner if they invest some money into the development and production of some great, new service. Like in many dramas, there are also some supporting roles important for the development of the story. In the second act, a kind of background choir appears, taking over responsibility for the proper evolution and integration into economic and social structures. These are public organizations building the legal framework around every new ecosystem as well as standard organizations building sets of rules and practices to make the new technology easier to unfold and to implement cross the borders of companies, countries and regions.

Common Patterns of Evolution

Since the time when the first civilizations were formed, division of labor increasingly took over and societies started to organize themselves towards a structure of specialized groups delivering services for others. Service ecosystems based on networks have existed for at least 250 years. Today's service ecosystems have its predecessors in network structures like water pipelines, railroad networks, electrical power networks, telegraph networks, telephone networks and radio and television broadcast networks. Drilling down into the technological basics and roots of network based service ecosystems and understanding the economic processes behind their creation, development and sometimes also their decline helps us to better understand the long-term evolution in many areas. We learn that patterns are often repeated and situations or trends we experience today are not new in many cases. Sometimes they follow already known basic models, which are triggered by specific initial situations, the status of technology and human behavior. Learning from the past helps to better understand current situations and enables us to foresee future developments.

Many of these technical and economical evolutions followed the same major path as we can see with IT and cloud computing during the last decades. Similarities between the development of information service ecosystems and other service ecosystems are the invention and application of a new technology, the development of first prototypes for a proof-of-concept, the creation of a nexus of enterprises, public or federal organizations for production and service delivery and evolution of technical and legal standards.

Crossing Borders with New Technologies

All the service ecosystems introduced during the last three centuries have something in common, in that they initiated the crossing of a significant border in the way we live and how our society is organized. Public transport networks introduced with the railway in the first half of the ninteenth century had a tremendous effect on the awareness of geography. Within less than 30 years from 1840 to 1870 it was possible for nearly everybody to travel to remote places and, in some cases, to cross continents within a couple of days at comparably low costs. The availability of the international telegraph connections around 1860 and, later, the telephone around 1880 completely changed the way people communicated, reducing the worldwide transport of messages and information from weeks to minutes to seconds. The introduction of broadcasting media radio in the 1920s and television in the 1950s introduced mass communication including its potential effects on education, marketing as well as politics.

All these new technologies and the successive economic structures represent service ecosystems based on their according networks. The invention of a new technology and its transition into an effective service ecosystem has always been a surprisingly new and exciting experience. None of these innovations had been predictable longer than a decade (or maybe two) before the event. At the time when the first electric telegraph was working in 1844, only very few people would have bet on a worldwide communication network 20 years later. Marconi—the inventor of the wireless communication—was considered as lunatic in 1895 by the Italian Post and Telegraph Minister. 20 years later, he had built a group of international enterprises providing and selling worldwide wireless communication systems. In 1990, nobody would have expected that 20 years later a good part of the world population would have mobile access to thousands of strange types of services like social networks, video calling or map and navigation applications, all from a device that fits into a shirt pocket. Technical innovation and the introduction of a new type of service ecosystem is something like a "Black Swan:" unexpected and unpredictable (Taleb, 2007).

New Borders Crossed by Cloud Computing

It already became obvious that information technology, cloud services and social networks were the means to cross many new borders. Due to the nature of the Internet as a basic structure, it evolved to an international global ecosystem soon after its start. When the first Internet connection was established between countries in 1982, the underlying structure became irreversibly global. The second important border crossed is the border between private and public space. No other service ecosystem has changed and is still changing the understanding about private and public space in such a dramatic way. Finally, globally interconnected cloud services contain another important major characteristic. They not only process data, but also allow for the storage of this data, which helped data to became the new currency of this ecosystem.

Cloud Computing has several similarities to other service ecosystems, but it also brought many new and sometimes unexpected economic and social effects with it. In comparison to other, older service ecosystems, the Internet and cloud computing intentionally has a global reach. It is also highly coupled within the network itself and to many other types of services outside the Internet. Today, a growing number of businesses not only use but need Internet and cloud services to develop, produce and provide their products and services. Internet and cloud services are a multipurpose media, they can be used for a growing number of different types of applications like information retrieval, shopping, social contacts, integration of users into business processes and many more. The Internet and the tools used for cloud computing are partly evolving from the open systems movement which is, again, a branch of the open society discussion started in the 1960s. The personal computer and, later, the Internet hype that started in the 1990s also created a new type of entrepreneurs who had the vision to change the world using information technology. Finally, the social impact on privacy and the question of ownership of personal data is one of the major discussions triggered and empowered by the growing usage and demand for Internet and cloud services.

Today Cloud Computing and many related technologies like cloud services, social networks, Internet of things, big data and analytics have gained a major and constantly growing impact on our daily life and trigger a huge change in the general information technology ecosystems. At the same time, we face a huge transition in how we perceive privacy and the borders between private and public space.

Creating Businesses and Industries
Service ecosystems created large, new industries. For some time, the US railroad business was the largest industrial sector after agriculture while the successful connection of the telegraph systems between Europe and US had a huge impact on the economy of both continents. Also, all the service ecosystems went through several developments in building different types of ecosystems. The building of the UK-railroad industry went through dramatic phases, including a stock market crash in 1846, which is comparable to the burst of the Internet bubble in 1999. The information service industry was created in the 1950s and experienced its first major change in the 1980s with the introduction of the personal computer. Soon after that, in the 1990s, an even much more dramatic transition was trigger by the World Wide Web, which is still ongoing. The current focus on cloud computing and services is once again changing all related ecosystems, and creating a plethora of new types of enterprises and consumers.

Information Technology as Logical Successor
Information technology and what we call cloud services today is a logical successor in this sequence. It is based on new technologies (computer, network and virtualization) and went through several partly dramatic technical evolutions and sometimes revolutions. It also created a large nexus of economically connected enterprises, as well as public and private organizations. Like some other service ecosystems (public transport or telegraph) information technology based on the Internet as connecting network moved from federal control to private enterprises operating the infrastructure and providing the services.

Structure of the Book

The book moves along two lines. The historical line will explore the evolution and creation of network based service ecosystems. It will show that our current experience with cloud computing and cloud services is the result of an at least 200-year evolution and will also project the current trends into the future.

The second line focuses on the interlock between technology, economy and social effects. We will narrate the epic stories about technology inventions, the fight for finding standards, and the economic processes creating successful business models and ecosystems, and the propelling effects of social impacts.

In these stories women and men played major roles as inventors, engineers, entrepreneurs and investors. We integrated their personal stories into our stories. Sometimes political situations influenced the introduction and application of new technology and we had to explore the effects on the overall evolution. In many cases, the regional culture of political and economic systems drove the creation of service ecosystems in a specific direction. Some of those early decisions have been changed over the decades; some of them are still defining today's economic structures. We uncovered these roots whenever possible.

Introducing Cloudiness

Our daily life is largely determined by using services and consuming products delivered by companies and organizations far beyond our horizon. The general term used today for these economic and technical structures is "cloud" and the number of these different clouds is constantly growing, which made us choose the title "Inventing the Cloud Century," to describe this phenomenon on a general basis. Cloudiness is an increasing phenomenon. In this book, we will focus on several fundamental findings and, as an introduction, we highlight these fundamental assumptions as our starting points.

A Short History of Service Ecosystems

In this initial chapter, we will focus on the history and development of service ecosystems based on networks. We will start with early evolutions and analyze the first public water supply networks created by the ancient Romans. Moving along the path of history we will see that early public transportation networks that appeared in Europe in the fifteenth century and developed into continent wide service operations by the nineteenth century. With these first service ecosystems, we can observe the evolution of ecosystems and the changing roles between private and federal control as well as the creation of the first business models. With the beginning of the industrial age in the eighteenth century, technology became the major trigger behind new network-based ecosystems. Supported by the innovation of steam technology, the railway became the first exponent of network centric services, leading to a revolution in mobility. The demand for public transport of goods and later persons formed the first, large-scale service ecosystems in Europe and US.

Early Information Network Services

The industrial age was also the starting point of messaging and information networks. Going back in history, we will find the first message networks using optical semaphores in the late eighteenth century in France and UK. The invention of the electrical telegraph triggered the first international message transportation networks. The introduction of the telegraph initiated the creation of large service ecosystems. Some of today's businesses, federal organizations and standard organizations have their roots in that epoch. At the end of the ninteenth century, the electrical telegraph was replaced by the telephone, again creating new service-based ecosystems. With the invention of wireless communication, a completely new technology appeared and, again, started to trigger the formation of new, huge service ecosystems based on networks, as seen in radio and later television. Telephone, radio and television are the predecessors of today's Internet.

Making of Digital Computers

Computers have always been and will remain the heart of information processing, their form factor and base technology change over time, making them useable for every kind of new application that we can imagine. We will tell the story from early computing devices to mainframes, servers, PCs and data centers.

Networks for Sharing and Connecting
Networking enables the necessary connectivity between devices covering small, local to wide distances by providing timely and appropriate technologies for fixed as well as mobile communications. We describe the evolution of local area and wide area networks driven by computing and highlight the usage of the Internet protocol suite for the success and breakthrough of the Internet.

Managing Virtual Storage
Storage Virtualization became the only way to cope with the ever-growing amounts of data for modern corporate IT and to handle the demands of storage re-configurations as well as demanding data migrations. We describe how storage virtualization frees applications from knowing details about storage systems through adding a new layer of software and hardware.

From Physical to Virtual Servers
Server Virtualization addresses the needs for more power in modern data centers resulting from growing server demands and scalability. We show how server virtualization allows for the partitioning of one physical server into multiple virtual machines and we also detail the most common server virtualization models.

Software Defined Virtual Networks
Network Virtualization is used for consolidation of infrastructure without compromising access controls protecting sensitive information.

Virtualization became the foundation and enabler of many new technology trends like cloud computing, Internet of Things, fog computing and big data and analytics and, hence, is one of the central chapters of this book. We describe why network virtualization, also known as Software Defined Networking (SDN), already has a long history going back to the mid 1990s. We also discuss what the actual limiting factors of legacy networking technologies are today and what we understand when we talk about architectures for future networking.

Building the Internet
The foundations of what we know today as the Internet were laid in 1960s by a group of researches in the USA working together in the federally funded Advanced Research Project Agency (ARPA). In the 1970s, several parallel evolutions empowered the making of an international global computer network based on a consensual protocol standard. We will also follow the history of operating system approaches, which led to the industry wide acceptance of the UNIX-family as server operating system, the evolution of standard programming languages and the importance of the open system movement. With the commercialization of the Internet in the 1980s, the foundation was established for a broad distribution of new applications and businesses.

World Wide Web
In the last decade of the twentieth century, the World Wide Web was introduced in addition to the Internet and started its dramatic transition of Internet usage. As a consequence of the "WWW"-introduction, the development of web services

followed as a standardized integration method. By 2007, the iPhone expanded the way consumers interacted with the service ecosystems and created a completely new part of the ecosystem. This also created a new industry that provided apps and service platforms, which delivered services to a larger community, thus creating a mass market. We will give an overview on the development of the market created by the Internet and the World Wide Web and the vibrant changes between 1992 and today. New players appeared and the information technology nexus went through a sequence of major transitions regarding the role of the "old companies" and the enterprises "born on the net."

Cloud Computing
The chapter gives a detailed overview on what we understand as "Cloud Computing" today. It concentrates on the technical levels and on standard definitions. We give brief introductions to the terms "cloud computing" and "cloud services" as they are defined today by describing the major players and roles, for example the consumers, the services and the resources. Following the timeline, we will explore the making of companies and services based on the Web and cloud computing.

Building Cloud Businesses and Ecosystems
Starting with the availability of the Internet and furthermore with the World Wide Web by 1992, completely new types of enterprises appeared. They were called new economy enterprises and were based on new types of business models using two different approaches: the mass distribution of Internet based services and the sharing of capacity and resources between a large group of customers. Since then, these new ecosystems have gone through an epic development of different pricing models, marketing approaches and partner strategies.

Creating Innovation
Innovation is and always was the major driver of economic, technical and social change. Innovation, as an effectual driver of change, seems not only to influence but sometimes also to dictate the transition into new economic and social structures. More than that, innovation became a kind of industry that tries to develop processes that identify, create or accelerate innovation to empower economic growth and change.

Security and Privacy Challenges
With the availability of international networks and with the wide distribution of personal computers, the occurrence of security threads and violations became a mass phenomenon. In parallel to the development of the new economies, a shadow industry of criminal organizations appeared. We will analyze the different types of threads and the technologies used today to break into computer systems and data collections and give an overview on security measures that are used for risk reduction.

Changes in Society and Politics
Like every other major technology leap, the Internet, the web and cloud computing was fueled by economic success and vice versa influenced ecosystems and social

behavior. With the increase of the global adoption of the World Wide Web and the accelerating digitalization of economies and our personal life, a huge challenge for governments and administrations developed in changing and building legal systems for this new world. Our perception of privacy and trust must be redefined. Borders are no longer geographical or political borders, but demarcated by virtual services and regions in the cyberspace.

Internet of Things
The Internet of Things (IoT) describes the connection of devices or physical objects with embedded sensors, actuators and software by networking technologies to interact with the internal states or the external environment. We see how the Internet protocol stack is reused for IoT and how cloud solutions create a new ecosystem of applications and instances of use.

Fog Computing
Fog Computing enables new applications and services by extending IoT and Cloud Computing to the edge of the network with the help of a distributed compute, network and storage infrastructure. These new IoT systems will mainly be self-efficient systems without any direct human interaction and be organized into types of macro endpoints. We will discuss a few of these systems and their requirements like smart traffic lights or smart connected vehicles, smart buildings and smart grid before we dig deeper into the Fog Platform requirements, the edge to cloud (core) relationship as well as the architecture.

Big Data and Analytics
Big Data Analytics describes how massive amounts of data are collected, organized and stored to allow efficient and timely analysis opens new areas of potential disruption in several different ecosystems. We will discuss what really is behind this new trend, what are the promised disruptions and what new ecosystems are inaugural. We will see how this massive amount of data is collected and organized appropriately to allow for efficient analysis, which is referred to as analytics, or big data analytics.

Future Technologies of the Cloud Century
All the technology areas we have discussed throughout this book like computing, networking, the Internet, virtualization, the IoT and big data analytics will see significant changes but also significant improvements, which will form the basis of how our lives might look some decades from now. Among all these evolutions, we will find some of the next big things that will surely revolutionize our future.

New Paradigms and Big Disruptive Things
Peer-to-peer, cognitive computing and quantum computing seem to be the major, disruptive trends for the near and distant future. Given that they are the candidates for disruption in economy, technology and society, we will analyze these three paradigms in more detail.

Arrival in the Cloud Century

All the evolutions discussed in this book will change our lives significantly and some of them will even provide the basis for the next big things to shape our economy, society and technology. Although, innovation is the trigger for improvement, we still carry the responsibility to make the right decisions as individuals as well as a society. We have arrived in the cloud century.

References

Chandler, A. D. (2005). *Inventing the electronic century. Harvard studies in business history*. New York: Free Press.

Russel, A. L. (2014). *Open standards and the digital age: History, ideology, and networks*. New York: Cambridge University Press.

Taleb, N. N. (2007). *The black swan: The impact of the highly improbable*. Prince Frederick, MD: Random House Publishing.

A Short History of Service Ecosystems

Where shall I begin, please your Majesty? Begin at the beginning, the King said gravely, and go on till you come to the end: then stop.

Alice's Adventures in Wonderland; Lewis Caroll

Typing "cloud" into a search engine will deliver more than 1 billion results within half a second. Many of the results refer to terms like cloud computing, cloud industry or cloud services. Within the IT industry, the word "Cloud" together with "cloud computing" and "cloud services" became one of the most used terms over the last 10 years. Since 2005 the number of queries related to cloud computing as an industry increased by a factor of 10 (Google trends "cloud computing"). "The Cloud," cloud computing and cloud services became the new paradigm of computing, based on the idea of sharing computing resources like hardware or software and using the Internet as the communication link and data highway between consumers and providers of IT resources. This increasing trend seems to move along a dramatic hype curve and simultaneously creates countless opportunities and challenges, but the basic idea is not new, even the word "cloud" has been used in science and technology for a long time. Long before the IT industry started to create the concept of cloud computing, clouds were used as a graphical symbol in drawings of technical systems to denote types of network structures that were used by the target architecture but whose internal structure was not visible and hidden in a "cloud." The origin of the term is still obscure but it is also claimed that science used the term "cloud" to describe a kind of nexus of elements far away, observable from an outside perspective, but at the same time not known or understood in its internal structure or mode of operation. Today, all this is true for cloud computing, at least as seen from the consumer or user of what is today called a cloud service.

It is no question that several restrictions must be accepted in parallel. Handing over ownership of tools and production methods to enterprises or public organizations leaves the consumer with a much less power of influence but also with no or very few responsibilities. The fact that middle class people in Europe and North America can make their food selections between more than 100,000 products in a super market, travel to remote places within hours or have access to tons of

M. Oppitz, P. Tomsu, *Inventing the Cloud Century*,
DOI 10.1007/978-3-319-61161-7_2

information and entertainment content within seconds comes at a price. The majority of people today are willing to pay that price, which is maybe a decreasing curve of quality and a changing aspect of privacy. It is also true that power and control over resources and networks has been always a major issue when implementing service ecosystems. What we experience today is the dramatic growth of the World Wide Web, with cloud services on top of it activating and provoking discussions about the freedom of network resources and the question of control and power. In fact, we are shifting more and more activities from "down on earth" to "the cloud" and we have been following that strategy for more than 200 years.

There is a major trend to create and consume products and services out of a "cloud," which was always triggered by new technologies. Implementation of these services offerings was put into effect by entrepreneurs trying to create businesses and ecosystems based on new technology. Development and availability of these service ecosystems was facilitated by the formation of standards created by engineers and standardization organizations. In all cases, the acceptance of service ecosystems was achieved through economic viability and through the desire to consume a type of service that was previously not available. Following this evolution, we must track technological progress as well as the rise of new business models helping to make services offerings available to the public consumers and to enterprises.

Cloud: An Old Concept

Cloud computing and cloud services seems to be a brand-new paradigm related to modern information technology and triggered by what we call today the World Wide Web, but this is fundamentally not true. Using a cloud service today is based on a well-defined agreement between the provider of the service and the consumer of that service. This agreement follows a set of simple principles: users agree with cloud providers about the essential characteristics of the service, the type of service they use and consume (the service model) and finally the level of privacy they want to be guaranteed.

This concept leads to many open questions, sometimes even to discussions and conflicts about the quality as well as detrimental effects of cloud services. None of these discussions are fundamentally new or related to modern IT. It is also true that, pushed by the rapid growth and use of cloud services and their global distribution, these discussions have a large impact on the acceptance of cloud services by organizations and private users. We will drill down into the history and development of services that compare to what we today call cloud services a little later. We will also see that most of the effects we experience today in terms of technology, social impact, payment and business models are not new.

Compared to other industries IT has a relatively short history of only a little more than 50 years. Comparing the development of other services to provider centric ownership models justifies our prediction that IT will follow that same trend. There are several examples of the creation and development of other services. Centralized water supply has existed for more than 2000 years. Public transport and mail was invented 400 years ago by companies like Thurn&Taxis in Europe and—later—Wells Fargo, Butterfield Overland in the US. Worldwide communication services developed in the middle of the nineteenth century using telegraph technology and were powered by the first successful transatlantic cable connection. Energy supply and distribution were industrialized more than 100 years ago by using electric power stations and gasworks together with the connected distribution networks. Music and video entertainment is becoming highly provider centric by using information technology.

Water

The ancient Romans were excellent engineers and besides their talents in building theatres, palaces and a huge empire, they also created one of the first network-based services. Building pipelines was brand-new technology at that time, the service—water for everybody—was based on a network of pipelines using bridges (aqueducts) and tunnels to transport the water dozens of kilometers from the mountains east and south of Rome to the consumer endpoints in the city. We have a complete documentation of this network provided by the Roman senator Sextus Iulius Frontinus in his book "De aquaeductu urbis Romae" (see Fig. 1).

Besides the desire for better quality of life for the ancient Romans technology was also a major trigger. It is true that water pipelines using aqueducts and tunnels had been built much earlier by the ancient Assyrians and Greeks. But it was the Romans who developed a new, basic building methodology to create pipeline networks on a larger scale. One of the key components was the Roman cement (Opus caementicium), which used gypsum, lime and volcanic dust as binders.

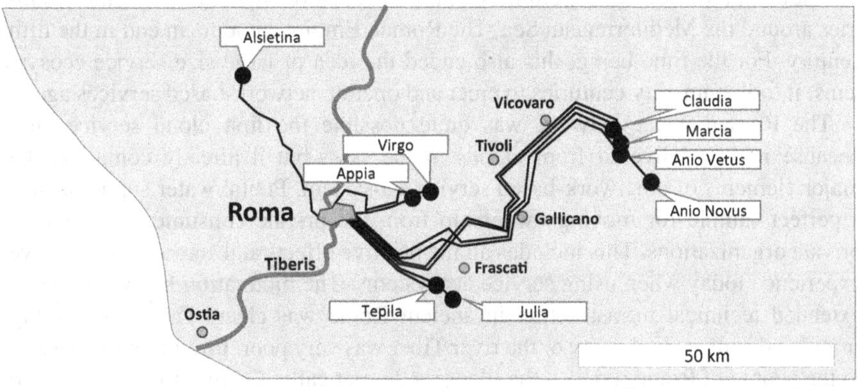

Fig. 1 The courses of the nine aqueducts at the time of Frontinus

Cement was used by the Romans for large buildings including arcs and domes. It enabled the builders to connect blocks of natural stone without exactly shaping the blocks, thus reducing the effort dramatically and increasing the stability of the construction. The magician who invented opus caementicium is unknown, but one of the first descriptions can be found in Vitruvius work "Ten Books on Architecture," 25 BC.

The service was free, as many cloud services today, but of course the rich Romans or the Roman senate, acting as investors for that service ecosystem, expected to receive the votes from the Roman citizens at the next senate election as payback. So, the price did not necessarily involve a monetary exchange, rather the value returned to the investor was delivered via the good conduct of a voting citizen. This business model was quite successful. The first Roman water pipeline was built in 312 BC by Appius Claudius Caecus, also known as the initiator of the "Via Appia," or the first paved "interstate highway" between Rome and Capua. In the following centuries, the pipeline network around Rome was extended. In the first century, a network of nine major supply lines transported water from springs in the mountains over distances of up to 150 km to central reservoirs in the city. A local network of pipes, some of them constructed as pressure pipelines, send the water to more than 1000 public fountains, numerous public baths and to private houses in Rome. Connecting a private house to the public network was subsidized by the government, but the consumers had to pay for the consumption. As with every new technology, people soon started to find ways to use the service to their own ends. Water was drawn off the overland pipelines by farmers to irrigate their land. Within the city, underground pipes were rerouted or illegal pipes were built by private homeowners to redirect water from official, regular pipes for the purpose of getting water without paying the bill. Network security is not an invention of the cloud age!

The provision of fresh water as a public service was a major characteristic of the Roman state and later became a fundamental feature of a modern civilized society. The technology of building water supply networks spread from Rome to all major cities within the Roman Empire. Today we can find the remnants of this huge pre-industrial service ecosystem in Spain, France, Germany and many other countries around the Mediterranean Sea. The Roman Empire came to an end in the fifth century. For the time being, this also ended the idea of large size service ecosystems. It took humanity centuries to erect and operate network-based services again.

The Roman water network was quite possible the first cloud service—not because it was delivered from clouds in the sky—but it already contained the major elements of a network-based service ecosystem. Public water supply is also a perfect sample for moving ownership from the private consumer to public or private organizations. This includes all the positive effects and some of the risks we experience today when using service ecosystems. The motivation for building this extended technical infrastructure in ancient Rome was clear. The water quality from local sources in the city or the river Tiber was very poor. Providing fresh water to the people of Rome reduced the illnesses dramatically. Getting fresh water at no costs and with low effort improved not only the quality of life for the Roman people, but also simple saved personal time and increased the working capacity. Maintaining a service through a public and centralized infrastructure has a huge potential to provide the service at a high-quality level. Getting something like fresh and clean water as a citizen of a community located in a wetland on the banks of a muddy river is otherwise quite impossible or very costly. Sharing these costs with all others to get a higher quality service is one of the principles of service ecosystems. As in every model of sharing resources, one part of the price is the loss of ownership and responsibility. As consumers of a public water supply, you are no longer the owner of your private water source and you even have no influence or responsibility on where your service is coming from and how it is delivered to the endpoint you are using. This also leads to certain risks. Water pipelines can be destroyed, tunnels blocked, the water itself could be poisoned by enemies and out of the reach of the end user. Security measures to guarantee a flawless supply had to be set in place by the public provider. Extending the network to more than 10 supply lines all leading to Rome was not only a question of capacity, but also of redundancy. Distributing the water via large reservoirs storing each of them up to 4 Mio liters of water secured the supply during the summer time.

Public Transport and Postal Services

Beyond the basic needs like shelter, water and food, transportation and message services were another basic requirement for cross border trade and business as well as cultural exchange between individuals. The history of public transport and postal services started 2000 years ago. The service has remained important to businesses and individuals for today, providing a network based ecosystem as part of the daily life as well as an important element for all types of businesses (see Table 1).

The First Transport Networks

27 BC for Federal Use Only

The Roman Empire was also forerunner to the first organized courier service the "cursus publicus." It was introduced by the Emperor Augustus (27 BC–14 AC) to connect Rome to the provinces and Italy. The service was kept alive for more than 500 years until the decline of the Western Roman Empire in the sixth century. The network consisted of the well-maintained Roman road network linking Rome with major cities in Europe and North Africa and is documented in detail in the Tabula Peutingeriana, a large map having its origin in the fifth century and showing all hubs and relay stations (see Fig. 2).

The relay stations provided horses and vehicles for change to guarantee a rapid flow of message relaying and travel. Use of the service was restricted to the

Table 1 Public transport timeline

−10	Roman Cursus Publicus starts operation
500	Cursus Publicus closed after 500 years
1400	
1490	Niederländischer Postkurs between Innsbruck (Austria) and the Netherlands
1500	Niederländischer Postkurs opened for public Service
1500	
1516	Franz von Taxis appointed as Main Postmaster of the Netherlands by Emperor Karl V
1561	The European post network covers connections between Germany, France, Italy and Spain
1596	Leonhard I appointed General Main Postmaster of the Holy Roman Empire
1600	
1661	Carrosses à cinq sols in Paris
1662	"Les carrosses a cinq sols" start operation in Pars
1680	"Les carrosses a cinq sols" is closed due to service level problems
1691	Thomas Neale receives a license for the "North American postal Service" from William III
1700	
1775	Benjamin Franklin appointed as first US Postmaster General
1792	Foundation of the US Post Office Department (USPOD)
1800	
1848	European Post Services taken over by Federal
1850	American Express founded as transportation company
1856	Wells&Fargo founded
1860	Pony Express operated between Missouri and California
1861	Pony Express closed
1900	
1907	UPS founded by James Casey
1919	Yamamato founded as local courier service between Tokio and Yokohame
1946	TNT founded by Ken Thomas
1969	DHL founded by Adrian Dalsey, Larry Hillblum and Rober Lynn
1971	Federal Express (FedEx) founded by Frederick Smith
1989	TNT acquired by Netherlands Post
2000	
2002	DHL acquired by Deutsche Post

government and public administration, a certificate issued by the emperor himself was necessary to use the services. The network covered Europe from Britain and Spain in the West to Greece in the East and extended to the Near East and North Africa (see Fig. 3). At its peak, it spanned 80,000 km.[1]

After the sixth century and the decline of the Western Roman Empire, travelling within Europe took a long time and the transport of goods and messages was

[1]Roman road map on Google maps: http://omnesviae.org/de/

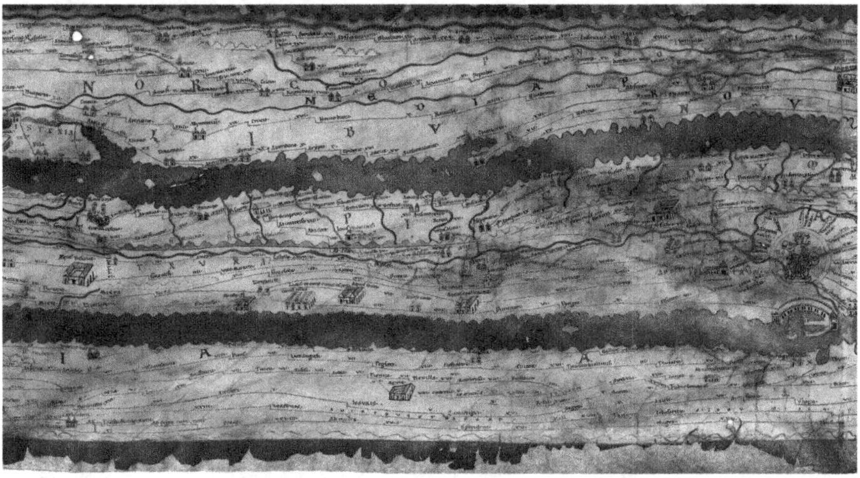

Fig. 2 Copy of the Tabula Peutingeriana from the thirteenth century (by courtesy of the Österreichische Nationalbibliothek)

associated with high costs. The originally well-maintained Roman roads were in a poor condition, interstate trade and communication decreased. With the upcoming new trading routes between Europe and Asia, Africa and—in the fifteenth century—the Americas, global commerce increased again. Trading was in the hand of trading companies, the transport of goods between the centers of production and the major cities were organized by private haulers. Messages were transmitted using private couriers and travelling by horse or coach had to be organized privately.

1490 Reinvention in the Fifteenth Century: The Emperor and the Entrepreneur

With the commercial development of European states and their growing trade volume the demand for transportation and voyages increased together with the demand for communication and message transfer. The first organized message transfer networks were implemented by private trading companies to connect their settlements and trading houses. Frequent courier services were also organized by the Catholic church in the fourteenth century to connect Rome and Avignon with the major dioceses. Private persons sending letters across borders was still either impossible or very expensive. It was the successful cooperation between a ruler and a private entrepreneur that triggered the first public mail service in Europe. Emperor Maximilian I appointed Franz von Taxis and his brother Janetto von Taxis to establish a postal service called the "Niederländische Postkurs" in 1490 between Innsbruck in Austria and the Netherlands. Later it was extended to Italy and Spain. Initially reserved for official mail only, the service was opened to the public in the sixteenth century. From the fifteenth century until the nineteenth century, the family of Turn and Taxis was the major driver and entrepreneur behind many European mail services. The right to organize and run a mail service was in the hands of the

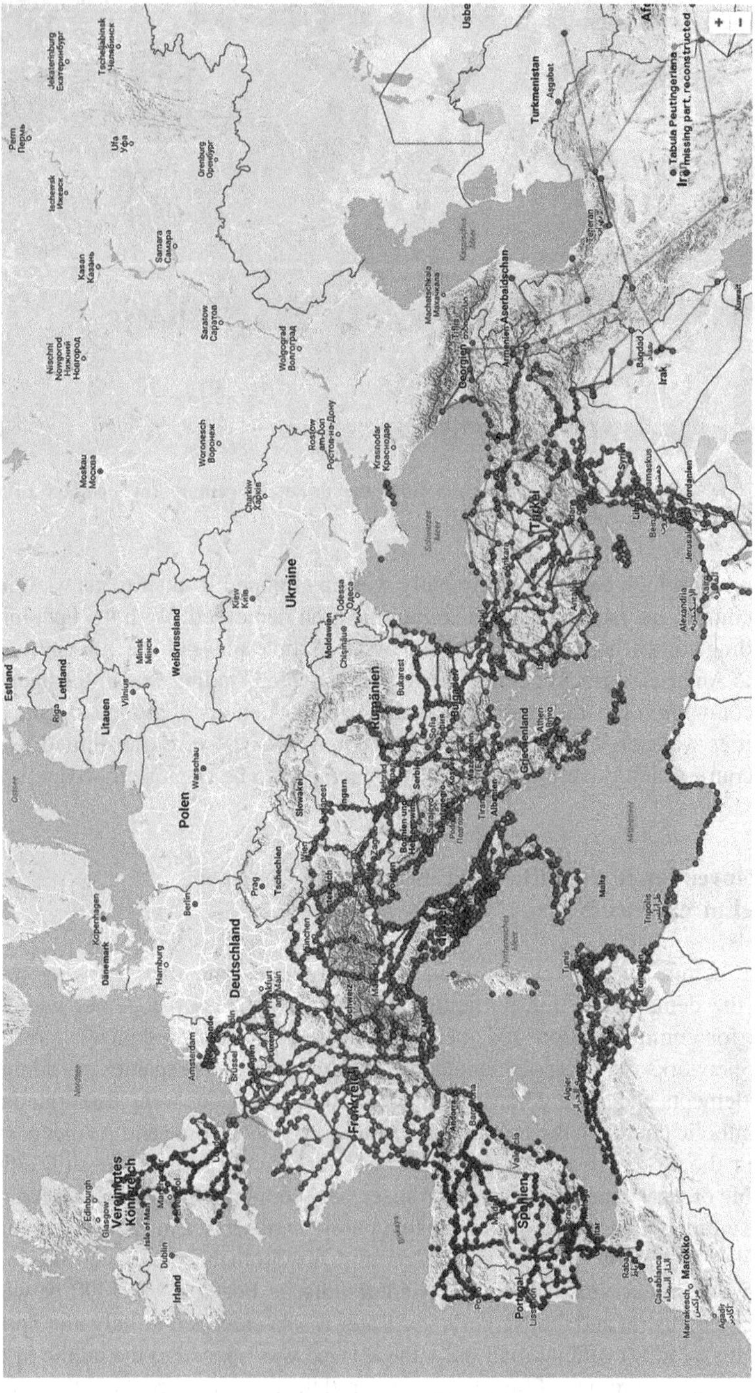

Fig. 3 Roman road network projected into a modern map (by courtesy of René Voorburg and omnesviae.org)

ruler, so many mail services at that time had the name "Kaiserliche Post." The rulers (emperors, kings or princes) had no real interest in organizing the mail service into a business. So, they appointed trusted men or families as postmasters. Franz von Taxis was appointed as Main Postmaster of the Netherlands in 1516 by Emperor Karl V, 1596 Leonhard I. von Taxis was appointed General Main Postmaster of the Holy Roman Empire. In the following centuries, the family of Taxis introduced and expanded the mail services in the countries of the Empire. The word "taxi" has its origin in the name of the Taxis-family. Around the same time official post businesses were founded in other European countries.

Since most of the European countries had been monarchies at that time, the basic right to run a postal service rested with the ruler of the state. In most cases, the ruler—like Karl V as Emperor of the Holy Roman Empire appointed a group of private entrepreneurs or an aristocratic family to operate the service. In the beginning the service was reserved for official mail only, but soon private consumers and businesses tried to use the services as well. At the beginning of the sixteenth century, the "Niederländische Postkurs" was opened for the public. Other postal services followed that approach, fees were calculated depending on the weight and the distance. The fee was collected on receiver's side thus making sure that payment was only collected after successful delivery. With the introduction of the stamp in the mid of the nineteenth century the billing method was standardized and changed to a prepaid service. In the mid of the seventeenth century the post also started to transport packages and people introducing stagecoaches.

1662 The First Metro in Paris: Les carrosses a cinq sols

In the seventeenth century cities began to grow. The population of Paris grew from 200,000 in 1500 to more than 500,000 in 1700. Transport within the comparably small and crowded cities began to become a problem. The French scientist, philosopher, mathematician and poet Blaise Pascal, who also constructed the first mechanical calculator, was also the first to try introducing public transport to the City of Paris. Pascal was motivated by the idea to provide a cheap and easy-to-use local transportation network for the public. Together with the aristocrats Duke de Roannez, Marquis de Crenan, Marquis de Pomponne, in 1661 he applied to King Louis XIV for the approval to organize a local network of carriages in the city of Paris. Approval was given and the private enterprise started operation in March 1662. The fare was 5 sol (les carrousses a cinq sol). The service consisted of five lines, one of them a circle line, and was operated from 7 in the morning until 8 in the evening.

Pascal tried to introduce a reduced ticket price for poor people but the idea was rejected. Later a discussion started about who could use the service and soon certain groups like soldiers and servants were banned from the usage which corrupted Pascal's initial idea about providing a service network for everybody. Discussions about funding, quality and the restriction of usage for certain groups of citizens led to violent demonstrations. Finally, "Les carouses a cinq sol" closed operation around 1680.

International Rollout

1691 First Private Post Service in America

The first postal service was introduced in the American British colonies in 1691 as "North American Postal Service." King William II and Queen Mary II granted a 21-year-license for postal services to Thomas Neale: "*to erect, settle, and establish within the chief parts of their majesties' colonies and plantations in America, an office or offices for receiving and dispatching letters and pacquets, and to receive, send, and deliver the same under such rates and sums of money as the planters shall agree to give, and to hold and enjoy the same for the term of twenty-one years.*"

1775 US Mail

In the United States the postal service (US Mail) has its origin in 1775 when Benjamin Franklin was appointed as the first postmaster general. The first official US post office was created in 1792 as the Post Office Department (USPOD). From that point onwards, the US Mail took over responsibility for the distribution of letters and packages within the growing United States, expanding the network to an organization of over 27,000 local post offices until the middle of the nineteenth century.

1820 Swing Back to Federal Control in Europe

In the nineteenth century the European political landscape was very colorful and Europe partly fragmented in many small and midsize states. Postal services were operated in Austria, Switzerland, many Italian cities and local princedoms and in many of the German princedoms. Those postal services were monopolies based on the approval of the local ruler or the emperor. This led to numerous different, partly regional postal service organizations, covering most of the European landscape. The business model was simple: a monopoly was granted either by the government of the state or the city to a private enterprise. Pricing was in the responsibility of the owner of the monopoly. After the political reorder of Europe between 1815 and 1848, the postal services were taken back from the private organizers and entrepreneurs and reorganized as public services offered and operated by the state itself.

1850 American Express and Wells Fargo

The US Mail Service had started as a public organization and later developed into an ecosystem of public and private companies working together. This led to the development of a second market segment for international postal services (express,

Table 2 Market share of global courier and delivery companies

Company	Founded	Country	Estimated market share 2015
UPS	1907	USA	19%
Yamato Holdings	1919	Japan	5%
TNT Express	1946	Australia	4%
DHL	1969	USA	9%
Fedex	1971	USA	14%
Others:			49%

courier services and parcel services). In the nineteenth century, a parallel ecosystem was created by entrepreneurs recognizing the demand for value added services like money transfers, travel and fast messaging. Some of the first movers were Henry Wells, William G. Fargo and John Butterfield, all of them already in the express industry and, joining forces in 1850, founded American Express. In 1856, Wells Fargo & Company was founded to introduce general forwarding and commissions, buying and selling of gold and freight service between New York and California. The company also participated in the famous Pony Express, which ran an express letter service from 1860 to 1861 between Missouri and California over a distance of 3000 km, reducing the delivery time to 10 days. This nexus of private held companies soon started to cooperate with the official US Mail by taking over freight delivery on the transcontinental routes.

Business and Market Today

Today postal services are provided in a combination between the federal post offices and many of private and global reaching delivery services. They are interconnected by international regulations and international agreements. The market size (revenue) of private global courier and delivery services is around $242 billion (2015), with an average annual growth rate of 0.6%. The six largest companies have more than 50% of the market.[2]

Following the first service providers in the US (Wells Fargo, American Express and others), several companies started operation in the beginning of the twentieth century (see Table 2).

UPS was founded in 1907 as *American Messenger Company* by James E. Casey and renamed in 1919 to *United Parcel Service* (UPS). Yamato was founded in 1919 as local courier service between Tokio and Yokohama. Today, it's the largest delivery service in Japan. DHL was founded in San Francisco in 1969 by Adrian Dalsey, Larry Hillblom, und Robert Lynn. It was acquired by Deutsche Post in 2002. FedEX (Federal Express) was founded in 1971 by Frederick W. Smith in

[2]IBISWorld, Global Courier & Delivery Services, 2016: https://www.ibisworld.com/industry/global/global-courier-delivery-services.html retrieved 2017-01-20.

Little Rock (Arkansas). TNT was founded in Australia in 1946 by Ken Thomas. After the privatization of the Netherlands Post (PTT Netherlands) in 1989, TNT was acquired by PTT Nederland in 1996, headquarters were moved to the Netherlands and later PTT was renamed to TNT Logistics. The courier and delivery market went through many transitions during the last few decades. During the nineteenth century, the volume of letters transported by the different postal services increased dramatically. In UK, the average number of letters per person went from 5 in 1834 to more than 100 in 1920. This trend continued until the end of the twentieth century and ended abruptly with the appearance of the Internet and email. In the first decade of the twenty first century the volume of letters sent via postal services decreased dramatically. The volume of paper mail sent through the US Postal Service has declined by more than 15% since its peak at 213 billion pieces per annum in 2006.[3] On the other side, an increasing volume of B2C (business to consumer) delivery services has led to still growing markets for postal services. This trend is powered by the increasing volume of Internet e-commerce orders using platforms like eBay or Amazon.com.

Social Impact

At the end of the eighteenth century all capitals and major cities were connected via postal services. With this service, sending letters within Europe was possible and the delivery—mostly—guaranteed. Still, the delivery time was not very impressive as it often took days or even weeks to arrive. The average daily distance for messengers riding a horse was not much more than 160 km. This performance could only be achieved through the well-organized changing of horses at relay stations. In any event to get a letter or small package from Berlin to Paris took 1 week, from Rome to London it took 2 weeks. Transport of persons or volume goods using carriages or freight cards took two to four times longer. Depending on the quality of the road, the average speed of a carriage was between 2 and 10 km/h, resulting in a daily average distance of between 20 km and 100 km. Thus, sending letters or freight was not much quicker as in the times of the Roman Empire but it was open to the public and ordinary people. Business correspondence between remote trading partners could be based on a reliable messaging network thus facilitating international trade. Exchange of private letters between European cities was possible and empowered cultural and familial connections.

In comparison to other network based ecosystems (railway, telegraph, telephone), public transport and postal services are not technology driven. Nevertheless, they can serve as long lasting samples for the economic development and social acceptance of network based, service ecospheres. Started as private initiatives by highly motivated entrepreneurs (Emperor Augustus, Blaise Pascal, Franz

[3]Postal Museum, Post Office Statistics, 2016 http://postalmuseum.org/discover/explore-online/history-of/statistics/ retrieved 2017-02-10.

von Taxis, Henry Wells, Williams Wells and others), these figures were empowered by the government, public transport and postal services to create these network-based services. In the middle of the nineteenth century most of the postal service was taken back under the federal control. The availability of a reliable postal service is still seen today as one of the responsibilities of the administration. Nonetheless, several additional services were created by private companies to widen the services offered to the citizens. In comparison to the federal post services, these services are mainly international services like money transfer, international delivery or express post services. Today, the service ecosystem, making up the network of postal messaging and package delivery, consists of a rather harmonic cooperation between the national post organizations and a small number of large international delivery services.

From a consumer point of view, the postal services are a network-based service that takes over all responsibilities from the end consumer and transfers them to an opaque network of organizations, infrastructures and processes. After more than 200 years of postal history, it is normal to give up ownership of a (private) letter or package and hand it over to a mainly unknown group of persons acting in an invisible organization.

At this point of time—at the beginning of the nineteenth century—there was no major difference in technology and delivery speed compared to the first message networks run by the ancient Romans. Still, message delivery was not much faster than 150 km per day and freight was transported at a maximum speed of 100 km per day overland. It was time for something new, it was time for *machines*.

Railway

Since the introduction of the wheel thousands of years ago, nothing had changed by the beginning of the nineteenth century. Speed and limitations of transport were still defined by the quality of the road and the power of the horses. Beginning in the 1920s, the first public transportation network based on a new technology was started. It soon became a worldwide service for logistics and personal travel. Railways went through several dramatic transitions of ecosystems in the US and Europe and created large, new industries that had a huge impact on daily life and business. In the twentieth century, the railway technology was again propelled by the invention of electric transmission (1920) and the introduction of high-speed trains in Europe and Asia (1980). Railway was the first network based ecosystem open for the public and thus had a major impact on society and daily life (see Table 3).

Table 3 Railway time line

1700	
1746	James Watt builds first efficient steam engine
1800	
1821	Royal Assent allows construction of first railway line
1825	First railway line opened between Stockton and Darlington
1825	Locomotive Nr.1. pulled a train composed out of 38 wagons
1830	First US railway line: South Carolina Canal and Rail Road Company
1833	Stockton Darlington Line makes first profit
1840	UK Act for Regulating Railways
1842	UK Railway Clearing House founded as association of railway companies to manage the allocation of revenue
1846	272 new application for railway lines in the UK
1846	Railway stock bubble burst in the UK
1850	Global railway network total size is 38,000 km
1850	Railway network in the US reaches 14,000 km
1850	Land Grant System introduced by the US government for railways
1861	Central Pacific founded
1862	Union Pacific founded
1869	First intercontinental railway finished at Promontory, Utah
1870	Union Pacific goes into bankruptcy
1871	US Land Grant System ended
1880	Railroad industry is the largest employer in the US outside agriculture
1885	Railway network in the UK reaches 30,000 km
1887	The ICC established by the Interstate Commerce Act of 1887
1887	US Interstate Commerce Act to control business activities of the railroad companies
1900	
1900	Global railway network total size is 450,000 km
1914	US government takes over control of the railway companies during WW I
1917	United States Railroad Administration (USRA) established 1917 and became law by the Railway Administration Act in 1918
1920	Federal control ends in the US
1920	Global railway network reaches 1.5 million km
1920	Diesel-engine and diesel-electric transmission introduced
1922	UIC (French: Union Internationale des Chemins de fer) was created on 20 October 1922.
1964	First high speed train "Shikansen" opened in Japan
1981	The French TGV ("Train a Grand Vitesse") starts operation between Paris and Lyon
1990	China plans a high-speed train between Beijing and Shanghai
1991	The German ICE starts operation between Hannover and Würzburg
1992	Union des Industries Ferroviaires Européennes (UNIFE) established as association of railway manufacturers
2000	
2015	China has the longest high-speed-rail network in operation (19,000 km)

Technology 1.0: The Steam Age

1825 The First Railway Line

The transformation from horsepower to machines started with the idea of creating power out of fire. Two major inventions triggered that new development and, in this case, we have the names of the two men who invented and improved a new class of machines and thereby revolutionized technology and businesses. James Watt built the first efficiently working steam engine around 1764 and George Stephenson improved already existing locomotives thereafter. Under his direction, the first operated railroad track between Stockton and Shildon was built and successfully operated. On September 27, 1825 Stephenson's "Locomotive Nr. 1" pulled a train composed of 38 wagons.

The new technology was first used for industrial purposes. Steam engines powered the pumps of mines, the mechanical looms or the forging hammers in iron works. Steam powered locomotives pulled transport vehicles in factories or mines. But soon the new technology was used for public transport. Rails for easier transport were already introduced as wooden rails in the seventeenth century. Later the wooden rails were changed to iron rails and the first railroad tracks were built and operated for goods transportation and persons over short distances. The wagons were pulled by horses.

The introduction of the first railroad track between Stockton and Darlington was a lucky cooperation between Stephenson as inventor and Edward Pease as entrepreneur. Pease was a member of a group of Darlington businessmen who were interested in preventing a new canal project that would not touch Darlington. Instead, they promoted the construction of a "tram road" between Darlington and Stockton following a route that touches Darlington and the local coal industry. Stephenson together with Nicholas Wood, manager of the coal mine where Stephenson had already constructed locomotives, managed to convince Edward Pease to change the design to a real railroad and to plan for an operation using steam locomotives.

Besides the technology, two prerequisites were necessary to create a railway as a successful enterprise: the routing of the track must be approved by the parliament issuing a "Royal Assent" and funding had to be secured. The royal assent was issued in 1821 and allowed for a railway "*that could be used by anyone with suitably built vehicles on payment of a toll, that was closed at night, and with which land owners within 5 miles (8 km) could build branches and make junctions*". Basically, this type of agreement has been used for the approval of each railway routing since then. Funding the whole project was also not easy and the selling of shares to private persons not very successful. Edward Pease took over the majority of shares thus also had influence on strategy and operation, but also carried the most risk. Edward Pease was the venture capitalist and the entrepreneur in one person. At the finalization of the line and beginning of operation, the real cost turned out to be far more than the initial estimation. Three years later—in 1825—the line was opened and started operation, transporting coal and other goods as well as people. The company now named Stockton & Darlington Railway started with debts of BP 60,000 It took 8 years and some of restructuring, mergers and acquisitions before S&D Railway turned a profit.

1840 Shaping the Ecosystem and First Regulations

In the following years, many railways were constructed in England and Scotland most of them connecting industrial centers, cities and harbors over distances not more than 50 km. The first long distance railway was the Grand Junction Railway stretching 132 km between Birmingham and Warrington. A little bit later, a line from Liverpool to Manchester was finished, connected to the Grand Junction railway and forming the first piece of a countrywide railway network. In 1840 railway lines in Britain were few and scattered but, within 10 years, a virtually complete network had been laid down and the clear majority of towns and villages had a rail connection. Until 1840, the government followed a "laisser-faire" approach issuing the approval for a new line without conditions. Starting in 1840, a first attempt at more regulation was made with the "Act for Regulating Railways," introducing the appoint of railway inspectors empowered by the UK Board of Trade.

The network expanded at high speed. Due to liberal political ideologies, it was operated by dozens of smaller or midsized companies owned by private investors. Each railway was a separate company with its own infrastructure, rolling stock and stations. Travelling meant that at each stage of the journey it was necessary to change trains and buy a new ticket.

1842 Building Standards: Gauge, Time and Processes

Stephenson and Pease were a successful team. In the following years, they created a business around railways consulting other groups in planning lines and constructing locomotives. They also introduced one of the most important standards of railways. The gauge used for the Stockton & Darlington line (4 ft 8 ½ or 1435 mm) is still the most widely used gauge for railways: the so-called standard gauge or Stephenson gauge. Many of the new built lines simply used that gauge because locomotives were delivered by Stephenson or built using the same design. According to folklore this was exactly the gauge of Roman chariots used on the ancient Roman railroad tracks in England. It's a good story but it's not true. There is no standard for Roman chariots. We know only about a Roman measure for roads, the "latitudo legitima" which was 2.4 m (8 Roman feet).

Standardization of processes was the next logical step. One of the major promoters of railways in Britain was George Hudson, called the "railway king" of

Britain. He invested in some of the companies and amalgamated numerous short lines. In 1842, he set up the Railway Clearing House as an organization to group railway companies together by providing a system of revenue allocation and management of fare prices paid by passengers for travelling over the lines of other companies. The Railway Clearing House providing uniform paperwork and standardized methods for transferring passengers and freight. In the absence of common regulations, Hudson followed the approach to create own standards, processes and best practices to grow business and reduce costs.

Interestingly, the railway and the railway schedules were a major trigger for the introduction of a global standard time based on defined time zones and Greenwich as the reference longitude for Greenwich Mean Time (GMT). The story goes that in June 1876 the chief engineer of the Canadian Pacific Railroad, Sandford Fleming, missed a train at an Irish train station between Londonderry and Belfast because his printed timetable mixed up p.m. and a.m. Fleming had to spent the night at the train station and began to think about an international time standard. It took him 8 years of convincing governments and lobbying until the International Meridian Conference of 1884 concluded the International Standard Time (Blaise, 2001).

1846 The First Technology Bubble

Shares of railway companies increased dramatically in the 1840s attracting new private investors, many of them from the middle class looking for easy money. In 1846, no fewer than 272 acts of Parliament were passed setting up new railway companies with a total of 15,000 km of new proposed routes. Only two thirds were built, all other companies collapsed before starting operation or were taken over by a competitor. Finally, in 1846, the bubble burst partially triggered by a raise of interest rates by the Bank of England. Investors, many of them middle class families, sunk their complete savings into railway companies. During the 1850s and 1860' the UK railway business saw a slight economic upturn. Around 1885, the British railway network had a length of more than 30,000 km. This was the peak of the network expansion. Until the 1940s the network stayed as this size and then started to decline in length. Today the UK railway network has a total size of around 15,000 km.

1850 Global Rollout and Transcontinental Networks

On an international level, the network was expanded at a high speed. The first railways powered by steam locomotives were introduced in the European countries and the United States. Until 1830, the first railway lines had been erected in Austria, France and the US. In the following decade, most of the European countries had

Table 4 Growth of global railroad network between 1830 and 1883

	Total length of railway network (km)		Growth rate per year (km)
1830	332	1830–1840	826
1840	8591	1841–1845	1767
1850	38,022	1846–1850	4120
1855	68,148	1851–1855	6025
1860	106,886	1856–1860	7748
1865	145,114	1861–1865	7646
1870	221,980	1866–1870	15,373
1875	294,400	1871–1875	14,484
1880	367,235	1876–1880	14,567
1881	393,232	1880–1881	24,515
1882	421,566	1881–1882	28,334
1883	443,441	1882–1883	21,875

Table 5 Length of global railroad network in 1900

	km
Europe	190,134
America	235,016
Asia	19,656
Australia	12,017
Africa	5996
Global	**463,000**

started to create their own network. Between 1830 and 1850 the total worldwide length of railway lines had expanded from 332 to 38,022 km[4] (see Table 4).

By the end of the nineteenth century, the global railroad network spanned a distance of more than 450,000 km. The US-network had expanded to 235,000 km followed by Europe with 190,000 km (see Table 5).

1869 Going Transcontinental

For the US, the new technology was a great gift to conquer the continent. The first line to use steam engines was the South Carolina Canal and Rail Road Company in 1830. The company followed the British railroad technology, initially importing locomotives built in UK and soon designing their own engines by local engineers like Matthias W. Baldwin. Within a short time, American entrepreneurship managed to introduce new technology for transportation at high speed. In 1850, the network had been expanded to 14,000 km. One major trigger for the rapid

[4]Length of Global Railroad Network, Meyers Konversationslexikon. 4. Edition, Vol 5. Bibliographisches Institut, Leipzig 1888, pp. 428–447: https://de.wikisource.org/wiki/MKL1888:Eisenbahn retrieved 2017-01-20.

expansion of railroad networks and enterprises was the land grant system started 1850 and operated by the federal government. The US government did not follow a "laisser-faire" political approach like the UK government, they pushed entrepreneurship by giving new land to railway companies. A total of 129 million acres (516,000 km^2) were granted to railway companies before the program ended in 1871. The companies could either sell the land or pledge it to shareholders. Thus, investing in railway companies was a safe thing. Many new enterprises were founded and started to build new lines across the continent.

The first transcontinental railway was built by two companies: The Union Pacific started to build the track from Omaha to the West, the Central Pacific started in San Francisco heading to the East. The project resulted in a transcontinental challenge where both companies tried everything to grab the larger portion of control over the line. In the end, the meeting of the two tracks happened in Promontory, Utah, on May 10th, 1869. The investors on the western side were the "Big Four" of California: Leland Stanford, Collis Huntington, Charles Crocker, and Mark Hopkins forming the Central Pacific. At the east coast the Union Pacific was founded 1862, collecting money from several private investors. 20 years after the completion of the first transcontinental railway, a network of railway lines crisscrossed the North American continent.

The lines were operated by numerous different companies thus creating not only a colorful map of connections, but also a rapidly changing and opaque nexus of enterprises and entanglements. In the 1870, the Union Pacific was shaken by bribery scandals and went into bankruptcy before being taken over by the businessman Jay Gould, who collected and reorganized a number railway companies. In the late nineteenth century, J.P. Morgan played a significant role in consolidating railway companies. Together with other industrialists like Cornelius Vanderbuilt and Jay Gould, Morgan amassed an enormous wealth out of railway enterprises and market consolidations which led to a handful of monopolies controlling the business.

1887 Fight Against Monopolies in the USA: The Interstate Commerce Act

The monopolistic practices of the railway companies forced the US government and congress to pass the Interstate Commerce Act in 1887 to control the business activities of the railroads. It introduced the ICC (Interstate Commerce Commission), which had the authority to set maximum freight rates and review a company's financial records. The success was limited: By 1906, two-thirds of the rail network in the US was controlled by seven entities with J.P. Morgan controlling the largest portions. By 1880, the railroad industry was the largest employer outside the agricultural sector. The industry operated over 17,000 freight locomotives and 22,000 passenger locomotives. The impact on the US economy was huge. The opening of hundreds of millions of acres of land for agriculture, lower costs for food, the creation of a large national market and finally the development of new system management practices and engineering excellence had been major drivers for industrial growth (Chandler, 1965).

1914 Federal or Private?

In most of the European countries, the first railways were built by private entrepreneurs but soon taken over by the federal organizations. States like the German Empire or the Austrian Hungarian Monarchy were convinced that a public infrastructure had to be controlled and owned by the government. There had been some private initiatives in the middle of the nineteenth century during public budget crises but, in the end, most the European railroad network was federal controlled. After World War II, the German, Austrian, French, Italian and other railroad networks became federal owned state railways and kept that status up to today. In the UK as well as in the US, the government began to take over control from the private or public railway companies towards the beginning of World War I.[5]

As in the UK the US governments took over management of railroads during World War I, the US introduced the United States Railroad Administration. This led to a standardization of equipment and better coordination of freight traffic. The federal control ended in 1920. In the following decades, the importance of rail transportation declined. Competition came from the new interstate highways used by a growing number of trucking businesses for freight traffic and the growing public airline services.

Technology 2.0: Electricity, Diesel and High-Speed

1920 Diesel Engines and Electric Power

There are two major technological steps in the development of railroad networks. Beginning in the 1880s, electricity was first used for railways. At the turn of the century, railroad lines in the UK, the US and in other European countries were changed to electrical operation. In the 1920s, the diesel engine and diesel-electric transmission became an alternative to steam locomotives. Both technologies created new industries that developed infrastructure for electric powered locomotives and the electrification of railway lines as well as developed and produced diesel-electric powered trains. The electrification of railway lines was the technology of choice for many European countries. In the US and partly in the UK diesel-electric transmission over long distances was the winning technology.

[5] UK Government, Rail usage, infrastructure and performance: https://www.gov.uk/government/statistical-data-sets/rai01-length-of-route-distance-travelled-age-of-stock retrieved 2017-01-20.

1922 Organizations to Support Standards

After World War I, the international railway network was the most important provider of public transport services. In the 1920s, it has reached a global expansion of 1.5 million km and was the backbone for personal travel and freight logistics in Europe, America and in some Asian countries. Interstate transport and logistics became more and more important and the need for international technical and processing standards was evident. In this environment, the UIC (Union Internationale des Chemins de fer) was created on October 20, 1922.

With the expansion of the network in Europe and Asia and the introduction of new technologies, the availability of standards for the new technologies became more and more important. The ecosystem supporting the railway networks expanded and became more complex. In 1992, a group of European providers of railway technology founded the Union des Industries Ferrovaires Europḗenes (UNIF) with the goal of creating a system of accepted technical and quality standards for railway equipment. Today, the members of the UNIFE have a market share of more than 80% in Europe and 50% worldwide. Associated members of the UNIFE are also the European national railway associations thus representing more than 1000 national and private railway enterprises. The UNIF has developed the "International Railway Industry Standard (IRIS) as a comprehensive set of quality management rules and specifications for railway infrastructure, maintenance and operation.

1945 Competition and Decline

After World War II, many railroads were pushed out of business, passenger train service was displaced by airlines and the growing usage of cars. Especially in the US, the ongoing regulation of freight rates by the Interstate Commerce Commission limited the railways flexibility to respond to the new market conditions. The global network reached a peak around 1930 followed by a lengthy decline before new technologies led to a revival of railway transport in the 1970s and 1980s[6] (see Fig. 4).

1970 High Speed Trains

In the 1970s and 1980s, the first high-speed-trains were tested and started operation in Japan (Shinkansen), and later in Germany (ICE), France (TGV) and Spain (ADIF). High-speed-trains reach a speed of more than 200 km/h and, in some cases, more than 300 km/h. The centers of high-speed trains are Europe and Asia.

---

[6]Union internationale des chemins de fer; http://www.uic.org/statistics#Railisa-Database retrieved 2016-08-15.

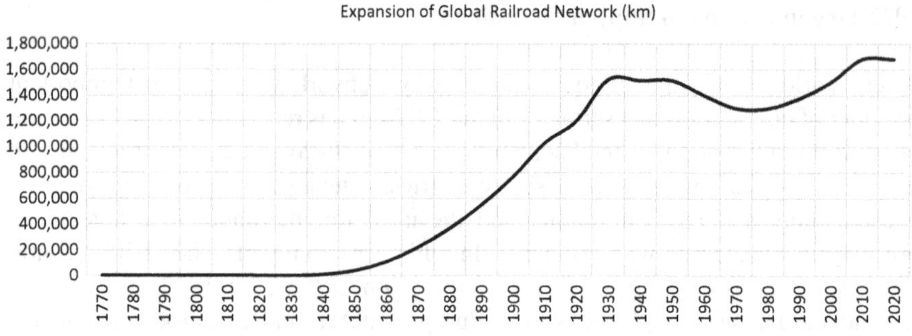

Fig. 4 Size of global railroad network 1830–2015

Table 6 Number of high-speed-trains operated

Year	Global	Europe	North America	Asia
2008	1737	1050	20	667
2010	2102	1243	20	839
2012	2777	1670	20	1087

Especially in Europe, the network of high-speed-railways is growing rapidly[7] (see Table 6).

Today more than 3000 high-speed trains are operated worldwide, most of them in Asia (Japan, China, Taiwan, South Korea) and Europe.

Social and Economic Impact

From the beginning, the railway caused a huge impact on society as a mass transportation service. The service itself had a major influence on the daily life of the people. For the first-time, travelling was achievable for everybody, distance and time took on a completely different meaning. Particularly the first transcontinental connection in America changed the perception of geography completely (Brown, 1977; Chandler, 1965). The reflection in all kinds of media and popular culture is tremendous. There are railway-songs, railway-movies, railway-novels, railway-poems and, in some places, large communities of railway-hobbyist's building their own small worlds of railway networks. With the movement of the railway ecosystems into more federally controlled entities in Europe, the railway services also became an important part of national identify. People in countries like Switzerland are proud to show their yearly subscription ticket for the Swiss Railroad. Citizens of countries operating high-speed trains are proud of the maximum speed achieved by the newest model. Running trains according to a timetable is still

[7]Union internationale des chemins de fer; http://www.uic.org/statistics#Railisa-Database retrieved 2016-08-15.

perceived as sign of proper administration by the federal authorities and the quality of public transport is one of the most discussed matters in evaluations of the quality of life in a specific country or city.

Railroads as the First Network Based Service Ecosystem

The development of railroads is the first example of a network based service ecosystem available to the public. Starting simultaneously in the US, UK and only a little bit later in other European countries, the railroad infrastructure created a huge industry, new jobs and job profiles. Railroads are based on new technologies (steam, electricity), a network structure, the sometimes-painful creation of standards and the sometimes-dramatic development of businesses. Railway as a public service is also a good example of moving ownership and responsibility to an invisible organization. Travelling by train means that the consumer has no influence on the resources, processes or technologies involved in the service provided. As a passenger on a train, you give away all types of ownership and you gain—if everything works properly—comfort, reliability and safety at a low price. That's the basic idea of a network-based ecosystem.

The development of railroad eco systems shows two different economic approaches. In the UK and Europe, railways started as private enterprises but were, in a second phase, taken over by federal control. In those countries, most railroad operations were part of the service provided by the state to its citizens. In a third phase, beginning with the deregulation in Europe in the 1980s, some—but smaller—parts of railroad services were run by private enterprises, although most of the systems were still operated by the federal administration. Public transport is an important part of the federal responsibilities. In the US, railroads also started as private enterprises and went through a heavy restructuring caused by the financial interests of a small group of people. Federal legislation tried to create a clear framework of rules and regulations to secure proper and safe operations. This had two consequences: the US railroads could not find an efficient strategy to compete in changing markets caused by the growing automotive and airline industry and the railroad technology was not developed at the same speed as in Europe and some Asian countries because of missing or declining learning bases and innovations. Some European (Germany, France, Spain) and some Asian countries (Japan, China, Taiwan, South Korea) could meet the challenge of developing an efficient mass transportation system for a growing population and a higher demand for mobility by investing in and building the next level of infrastructure creating a high-speed mass transportation system. Those countries are leading in those technologies today.

References

Blaise, C. (2001). *Time lord. Sir Sandford Fleming and the creation of standard time*. New York: Pantheon Books.
Brown, D. (1977). *Hear that lonesome whistle blow*. New York: Henry Holt & Company.
Chandler, A. (1965). *The railroads: The Nation's first big business*. New York: Arno Press.

Early Information Network Services

"If you succeed, you will bask in glory"
The first message over Chappe's Semaphore system

Sending messages over long distances in a short time is one of the major achievements of the last 200 years and had a huge influence on technical, economic and military development. There are four major phases: the optical telegraph invented in the eighteenth century and used until mid of the nineteenth century, the electrical telegraph based on Morse's code, the telephone and wireless communication. All four major technical steps created communication networks and services based on these networks, being predecessors to the Internet and the Web.

The First Optical Communication Network

The invention of the optical telegraph can be connected back to two brothers in France. Around 1792, the French engineer Claude Chappe together with his brother Ignace started to experiment with a system consisting of black wooden arms, the position of which indicated alphabetic letters. The whole construction was mounted on a tower and could be controlled with a system of ropes and two handles from the ground. It became the first messaging system based on technology not relying on horses. The idea was to construct a line of semaphores each of them within visible distance of the next semaphore station. Each station should receive a message via optical observation of its predecessor station and then use its own semaphore system to signal the message to the next station (see Table 1).

1792 The First Mechanical Communication Line
In 1792, the Chappe brothers could convince the French National Assembly to construct the first line of semaphore stations from Paris to Lille over a distance of 230 km using 15 stations. Given a distance of 230 km, a courier would have needed at least 24 hours. The semaphore line increased the speed of message transport by a factor of 50. Depending on the weather conditions, a symbol could be received by a

M. Oppitz, P. Tomsu, *Inventing the Cloud Century*,
DOI 10.1007/978-3-319-61161-7_3

Table 1 Optical telegraph timeline

1791	First optical telegraph line in France from Paris to Lille
1794	Optical Telegraph lines in Sweden
1795	Optical Telegraph network in the UK, Murray system
1816	Murray system replaced by Popham system
1825	Private line between Liverpool and Holyhead for signaling ships arrivals
1837	French government banns private communication
1849	Telegraph Hill line in San Francisco erected, in operation until 1862
1850	Optical telegraph network in France expanded to 5000 km
1854	Admiralty line between London and Portsmouth closed and replaced by electrical telegraph
1858	Liverpool—Holyhead line replaced by electric telegraph

station and passed to the next station within 30 sec. It is reported that a 36 symbol-message sent from Lille to Paris took 32 minutes from start to finish. The French government operated the network and expanded it throughout France until 1850. At its most extensive, it consisted of 534 stations covering more than 5000 km. In the beginning, the French optical telegraph network was used for military and public administration purposes only. Later, it was also used to broadcast stock prices and lottery results to avoid bribery with the lottery numbers. In 1837, the French government passed a bill banning private networks to keep control over the usage of telegraph technology.

1794 Networks in the UK, Sweden and Germany
After the first successful line between Paris and Lille, optical telegraph lines were soon introduced in Sweden (1794) and the UK (1795). The British system was slightly different from the Chappe system and was proposed by Lord George Murray. It used a matrix of 6 shutters each capable of being switched to a horizontal or vertical position thus implementing a 6 bit-code system for the transmission of 64 different symbols. The UK lines were operated by the Admiralty and expanded to a network between London and Dear, Portsmouth, Plymouth, and Yarmouth until the end of the Napoleanic wars in 1816, used for military messages only. By 1816, the Murray system was replaced by a simpler construction designed by Sir Home Popham. A new telegraph line dedicated to the Admiralty was erected between London and Portsmouth and stayed in operation until 1854 (see Fig. 1).

1825 Improving Business by Fast Communication
The economic communities in England soon recognized that fast message broadcasting could also speed up their business. In 1825, the Board of Port of Liverpool applied to Parliament for approval of a telegraph line between Liverpool and Holyhead at the Isle of Anglesey, a distance of over 150 km: *"establish a speedy Mode of Communication to the Ship-owners and Merchants at Liverpool of the arrival of Ships and Vessels off the Port of Liverpool or the Coast of Wales, by building, erecting and maintaining Signal Houses, Telegraphs or such other Modes of Communication as to them shall seem expedient, between Liverpool and*

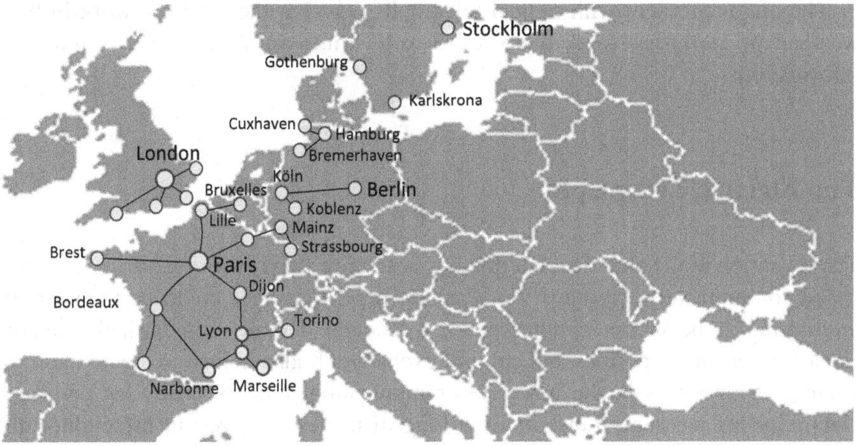

Fig. 1 Optical semaphore lines in Europe around 1800

Hoylake, or between Liverpool and the Isle of Anglesea." The purpose of the line was to communicate the arrival of ships to the Liverpool merchants as quick as possible. The line was in operation until 1858 and then replaced by an electric telegraph line.

Social and Economic Impact

Optical telegraphs formed the first communication networks, and initiated the first service ecosystems for information distribution. There are a lot of similarities between optical telegraph services between 1790 and 1850 and cloud services found on the Internet later. Optical telegraphs used a code system for transporting messages, they connected major and geographically distant points and formed the first network connecting hot spots of information sources with information consumers. In both cases—optical telegraphy and Internet—the first movers were military organizations: The army in France and the navy in Britain, soon followed by commercial oriented interest groups, that recognized the importance of quick information distribution.

The name "telegraph" was used for the new technology beginning in France in 1791. The story goes that Chappe wanted to call his invention *tachygraphe,* the Greek word for "fast writer." Chappe's friend Miot de Melito convinced him to change to *telegraph*—"far writer." Even today many places are known as "telegraph hill" having been the sites were semaphore stations were erected. One of the most famous telegraph hills is the one in San Francisco. It was the place for a semaphore station broadcasting information to the San Francisco merchants about ships passing the Golden Gate Bridge. This information service was provided from 1849 until 1862 and then replaced by electric telegraphs.

The major problem with optical telegraphs was that they did not work in bad weather or darkness—and there were still limits in speed and volume of information.

The Electric Telegraph

The invention of the electrical telegraph was a major step in the creation of network based communication systems. For the first time, messages could be transported around the globe within minutes or even seconds. The new media had a major impact on business, politics and cultural exchange. It also helped to create the first business ecosystems in the new telecommunication industry. The idea of using electricity for message transport over long distances appeared in the eighteenth century. In 1753, an anonymous writer in the *Scots Magazine* suggested using an electrostatic telegraph with one wire for each letter of the alphabet and pith balls to indicate the active line. Unfortunately, the lack of a reliable source of continuous current doomed the system to failure. 50 years later in 1800, the Italian physicist, Allesandro Volta, solved this problem by inventing the electrical battery as a steady and reloadable source of electric power. In the following years, several attempts were made to design a working system. Successful prototypes of electric and magnetic telegraph systems were built and demonstrated by Samuel Thomas von Sömmering (Germany) 1809, Francis Ronalds (UK) 1816, Joseph Henry (USA) 1828, Baron Schilling von Canstatt (Russia) 1832, Carl Friedrich Gauss (Germany) 1833, David Alter (USA) 1836. The first working telegraph systems were introduced in England and USA in 1838 (see Table 2).

1838 First Prototypes in the UK
It was the cooperation between an inventor—William Cooke and a scientist—Charles Wheatstone that led to the first practical telegraph system that found its way to commercial use. In this partnership, Cook provided the basic idea of using a set of five needles to indicate a 5-bit signal transmitted via a 6-wire telegraph line. Wheatstone added his know-how about electricity, and together they achieved the installation of the first test line between Euston and Camden Town along the railroad track. By 1838, the first commercial line was installed on the Great Western Railway between Paddington Station to West Drayton, a distance of over 21 km. These first lines proved that message transmission using electricity was possible but still were not reliable. Problems occurred due to faulty isolation between the 6 wires, forcing the engineers to find solutions with a reduced number of wires.

1838 Morse's Idea
At the same time, the American, Samuel Morse, started to think about a better solution for transmitting different symbols over a single wire. During an Atlantic crossing, he developed the idea of a code system consisting of long and short signals. Morse created a table referencing each symbol (character or digit) to a combination of dots and dashes. Morse also ordered the characters along their

Table 2 Electrical telegraph time line

1830	
1837	Cooke and Wheatstone patent telegraph in England.
1838	First Telegram sent over a distance of 3 miles at Speedwell Ironworks, Morristown, NJ
1838	Morse's Electro-Magnetic Telegraph patent approved.
1840	
1843	First message sent between Washington and Baltimore.
1846	House's Printing Telegraph patent approved.
1846	First commercial telegraph line completed. The Magnetic Telegraph Company's lines ran from New York to Washington.
1848	Associated Press formed to pool telegraph traffic.
1849	Bain's Electro-Chemical patent approved.
1851	Hiram Sibley and associates incorporate New York and Mississippi Valley Printing Telegraph Company.
1851	Telegraph first used to coordinate train departures.
1857	Transatlantic cable laid
1857	Treaty of Six Nations is signed, creating a national cartel
1859	First transatlantic cable is laid from Newfoundland to Valentia, Ireland, fails after 23 days.
1861	First Transcontinental telegraph completed.
1865	International Telegraph Union (ITU) founded at the Conference Telegraphique , Paris
1866	Western Union merges with major remaining rivals.
1866	First successful transatlantic telegraph laid
1867	Stock ticker service inaugurated.
1870	Western Union introduces the money order service.
1881	Jay Gould gains control over the Atlantic & Pacific and of Western Union.
1900	
1908	AT&T gains control of Western Union. Divests itself of Western Union in 1913.
1924	AT&T offers Teletype system.
1926	Inauguration of the direct stock ticker circuit from New York to San Francisco.
1930	High-speed tickers can print 500 words per minute.
1945	Western Union and Postal Telegraph Company merge.
1988	Western Union Telegraph Company reorganized as Western Union Corporation
2000	
2006	Western Union closes the telegraph service

frequency thus coding the character "E" as most frequent character with one dot, the character "T" with one dash and so on. Morse also designed the first writing device consisting of a pendulum with a pen drawing marks on a strip of paper indicating the long and short signals. Morse's first concept of a sending device was a contact table with small long and short copper tiles. Moving a conductive pen over the table indicated long and short impulses on the wire. Later, the construction was replaced by a hand-operated key, the so-called Morse-key.

1844 First Commercial Line in USA
At the same time as Cooke and Wheatstone in England were collecting their first experiences, Samuel Morse achieved the installation of the first test connection at the Speedwell Ironworks near Morristown, NJ in 1838. Nevertheless, it took him and his partners 6 years before they received a grant of $30,000 and permission from the government to erect the first commercial line between Washington and Baltimore in 1844. The first words communicated were "What hath God wrought?". Morse split the patent rights between himself and his partners: Amos Kendall, Leonard Gale, and Alfred Vail. Discussions and discord between the partners led to a split of the patent right geographically.

1846 Competing Systems
In 1846, a rival US patent for the telegraph was introduced by Royal House and Alexander Bain. The patent was used by competing enterprises building numerous private firms and telegraph lines. By 1851, ten separate lines ran to New York City connecting New York to Philadelphia, Boston and Buffalo. The report by the Bureau of Census listed 75 different companies operating 21,147 miles of wire. The market was highly competitive and rates fell dramatically. Quality was a major problem; messages were garbled or lost due to transmission problems. A message sent from Boston to St. Louis could have passed over the lines of five different companies.

1851 Shaping Business in the USA and the UK
In 1851, a group of business men decided to use the new technology based on Morse's and Vail's patent to create a new business. Hiram Sibly and Samuel Selden founded the "New York and Mississippi Valley Printing Telegraph Company." At the same time, Ezra Cornell (who also found Cornell University in 1865) created the "New York & Western Union Telegraph Company" using one of his bankrupt companies. All three businessmen decided to merge under the name of "Western Union." The business grew rapidly, by 1851 the US telegraph network had a size of 32,000 km. In 1852, the Supreme Court declared Bain's patent as an infringement of Morse's patent and lines operated on Bain's technology were merged with Morse lines. The next step of integration took place when the six largest companies signed the "Treaty of Six Nations" as a pooling agreement. By 1957, the telegraph market still consisted of several regional providers, but most of them owned and used both patents: Morse and Royal House. On October 24th, 1861, the first intercontinental connection between east and west coast was completed by Western Union, proving that the electric telegraph can be used for long distance communication.

In a final phase of integration, the pool members consolidated into a national monopoly. By 1864, only Western Union, the American Telegraph Company and The United States Telegraph Company remained as telegraph providers. Western Union absorbed its last two competitors in 1866 and thereby reached its position of market dominance. Western Union grew rapidly from a capitalization of $385,700 in 1858 to $41 million in 1876.

In the UK, the first telegraph business was founded by Sir William Fothergill Cooke and John Lewis Ricardo as the "The Electric Telegraph Company" in 1846.

The company bought all patents held by Cooke and Wheatstone. In 1855, it merged with the "International Telegraph Company," another private provider, to become the "Electric and International Telegraph Company." By 1850, around 20 private telegraph companies operated lines between more than 1100 cities and towns in England, Scotland and Wales. Telegraph lines were operated by those private companies or by railroad companies installing telegraph lines along the railway lines. Soon, additional services were offered. In 1852, time signals were first transmitted from the Royal Greenwich Observatory. In 1872, the London Stock Exchange introduced the first stock ticker provided for and operated by the Exchange Telegraph Company of London. Of all these different startups, it was the Electric Telegraph Company that achieved establishing a learning base in technology, operation and standards, which made it the dominating provider in the UK. Renamed to the "Electric and International Telegraph Company," it was nationalized by the government in 1870 and taken over by the General Post Office (GPO) to become the primary supplier of communication services in the UK. Since 1981, it has operated under the name of "British Telecommunication", later "BT."

1858 Cyrus Field and the Transatlantic Adventure

The project of connecting Europe and America by cable is related to the name Cyrus Field, an American business man and entrepreneur. Field succeeded in manufacturing business between the 1830s and 1850s. His business earnings allowed him to retire with a fortune of $250,000 at the age of 34. He turned his attention to the telegraph and got in contact with Canadian engineer, Frederick Gisborne, who, in 1851, was involved in a project constructing a landline crossing in Nova Scotia. Gisborne and Field considered the idea of extending that connection across the Atlantic to Europe. At that time, a submarine cable was already operated between England and France. Field started to get investors from both sides of the Atlantic and promoted the "New York, Newfoundland and London Telegraph Company" to organize funding.

The first attempt was undertaken in 1857 and failed, the cable broke on the first day and had to be repaired. The cable broke again in the second trial in the middle of the Atlantic. The next year, new attempts were made and this time with two vessels meeting in the middle of the Atlantic, splicing the cable and then sailing east- and westwards. After three incidents with broken cables, the expedition made landfall with the cable in Trinity Bay, Newfoundland and in Knightstown, Ireland in August 1858. Later that month, the first message could be transmitted across the ocean (see Fig. 2). Unfortunately, the electric engineers had different ideas about the proper operation of the cable and the level of voltage to use. It turned out that the voltage used was too high (several thousand volts) and destroyed the cable after 3 weeks of test operation. The total volume of messages sent during that period was not more than 4359 words, while the cost of laying the cable was $1.2 million.

Although public opinion immediately switched from enthusiasm to disappointment and disbelief with respect to the success of the idea, Field achieved in creating a new pitch for his transatlantic cable project and found new investors forming the Anglo-American Telegraph Company with an initial capital investment of

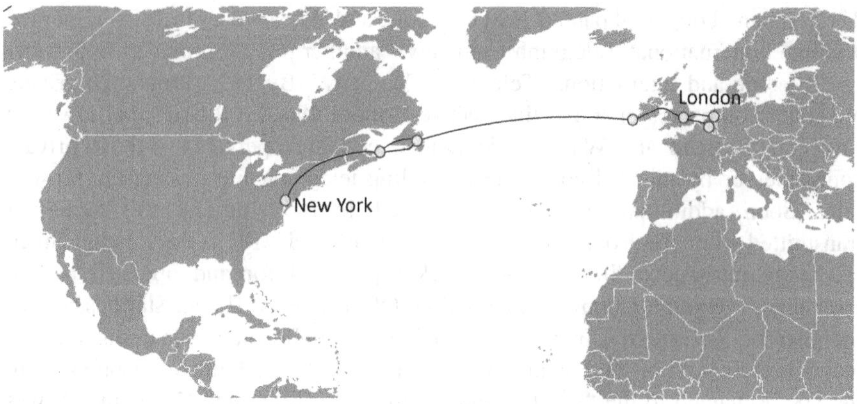

Fig. 2 The first transatlantic cable

£350,000. In the summer of 1865, a new attempt was made using the "Great Eastern," the largest ship at that time. Although the cable broke again two times and was also lost in the 1856 expedition, the next trial in the summer of 1866 was a success. The cable was landed safely on September 7th, 1866. The new connection was also a vastly improved version of the 1858 cable. Message transmission was much more reliable and quicker. The first operable transatlantic cable could transmit eight words per minute, which relates to 1 byte per second or 8 bit/s. It took more than 50 years to expand that bandwidth to 120 words per minute (120 bit/s). Today modern fiber optic submarine cables have a bandwidth of 2 $\times$ 2.5 Gbit/s.

1865 Creating Standards: The International Telegraph Union

The growth of national and international telegraph networks soon led to many compatibility problems between the different systems and how they were operated. Though the Morse code was accepted as the most efficient coding system, several different variations of the Morse code system were introduced in the US and Europe. In the US, the "American Morse Code" was used which specified different length of gaps between symbols and words and presented three different lengths of dashes with different meanings. A modified version was adopted as the European standard in 1865, called at first the "Continental Morse" which eliminated the different lengths of gaps and dashes. This version became later known as the "International Morse." Similarly, telegraph operators tended to use abbreviations for frequently used words and companies using the telegraph for internal messages started to use codebooks to encrypt their communication. The production of codebooks became an important business and was one of the first commercially used encryption methods. Using of codebooks turned out to become a major problem because the failure rate of telegraph operators was much higher when they send or received encrypted messages composed out of meaningless character sequences (Standage, 1998).

In Europe, the pressure to introduce a framework of standards was especially high because of the growing international traffic between national telegraph operators. When messages had to pass borders between two incompatible telegraph systems it was necessary to decode and encode every message, which costs time, and lead to the garbling of messages. Different national pricing schemes made the calculation of international message costs even more difficult and time consuming. One first attempt at standardization was made within the German Federation in 1850. A conference of representatives of Prussia, Austria-Hungary, Bavaria and Saxony was held in Dresden to discuss those matters. Thus, the "Der Deutsch-Österreichische Telegraphenverein—DÖTV" was founded to harmonize the technical, operational and tariff standards between the national telegraph administrations. In 1855, the "Western European Telegraph Union" was founded by France, Belgium, Spain, Sardinia and Switzerland. Both unions cooperated from the beginning. Finally, in 1865, the representatives of 20 European states met at the "Conference Telegraphique" in Paris to establish the "International Telegraph Union—ITU" which became the first international organization for telegraph and, later, communication standards. The ITU still exists and has created since then standards for all types of communication methods, one of them being the "ITU definition of cloud architecture" (ITU, www.itu.int 2014; Wenzlhuemer, 2010).

1890 Intercontinental Connections and British Dominance

In the following years submarine cables were laid between all continents by many new enterprises. In 1891, submarine cable connections existed between Europe, North America, South America, Asia and Australia (Standage, 1998) (see Fig. 3).

British submarine cables became dominant from the 1850s into the twentieth century. For Britain, with all its overseas territories, it was a strategic goal not only to "rule the waves" but also to have control over the communication lines to the overseas territories. This political strategy was called the "All Red Line" and was comprised of submarine cable connections between Great Britain, Canada, Australia, South Africa and India all under British control, thus forming the first international network that spanned the globe (see Fig. 4). By the end of the nineteenth century, British companies owned and operated more than 60% of the international cable connections, in 1923 the share was still more than 40%. During World War I, the German connections were quickly cut, while the British lines stayed reliable and uninterrupted.

1876 Western Union Becomes the Major US Provider and Misses the Next Train

By the end of the nineteenth century, Western Union dominated the US communication industry based on the electrical telegraph technology and with a 90% market share. It introduced new types of services such as the stock ticker in 1866, a standardized time service in 1870 and a telegraph based money transfer service in 1871. By end of the nineteenth century, Western Union operated a million miles of telegraph lines and two international undersea cables. By 1881, the railroad tycoon Jay Gould had gained control over the Atlantic & Pacific and of Western Union. In 1909, AT&T gained control over Western Union, but it turned out soon after that both companies went in different directions. Additional pressure from the Department of Justice using the Antitrust Act forced the two companies to split again.

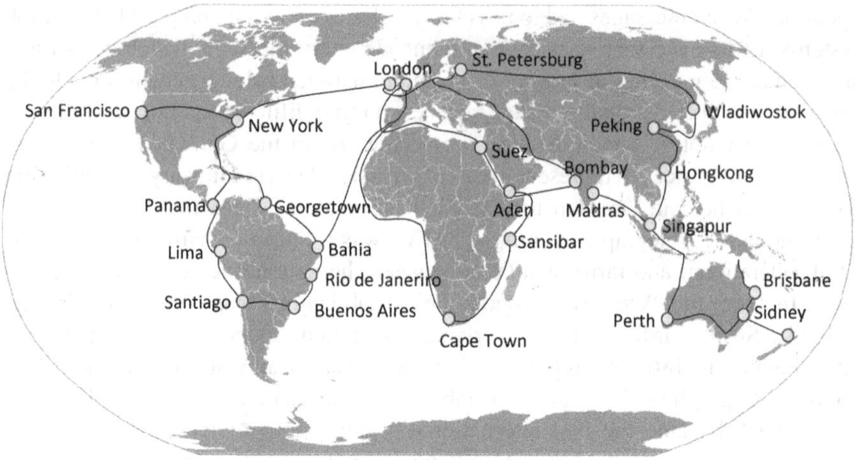

Fig. 3 Major intercontinental telegraph lines in 1891

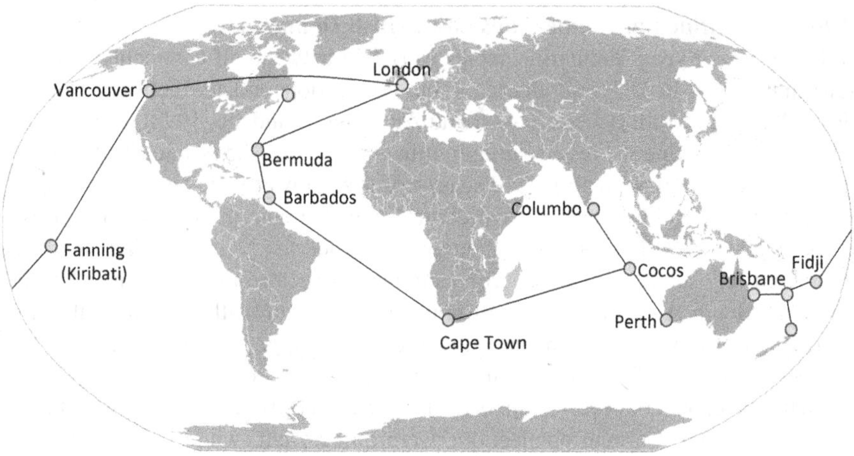

Fig. 4 The "All Red Line" in 1903

In the late nineteenth century, the telephone was the logical successor of the electric telegraph. Unfortunately, Western Union was drawn into a patent lawsuit between its partner company Western Electric and Alexander Graham Bell and decided not to invest in the telephone. In an internal memo in 1876, Western Union's president William Orton said: "*This 'telephone' has too many shortcomings to be seriously considered as a means of communication.*" Therefore, the US telephone business was taken over by the Bell Telephone Company, which was later acquired by AT&T. As a result, Western Union was left with a declining telegraph business and concentrated on money transfer services.

Table 3 Number of telegraph messages between 1870 and 1970

Date	Messages handled	Date	Messages handled
1870	9,158,000	1930	211,971,000
1880	29,216,000	1940	191,645,000
1890	55,879,000	1945	236,169,000
1900	63,168,000	1950	178,904,000
1910	75,135,000	1960	124,319,000
1920	155,884,000	1970	69,679,000

Table 4 Telegraph and telephone rates (USD)

Date	Telegraph	Telephone
1850	1.55	
1870	1.00	
1890	0.40	
1902		5.45
1919	0.60	4.65
1950	0.75	1.50
1960	1.45	1.45
1970	2.25	1.05

During the first decades of the twentieth century, the telegraph business declined, superseded by telephone[1] (see Table 3).

The rates for telephone calls decreased and, in 1960, the price for a 3-minute phone call was equal to a 10-word telegraph message[2] (see Table 4).

AT&T introduced the teletype writer in 1931, again drawing telegraph business from Western Union. In the following decades, Western Union invested successfully in numerous types of communication services including intercontinental connections and satellite networks. Western Union was the first company to launch a fleet of communication satellites in 1974 called Weststar. The telegraph service was operated until February 2006. On this day, the Western Union website showed this notice: *"Effective 2006-01-27, Western Union will discontinue all Telegram and Commercial Messaging services. We regret any inconvenience this may cause you, and we thank you for your loyal patronage. If you have any questions or concerns, please contact a customer service representative."*[3]

[1]Economic History Association, History of the U.S. Telegraph Industry: https://eh.net/encyclopedia/history-of-the-u-s-telegraph-industry/ retrieved 2016-12-10.

[2]Economic History Association, History of the U.S. Telegraph Industry: https://eh.net/encyclopedia/history-of-the-u-s-telegraph-industry/ retrieved 2016-12-10.

[3]Washington Times, 2006, http://www.washingtontimes.com/news/2006/feb/2/20060202-120246-7642r/ retrieved 2016-10-12.

Social and Economic Impact

Around 1800, before the introduction of the railway, the travel time between Paris and Berlin was 1 week, and from New York to Cleveland it took 2 weeks. The railroad cut this time by the half. Transatlantic crossing time by an ocean liner was around 10 days in 1850. However, using the telegraph, messages could be passed between all these locations within minutes and at comparable low costs.

The social influence of the telegraph is not limited to the personal use of the new technology. Sending or receiving private messages within minutes or hours is one thing, but the effect on the economy was much more dramatic. The telegraph was the first international network based ecosystem that had a broadband effect on the efficiency of business operations, the speed of business transactions and financial markets and the capability of delivering goods from remote production places in short time. These effects had a much broader and deeper influence on the daily life of the citizens in Europe and North America.

Creating an Ecosystem with Railroads

One of the first business users of the new telegraph systems were the railroad companies. Building new telegraph lines was comparably easy along already existing railway tracks. After a while, the railroad companies also started to use the telegraph connection to improve their technical operations. At that time, most of the railway lines were single-track lines, which needed a perfect coordination of the train schedule to avoid incidents. Even when operating on double-track lines, the occupation of each track by trains must be controlled. Using the telegraph to coordinate and signal departure times and the occupations or reservations of route-sections is an efficient method and helps to reduce costs and risk and improves operational efficiency.

Impact on Politics, Military and Police

Telegraphs also changed the work of police and military forces. John Tawell, murderer of his mistress, was the first criminal captured thanks to the telegraph. Trying to escape, Tawell jumped on a train in Slough heading for Paddington. The police used the telegraph to send a quick notice to Paddington and Tawell was expected by the police on arrival. The telegraph during wartime appeared for the first time in the Crimea War between Russia and Britain. A telegraph department was part of the British forces and started to build a local network in the field. The much more serious effect at that point of time affected the European telegraph network which already reached from England to St. Petersburg. That resulted in the Russian military and government having access to all political and military news from their enemy. As usual, the War Ministry in London proudly issued the precise details about numbers of troops deployed for the expedition. In older times, the troops would have outstripped the news of their arrival. Now the Russian forces received that information and knew exactly what they had to expect weeks before the arrival of the British forces. The telegraph also collapsed the distance between the soldiers, the citizens and families back home and between the commanders in the field and the government. Executing long term plans and making onsite decisions is followed completely different rules with temporary communication between board-level and execution officers.

Changing Economic Processes

Another economic segment depends on quick messaging and near-real-time information: agricultural production. The invention of refrigerated cars around 1875 enabled the delivery of fresh meat and vegetables over long distances on time. This is only possible when advertising, ordering, packing and delivery can be coordinated over long distances as well. The telegraph had a tremendous impact on the financial markets. Stock prices could be communicated from the central stock exchanges in New York, London and other places within minutes to traders somewhere in the same city or at more remote locations. The introduction of the stock ticker in 1867 improved that service and created the first one-to-many automatic news broadcasting system. Stock tickers became a standard office equipment for traders as well as large enterprises to follow the market in real-time. An additional effect was the rapid news distribution about political events, weather conditions, ship accidents and other relevant events, which influenced the market. Within only a few years, the financial market places and traders were connected to each other and the rest of the world in near real time thus completely changing the operation of stock exchanges and the mechanics of the financial markets. Financial market places were centralized (New York, London) which led to the multiplication of buying and selling opportunities, more liquidity and a rapidly growing trading volume.

Decline and Fall of a Disruptive Technology

Within 20 years (between 1840 and 1860), the world has shrunk dramatically. Starting with the first commercial use of the telegraph in the 1840 and progressing with the transatlantic connection in the 1860s, the electrical telegraph created a completely new perception of the world and multiplied the opportunities for making business, delivering goods and services and planning and executing political, diplomatic and military missions. The electrical telegraph was the first international communication system. It created a new industry segment in the US and Europe and established many new types of enterprises building a nexus of communication businesses. In the US and UK, private companies took over from the beginning, struggling through some decades of transition and market restructuring. In the end, it was one company in the US (Western Union) that created a de facto monopoly on telegraph services. In UK and the other European countries, the telegraph business was soon taken under federal control and merged into the national post administration structures. In all cases, the telegraph technology played a major role as forerunner of other communication networks (telephone, radio, TV, Internet). It played this role through the first half of the twentieth century. Declining because of the advent of the telephone and other services, the electric telegraph died a silent death at the end of the twentieth century.

Telephone

The telephone was the first globally available communication network and thus is also the first example of a public network based service ecosystem. It is based on mass produced end devices, a standardized global network, and a set of developed business models and it had a major social impact on worldwide personal and business communication. Using phone communication today is an indispensable element of daily life for individual communication and business relationships (see Table 5).

Technology

The idea of transmitting speech and sound over electrical lines had been discussed by many technicians since the middle of the nineteenth century. In 1854, the Frenchman Charles Boureul, published an article "Transmission électrique de la parole" in the Paris newspaper "L'Illustration" on the principles of speech transmission. In Germany, Philip Reis constructed the first speech transmitting telephone in 1861. The Italian, Antonio Meucci, filed a first intent, but not a formal patent application at the US Patent Office for a "Sound Telegraph" in 1871 and Thomas Edison experimented with "Acoustic Telegraphs" in 1875.

1876 Graham Bells Patent
Finally, it was the American Graham Bell who made him to finish line first and on January 20th, 1876, he signed his patent application for the telephone, which was granted on March 7th, 1876. There was only one problem: the device did not work. Simultaneously, the American inventor Elisha Grey was working on the same problem and applied for a telephone patent using a so-called liquid transmitter. Grey's application was filed on February 14th, 1876. On March 10th, 1876—3 days after his patent was granted—Bell could transmit a sentence from one room to the other for the first time: "*Mr. Watson, come here! I want to see you*". As part of the lab configuration, a liquid transmitter comparable to Grey's design was used by Bell. The discussion about the origin of Bell's design has continued since then. Bell himself improved his design within a couple of weeks and never used the liquid-transmitter design again.

1878 Telephone Network and Switching
With the introduction of the working telephone, the first element of a network based service ecosystem was available. Using Bells invention made it possible to build a point-to-point connection between two end devices. For the creation of a network of telephone users an important building block was missing: the telephone exchange. A telephone exchange is necessary to provide connections between two or more individual telephone subscribers. The exchange device makes it possible to call each other and makes telephone service an everyday communication tool. The first

Table 5 Telephone time line

1600	
1667	Robert Hooke invented a string telephone that conveyed sounds over an extended wire by mechanical vibrations. It was to be termed an 'acoustic' or 'mechanical' (non-electrical) telephone.
1700	
1753	Charles Morrison proposes the idea that electricity can be used to transmit messages, by using different wires for each letter
1800	
1844	Innocenzo Manzetti first mooted the idea of a "speaking telegraph" (telephone).
1854	Charles Bourseul writes a memorandum on the principles of the telephone
1854	Antonio Meucci demonstrates an electric voice-operated device in New York; it is not clear what kind of device he demonstrated.
1861	Philipp Reis constructs the first speech-transmitting telephone
1871	Antonio Meucci files a patent caveat at the US Patent Office for a device he named "Sound Telegraph"
1872	Elisha Gray establishes Western Electric Manufacturing Company.
1875	Bell uses a bi-directional "gallows" telephone that could transmit "voice like sounds," but not clear speech.
1875	Bell's US Patent 161,739 "Transmitters and Receivers for Electric Telegraphs" is granted.
1875	Thomas Edison experiments with acoustic telegraphy and in November builds an electro-dynamic receiver, but does not exploit it.
1875	Hungarian Tivadar Puskas (the inventor of telephone exchange) arrived in the USA.
1876	Alexander Graham Bell patents the telephone.
1876	Bell signs and notarizes his patent application for the telephone.
1876	Bell's US patent No. 174,465 for the telephone is granted.
1876	Bell transmits the sentence *"Mr. Watson, come here! I want to see you!"* using a liquid transmitter and an electromagnetic receiver
1876	Elisha Gray designs a liquid transmitter for use with a telephone, but does not build one.
1877	Bell's US patent No. 186,787 is granted for an electromagnetic telephone using permanent magnets, iron diaphragms, and a call bell.
1877	Emile Berliner invents the telephone transmitter.
1877	Edison files for a patent on a carbon (graphite) transmitter. Patent No. 474,230 was granted on 3 May 1892, after a 15-year delay because of litigation. Edison was granted patent No. 222,390 for a carbon granules transmitter in 1879.
1877	the Scientific American publishes the invention from Bell
1877	The article in the Scientific American is discussed at the Telegraphenamt in Berlin
1877	The first commercial telephone company enters telephone business in Friedrichsberg
1877	The first experimental Telephone Exchange in Boston.
1877	First long-distance telephone line
1878	The first commercial US telephone exchange opened in New Haven, Connecticut.
1887	Tivadar Puskás introduced the multiplex switchboard.
1900	
1915	First US coast-to-coast long-distance telephone call
1920	20 million fixed line telephones worldwide installed

(continued)

Table 5 (continued)

1974	Western Union places Weststar satellite in operation.
1974	Western Union launches the first Weststar Satellite
1990	1000 million fixed line telephones installed worldwide

telephone exchange was developed by the Hungarian engineer, Tivadar Puskas, while working on a telegraph exchange for Thomas Edison. Puskas design was first used in New Haven, Connecticut, in 1878 and served 21 subscribers. The design was improved later by George W. Coy. These first exchange switchboards, although manually operated, were the forerunner of today's Internet switching and routing devices.

Building an Ecosystem

1877 AT&T and the Bell System in USA
When Graham Bell started to work on the telephone in 1873 he was a professor at the Boston University. He became introduced to Gardiner Hubbard and Thomas Sanders—two wealthy patrons who decided to fund Bell's work on the "multiple tone telegraph." After the successful patent filling, Gardiner Hubbard organized the Bell Telephone Company to hold Bell's patents. The company was merged with a sister company in 1879 and formed two new companies: the National Bell Telephone Company and the American Telephone & Telegraph Company. By the end of 1899 both companies merged and became American Telephone & Telegraph Company—AT&T. AT&T became, within a short time, the monopolized provider of telephone services in the US and the head of the so-called Bell System, introducing and implementing most of the telephone related standards in the US.

The Bell System was a group of companies led by the Bell Telephone Company and from 1899 led by AT&T. AT&T provided telephone service to the US and Canada from 1877 to 1984 as de facto monopoly. In 1984, the system was broken into independent companies by a US Justice Department mandate.

1880 Creating Standards: NTEA, AIEE and ASA
The Bell System was the driving force behind the creation of the US telephone industry and the standardization framework. In 1880, there were no telephone standards. In the US, the first networks were built by entrepreneurs licensing the Bell patents, the infrastructure and components were delivered by up to five different manufacturers following different ideas about design and procedures. The situation began to change when the Bell managers, led by Theodore N. Vail and Gardiner Hubbard, started to consolidate the production of telephone equipment within one company, Western Electric. Vail served as president of AT&T after it's founding in 1885 and was one of the drivers behind telephone standardization until his retirement in 1919. As the first platform, the National Telephone

Exchange Association (NTEA) was used. The NTEA was a typical US association founded by a group of telephone companies and equipment producers. In the committees of the NTEA, engineers proposed standards for construction, production and operation of telephone networks based on their daily working experience. These proposals could then have turned into a proposal or approved standard accepted by the telephone industry. AT&T supported that strategy to create efficient production and operation methods for their network and the components delivered by Western Electric. Standardization became the sustaining ideology of AT&T and Bell, as holder of the telephone patents. Within AT&T, Bancroft Gherardi, as vice president and chief engineer, was the main advocate for the standardization strategy. Together with John Carty and Frank Jewett, Gherardi developed the so-called "Bell System" as a framework of standardized development, production and operation of the AT&T telephone network in the US. He was an engineer by heart and used standardization primarily as a tool to manage complexity.

Besides the standardization initiatives within Bell, AT&T and other telephone companies using the NTEA platform, Gherardi also understood that AT&T must take part in other standard organizations. In 1920, AT&T engineers reached out to the world outside the Bell System and joined standard organization like the AIEE (American Institute for Electrical Engineering) and the ASA (American Standard Association). The AIEE was founded 1884 by some of the most prominent inventors and engineers like Niklaus Tesla, Thomas A. Edison, Elihu Thomson, Edwin Houston and Edward Weston. Its goal was to: *"promote the Arts and Sciences connected with the production and utilization of electricity and the welfare of those employed in these Industries: by means of social intercourse, the reading and discussion of professional papers and the circulation by means of publication among members and associates of information thus obtained."* The AIEE and the IRE (Institute of Radio Engineering) merged in 1963 to form the "Institute of Electrical and Electronics Engineers" (IEEE), becoming the world's largest technical society. Gherardi himself served as president of the AIEE from 1927 to 1928 and ASA from 1931 to 1932. The role of the AIEE was to concentrate on standards regarding electrical engineering thus covering all aspects of the electric industry. The ASA was originally formed in 1918 as the American Engineering Standards Committee (AESC) by five engineering societies and three government agencies to create a platform for the coordination of industry standards between industry associations and the government. In 1966, ASA was reorganized and became, in 1969, the American National Standards Institute (ANSI). Even today, it serves as the national umbrella organization to coordinate national industry and government interests as a national standard organization and acts as hub to international standard organizations like ISO.

Due to the US government's regulators interventions, AT&T divested almost all its international interest in 1925 with the exception of the Canadian Bell Telephone Company. The European divisions were acquired by the International Telephone & Telegraph Company (today ITT). The research and development activities were concentrated in the Bell Telephone Laboratories, which were later, renamed the

Bell Labs. As the Bell Labs, they became one of the most important research institutions between 1940 and 1980. Disruptive inventions and technologies like the transistor, the laser, the UNIX operating system and the C programming language were developed by researchers in the Bell Labs. In 1996, AT&T decided to move the Bell Labs to the new founded Lucent Technologies, which was merged, with the French Alcatel in 2006. AT&T itself is still the largest provider of telephone services in the US with revenue of around $14.8 billion.

1877 Germany
In Germany, the approach was different. Although the German, Philipp Reis, built the first working telephone in 1861, there were no further efforts to use the new technology. Following the breaking news from the US, the Berlin "Generalpostmeister" Heinrich von Stephan decided to build an experimental line over two km in 1877. The test operation was successful and Siemens & Halske was commissioned to produce additional telephones. The first telephone exchange was erected near Berlin in 1877 and the daily production of telephones by Siemens grew rapidly to 200 per day. In the following 5 years, several local telephone networks were constructed. Beginning in 1883, those local networks were connected which led to a high coverage of telephone connection within Germany. On August 6th, 1900, the remote connection between Berlin and Paris was opened. By 1936, the German telephone network served 3.39 million subscribers and had a range of 25 million km. The telephone network was built and operated by the federal Deutsche Post until the 1990s when the telephone branch was split and Deutsche Telekom was formed. With the deregulation of the communication industry in 1998, the monopoly by the German telephone provider Telekom disappeared.

1878 Great Britain
In January 1878, Alexander Graham Bell had the honor of demonstrating the telephone to Queen Victoria. In the same year, the first telephone line in the UK was installed connecting the Manchester premises of Mr. John Hudson with his other premises. The line was erected by David Moseley and Sons. Soon, the number of installed telephones across the country grew and it became necessary to construct telephone exchanges. The first telephone exchange was opened in London in August 1879.

In the beginning, the telephone service in the UK was provided by private companies and later by the General Post Office (GPO). In 1912, the GPO took over the private sector telephone companies. Telegraph and telephone services became the exclusive responsibility of the Post Office Engineering Department. British Telecommunications (BT) was formed in 1980, and became independent of the Post Office in 1981. British Telecommunications was privatized in 1984, with some 50 percent of its shares sold to investors. The government sold its remaining stake in further sales in 1991 and 1993.

Other European countries followed the same approach in creating federal controlled companies to provide telephone services. Most of these organizations were part of the Federal Postal Administration of the country. In nearly all cases, the

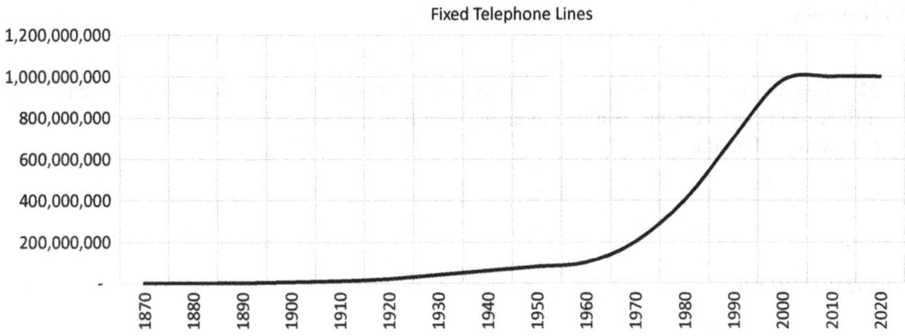

Fig. 5 Number of installed telephones

deregulation phase in the 1980s lead to a split up of telecommunication enterprises and the subsequent privatization.

Social and Economic Impact

The impact of the telephone on society was much broader than with the telegraph. Telephone was the first international communication service and thus also the first publicly available network based service ecosystem. The number of telephone connections was low in the beginning but by the 1950s, having a telephone connection at home was achievable. In tandem, the telephone companies tried to provide numerous public telephone booths to deliver the service to a broad range of customers. The payphone was invented by William Gray in 1889. The first public phone booths were installed at the beginning of the twentieth century (London 1903). At the same time, public phone booths were provided by the national post offices in their branches and by other public places like restaurants. All those public phones were coin operated.

By 1920, there were 20 million telephones connected worldwide, by 1960 the number increased to 100 million reaching its peak in 1990 with 1000 million fixed line connections.[4] At the beginning of the twenty first century the number of fixed telephone connections stagnated due to the competition from mobile phone technology (see Fig. 5).

[4]ITU Statistics, 2015: https://www.itu.int/en/ITU-D/Statistics/Pages/stat/default.aspx retrieved 2017-01-20.

Wireless

> My chief trouble was that the idea was so elementary, so simple in logic that it seemed difficult to believe no one else had thought of putting it in practice.
> Guglielmo Marconi

Technology

1895 Marconi

The idea of wireless transmission of electric signals was developed in the middle of the nineteenth century by (amongst others) the German scientist Heinrich Hertz. Hertz could prove the existence of electromagnetic radiation, called at that time "Hertzian waves." Hertz died in 1894 and the standard unit for frequency is named after him. The practical application of radio waves is attributed to the Italian, Guglielmo Marconi, and was first demonstrated 1899. Marconi, son of the Italian aristocrat and landowner Giuseppe Marconi and Annie Jameson (granddaughter of the founder of the Irish whiskey distillers Jameson & Sons), started as a young man with electrical experiments at the University of Bologna, following Hertz's ideas and searching for a method for wireless transmission of signals. Marconi's transmission technology was based on a spark transmitter, used to send telegraph-type signals, not voice or any other type of sound. In 1895, he could send a signal over 2.4 km across a hill at his father's estate. He first tried to find funding in Italy, but received no answers from the officials. The story goes that the Italian Minister of Post and Telegraphs wrote the remark "to the Longara" on Marconi's letter referring to a famous insane asylum in Rome.

1897 Proof of Concept in the UK

Finding no support and interest from Italian officials, he decided to move to England. Within a short time, he gained the interest and support of William Preece, the Chief Electrical Engineer of the British Post Office. A series of test transmissions were organized and on May 13th, 1897 Marconi send the first message over open water across the Bristol Channel over 6 km. A series of additional demonstrations followed including tests in La Spezia, Italy in 1897, a test transmission between England and Ireland in 1888 and the crossing of the Channel in 1899. In 1899, Marconi was invited to the US and could demonstrate the first ship-to-land transmission, also reporting the results of the Americas Cup yacht races to New York. In 1897, Marconi was granted a British patent and started to work on the first transatlantic transmission in 1901. Failure or ambiguous first results led to a structured test scenario. In 1902, Marconi installed a wireless station on the ship SS Philadelphia sailing west from Great Britain and exchanging signals with the Marconi high-power station at Poldhu, Cornwall. The test scenario was successful. One result was that radio signals travel much farther at night than in the day. On December 17th, 1902, the first transatlantic wireless transmission was send and received successfully between the Marconi station in Glace Bay, Nova Scotia, Canada and Cornwall.

1897 First Business and Marconi's Wireless Telegraph and Signal Company
On July 1897, the "Wireless Telegraph and Signal Company" was formed by Marconi, renamed in 1900 to "Marconi's Wireless Company" with headquarters in London. The first regular transatlantic wireless service was introduced in 1907, but for many years the company had problems providing stable connections. The Marconi company started to build high-power stations on both sides of the Atlantic offering wireless services to ships crossing the ocean thus creating the first "mobile communication" services. In 1900, Marconi founded the Marconi International Marine Communication Co. to provide and operate that segment of the business and the Marconi Wireless Telegraph Company of America. The operators on board the ships were not employed by the shipping company but by the Marconi company, offering the wireless service as a kind of shop-in-shop to the passengers. The wireless telegraph played a major role in the "Titanic" catastrophe in 1912. The ship received several ice warnings hours before the collision with the iceberg. The discussion about how serious those messages were taken fills libraries. During the sinking of the ship, the wireless Marconi operator sent the first "S.O.S." signals finally received by the SS Carpathia that arrived in time to rescue 700 of the 2200 people on board.

Wireless communication turned out to be vital for the Navy and for administrational overseas communication. In 1912, Marconi received an order from the British government to build a wireless communication network between London and the British Overseas Territories. The project was overshadowed by the Marconi scandal. Members of the British government were accused having used insider information and had bought Marconi shares in an American subsidiary. In the following decades, the different Marconi companies in Europe and the US went in different directions. In the US, Marconi was acquired by RCA in 1920. The UK based, Marconi's Wireless Telegraph Company, was one on the founding members of the BBC. By 2000, the Marconi companies went through numerous transitions and were spilt into several different branches. In 1946, the UK Marconi Electronic Company was acquired by English Electric. English Electric was itself acquired by the UK GEC (General Electric Company) in 1968. Parts of the remaining Marconi Company were acquired by Ericsson in 2006.

Today, several enterprises still have "Marconi" in their company name, most of them having only very thin relations to the original Marconi Wireless Telegraph Company. Guglielmo Marconi died in 1937.

1904 The Next Step: Electronic Tubes
One problem of Marconi's invention was that they continued to use the spark transmitter technology for a long time despite the fact that a new and much more efficient device was already available. Many scientists had experimented with vacuum tubes since the 1880s, including Thomas A. Edison and Nikolaus Tesla. Vacuum tubes can be used to control the electric currents between electrodes in an evacuated glass container.

Based on earlier works by Thomas A. Edison, Frederic Guthrie, Nikolaus Tesla and others, John Ambrose Fleming of University College London succeeded in constructing the first practical application of a vacuum tube (or "thermionic valve") in 1904. It soon turned out that this device could be used as a detector in electrical wireless telegraphs. Fleming received a patent in 1904 (US Patent 803,684—Instrument for converting alternating electric currents into continuous currents) and worked also as a consultant for Marconi and the Edison Electric Light Company.

Following the ideas of Fleming, Lee Deforest, in the US, added a control grid to the vacuum tube turning Fleming's "Diode" into a "Triode" which made it possible to use the so called "Audion-Tube" as an amplifier for received radio signals. Deforest also received a patent (US Patent 879,532) for his Audion. In 1906, though both inventions were granted UK and US patents, a long-lasting court battle started with many different stages of victory for both sides. Earlier in 1902, Deforest founded, together with his sponsor Abraham White, the "American Deforest Wireless Telegraph Company". The vision was to create the first "world-wide wireless." One of its major contracts was the construction of a first wireless network for the US Navy. After some internal conflicts with his partner, Deforest decided to resign from the company and cashed in his stock for $1000 in November 1906. American Deforest was then reorganized as the United Wireless Telegraph Company. For some years, it became the largest and dominant US radio communication enterprise until its bankruptcy in 1912.

Building Business: The Begin of the Electronic Industry

At the begin of the twentieth century, the spark transmitter was replaced by the electronic vacuum tube. It became the first "electronic" component of the electronic industry and was the basic building block for radios, televisions and the first computers until the invention of the transistor in 1948. The triode was the basic device for the construction of amplifiers for long-distance telephone and radio communications, radars, and later the early electronic digital computers.

In tandem with developing their inventions, Marconi, Deforest and Fleming built their own businesses while the existing electrical industry in the US and Europe started to take over the new technology to create markets and business. Those companies entered the market with some decades of experience in building products and in the transition of prototypes to mass production.

1904 Telefunken in Germany

In Germany, Siemens & Halske AG (Siemens) and Allgemeine Elektrizitätgesellschaft (AEG) formed Telefunken to commercialize the new technology in 1904. In the beginning of the twentieth century, both firms already belonged to the largest electrical industries worldwide. Siemens was founded in 1847 by Werner Siemens. It had a long history in electrical engineering and the production of electrical telegraph equipment. AEG was founded in 1883 in Berlin as "Deutsche Edison-Gesellschaft für angewandte Elektricität." Both partners provided knowledge and organizational structures to create new industries based on the electronic technology. In the following years, Telefunken played a major role not only in developing the new vacuum tube technology but also in building a new business ecosystem.

1906 AT&T and Lee Deforest's Audion Tube
In the US, it was General Electric (GE), America Telephone and Telegraph (AT&T) and Westinghouse that took over the electronic industry. All three companies had been major players in the new electrical industry since the end of the nineteenth century. General electric had produced and improved the light bulb and X-ray tubes since 1890. AT&T, together with Lee Deforest, improved the line amplifier and commercialized the new technology with the mass production of the Audion tube. Westinghouse had the capabilities to produce electronic tubes during World War I.

Between 1900 and 1920, wireless radio communication became the first application of electronics and was used mainly for military purposes (Navy), federal communication networks and civil applications on passenger ships. But the next big thing was already waiting.

Standards

1906 International Radiotelegraph Convention
In the first decade of the twentieth century, it became obvious that the new technology began to spread and that international regulations about the usage should be settled. In 1906, the first International Radiotelegraph Conventions was signed in Berlin. It concentrated on the service regulations and the intercommunication between ships. It also introduced the SOS-signal and established the governmental licensing requirements to operate ship stations. Also, a first official list of radiotelegraph stations was established and an agreement about the charges for radio telegrams was signed.

1912 Standards for Sending Stations
At the International Radiotelegraph Conference in London in 1912, the International Radiotelegraph Convention was signed establishing additional standards for call signs as unique identification of sending stations. A system of prefixes for certain major countries was defined (A-Germany, F-France, B-Great Britain) thus introducing the first global system of unique addressing. To improve the efficiency and quality of operation, the 1927 Washington conference allocated frequency bands to the various radio services (fixed, maritime and aeronautical mobile, broadcasting, amateur, and experimental) (ITU History, 2016). Up to today, World Radio Communication Conferences were held frequently—every

3 years—and produce the ITU Radio Regulations to provide the framework for global harmonization in the use of the radio-frequency spectrum.

Broadcasting

One major difference between the electrical telephone and radio communication is that a signal sent could be received by everybody who has a receiver device. Based on the mass production processes developed during the first two decades of the twentieth century, it became clear that broadcasting information to a large audience is the next step. Comparable to the telephone, radio broadcasting is also a mass market but has a completely different range of possible services including information, entertainment and advertising. Additionally, it soon turned out that radio broadcasting is the most efficient media channel for political messaging. The 1920s were the starting point of mass media. The missing element was the business ecosystems moving that huge possibility to a global industry.

Technology

1906 Amplitude Modulation
At the beginning of the 1920s, the basic elements of technology for radio broadcasting technology were already there. One of the last gaps was the necessary split of frequencies so multiple broadcasts from different senders would not interfere. The Canadian inventor Reginald Fessenden together with Lee Deforest solved that problem in 1906 inventing the amplitude modulation (AM). Using amplitude modulation each sender could use his band of frequency as opposed to the spark radio, which covers the entire bandwidth. Fessenden could demonstrate the functionality of the new transmitter on December 21st, 1906 at Brant Rock, Massachusetts.

1909 The First Radio Program
After World War I, the new application of wireless communication found its first prototypes. In the US, first experiments took place as early as April 1909. Charles "Doc" Herold started a daily program from his Herold electronics school in San Jose ("San Jose calling"). This first radio operation was interrupted by World War I. Starting with 1915, the experimental radio station 2XG in New York City ("Highbridge Station") licensed to Deforest Radio Telephone and Telegraph Company sent a news and entertainment program on a regular schedule. In 1916, 2XG was the first radio station using a vacuum-tube transmitter. In November 1919, the Dutch entrepreneur and manufacturer with the long name Hanso Schotanus a Steringa Idzerda, sent the first Dutch broadcasting from his private apartment in Den Haag. In Germany, the first broadcast was a Christmas concert sent from

Königs Wusterhausen and operated by the German Reichspost in December 1920. In the UK, the first experimental radio broadcasting was organized from Marconi's factory in Chelmsford in 1920. It was a short entertainment program featuring Dame Nellie Melba, a famous Australian opera soprano, and sponsored by the Daily Mail's Lord Northcliffe. Other European countries followed with experimental broadcasts some of them organized by private persons, some of the already produced by federal post offices.

Building Radio Business

1921 The Radio Corporation of America

Within short time, the industry recognized the huge acceptance of the new media and started to enter the ecosystem. In the US, it was General Electric (GE), America Telephone and Telegraph (AT&T) and Westinghouse that joined forces to enter the market. In 1921, these three companies formed the Radio Corporation of America (RCA) as patent pool for radio-related patents. From that point of time until the 1960s RCA was the driving power behind the creation of radio broadcasting industry in the US (ETHW, 2016). David Sarnoff was promoted to general manager and immediately started to build a nationwide network of more the 200 distribution outlets. By 1925, RCA had also built and operated four broadcasting networks. Within 3 years, RCAs revenue reached $50 million. Within a short time, many private radio networks were founded, a part of them owned by manufacturers and department stores interested in selling radios, others owned by newspapers to sell newspapers or express their opinions. The ecosystem created consisted of RCA as the patent holder of the technology and distributer of the radios to the consumers and GE, AT&T and Westinghouse as production platforms for the equipment. This structure forming a de facto monopoly of RCA proved to be stable until the 1960s when, within a few years, the US consumer electronics industry lost its market to Japan (Chandler, 2005).

1922 The British Broadcasting Company

In the UK, a consortium of 6 companies decided to follow the same path as in the US. On October 18th, 1922, the British Broadcasting Company (BBC) was founded by a group of UK and US electrical equipment manufacturers: Marconi's Wireless Telegraph Company, Vickers Electrical Company, Radio Communication Company, The British Thomson-Houston Company, The General Electric Company and The Western Electric Company. The target of the new enterprise was to offer a broadcasting program to grow sales of radio receivers. It turned out soon that this setting was more a public service than a commercial enterprise. Sales of receivers were interfered with by rival non-licensed devices and the program schedule had to follow governmental guidelines. Discussions between the BBC and the General Post Office (GPO) about the future of the BBC lasted until 1926 when a general strike broke out in May, interrupting newspaper production. Within a couple of hours, the BBC became the single source of information for the UK citizens. The BBC director and co-founder, John Reith, used the BBC's information monopoly to manage the crises together with the PM Stanley Baldwin, including giving the PM the chance to speak over the radio from Reith' private home. In the following years, the BBC was settled as the radio monopoly in the UK. There was no advertising and income was generated from a tax for receiving sets. The monopoly of the BBC lasted until the 1980s when, due to the deregulation of the UK media market, the BBC faced increased competition from the commercial sector.

The UK model was the blueprint for most of the European countries. Germany (ARD, ZDF), France (Radio France), Italy (RAI), Austria (RAVAG), Spain (RNE) and others organized the radio—and later television—ecosystem in the same way. With the founding of federal controlled radio companies, they created a monopoly that lasted until the European deregulation phase in the 1980s.

Standards and Regulation

Compared to other network based technologies, radio broadcasting was also based on new paradigm regarding the usage of the resource of public "air space." Providing a broadcast service means simply that the sending network is occupying a certain frequency band within its reach. It soon became clear that the allocation of bandwidth is one of the basic regulation requirements when organizing a nexus of competing radio broadcasting networks.

1912 The Institute of Radio Engineers
From the beginning, technical standards had to be developed. In 1912, the Institute of Radio Engineers (IRE) was founded as a platform for engineers and producers of radio equipment to discuss and approve standards. The IRE was merged in 1962 with the American Institute of Electrical Engineering (AIEE), which became the Institute of Electrical and Electronics Engineering (IEEE).

1919 Federal Rules and Regulations in USA, UK and Germany
It was also clear that the ownership of the frequencies should be in the hands of the state to allow for clear regulations. In the US, radio regulation was introduced in 1910 passing an initial, modest legislation about the usage of radio technology. Following the rapid growth of broadcasting during 1920s, the US Radio Act was passed in 1927. In 1934, the Federal Communication Commission was established as permanent body to determine regulations on radio and television. In Germany, the government declared in 1919 the so-called "Funkhoheit" as principle right of the state to operate radio stations as well as receivers. A ban on receiving radio transmissions by private persons was abolished in 1923. The broadcasting fee was introduced in 1923 thus also providing funding for the federal radio network. In the UK, broadcasting was regulated by the Radio Authority, which was moved into the Office of Communication (Ofcom) in 2002.

Social Impact

The impact of radio broadcasting on society and culture was immense. For the first time, a mass media was available to distribute news, entertainment and political expression to everybody.

In Britain, the first broadcast in 1920 immediately caught the peoples' imagination and created public enthusiasm forcing the government to regulate the new media. In the US, radio broadcasting was quickly adopted and reached a first climax in the 1930s. Its power and social influence became most obvious in the effect of Orson Wells' War of the Worlds broadcast. On Halloween night in 1938, Orson Wells used an adaption of H.G.Wells "War of the Worlds" as script for the "Mercury Theatre on the Air" radio show. The presentation was so realistic that listeners were caught up by the realism, believing the story to be true. In the 1930s and 1940s radio became the most important media for political messaging. The German regime used radio to influence the public opinion by broadcasting major political speeches. In the UK, King George VI speech on Britain's entry into World War II on September 3rd, 1939 had a tremendous influence on the spirit of the people. During the war, radio was the major source of information and it kept this role for the following decades. Today, radio is still a strong part of the media landscape, although TV in the 1950s and 1960s and later the Internet news services have taken over.

Status Before Internet and Cloud Computing

Let us stop here for a moment. By 1950 and much more by 1980, many network-based service ecosystems are part of the daily life of a major part of the world's population. Many people in the Western world and a growing number of the population in the less developed countries are using network based service ecosystems now. And they are not only using them, many of these services are a vital part of their daily private or business life.

In 1990, 76% of the world population has water at home. There are 700 million fixed phones used worldwide. More than 75% of the households have electricity. Public transport systems like railway, metro lines or bus cover most of the globes populated surface, radio and television has reached nearly 100% coverage. All these services are network based service ecosystems. By 2014, the number of mobile phones exceeded the world population with more than 7000 Mio mobile phones. We have already moved ownership to those cloudy services and gave up responsibility in exchange to a much higher level of comfort and quality. Imagining that only one of these services is interrupted, it would cause chaos and instability. We are used to accepting network based service ecosystems for more than 100 years or three generations.

So, what's the problem when we take the next step and accept cloudiness and Internet cloud services as natural expansion of our comfort? Comparing other service ecosystems with today's cloud services may lead to the assumption that cloud services are the same kind of service ecosystems, only based on another new technology. Although all the major elements of service ecosystems (network, end-devices, business models, standards) can be identified within Internet, The Web and cloud services, there are some huge differences as well.

- Internet and cloud services are multi-purpose systems: todays' cloud services provide a large and rapidly growing number of different solutions and applications, we are dealing with a multi-purpose service ecosphere.
- Change and development of new aspects of network technology (e.g. IoT) is accelerating dramatically.
- The number of players in terms of providers of technology, providers of services is much larger than in other service ecosystems.
- Cloud services and other types of Internet based services are distributed worldwide, regional borders do no longer exist.
- The integration and coupling with other industries and services is also increasing very fast. A growing number of businesses are going the way of digitalization, which means that the usage of Internet, Web and cloud services is becoming a major part for their production. Operation and customer relationship.
- New technologies like the Internet of Things or Cognitive Computing are built partly on the platform of Internet and cloud computing. It could be the case that these new technologies have a much bigger impact on our daily life than cloud computing.

To understand the new world of cloud computing we must distinguish between cloud computing as the set of technologies used (Internet, WWW, computing, virtualization, etc.) and the term "cloud services" which describes the distribution of services and solutions delivered to the customer and based on those technologies (Google search, Facebook, Ebay, Amazon Web Services and thousands of others). Cloud services are based on cloud computing, which can be described as a layered model of technologies, having the network at the bottom and the most advanced technologies like Internet of Things on top of that stack. One target of this book is to accompany the reader through these layers' step by step. Starting at the bottom, the first elements are network, computing (or computers) and virtualization. These are the basic technologies used to build the Internet and later the cloud computing industry. In the next chapters, we will focus on the making of these technologies.

References

Chandler, A. D. (2005). *Inventing the electronic century. Harvard studies in business history*. New York: Free Press.

ETHW. (2016). *Engineering and technology history Wiki – RCA*. Retrieved 2016, from http://ethw.org/RCA_%28Radio_Corporation_of_America%29

ITU. (2014). *www.itu.int*. Retrieved from Information technology – Cloud computing – Reference architecture: http://www.itu.int/rec/T-REC-Y.3502-201408-I
ITU History. (2016). Retrieved 2016, from http://www.itu.int/en/history/Pages/ITUsHistory-page-2.aspx
Standage, T. (1998). *The Victorian internet*. London: Weidenfeld & Nicolson.
Wenzlhuemer, R. (2010). *The history of standardisation in Europe*. Retrieved from http://ieg-ego.eu: http://ieg-ego.eu/en/threads/transnational-movements-and-organisations/internationalism/roland-wenzlhuemer-the-history-of-standardisation-in-europe

Making of Digital Computers

Computers are useless. They can only give you answers.
Pablo Picasso
Man is still the most extraordinary computer of all
John F. Kennedy

Computing has gone through many phases since it was first commercially introduced in the 1960s. It started with mainframe computers, which almost disappeared as personal computers and servers became popular at the end of the 1980s. This and the invention of capable networking technologies like Ethernet, made computing accessible for more people and businesses in a much more efficient and cheap way than before.

This initiated a major change in the computing ecosystem by moving away from centralized to distributed computing. Personal computing soon got widely accepted and we will describe how two ecosystems based on Windows and Apple evolved over the past 25 years and still dominate modern computing.

We will discuss how workstations evolved and how servers became the building blocks of modern data centers as the basis of virtualization, before we consider one of the latest evolutions in computing: the renaissance of mainframes at the beginning of the 2010s.

Computing already began evolving long before the twenty-first century, when most of the modern computing we know and remember today began to appear and started to transform our lives. The cradle of computing dates back several thousand years and started with mechanical machines that were used for simple calculation tasks (see Table 1).

M. Oppitz, P. Tomsu, *Inventing the Cloud Century*,
DOI 10.1007/978-3-319-61161-7_4

Table 1 Computing time line

−2300	Abacus as first calculating device
−100	Antikythera mechanism
1830	
1837	Babbage publishes description of analytical engine
1890	
1890	Hollerits tabulating machines used for the US census
1910	
1911	IBM founded
1930	
1931	Kurt Gödel: "Über formal unentscheidbare Sätze der Principia Mathematica und verwandter Systeme"
1936	Alan Turing "On Computable Numbers"
1940	
1940	Claude Shannon publishes his theory about computing
1941	Zuse constructs his Z3 as first programmable electromechanical computer
1945	First Draft of a Report on the EDVAC
1946	ENIAC
1948	Design of the Princeton Computer
1948	The "Manchester Baby" is operated as Small Scale Experimental Machine
1949	EDVAC goes into operation
1950	
1951	Sperry Rand delivers the first UNIVAC to the US Bureau of Census
1957	Ken Olsen founds Digital Equipment Corporation (DEC)
1957	William Norris founds Control Data Corporation (CDC)
1957	Honeywell introduces D-1000
1960	
1961	Burroughs introduces BS5000
1962	IBM introduces System 360
1962	NCR introduces NCR-315
1964	DEC introduces ist PDP-8
1969	VDU (Video Display Unit)
1970	
1971	DEC VT100
1971	IBM 3270
1972	Intel introduces Intel 4004 processor
1973	First Graphical User Interface (Xerox Alto)
1977	First PCs: Apple II
1977	First PCs: Commodore
1978	DEC introduces VAX 11
1980	
1980	Data General introduces Eclipse MV/8000 as on of the first 32-bit architectures
1981	IBM PC+"Clones"
1981	Terminal Emulations on PCs (IBM 3270, DEC VT100)

(continued)

Table 1 (continued)

1982	Apple Macintosh
1982	Microsoft Windows
1990	
1990	Next introduces nextStation
1993	Apple introduces Power Mac G3
2000	
2003	First 64-bit architectures introduced by Compaq, HP, Apple
2007	Apple introduces iPhone
2008	The MacBook Air and other laptops rely totally on flash memory for storage
2010	
2010	Apple introduces iPad
2015	IBM introduces ist z13 mainframe architecture

History of Computing

The earliest computation tool we know was the abacus, which was developed between 2700 and 2300 BCE by the Sumerians. The abacus was actually a table of successive columns that delimited the successive orders of magnitude of the sexagesimal number system.

There were several, other mechanical approaches over the following centuries like the Antikythera mechanism (Christian C. Carman, 2014), which was used to calculate astronomical positions and was developed around 100 BC by the ancient Greeks.

From Mechanical to Electrical Computing

The evolution of computing took until the nineteenth century before it finally reached the status of what we call a modern computer, albeit still based on pure mechanics. In 1837, the Analytical Engine (Freiberger, 2014) was a proposed mechanical general-purpose computer designed by the English mathematician and computer pioneer, Charles Babbage. The Analytical Engine allowed for the expansion of memory, used an arithmetic unit and logic processing capabilities and interpreted a programming language with loops and conditional branching.

In the twentieth century, the evolution of computers went from mechanical to electrical and later electronic elements but already, all these digital machines were able to render states of numeric values and store individual digits at first using relays until the 1940s and from then on transistors.

1928–1936: Mathematical Theory—Gödel, Turing and von Neumann

In 1928, the "Entscheidunsproblem" (German for decision problem) was a challenge posted by the German mathematician David Hilbert that asked a simple question: Is it possible to decide if a given logical statement is valid or not[1]?

In 1931, the Austrian mathematician, Kurt Gödel, published a paper on the "Entscheidungsproblem," stating that no system of axioms is capable of proving all truths. For any such system, there will always be statements that are true, but that are improvable within the system. The second incompleteness theorem, an extension of the first, shows that such a system cannot demonstrate its own consistency.

This was bad news for the so-called Hilbert-program,[2] a challenge introduced by David Hilbert consisting of 23 problems, where the decision problem was listed as number 10. The basic idea behind the Hilbert program was to create a solid foundation of proven axioms, which can be used as the basis for the related axioms and theorems in mathematics. Gödel struck a heavy blow to the Hilbert program and the hope in creating a solid base for theoretical mathematics with his paper "*Über formal unentscheidbare Sätze der Principia Mathematica und verwandter Systeme I*" (Gödel, 1931). Gödel simply proved that complete certainty in a decision is not possible there will always be problems and theses that cannot be decided.

1936 The Turing Machine

In 1936, Alan Turing published his paper "On computable numbers, with an application to the Entscheidungsproblem".[3] From a theoretical point of view, this was the starting point of the digital age. Turing concentrated his paper on a very specific question related to the Entscheidungsproblem (Turing A. M., 1936) to answer this persistent problem: Is it possible to decide if an algorithm, once started and working on a given input, will stop and produce a result or run forever?

Turing's paper and approach was dramatically fundamental for the theory of mathematics as well as for the dawning of the digital age. It contains two revolutionary assumptions: the Entscheidungsproblem is unsolvable in the general case

[1]DecodedScience, DeHaan, The Turing Machine versus the Decision Problem of Hilbert, https://www.decodedscience.org/the-turing-machine-versus-the-decision-problem-of-hilbert/14072, 2012, retrieved 2017-01-27.

[2]philsci, Zach Robert, Hilbert's Program Then and Now, http://philsci-archive.pitt.edu/2547/1/hptn.pdf, 2005, retrieved 2017-01-27.

[3]cs.virginia.edu, Turing Alan, https://www.cs.virginia.edu/~robins/Turing_Paper_1936.pdf, 1936, retrieved 2017-01-27.

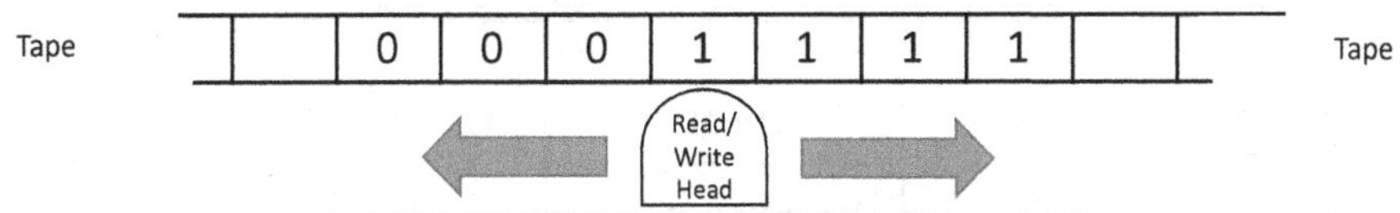

Program:

Status	Symbol read		Write Symbol	New Status	Move
A	0	→	1	B	R
A	1	→	1	C	L
B	0	→	1	A	L
B	1	→	1	B	R
C	0	→	1	B	L
C	1	→	1	HALT	R

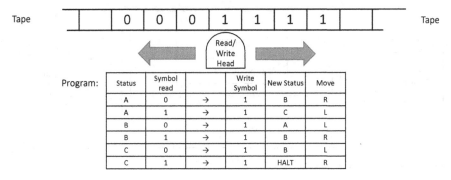

Fig. 1 Turing machine

and a hypothetical machine can be constructed that can solve any solvable problem using simple rules! This hypothetical machine is called today the Turing Machine.[4]

Turing decided to describe a machine consisting of four simple components: a (potentially endless) tape divided into cells which may contain symbols (letters, digits), a head that is able to read, write or delete symbols from the current cell on the tape and also can move the tape to the left or the right, a register storing the current state of the machine and a finite table of instructions as a program that, given the state the machine is currently in and the symbol it is reading on the tape, tells the machine to do the following in sequence (see Fig. 1):

- Either erase or write a symbol and then
- Move the head to the right or left or stay in the same place
- Assume the same or a new state, as prescribed

Turing also extended the approach from the specific machine calculating one specific function and using a specific instruction table as a program to the universal machine. The instruction table is printed as a sequence of symbols on the tape and read by the machine in an initial sequence before processing the calculation: "*It is possible to invent a single machine which can be used to compute any computable sequence. If this machine **U** is supplied with the tape on the beginning of which is written the string of quintuples separated by semicolons of some computing machine **M**, then **U** will compute the same sequence as **M**.*" 'On computable numbers, with an application to the Entscheidungsproblem' from *Proceedings of the London Mathematical Society*, (Ser. 2, Vol. 42, 1937) (Turing, 1937).

John von Neumann used this hypothetical design for the actual design of the Princeton Computer in 1948, at a time when a number of computer projects had already been under way. The most advanced machine was built in Philadelphia in 1946—ENIAC, followed by the EDVAC in 1949.

[4]AlanTuring.net, Copeland Jack, http://www.alanturing.net/turing_archive/pages/reference%20articles/what%20is%20a%20turing%20machine.html, 2000, retrieved 2017-01-26.

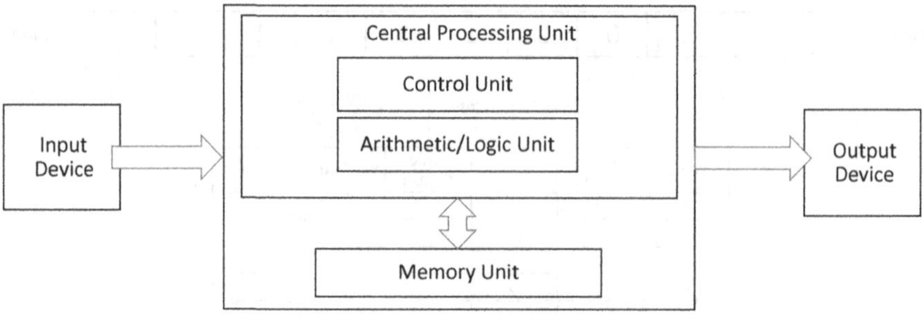

Fig. 2 Von Neumann architecture

Von Neumann Architecture

In an internal report "First Draft of a Report on the EDVAC,"[5] Neumann describes the basic elements and functions of a digital computer, today known as the "von Neumann architecture."[6] The von Neumann architecture (Macrae, 1992) consists of a processing unit containing an arithmetic logic unit and processor registers, a control unit containing an instruction register and program counter, a memory to store both data and instructions, external mass storage, and input and output mechanisms (see Fig. 2).

All digital computers since then have been based on the von Neumann Architecture. With the von Neumann architecture, a number of basic decisions were made that still apply today, several decades later:

- Computers are digital machines (0/1)
- Digital computers are programmed using a simple structure: "do something with a given data"—INSTRUCTION (ADDRESS)
- Digital computers are working asynchronous, the sequence of instructions processed is defined by the clock rate of the machine

These design principles still follow the hypothetical Turing Machine and lead to a number of restrictions and obstacles. Although some optimizations were introduced during the following decades, the basic assumptions are still the same today.

The von Neumann architecture is the basis for all subsequent computer designs up to today and introduced the idea of allowing machine instructions and data to share the same space which consisted of three major parts: the Arithmetic Logical Unit (ALU), the memory and the Instruction Processing Unit (IPU). The von Neumann machine design uses the Reduced Instruction Set Computing (RISC) architecture with a total of 21 instructions for performing all tasks. It is important to

[5]John von Neumann: "First Draft of a report on the EDVAC" http://library.si.edu/digital-library/book/firstdraftofrepo00vonn, retrieved 2016-11-14.

[6]Quora.com, Can someone explain the Von Neumann Architecture?, https://www.quora.com/Can-some-one-explain-the-Von-Neumann-architecture, 2017, retrieved 2017-01-26.

mention that there is another prominent architecture based on Complex Instruction Set Computing (CISC), which uses more instructions for computing.

Between 1940 and the 1950s, a number of experimental computing machines were developed in Europe and US. The German Konrad Zuse constructed the Z3,[7] one of the world's first functional, programmable computers in 1941 (Zuse, 2013). Other milestones were the completion of the Manchester Baby in 1948[8] (Moore S., 2010), Pilot ACE (Automatic Computing Engine)[9] by Turing in the early 1950s (Yates, 1997), or the Mathematical Theory of Communication found by Shannon in the 1940s (Claude and Warren, 1948), plus many more.

Von Neumann Bottleneck: Lack of Parallel Processing
The von Neumann architecture is based on a defined sequence of processing steps executed by the control unit and thus not able to implement types of parallel interactions between elements in more complex structures like social networks, processes in the atmosphere, processes analyzed in atomic physics and many others.

The disadvantage was already recognized by von Neumann himself and is known as the Neumann Bottleneck. Back in 1951, Neumann already discussed the notion of creating architectures consisting of digital machines interacting with each other on a more analog basis. These architectures would follow the idea of networks much more, later described as neuronal networks, and would find a first simple model in the description of semantic networks. Today, artificial intelligence and cognitive computing have developed these ideas into the first running applications. Also, social networks and search engines today can be described as digital & analog architectures consisting of a large number of "von-Neumann" nodes interacting with each other and continuously creating a growing number of references (Dyson, 2012).

Women and the Development of Computers

There is still a debate today as to who built the first computer. One of the first machines, the Eniac 1, was developed in 1942. It was built to help the US army calculate complex flight paths and two men are generally credited as the inventors of that machine, while mostly women did the programming. Francis Betty Snyder Holberton, Betty Jean Jennings Bartik, Kathleen McNulty Mauchly Antonelli, Marlyn Wescoff Meltzer, Ruth Lichterman Teitelbaum and Frances Bilas Spence

---

[7]Inverse.com, Brown Mike, Konrad Zuse's Z3, the World's First Programmable Computer, Was Unveiled 75 Years Ago, https://www.inverse.com/article/15542-konrad-zuse-s-z3-the-world-s-first-programmable-computer-was-unveiled-75-years-ago, 2016, retrieved 2017-01-26.

[8]Manchester.ac.uk, University celebrates Baby's 65th birthday, http://www.manchester.ac.uk/discover/news/university-celebrates-babys-65th-birthday, 2013, retrieved 2017-01-26.

[9]AlanTuring.net, Turing's Automatic Computing Engine, http://www.alanturing.net/turing_archive/archive/infopages/london1st.html, 2017, retrieved 2017-01-26.

were the programmers and they were not taken seriously at this time but 55 years later, they were finally seen as pioneers[10] because of another woman, Kathy Klein.

Kathy Klein, a female Harvard student of information technology, saw pictures of these six women working but there were no names shown. She did not believe that she was told by the Computer History Museum that these six women were models and found their addresses. By interviewing them, the truth came to the surface and finally in 1997, the six Eniac-women were honored during a ceremony in Silicon Valley for their contributions.

The Birth of IBM: The Mother of Mainframes

A guy named Herman Hollerith was the first person who typed data on cards, sorted them and printed them mechanically, which was a major improvement over the manual methods used before.[11] With this invention, Hollerith could improve the processing of the US Census in 1890 from 8 years down to 1 year.

Hollerith also founded a company based in Georgetown (Columbia), Maryland that built punched card equipment. This company was called Tabulating Machine Company and became known as IBM[12] in 1911. Calculators used at these times were purely mechanical, which later changed from mechanical to relay and then to tube, making calculations remarkably faster. In the beginning, arithmetic systems were decimal, but changed to binary in order to better represent the two-state states of the new switching elements. As a next step, a "computer" was put between the card reader and the printer to crunch the data, which marked the beginning of the era of computing (Nallur Prasad, 1994).

1960 Mainframes and Early Computing

Obviously the most important influencers in the area of modern computing, especially at its commercial start, were mainframe computers[13] also called big irons. After dominating these early times of computing, they almost disappeared between 1990 till 2010 due to several inventions like the personal computer (PC) as well as

[10]Der Standard, Hagen Lara, Frauen waren maßgeblich an der Entwicklung von Computern beteiligt, ..., http://mobil.derstandard.at/2000033996437/Die-vergessenen-Wegbereiterinnen-des-Computers, 2016, retrieved 2017-01-26.

[11]IBM100, The Punched Card Tabulator, http://www-03.ibm.com/ibm/history/ibm100/us/en/icons/tabulator/, 2017, retrieved 2017-01-26.

[12]IBM100, IBM Is Founded, http://www-03.ibm.com/ibm/history/ibm100/us/en/icons/founded/, 2017, retrieved 2017-01-26.

[13]IBM, IBM Mainframes, https://www-03.ibm.com/ibm/history/exhibits/mainframe/mainframe_intro.html, 2017, retrieved 2017-01-26.

the evolution of capable networking technologies. We see a renaissance of these mainframes at the beginning of the 2010s. We will discuss the renaissance of mainframes later in this chapter, first we look into what mainframe computers, also commonly called mainframes or big irons, are.

Mainframe (Stephens, 2008) refers to the large enclosures or cabinets housing the central processing unit, as well as the main memory of early computers that dominated computing during the 1960s and 1970s (Stakem, 2015). Big customers like banks, enterprises, airlines, large universities etc. bought these large, room-filling computing systems. Computers at these times were built for two main purposes: business accounting and scientific calculations and, additionally, research.

Early User Interfaces

At these times even writing the simplest types of programs were still cumbersome because of typing the cards. I remember during my time at university (this was in the 1970s), many of us attended a class just to make these computer—beasts actually do something. You had to write down the problems to be solved in programming languages like Fortran (formerly FORTRAN derived from Formula Translation) or COBOL (Common Business Oriented Language) in a series of line-by-line instructions for the process of calculations or analytics you wanted to achieve.

You then had to type each individual line of the handwritten program onto punch cards, which were perforated in a way that the computer could read. These cards had to be carefully arranged in the right order and handed to the operator, who put your cards in the queue behind many others before they were fed into the mainframe. Finally, the machine would spit out the results on accordion-folded paper and, in many cases, the programs had to be corrected several times before you got the desired results, so the whole cumbersome process had to be reiterated again and again.

Mainframe and Virtualization

Besides punch card interfaces there existed two other important ideas revolutionizing and optimizing computing. Since the big irons were not computing and waiting for the next input most of the time, the concept of time-sharing[14] evolved. A special control program had to juggle the resources in order to allow useful

[14]tutorialspoint, Types of Operating Systems, https://www.tutorialspoint.com/operating_system/os_types.htm, 2017, retrieved 2017-01-26.

programs to always run. Another concept for improved usage was to connect remote terminals to the mainframe in order to allow users access from multiple different locations, which later led to the first networking approaches used in computing, marking the invention of LANs (Local Area Networks) in order to connect the terminals to the mainframe. These two inventions were actually the first virtualization mechanisms in computing as they allowed each user to use their own machine at least for some periods of time.

The Big Mainframe Players

A few big, technology companies dominated the mainframe market which was clearly led by IBM[15] (International Business Machines), followed by some smaller companies the biggest being RCA[16] and General Electric, who actually gave up in the 1970s. This left the rest of the market to Burroughs, Control Data Corporation, DEC (Digital Equipment Corporation), Honeywell, NCR[17] and Univac.[18]

Interestingly enough, these companies were headquartered and based primarily in eastern USA, with IBM around New York City, DEC and Wang in the Boston area, Burroughs in Detroit, Univac in Philadelphia, NCR in Dayton and Cray, Honeywell and Control Data around Minneapolis. The only exception was Hewlett-Packard (HP) based in Silicon Valley, but HP concentrated rather on scientific testing and measurement instruments and not computers. The computer industry in the 1970s was mainly developing and building their products for a few, selected big customers, who were able to pay a significant amount of money for their equipment and its maintenance. The customers numbered in the hundreds, so this was far from being a mass market like computing has become over the following years and still is today.

Later, the term mainframe was used to distinguish high-end commercial machines from less powerful units. Most large-scale computer system architectures were established in the 1960s, but continue to evolve today, as we will see when discussing the mainframe renaissance in the middle of the 2015s.

---

[15]IBM, International Business Machines, http://www.ibm.com/us-en/, 2017, retrieved 2017-01-26.

[16]ETHW.org, RCA (Radio Corporation of America), http://ethw.org/RCA_(Radio_Corporation_of_America), 2017, retrieved 2017-01-26.

[17]Computerhistory.org, Company: National Cash Register (NCR) company, http://www.computerhistory.org/brochures/companies.php?alpha=m-p&company=com-42bc20992be7c, 2017, retrieved 2017-01-26.

[18]History.com, 1951 Univac computer dedicated, http://www.history.com/this-day-in-history/univac-computer-dedicated, 2017, retrieved 2017-01-26.

1970 The Rise of Minicomputers

DEC could dominate a new segment of minicomputers,[19] which were used by smaller businesses or departments within larger corporations. DEC was also one of the leaders in paving the way to connecting their computers and servers via an upcoming network technology called Ethernet, which allowed for the building of simpler but, for these times, still very efficient computing systems. Many programmers could now use these systems simultaneously in a much more efficient and less cumbersome way than the big old mainframes with their punch cards. On the upper end of computers was Cray Research[20] a company founded in 1972 that built supercomputers mainly used for scientific purposes and mathematical modeling (Cray Research, 1977). On the lower and cheaper end of computer systems was Wang also founded in the early 1970s, that became famous for their Word Processing System (Wang WPS),[21] which was still not a personal computer, but pretty close to it.

In the 1980s, the area of information technology was changing on many fronts. First minicomputer based systems that were often called departmental computers became sophisticated enough to replace low-end mainframes and one of the most prominent examples was the DEC VAX[22] from Digital Equipment.

1980 Personal Computers

At the end of the 1970s, the new microcomputer industry started to establish itself in Silicon Valley, which, over time, gave birth to the personal computer. All of this started with hobbyist groups forming computer clubs, like the Homebrew Computer Club.[23] Meetings were often held in private garages and during one of these meetings, for example, one of the first microcomputers was shown as a hand assembled kit for under $500. These first microcomputers could not do much but they were the first computers you could own personally, program as often and as long as you wanted without the need to punch your cards before handing the final stack to the operator for input to the machine. This was the time when Bill Gates

[19]Centre for Computing History, 1960, DEC released ist first mini computer: PDP-1, http://www.computinghistory.org.uk/det/5495/DEC-released-its-first-mini-computer-PDP-1/, 2017, retrieved 2017-01-26.

[20]Computerhistory.org, Cray Research, Inc., http://www.computerhistory.org/brochures/companies.php?alpha=a-c&company=com-42b9d5d68b216, 2017, retrieved 2017-01-26.

[21]Wang2200.org, Word Processing on Wang 2200 Systems, http://www.wang2200.org/docs/brochure/2200WP_WordProcessing_Brochure.700-6450.3-81.pdf, 2017, retrieved 2017-01-26.

[22]old-computers.org, VAX 11/780, The First VAX System, http://www.old-computers.com/history/detail.asp?n=20&t=3, 1977, retrieved 2017-01-26.

[23]Computer History Museum, The Homebrew Computer Club, http://www.computerhistory.org/revolution/personal-computers/17/312, 2017, retrieved 2017-01-26.

founded Microsoft in 1975 to develop programming languages for the Altair[24] microcomputer, which was the first of its breed.

In 1971, Intel introduced the Intel 4004[25] microprocessor, which marked the decline of microprocessor costs. The Intel 8008[26] followed in 1972, which was already a pretty powerful and easy to use microprocessor and allowed the building of the first personal computers, or maybe better called the ancestors of PCs. The first software to run on these machines was a Microsoft developed BASIC interpreter that allowed users to develop programs in higher-level languages. This also marked the first alternatives to hand assembled machine code that could be directly loaded into the microcomputers memory using a front panel of toggle switches, push-buttons and LED displays. While this still emulated the input method of early mainframe and minicomputers, after a very short period of time, this very cumbersome method was replaced by input/output (I/O) through a terminal. This became the preferred human to machine interface, as it still is today.

The Homebrew Computer Club

Although the Altair spawned an entire business, another side effect it had was to demonstrate that the microprocessor had reduced the cost and complexity of building a microcomputer, and that anyone with interest could build his or her own. Many such hobbyists met and traded notes at the meetings of the Homebrew Computer Club (HCC)[27] in Silicon Valley. It was a group of geeks, regularly meeting in the HCC, who advanced the designs of these machines over the coming years, one of the best known was Woz (Steve Woznak),[28] who later worked closely with Steve Jobs[29] at the heart of early Apple computers company.

The development goal at this time was to build microcomputers that were easier to program, operate and control. So, one of the first steps was to add a keyboard in order to allow more direct input of commands and data, a TV monitor to visualize the typing results and a cassette tape recorder to store data and programs. Woz was a very talented young man and this allowed him to design new computers that were

---

[24]history-computer, The Altair 8800 of Ed Roberts, http://history-computer.com/ModernComputer/Personal/Altair.html, 2017, retrieved 2017-01-26.

[25]Intel.com, The Story of the Intel 4004, http://www.intel.com/content/www/us/en/history/museum-story-of-intel-4004.html, 1971, retrieved 2017-01-26.

[26]old-computers, 8008 Microprocessor released by Intel, http://www.old-computers.com/history/detail.asp?n=69&t=3, 1972, retrieved 2017-01-26.

[27]Computer History Museum, The Homebrew Computer Club, http://www.computerhistory.org/revolution/personal-computers/17/312, 2017, retrieved 2017-01-26.

[28]biography.com, Steve Wozniak, http://www.biography.com/people/steve-wozniak-9537334#synopsis, 2017, retrieved 2017-01-26.

[29]biography.com, Steve Jobs, http://www.biography.com/people/steve-jobs-9354805, 2017, retrieved 2017-01-26.

easy to program, reliable and cheap. Steve Jobs started working with Woz to sell the parts of these computers to other club members, so they could build them by themselves. Although this strategy allowed Jobs and Woznak to have some limited financial success, it was not the big breakthrough, as the machines they advertised and sold were mainly for technical geeks and not designed for running practical day-to-day business affairs, but it was the start of the first microcomputers.

Computers from the Starting Period

During the first years of the PC industry that lasted roughly from 1972 to 1977, many computers appeared based on the idea of building smaller and more personal machines, like the Datapoint 2200,[30] the French Micral N,[31] the Xerox Alto,[32] the IBM 5100[33] or the Altair 8800 (that was already based on a single microprocessor), the Intel 8080,[34] Zilog Z80[35] and finally the Intel 8085[36] microprocessor chips. The operating system of the Altair was CP/M-80,[37] which turned out to become one of the first, popular microcomputer operating systems to be used by many different vendors, and many software packages were written for it, such as WordStar and dBase II.

1977 Apple II: The First Personal Computer for Everyone

It was the Apple II,[38] introduced in 1977 that promised to be a new machine finally offering versatility and usefulness never seen before in a microcomputer. This was the first complete computer, which only lacked a TV monitor, to be the perfect tool

[30]old-computers, Datapoint 2200, http://www.old-computers.com/museum/computer.asp?st=1&c=596, 1972, retrieved 2017-01-26.

[31]history-computer, Micral N, http://history-computer.com/ModernComputer/Personal/Micral.html, 1973, retrieved 2017-01-26.

[32]computerhistory.org, Xerox Alto, http://www.computerhistory.org/revolution/input-output/14/347, 1973, retrieved 2017-01-26.

[33]oldcomputers, IBM 5100 Portable PC, http://oldcomputers.net/ibm5100.html, 1973, retrieved 2017-01-26.

[34]Centre of Computing History, Introduction of the Intel 8080 2MHz Microprocessor, http://www.computinghistory.org.uk/det/6184/introduction-of-Intel-8080-2MHz-microprocessor/, 1974, retrieved 2017-01-26.

[35]Z80, Z80 CPU Introduction, http://www.z80.info/z80brief.htm, 1976, retrieved 2016-01-26.

[36]cpu-world, 8085 family, http://www.cpu-world.com/CPUs/8085/, 1976, retrieved 2017-01-26.

[37]cpm80.com, CP/M-80 Information and Download Page, http://www.cpm80.com, 1974, retrieved 2017-01-26.

[38]oldcomputers, Apple II, http://oldcomputers.net/appleii.html, 1977, retrieved 2017-01-26.

for home, school or office with an advanced microprocessor, enough memory, and significantly improved performance over the Apple I.[39] It even included an audio amplifier and speaker and allowed for the use of a joystick for playing games and a cassette tape drive for data storage. It had a built in BASIC programming language to make it useful to hobby programmers as well.

On top of all of this, it allowed for future hardware modifications for performance optimization for special applications like number crunching, gaming, or writing programs via expansion slots. This was the reason why the Apple II had the potential to become a very capable computer as soon as professionally designed software applications and special expansion cards became available. The Apple II already had a specially designed power supply that generated much less heat; hence no permanent running fan was required. Finally, for a lot of consumers, this computer was the first design that made it acceptable for everyday life usage, which turned out to be one of the biggest success factors behind this machine and its design.

The Rise of PCs

Like the Apple II, soon a number of other, so-called home computers saw the light of day, like the Atari 400/800,[40] Sinclair ZX80[41] and ZX81,[42] Commodore 64,[43] BBC Micro[44] and others. In 1981, the first personal computer to see widespread use was the famous IBM PC.[45] It used the Intel 8088 CPU[46] and was based on an expandable and open card based architecture allowing third parties to develop software for it. The IBM PC typically used IBM's PC-DOS[47] operating system that was a rebranded version of Microsoft MS-DOS.[48] One can see the impact of the Apple II and the IBM-PC when Time magazine named the home computer "Machine of the Year,"[49] the first time in the history of the magazine that an object was given this award instead of a human.

[39]oldcomputers, Apple I, http://www.oldcomputers.net/applei.html, 1976, retrieved 2017-01-26.

[40]oldcomputers, Atari 400, http://oldcomputers.net/atari400.html, 1978, retrieved 2017-01-26.

[41]oldcomputers, Sinclair Z80, http://oldcomputers.net/zx80.html, 1980, retrieved 2016-01-26.

[42]oldcomputers, Sinclair Z81, http://oldcomputers.net/zx81.html, 1981, retrieved 2017-01-26.

[43]oldcomputers, Commodore 64, http://oldcomputers.net/c64.html, 1982, retrieved 2017-01-26.

[44]old-computers, BBC Micro, http://www.old-computers.com/museum/computer.asp?c=29, 1981, retrieved 2017-01-26.

[45]oldcomputers, IBM PC, http://oldcomputers.net/ibm5150.html, 1981, retrieved 2017-01-26.

[46]CPU-World, Intel 8088, http://www.cpu-world.com/CPUs/8088/, 1979, retrieved 2017-01-26.

[47]Computer Hope, PC-DOS, http://www.computerhope.com/jargon/p/pcdos.htm, 1981, retrieved 2016-01-26.

[48]Computer Hope, MS-DOS, http://www.computerhope.com/jargon/m/msdos.htm, 1981, retrieved 2016-01-26.

[49]Time, Machine of the Year, http://content.time.com/time/covers/0,16641,19830103,00.html, 1983, retrieved 2017-01-26.

Over the next decade, PCs and several clones dominated the market, with ever increasing processing power, enabled by the introduction of the first 32-bit architectures through the Intel 80386[50] in 1985. During this period, PC clones slowly outpaced sales of all other machines and, by the end of the 1980s, PC XT[51] clones began to take over the home computer market. They were typically sold under $1000 and could be ordered by email. This cheap price was also made possible by using older, 16-bit technologies manufactured by Cyrix[52] and AMD[53] instead of Intel.[54]

1980 From Personal Computers to Workstations

The early 1980s saw the birth of high-end personal computers, subsequently called workstations, like the SUN-1 Unix workstation in 1982. Andy Bechtoldsheim, a founding member of Sun Microsystems, originally designed the Sun workstation at the Stanford University as a personal, CAD workstation. It was based on a Motorola 68000 processor with an advanced memory management unit (MMU) and supported a Unix operating system with virtual memory support. Sun Microsystems[55] was acquired almost three decades later by Oracle.

There followed a number of further evolutions in the workstation area such as the NeXT Cube[56] in 1988 that was developed by Steve Jobs, who had left Apple by that time. The NeXT pioneered the first object-oriented programming concepts but the whole concept turned out to be a failure and NeXT had to shut down hardware operations in 1993.

Another evolution changing computing was the introduction of the CD-ROM, which became an industry standard by the middle of the 1990s. The CD-ROM, originally developed for audio in the 1980s, could now store a lot of important computer data, like an operating system and applications and started serving as storage for multimedia. This meant that most desktop computers now came with built-in stereo speakers and were capable of playing CD quality music and sounds with so called sound cards.

Power workstations based on 32-bit architectures were released on the market, used for desktop publishing and graphic design. In December 1996, Apple bought

[50]CPU-World, Intel 80386, http://www.cpu-world.com/CPUs/80386/, 1985, retrieved 2017-01-26.

[51]old-computers, PC XT, http://www.old-computers.com/museum/computer.asp?c=286, 1983, retrieved 2017-01-27.

[52]pctechguide, Cyrix CPUs, https://www.pctechguide.com/cyrix-cpus, 1989, retrieved 2017-01-26.

[53]AMD, Advanced Micro Devices, http://www.amd.com/en-gb, 2017, retrieved 2017-01-26.

[54]cpushack, the Life Cycle of a CPU, http://www.cpushack.com/life-cycle-of-cpu.html, 2017, retrieved 2017-01-26.

[55]Oracle, Oracle and Sun Microsystems, https://www.oracle.com/sun/index.html, 2016, retrieved 2016-12-09.

[56]oldcomputers, Next Cube, http://oldcomputers.net/next-cube.html, 1988, retrieved 2017-01-27.

NeXT[57] during a time when they were almost bankrupt and Steve Jobs returned to Apple. Jobs brought Apple back to turning a profit with the release of the Mac OS 8[58] operating system, the PowerMac G3[59] and finally with the famous iMac[60] computers. The iMacs sold several million units and helped make Apple profitable again along with the introduction of the newly designed, next generation operating system Mac OS X[61] in 2000. The iMac, now in a different form, is still on sale today and has not lost any of its charm nor its superiority.

These examples show that revolutionary ideas are very often much more necessary than just the evolution of old and established paths, but also new technologies must be available at the right time to initiate new waves of products and business. Workstations, CD-ROMs, integrated sound systems all mark great examples of how the idea of the personal computer was successfully developed and perfected over time, but only the perfect interplay between hardware, software and usability in combination with new ways of thinking outside of the box that finally enabled the creation of stunning, revolutionary products. In the end, this perfect mix allowed the disruption of existing structures and in turn also enabled companies, like Apple, to exceed traditional business expectations.

1990 PCs Getting Mature

Since the late 1990s and early 2000s, further improvements in computing were writeable CDs,[62] MP3s[63] and P2P file sharing. USB[64] (Universal Serial Bus) ports were added for easy plug and play connectivity to devices and DVD (Digital Versatile Disc) players.[65]

[57]computerworld, Apple buys Next for $400m to create the new MacOS, http://www.computerworld.co.nz/article/518966/apple_buys_next_400m_create_new_macos/, 1996, retrieved 2017-01-27.

[58]cnet, Mac OS 8 has arrived, https://www.cnet.com/news/mac-os-8-has-arrived/, 1997, retrieved 2017-01-27.

[59]apple-history, Power Macintosh G3, http://apple-history.com/g3, 1997, retrieved 2017-01-27.

[60]YouTube, Steve Jobs introduces the Original iMac – Apple Special Event 1998, https://www.youtube.com/watch?v=oxwmF0OJ0vg, 1998, retrieved 2017-01-27.

[61]YouTube, Macworld San Francisco 2000 – The Mac OS X Introduction, https://www.youtube.com/watch?v=Ko4V3G4NqII, 2000, retrieved 2017-01-27.

[62]Philips, The history of the CD – The CD family, http://www.philips.com/a-w/research/technologies/cd/cd-family.html, 2017, retrieved 2017-01-27.

[63]inventors.about, The History of MP3, Fraunhofer Gesellschaft and MP3, http://inventors.about.com/od/mstartinventions/a/MPThree.htm, 1996, retrieved 2017-01-27.

[64]eetimes, Legare Christian, Introduction to USB Part I, http://www.eetimes.com/document.asp?doc_id=1280130, 2012, retrieved 2017-01-27.

[65]didyouknow.org, History of DVD, https://didyouknow.org/dvdhistory/, 1996, retrieved 2017-01-27.

In 2003, the first personal computers based on 64-bit architectures from AMD (Advanced Micro Devices)[66] were introduces by Compaq and HP, while Apple released their PowerMac G5[67] system based on IBM 64-bit architectures.[68] Intel countered a year later with their own 64-bit architectures[69] in their newly released Xeon and Pentium 4 lines, first mainly used in high end systems, servers and workstations, but later gradually replaced by 32-bit architectures in consumer desktops and laptops in the second half of the 2000s.

Early networking and connectivity of computers, personal computers and laptops was achieved via different Local Area Networks (LANs) where Ethernet clearly succeeded, which will be discussed in detail in the chapter called "Networking Technologies and Standards". The speeds supported by these LANs became, of course, much higher than the original 10 Mbps and are now in the range of several Gbps. The 2000s also saw the birth and evolution of wireless networks, called Wireless LANs or WLANs, which are capable of supporting a few hundred Mbps in combination with huge numbers of users per domain.

Further advancements in computing and personal computing were the introduction of multi-core processors,[70] flash memory[71] and the replacement of CRT monitors by LCD screens. Once limited to high-end industrial use due to high costs, these technologies are now mainstream and available to consumers. In 2008, the MacBook Air[72] and some PC-based laptops started to work solely without internal optical drives (DVD-ROM) and hard-drives, relying totally on flash memory for storage and high speed ports for connecting to external, peripheral devices.

---

[66]hardwaresecrets, AMD 64-bit architecture (x86-64), http://www.hardwaresecrets.com/amd-64-bit-architecture-x86-64/, 2004, retrieved 2017-01-27.

[67]Macworld, Edwards Benji, Ten years in the shadow of the Power Mac G5, http://www.macworld.com/article/2042702/ten-years-in-the-shadow-of-the-power-mac-g5.html, 2013, 2003, retrieved 2017-01-27.

[68]Apple, Apple and IBM Introduce the PowerPC G5 Processor, https://www.apple.com/pr/library/2003/06/23Apple-and-IBM-Introduce-the-PowerPC-G5-Processor.html, 2003, retrieved 2017-01-27.

[69]PCWorld, Intel Readies First 64-Bit Chip, http://www.pcworld.com/article/116631/article.html, 2004, retrieved 2017-01-27.

[70]PCMech, De Looper Christian, All About Multi-Core Processors, https://www.pcmech.com/article/all-about-multi-core-processors-what-they-are-how-they-work-and-where-they-came-from/, 2015, retrieved 2017-01-27.

[71]TutorialsWeb, Flash Memory: Theory and Applications, http://www.tutorialsweb.com/computers/flash-memory/, 2017, retrieved 2017-01-27.

[72]YouTube, MacBook Air Introduction by Steve Jobs, https://www.youtube.com/watch?v=kvfrVrh76Mk, 2008, retrieved 2017-01-27.

1990s Servers Replacing Mainframes

Time had come when servers, based on microcomputer designs, could be deployed much cheaper than even the cost of low end mainframes, while offering connected users much greater possibilities and control over their systems. This was the time when terminals were generally replaced by personal computers. As a result, demand for mainframes decreased significantly and the installations remaining were in government and financial services. It quickly became the consensus that the mainframe was a dying breed, because they were replaced by personal computers connected via local area networks and the Internet. There were several analysts in the middle of the 1990s who said that the mainframes would be dead by 2000.

2010 Mainframes Renaissance

As it is often in life, the reality was different from what was predicted. In the late 1990s, several enterprises discovered new uses for their existing mainframes resulting from collapsing data networking prices, which made centralized computing more appealing again. Meanwhile e-businesses asked for large numbers of back-end transactions typically best processed by mainframe software, requiring a large size and the use of databases. In addition, the demand increased for batch processing like billing, fueled by the growth of e-businesses, which is naturally the home territory of mainframes.

When Linux became available as an operating system on IBM mainframes in 1999,[73] it allowed running hundreds of virtual machines on a single mainframe, this additionally helped it to stay popular and raise awareness as Linux allowed users to run open source software in combination with mainframe hardware. On top of this, came demands in emerging markets like China, where mainframes were used to solve exceptionally difficult computing problems requiring high amounts of transaction processing databases for a large amount of consumers across several industries such as insurance, government services, banking and credit reporting.

This was the reason why IBM finally could introduce the 64-bit z-architecture[74] (Stephens, 2008) and added a number of software products to the mainframe through the acquisition of different software companies. While this rebirth of

---

[73]IBM, IBM is Committed to Linux and Open Source, ftp://ftp.software.ibm.com/linux/pdfs/IBM_and_Linux.pdf, 1999, retrieved 2017-01-27.

[74]IBM, Greiner Dan, IBM z/Architecture CPU Features, A Historical Perspective, https://www.google.at/url?sa=t&rct=j&q=&esrc=s&source=web&cd=&ved=0ahUKEwiCpbvChOLRAhVMDJoKHYz1A-8QFghEMAc&url=https%3A%2F%2Fshare.confex.com%2Fshare%2F117%2Fwebprogram%2FHandout%2FSession9220%2FIBM%2520zArchitecture%2520CPU%2520History.pdf&usg=AFQjCNG8i9LjpD2bf8Qj1O7E3DW0Xf38Eg&sig2=A3Ob5xtPgkKckIuIFjsVow, 2011, retrieved 2017-01-27.

mainframes was not without hiccups resulting from overall downturns at the end of the 2000s, the death of the mainframe as predicted a decade earlier had clearly not happened.

Supercomputers Versus Modern Mainframes

It is important to distinguish supercomputers from mainframes. Supercomputers are usually at the forefront of processing capacity, especially when it comes to the speed of calculation and are used to solve engineering and scientific problems. A prominent example of the use of supercomputers is High Performance Computing (HPC),[75] as, for example, calculating the results of the Large Hadron Collider[76] in Geneva (Robert D. Kent, 2003). Supercomputers (Scientific_American, 2003) are optimized and built for data and number crunching, on the contrary the strength of mainframes is transaction processing.

The performance of mainframes is often measured in Millions of Instructions Per Second (MIPS), while supercomputers are usually measured in Floating Point Operations Per Second (FLOPS), which became more recently Traversed Edges Per Second (TEPS). Floating point operations are, in many cases, additions, subtractions and multiplications with a large number of digits in order to allow for precise modeling of, for example, weather predictions or nuclear simulations, making supercomputers extremely powerful machines in terms of computational abilities.

Contrary to these computational abilities, mainframes are optimized for transaction processing in the business world, where a typical transaction includes database system updates like airline reservation services, inventory control or banking. Thus, a transaction refers to certain operations like disk reading/writing, operating system calls, or data transfers between different subsystems.

The design of modern mainframes is typically optimized beyond MIPS for single task computations or FLOPS for floating point calculations by also taking into account redundant internal engineering. This results in higher reliability and security, stricter backward compatibility with older software, high hardware and computational utilization rates achieved by virtualization in order to achieve massive throughput and finally extensive input/output facilities including offloading to other machines.

[75] inside HPC, What is high performance computing?, http://insidehpc.com/hpc-basic-training/what-is-hpc/, 2017, retrieved 2017-01-27.

[76] Cern, The Large Hardon Collider, http://home.cern/topics/large-hadron-collider, 2017, retrieved 2017-01-27.

Mainframes in the Middle of the 2010s

In 2014, IBM announced its latest mainframe, the z13[77] (IBM, 2015) that took 5 years of design for an overall cost of $1 billion. The goal of this development was to build the latest and greatest, most powerful and most secure mainframe ever. The capacity is huge, as it can process 2.5 billion transactions a day, combined with real time encryption and embedded analytics, aimed at providing the necessary processing power required by the increasing demand of mobile devices. This shows that today a market still exists for modern mainframes and clearly indicates a step towards centralized computing as opposed to the very popular distributed computing running across large numbers of server based machines in data centers, which was the standard by 2015.

The Economic Cloud Solution

There are a lot of customers who still rely on mainframes because they need to run mission critical applications with requirements for reliability and security as, for example, in banking transactions. These modern mainframes offer a higher level of reliability, such as the mean time between failures measured in decades. Linux and OpenStack support guaranteed open source cloud computing, meaning that a private cloud can be directly run on such a mainframe, supporting up to 8000 virtual Linux machines simultaneously. This should allow for substantially lower costs of ownership compared to the same private cloud running on virtual machines in data centers. Implementations have already shown that total costs of ownership of these new mainframes are around 50% lower compared to traditional server farms and data centers. Additionally, cost savings come through the reduced need for floor space etc.

The typical customers for mainframes are those running mission critical applications on private clouds, as opposed to customers who outsource their own infrastructure to public clouds such as Amazon.com. This is clearly a different market, but it is here to stay or, perhaps, get bigger with future service requirements needing highly reliable, secure and fast transactions, something that cannot be guaranteed on typical public clouds.

[77]IBM, IBM z13, http://www-03.ibm.com/systems/z/hardware/z13.html, 2017, retrieved 2017-01-27.

References

Christian C. Carman, J. E. (2014, November 15). On the epoch of the Antikythera mechanism and its eclipse predictor. *Archive for History of Exact Sciences, 68*(6), 693–774.

Claude, S., & Warren, W. (1948). *The mathematical theory of communication.* Urbana: The University of Illinois.

Cray Research, I. (1977). *The Cray-1 computer system.* New York: Aladdin Paperbacks.

Dyson, G. (2012). *Turings cathedral.* New York: Pantheon.

Freiberger, P. A. (2014, August 21). *Enzyklopedia Britannica.* Retrieved from www.britannica.com/technology/Analytical-Engine

Gödel, K. (1931). Über formal unentscheidbare Sätze der Principia Mathematika und verwandter Systeme I. *Monatshefte für Mathematik und Physik, 38*, 173–198.

IBM. (2015). *IBM z13 Technical introduction.* IBM.

Macrae, N. (1992). *John von Neumann: The scientific genius who pioneered the modern computer, game theory, nuclear deterrance, and much more.* Providence, RI: American Mathematical Society.

Moore, S. (2010). The Manchester baby. *ECAD and Architectural Practical Classes.*

Nallur Prasad, J. S. (1994). *IBM Mainframes – Architecture and design* (2nd ed.). New York: McGraw-Hill.

Robert D. Kent, T. W. (2003). In T. W. Robert & D. Kent (Eds.), *High performance computing systems and applications.* Windsor, ON: Kluwer Academic Publishers.

Scientific_American. (2003). *Understanding supercomputing.* New York: Little, Brown & Company.

Stakem, P. (2015, February 22). *Mainframes, computing on big iron (Computer architecture book 15).* p. 119.

Stephens, D. (2008). *What on earth is a mainframe.* Perth: Longpela Expertise.

Turing, A. M. (1936, May 28). On computable numbers, with an application to the entscheidungsproblem. *Monatshefte Mathematik Physik, 37*, 349–360.

Turing, A. M. (1937). On computable numbers, with an application to the enscheidungsproblem. *Proceedings of the London Mathematikal Society, s2–42*(1), 230–265.

Yates, D. M. (1997). *Turing's legacy: A hisory of computing at the National Physics Laboratory 1945–1995.* London: National Museum of Science and Industry.

Zuse, H. (2013, June 17–18). Reconstruction of Konrad Zuse's Z3. *IFIP Advances in Information and Communication Technology, 416*, 287–296

Networks for Sharing and Connecting

Imagination is more important than knowledge. Knowledge is limited. Imagination encircles the world.

Albert Einstein

Computers always required some sort of networking and, of course, these networks changed from simple and slow point-to-point connections to more and more sophisticated technologies for sharing a common network medium. The networks had to cover buildings via local area networks (LANs) and also cities, countries and the whole globe via wide area networks (WANs).

We describe the major breakthroughs in computer networking through the successful introduction of local area networking technologies like Ethernet, besides many other economically unsuccessful attempts. This will help us to see the evolution from structured cabling with limited scalability, to bridging, switching and routing enabling the evolution of highly scalable while also fast networking.

An overview of networking cannot be complete without referring to standards, therefore we discuss how the Internet Protocol (IP) Suite successfully won the battle against the OSI Reference Model and how finally packet based networks based on the IP won the fight against cell based Asynchronous Transfer Mode.

We give examples of how the consequent development and deployment of the Internet protocol suite formed the basis of today's most successful communication ecosystem, the Internet, before highlighting IPv6 and what made it the ground-breaking foundation for the Internet and IoT (see Table 1).

Evolution of Computer Networks

Computer networks began to emerge in the late 1960s and inherited many useful properties from older and widely deployed telephone networks. One of the most exciting facts about computer networks is that they allow for the creation of stores of information accumulated by the human civilization since the existence of the human race and spanning over many thousands of years. This information store

M. Oppitz, P. Tomsu, *Inventing the Cloud Century*,
DOI 10.1007/978-3-319-61161-7_5

Table 1 Networking time line

1950	
1956	Comité Consultatif International Téléphonique et Télégraphique (CCITT) founded
1960	
1961	First packet-switching papers
1966	ARPANET planning starts
1966	Merit Network founded
1968	ALOHANet first experimental computer network in Hawaii
1969	BBN develops the Interface Message Processor for ARPANet—the first router
1969	ARPANET carries its first packets
1970	
1970	Mark I network at NPL (UK)
1970	Network Information Center (NIC)
1971	Merit Network's packet-switched network operational
1971	Tymnet packet-switched network
1973	CYCLADES network demonstrated
1974	Telenet packet-switched network
1975	Ethernet patent by XEROX
1975	The first TCP/IP communication tests between Stanford University and University College of London
1976	X.25 protocol approved
1978	Minitel introduced
1979	Internet Activities Board (IAB)
1979	3com as one of the first networking companies founded in Silicon Valley by Robert Metcalfe
1980	Ethernet standard introduced
1980	USENET news using UUCP
1980	
1980	Ethernet DIX standard published by Digital, Intel and Xerox
1980	Asynchronous Transfer Mode (ATM) development started
1981	BITNET established
1982	Simple Mail Transfer Protocol (SMTP)
1982	US Department of Defense declares TCP/IP as standard for all military computer networking
1983	MILNET split off from ARPANET
1983	Domain Name System (DNS)
1983	IEEE 802.3 Ethernet standard published
1983	Official migration of ARPANet to TCP/IP completed and activated
1984	IEEE 802.5 Token Ring standard introduced by IBM
1984	IEEE 802.4 Token Bus network introduced by General Motors
1984	CISCO founded by Len Bosnak and Sandy Lerner
1984	OSI model and standards by ITU and ISO finished
1985	FDDI (Fiber Distributed Data Interface) standardization work begins
1985	First InterOp conference held in San Jose, CA organized by Dan Lynch
1986	NSFNET with 56 kbit/s links

(continued)

Table 1 (continued)

1988	David Clark on the design philosophy of the DARPA Internet protocols "We reject kings, presidents and votings..."
1988	OSI Reference Model released
1988	NSFNET upgraded to 1.5 Mbit/s (T1)
1989	PSINet founded, allows commercial traffic
1989	Border Gateway Protocol (BGP)
1989	Federal Internet Exchanges (FIXes)
1989	First multiport Ethernet switches introduced by Kalpana
1989	IETF (Internet Engineering Task Force) RFC 1105 describes Border Gateway Protocol (BGP) version 1
1989	Tim Berners-Lee writes first online message eventually leading to the World Wide Web
1990	GOSIP (without TCP/IP)
1990	ARPANET decommissioned
1990	Advanced Network and Services (ANS)
1990	
1991	Wide area information server (WAIS)
1991	Gopher
1991	Commercial Internet eXchange (CIX)
1991	ANS CO+RE allows commercial traffic
1991	ATM Forum founded by Fred Sammartino of Sun Microsystems
1993	CISCO acquires switch manufacturer Crescendo
1996	Advanced gateway Server (AGS)—first commercial router
1996	Tag Switching introduced by CISCO
1997	Multi Protocol Label Switching (MPLS) standardization started by IETF
2000	
2003	FCOE (Fiber Channel over Ethernet) early version developed by Azul Technology
2010	
2010	Driverless cars experiments started by Google, Apple and BMW, etc.
2013	ETSI start specification of Network Function Virtualization (NFW)
2015	Three billion users on the Internet
2018	More than 400 Zetabytes per year data created by the Internet of Everything, Cisco projects
2019	Global mobile data traffic will reach 24.3 Exabytes per month, CISCO says
2025	US Exascale computer is expected to overtake leading Chinese supercomputers
2040	Above 1 Terabit/second access speed expected
2050	Storage capacity of micro SD card is expected to be three times the brain capacity of entire human race

keeps constantly growing with ever increasing rates that is also highly reflective of the recent popularity in big data and analytics, the latest technologies in the area of information storage and data mining.

Data communications used for computer networking covers many areas including interface standards, asynchronous versus synchronous communication, telephone switching systems, modems, and protocols. We can only discuss some major topics throughout this chapter and, for further reading, recommend numerous references found throughout the text. For a great start to understand data communications basics see (McNamara, 1982).

The evolution of computer networking is naturally tightly intertwined with the evolution of computing (Ceruzzi, 1998). The first computers originated in the 1950s and were used for batch processing instead of using them interactively as this is typically the case with computers today. Hence, these first computers were only built for a small number of privileged users.

Batch processing systems were based on mainframe architectures and were not designed for interactive operating modes (Larry L. Peterson, 2012). This was due to the fact that the heart of the computer, the processor, was very expensive and this cost was much greater than the costs to the user's labor time. During the 1960s, processors became cheaper and allowed for the building of multi-terminal systems offering the end users more convenient access to the computing resources (see Fig. 1).

These systems used time-sharing of the central processor, where the users could communicate from their own terminals with the central computer (often called the mainframe). At this stage, processing power was still centralized but data input and output was already distributed.

Ethernet: The Epic Foundation for Local Area Networks

Distributing data input and output became the typical approach used for accessing computers in the early 1980s. By then, the terminals had been moved out of the computing centers to reside on the desks of the end users, marking a significant improvement in usability. During this time, early Local Area Networks (LANs) were introduced to connect terminal servers to the mainframes, with the terminals directly connected to the terminal servers in a star shaped, point-to-point configuration. Ethernet (Held, 2003) was used as one of the first and most successful implementations for connecting the terminal servers to the mainframes, but soon

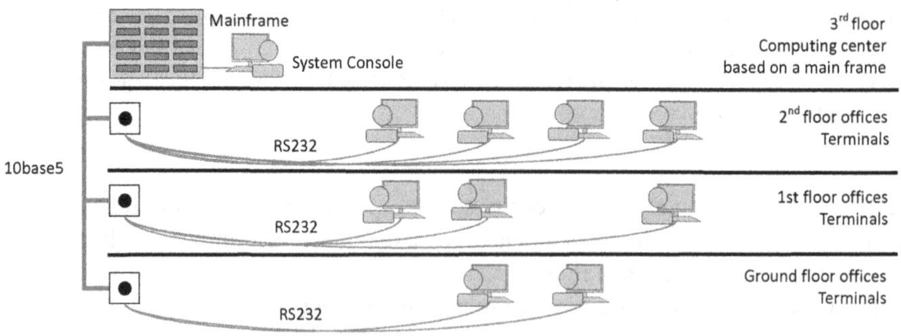

Fig. 1 Computer network based on multi-terminal system

more LAN examples followed (Madron, 1988; Minoli, 1st, 2nd & Next Generation LANs, 1993).

Beginnings of Ethernet

Ethernet had been developed at Xerox PARC around 1974. It had its origins in ALOHAnet (Tanenbaum, 2003), which Robert Metcalfe had worked on as part of his PhD dissertation. Robert Metcalfe named the Ethernet after the disproven luminiferous ether in 1973 as an "omnipresent, completely passive medium for propagation of electromagnetic waves." The Ethernet system was successfully deployed at Xerox PARC,[1] before Xerox filed the patent for Ethernet in 1975.

In 1979, Metcalfe left Xerox to form 3Com, one of the first networking companies in Silicon Valley and he managed to convince Digital Equipment Corporation (DEC), Intel and Xerox to collaborate on an Ethernet standard that also became known as the DIX standard, for Digital, Intel, and Xerox. The first Ethernet standard specified a 10 Mbps transmission speed, 48-bit destination and source addresses for identifying communicating stations on the medium and a global 18-bit Ether-type field for further control purposes. It was published in September 1980 as "The Ethernet, A Local Area Network. Data Link Layer and Physical Layer Specifications."[2] Version II of this standard was published 2 years later and became commonly known as Ethernet II, which finally resulted in the publication of the first 802-IEEE standard for local area networks, IEEE 802.3, in June 1983 (ANSI/IEEE, 802.3 Carrier Sense Multiple Access with Collision Detection, 1985a; Burg, 2001).

The first LAN cable used as a shared medium was Thick Ethernet Yellow Cable (RG-8), also commonly referred to as 10Base5[3] cable, because it allowed 10 Mbps baseband transmission over a 500-m segment length. This implementation needed physical repeaters in order to extend beyond 500 m. The protocol to access this medium—the Ethernet—was called CSMA/CD (Carrier Sense Multiple Access with Collision Detection). Around this time, Digital Equipment used Ethernet Version II for connecting their terminal servers to their host computers.

Ethernet Details

IEEE 802.3 defines both a physical layer as well as a media access control (MAC) layer. The physical layer is separated yet again into two sub-layers: the lower being the physical medium attachment (PMA) and the upper the physical layer signaling (PLS). The beauty of this approach is that the PLS sublayer is a medium, independent layer responsible for coding, while the PMA sublayer defines the medium

[1]etherhistory.typepad, Metcalfe, R.M. and Boggs, D.R., Ethernet: Distributed Packet Switching for Local Computer Networks, http://ethernethistory.typepad.com/papers/EthernetPaper.pdf, 1976, retrieved 2017-02-01.

[2]decnet.ipv7.net, The Ethernet—A Local Area Network, Data Link Layer and Physical Layer Specifications, http://decnet.ipv7.net/docs/dundas/aa-k759b-tk.pdf, 1980, retrieved 2017-02-01.

[3]The Network Encyclopedia, 10Base5, http://www.thenetworkencyclopedia.com/entry/10base5/, 2017, retrieved 2017-02-02.

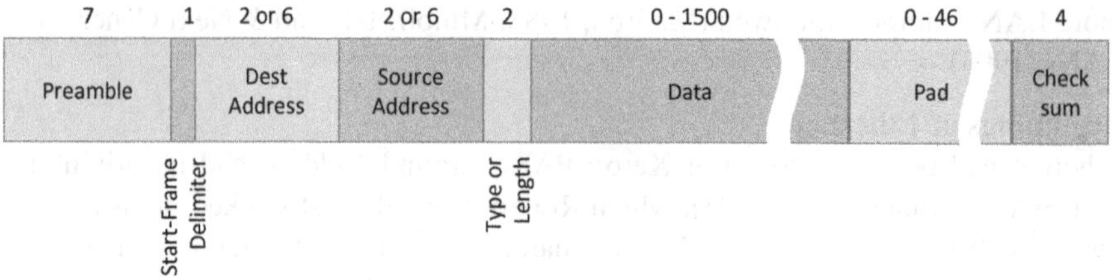

Fig. 2 Ethernet frame

access unit (MAU), attaching directly to the medium and providing for the transmission and reception of medium signals as well as collisions. Finally, Ethernet uses the carrier sense, multiple access and collision detection protocol to share the same medium (cable) between multiple end stations.

There are several mediums associated with different 802.3 implementations like 10Base5, 10Base2 or 10BaseT that will be discussed in detail in the section "From Ethernet to Structured Cabling." The medium is split into the PMA layer that determines how to attach to the specific medium, the PLS sublayer and the MAC sublayer, which are all common to variants of 10 Mbps 802.3. The simple structure of the MAC frame can be seen in Fig. 2.

The frame starts with a Preamble that is used by the receiver to establish bit synchronization, as there is no separate clocking information available on the medium (Ether) when no data has been sent. A Start Frame delimiter is a kind of "flag" following the preamble and indicates the start of the frame. Then come, the Destination MAC (Media Access Control) address followed by the Source MAC address, both 48-bits long.

The Type or Length field is the field that is different in 802.3 and Ethernet II implementations. In 802.3, it indicates the number of bytes of data in the payload (data) field and can range from 0 to 1500 bytes. Frames must be at least 64 bytes long (excluding the preamble) as required by the CSMA/CD mechanism. The Pad field is used for compensating the missing bytes. In Ethernet II, the Type/Length field is used to indicate the type of payload carried by the frame, like an IP payload. A special feature of the protocol allows both 802.3 as well as Ethernet II frames to coexist on the same medium (the network). Finally, the obligatory Checksum field ends the frame.

Other Local Area Network Standards

The IEEE 802 LAN Standards Family

Ethernet became the first successfully implemented and widely deployed standard LAN technology at the beginning of the 1980s. Many LAN technologies competed at this time including:

- Ethernet[4] (IEEE 802.3, driven by Digital Equipment Corporation, Intel and Xerox; ANSI/IEEE, 802.3 Carrier Sense Multiple Access with Collision Detection, 1985a; Burg, 2001)
- Token Passing Bus[5] (IEEE 802.4, driven by General Motors for their Manufacturing Automation Protocol also abbreviated as MAP; ANSI/IEEE, 802.4 Token Passing Bus Access Method, 1985b)
- Token Ring[6] (IEEE 802.5, introduced by IBM in 1984; ANSI/IEEE, 802.5 Token Ring Access Method, 1985c; Hans-Georg Göhring, 1990)

An excellent overview on Standards Organizations (SDOs) especially in the area of Local Area Networks can be found in (Stallings, 1990b).

FDDI: Fiber Distributed Data Interface

In 1989, the Fiber Distributed Data Interface (FDDI) (ANSI, 1987) standard joined the family of LAN standards, extending the reach to 100 km, or 62 miles, using high-speed fiber rings. FDDI[7] allows for speeds of up to 100 Mbps when using one fiber ring and up to 200 Mbps when using the second (backup) ring. FDDI used a larger, maximum-frame size (4352 bytes) than the standard Ethernet family, which only supported 1500 bytes of maximum-frame size, improving effective data rates (Bernhard Albert, 1994).

Sure enough, there was big competition between all these different approaches or, more precisely, between the vendors trying to make their LAN technology the most used, standard solution. It would take many years before Ethernet became finally the de facto LAN standard at the end of 1990s.

Later in this chapter, we will look into some of the most significant evolutionary steps of Ethernet and describe the most important changes to local area computer networks based on Ethernet, which led to structured cabling and was followed by bridging, switching and finally routing. Although the other LAN technologies were important to the overall evolution of local area networking, Ethernet finally won the competition and the basic principles are still valid today.

[4]The Network Encyclopedia, Ethernet, http://www.thenetworkencyclopedia.com/entry/ethernet/, 2017, retrieved 2017-02-02.

[5]computer.org, Strohl Mary Jane, Token passing local area networks, National Computer Conference 1985, https://www.computer.org/csdl/proceedings/afips/1985/5092/00/50920605.pdf, 1985, retrieved 2017-02-02.

[6]The Network Encyclopedia, Token Ring, http://www.thenetworkencyclopedia.com/entry/token-ring/, 2017, retrieved 2017-02-02.

[7]Cisco, Fiber Distributed Data Interface (FDDI), https://www.cisco.com/cpress/cc/td/cpress/fund/ith2nd/it2408.htm, 1999, retrieved 2017-02-02.

From Ethernet to Structured Cabling

As soon as personal computers started to replace terminals at the user desk, the LAN was extended to each desktop and the yellow cable, that was previously expensive, cumbersome and difficult to handle, was replaced by a much cheaper and easier to install coaxial cable solution based on 10Base2. 10Base2[8] allows 10 Mbps, baseband transmission of an over 200-m segment length and became commonly known as Thinnet Bus because of the thinner coaxial cable type RG-58A/U.

10Base2 used the same CSMA/CD access method as 10Base5, but the computers were connected via BNC connectors to the medium based on thin coaxial cable, allowing for the extension of the local area network to each desktop with the same type of cable as was used for the backbone. Segments could be extended and linked together by repeaters (see Fig. 3).

The next groundbreaking step in local area networking came with the introduction of twisted pair cables specified as 10BaseT[9] (for 10 Mbps, baseband transmission over twisted pair cables) that were used to transport the signals between the single machines. For the first time, this allowed designing structured cabling schemes instead of bus systems (Elliott, 2002). In order to connect the single machines so-called hubs were used, which were boxes allowing for the connection of up to a certain number of end systems via twisted pair cable while leaving the medium access protocol the same CSMA/CD mechanism. This was actually achieved by shrinking the bus into the backplane of the hubs. Hubs could either be connected together via twisted pair structured cabling or via the classic coaxial cable solutions 10Base5 or 10Base2 (see Fig. 4). Hubs could also have multiple interface cards with multiple twisted pair interfaces per card, thus concentrating and connecting very high numbers of end systems.

Principles of Layered Networking

As we are now moving into a bit more sophisticated networking areas other than just connecting computers via cable either locally or via the wide area, it is time to briefly introduce one of the basics of modern networking—the seven-layer networking model, which we also describe later in this section under OSI Reference Model. Thus, here is a quick introduction to the seven layers of networking (for dummies):

---

[8]The Network Encyclopedia, 10Base2, http://www.thenetworkencyclopedia.com/entry/10base2/, 2017, retrieved 2017-02-02.

[9]The Network Encyclopedia, 10BaseT, http://www.thenetworkencyclopedia.com/entry/10baset/, 2017, retrieved 2017-02-02.

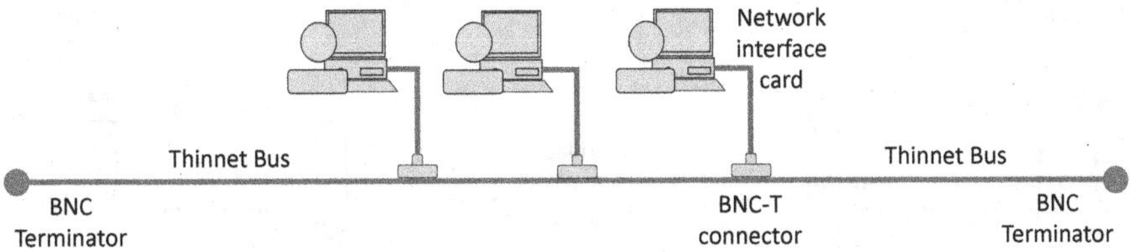

Fig. 3 Base 2 installations

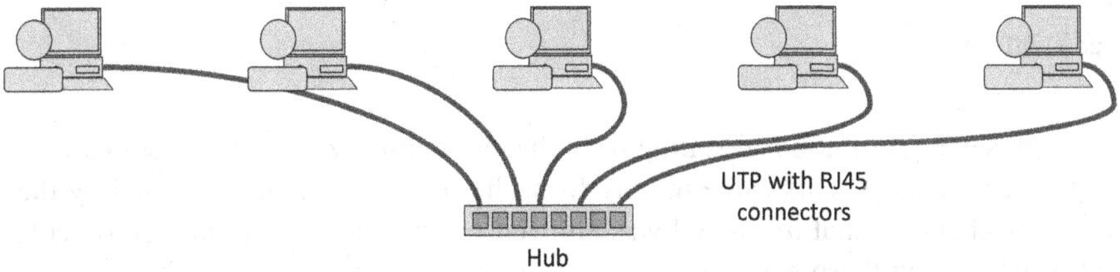

Fig. 4 Structured cabling with 10BaseT

Table 2 OSI layers

Layer 7	Application layer	Message format, human machine interface
Layer 6	Presentation layer	Coding into 1s and 0s, encryption, compression
Layer 5	Session layer	Authentication, permissions, session restoration
Layer 4	Transport layer	End-to-end error control
Layer 3	Network layer	Network addressing, routing of packets
Layer 2	Data link layer	Error detection and flow control on physical layer, switching of frames
Layer 1	Physical layer	Bit stream, physical medium, how to represent bits

Networking can be best described by separating its functions into layers. Several decades ago, these functions were split into seven layers by OSI. These layers, their names and the basic functions they perform are shown in the following table.

We will discuss many of the details of these layers throughout the course of this chapter, but it is helpful to see an overview now, before getting into the next sections (see Table 2).

Bridges Expanding LANs Beyond Cabling Limitations

Suddenly it became easily possible to wire complete buildings with structured cabling and prepare for later, flexible use of distributed computing, but there

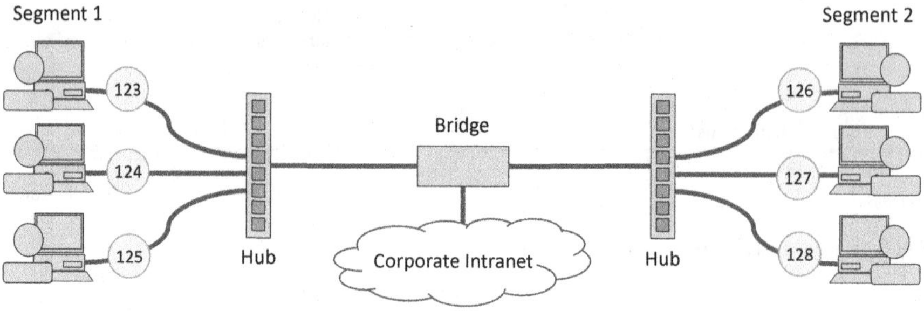

Fig. 5 Bridges

were still severe limitations resulting from the maximum size that could be spanned by physical cabling. This came mainly from the time constraints, imposed by the access mechanism, that the signal was allowed to travel as well as some protocols that were used on these networks.

To overcome these limitations, a new networking component, the so-called bridges, had to be introduced in order to allow for the extension of the physical segments based either on coax or twisted pair cable. This extension worked on the Data Link Layer, the layer above the Physical Layer, where all the cabling we have been dealing so far is described. We will further discuss the different layers and the layer model in the sections about Internet protocol suite and OSI Reference Model, but for now we should understand that the main reason to introduce bridges was due to the limitations of single segments and access domains of Ethernet.

Bridges[10] were invented in the mid 1980s and one of the key people behind this invention as well as the Spanning Tree Protocol (STP),[11] which is a key protocol for network bridges, was Radia Perlman (Perlman, 1992). Bridges allowed for extensions beyond the physical limitations of cabling systems and also allowed the bridging of LANs via wide area network (WANs) (see Fig. 5). This enabled the connection of computers in different buildings, cities, and countries or even on different continents building corporate networks or Intranets, as if they were on the same shared medium. The only limiting factor remained the speed of the interconnection links, mostly the WAN links.

Bridges also had important advantages other than just extending the physical reach of local area networks. They use a store and forward mechanism for the data sent between two end systems (called frames when operated on the bridge level) that require the usage of unique hardware addresses, the so-called Media Access Control (MAC) addresses. MAC addresses allow the filtering of traffic that would

[10]IEEE, IEEE 802.1D, Standard for Local and Metropolitan Area Networks, Media Access Control (MAC) Bridges, http://standards.ieee.org/getieee802/download/802.1D-2004.pdf, 2004, Revision of IEEE Std 802.1D 1998, retrieved 2016-12-05.

[11]Force10, White Paper, Evolution of the Spanning Tree Protocol, http://www.force10networks.com/whitepapers/pdf/F10_wp19_v1%201.pdf, 2007, retrieved 2016-12-05.

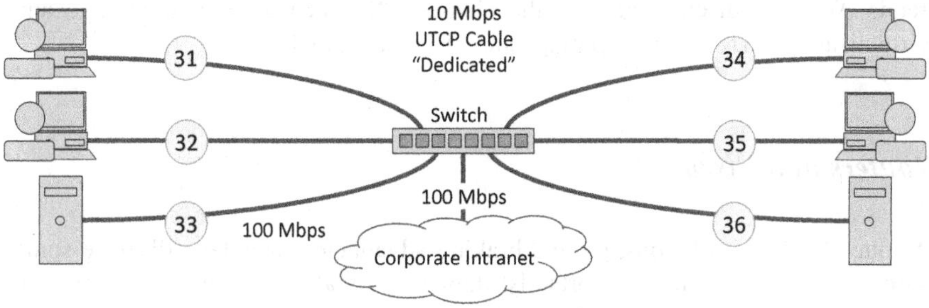

Fig. 6 Switches

otherwise be forwarded from one segment to another, but only if physical layer repeaters are used. By reading the destination address of incoming frames and using recorded addresses to determine the appropriate outbound port for the frame, traffic can be forwarded in a much more efficient way. Of course, bridges could be operated in combination with hubs in order to increase the port density on each side of the Ethernet LAN segments.

Switches Enabling Scalable Fast Networking

In the early 1990s, a new technology, called switching, started to revolutionize networking. Switches use bridging principles to forward traffic between ports, but instead of the store and forward of frames between input and output ports, they allow for multiple, dedicated transmissions between end stations directly connected to the switch ports without necessitating storing and forwarding (Charles E. Spurgeon, 2013). In case of Ethernet, switches still use Ethernet framing, build and maintain address tables just like bridges do, but all together this technology is more scalable because of the possibility of multiple, simultaneous conversations (see Fig. 6). Switches replacing hubs and bridges, result in each station that is connected to the switch getting the full 10 Mbps bandwidth of Ethernet.

Over time, the speed of Ethernet got faster from 100 Mbps, also known as Fast Ethernet[12] or 10Base100, to 1 Gbps, also called Gigabit Ethernet,[13] to later 10 Gbps and then speeds in the range of several 100 Gbps. Switches supporting this flexibility can be used to build highly scalable and ultrafast networks for many computers connected in dedicated places, which are called data centers. Data centers appeared in the 2010s and became the heart of many modern technology

[12]The Network Encyclopedia, Fast Ethernet, http://www.thenetworkencyclopedia.com/entry/fast-ethernet/, 2017, retrieved 2017-02-02.

[13]The Network Encyclopedia, Gigabit Ethernet, http://www.thenetworkencyclopedia.com/entry/gigabit-ethernet/, 2017, retrieved 2017-02-02.

trends. We will discuss them in the chapters "Cloud Computing, Storage and Virtualization, IoT, Fog Computing" and "Big Data Analytics."

Routers and Cisco

Despite all of their obvious, ground-braking advantages, switches still have some serious limitations, most importantly they do not allow for logical, separated networks. In order to better understand this, we can consider two enterprises who want to run separate networks in order to guaranty privacy and security. This requires an additional layer of networking above the data link layer that is commonly known as the network or routing layer. By introducing this layer and unique network addresses, a network layer element (the router) can use path determination throughout the network with a variety of metrics in order to forward packets from one network to the other. The data sent between end systems on the network layer (OSI Layer 3), using network addresses, is called packets. To recap our short networking crash course, data units sent on layer 2 (the data link layer) are called frames and data units sent on layer 3 (the network or routing layer) are called packets (Perlman, 1992).

The inventor of the router was Bill Yeager, who developed the first multiprotocol router at Stanford University in 1980. Len Bosnack and Sandy Lerner, a married couple who worked at Stanford University as computer operations staff, were interested in this project. They were joined by Richard Traiono, a computer engineer, and, in 1984, formed Cisco[14] (the name comes from San FranCISCO) to sell computer-networking products. Cisco is the only company that started to pioneer routing in the 1980s and, in 1986, Cisco already changed networking and the networking industry with the introduction of its first commercially available multi-protocol router.

The basic function of a router is simple as all it does is forward traffic—the data packets—based on the instructions or metrics it is given. Each network has its own unique addresses and works with a routing protocol. A router combines the network addresses and routing protocols and redirects traffic like email, WEB pages, files or documents across the network and is able to send the traffic from one location to another or even from one continent to another.

There were many different networking protocols and, accordingly, routing protocols (Dickie, 1994; Malamud, 1992) used at the beginning of routing, hence the term "multi-protocol router." Each had their own addressing schemes and dedicated functionalities. Finally, the Internet Protocol (IP) rose from the others and became the commonly used protocol today, as we will discuss in the section "Internet Protocol Suite." Addresses used on the IP networking layer are called IP addresses.

[14]Cisco, https://www.cisco.com, 2017, retrieved 2017-02-02.

Routing metrics are used by the routers to determine specific paths through the network based on load of the links, delay, bandwidth, reliability, hop count, Quality of Service (QoS) (Paul Ferguson, 1998) and much more. Since the beginning, IP routers sit at the heart of the Internet and forward packets on behalf of IP addresses throughout the network, thereby using sophisticated hardware and software technologies. We will look much deeper into the secrets of routing in the chapter "Network Virtualization"—where we will discuss control planes in detail. Routers can perform all the functions that bridges and switches can perform, plus enable many more features, as they can look deeper into the data sent and apply network services based on the destination IP address (Halabi, 1997).

There are several advantages that come with layer 3 router-based networks and, though some people still keep thinking they could overcome some of the basic network design rules, these rules have not changed up to today and will not in the future. One of the biggest advantages of router networks is the fact that, by their own nature, they impose a layer 3 network hierarchy, thus limiting necessary layer 2 mechanisms like layer 2 address resolution and broadcast to small and controllable layer 2 domains. This allows the design and deployment of much larger and more scalable layer 3 router networks, while keeping layer 2 overhead limited to layer 2 domains. The development of specialized routing protocols like Border Gateway Protocol (BGP)[15] and many other advanced routing mechanisms finally brought the scalability needed to build huge networks like the Internet.

Networking Standards

In this section, we will look deeper into different networking standards and architectures that all emerged around the 1980s, but only one of them lasted. This was the Internet protocol suite and it remains today the basis for all modern networking as well as the Internet. It is known today and probably will be for the next, several decades. At the beginning of networking there existed many different proposals and architectures for both local area networks (LANs) and the wide area networks (WANs). We already discussed the evolution of LANs and the victory of Ethernet, but it was actually the evolution of WANs and their different approaches, which, during the 1980s and 1990s, were responsible for an exciting fight between the different approaches, supported and driven by different interest groups (Malamud, 1992).

[15]IETF, RFC 4271, A Border Gateway Protocol 4 (BGP-4), https://tools.ietf.org/html/rfc4271, 2006, retrieved 2016-12-02.

The Birth of Modern Networking

Once the first successful computer networking technologies became commercially available at the beginning of the 1990s, the era of modern networks covering corporate sites and wide area networks (WANs) started. This was enabled by the introduction of public packet networks as an alternative to leased line based WAN connections. The upcoming WAN networks relied on optical communication technologies where the optical media was shared between many users via multiplexing.

These multiplexing technologies were mainly defined by ANSI in Synchronous Optical NETwork (SONET)[16] or the equivalent by the ITU in Synchronous Digital Hierarchy (SDH).[17] Both specifications are identical with the exception of some small deviations in the header of the used frames[18] and were specified up to transmission speeds of 160 Gbps. Due to the nature of optical transmission technologies used it was easily possible to span any physical distance ranging from metropolitan areas (MAN), to countries and overseas to worldwide optical networks (WAN). For more details also describing broadband business services used in these times and their strategic impact please refer to (Tomsu & Schmutzer, 2002; Wright, 1993).

Frame Relay

It was a technology called Frame Relay[19] that was the first successful implementation of packet-based WANs. Its popularity rose in the early 1990s when businesses and enterprises were mainly running multiple, vendor-specific networks supporting mainframes and client/server environments. From the beginning, Frame Relay was designed to be protocol transparent and was used to replace leased line WANs with a single protocol packet network (Buckwalter, 1999).

Distributed Queue Dual Bus

Another promising standard came in the middle of the 1980s, called Distributed Queue Dual Bus (DQDB), a technology specifically developed to span metropolitan areas typically up to a few tens of kilometers in their first incarnations (David

---

[16]Cisco Systems: SONET Primer, http://www.cisco.com/c/en/us/td/docs/optical/15000r5_0/planning/guide/r50engpl/r50appb.pdf, 2009, retrieved 2016-12-02.

[17]ITU-T, Rec. G783 (03/2006), Characteristics of Synchronous Digital Hierarchy, https://www.google.at/url?sa=t&rct=j&q=&esrc=s&source=web&cd=3&cad=rja&uact=8&ved=0ahUKEwj-yYyWi_nRAhVjEpoKHU-AAX0QFggqMAI&url=https%3A%2F%2Fwww.itu.int%2Frec%2Fdologin_pub.asp%3Flang%3De%26id%3DT-REC-G.783-200603-I!!PDF-E%26type%3Ditems&usg=AFQjCNEvV0GtaRh6isGgkBB1DrzvB4voVA&sig2=Q6fHx3ORGOd4ve1FnCRCXg, 2006, retrieved 2017-02-05.

[18]Cisco Systems, Understanding the Basic Differences Between SONET and SDH Framing in Optical Networks, Document ID 16180, 2006.

[19]Computer Networking Notes, Basic Concepts of Frame Relay Explained in Easy Language, http://www.computernetworkingnotes.com/ccna-study-guide/basic-concepts-of-frame-relay-explained-in-easy-language.html, 2017, retrieved 2017-02-05.

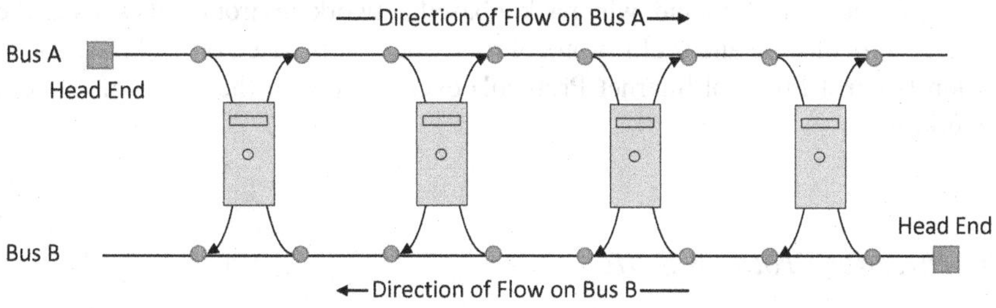

Fig. 7 Distributed queue dual bus

M. Piscitello, 1993). Later, in the beginning of the 1990s, these Metropolitan Area Networks (MANs) marked the convergence between computer and telecommunications industries and allowed simultaneous voice and data communications over distances of several hundred kilometers, expanding well beyond city areas (see Fig. 7).

The operation principle of MANs was actually the same as that of LAN operation by sharing the bandwidth of a bus (a dual bus in case of DQDB) between several users. Data streams were packetized by a Media Access Control Layer (MAC) handling efficient medium access control for voice, data and video transmissions, enabling the reservation of bandwidth for critical traffic types such as video or voice. The main difference between LAN and MAN standards was that the MAC layer was optimized for larger geographical distances. The Institute of Electrical and Electronic Engineers (IEEE) defined this standard under IEEE 802.6 DQDB[20] (ANSI/IEEE, 802.6 Local and Metropolitan Area Networks: Distributed Queue Dual Bus (DQDB) Subnetwork of a Metropolitan Area Network (MAN), 1990; Wright, 1993).

Asynchronous Transfer Mode

A very important contender for becoming the dominant networking technology was, for a while, a cell based approach called Asynchronous Transfer Mode (ATM). ATM was an evolution of the connection oriented voice networks, used as the main wide area networking, and interconnection technology before modern networking started (Wright, 1993). ATM will be discussed in detail in the section about ATM—Attempt to Integrate Data and Voice," but it is important to understand that it was based on fixed length cells as opposed to variable length packets of packet-based networks like Frame Relay. This marked the era of cell based versus packet-based networking. This competition lasted for at least 10 years before networks based on variable length packet technologies were finally established as the de-facto standard.

The two packet based solutions were mainly based on the Internet protocol suite and the OSI reference Model. We will highlight both in more detail throughout this

[20]ANSI/IEEE, Std 802.6, Distributed Queue Dual Bus (DQDB) access method and physical layer specifications, http://labit301.upct.es/LIPS/DQDB/802.6-1994.pdf, 1994, retrieved 2016-16-12.

chapter in order to understand, why packet based network environments such as the Internet protocol suite and multi-protocol router networks based on TCP/IP (Transmission Control Protocol/Internet Protocol) eventually won the battle of modern networking.

The Internet Protocol Suite

Internet Protocol Suite History

The Internet protocol suite is based on research and development from the Defense Advanced Research Projects Agency (DARPA) in the late 1960s and the first network running these protocols was the ARPANET[21] in 1969 (Stallings, 1989). In the early 1970s, Robert E. Kahn, who became the DARPA Information Processing Technology Officer, and Vinton Cerf, who was the developer of the existing ARPANET Network Control Program (NCP), started work on open-architecture interconnection models in order to design the next protocol generation of the ARPANET.

Soon Kahn and Cerf had worked out a fundamental change, hiding the differences between network protocols through the use of a common internetwork protocol, making the hosts themselves responsible for reliable communication as opposed to the network being responsible for the reliability in the ARPANET. There were, of course, many other names involved in this evolution, but the most important ones were Kahn and Cerf, who eventually created the first TCP/IP specification and the Internet protocol suite.

In 1975, the first TCP/IP communication tests between two networks were performed between Stanford and the University College of London (UCL) and in 1977, these tests were extended to Norway. In the following years, several other TCP/IP prototypes were developed at multiple research centers. Finally, on January 1st, 1983 the migration of ARPANET to TCP/IP was officially completed and the new protocols were permanently activated.

It was in 1982 before the US Department of Defense (DOD) declared TCP/IP as the standard for all military computer networking. In 1985, the Internet Advisory Board (called later the Internet Architecture Board) held its first 3-day workshop on TCP/IP for the computer industry, which was attended by 250 vendors' representatives. Also in 1985, the first Interop conference was held, focusing on network interoperability. This conference was typically attended by large vendors such as IBM and DEC, who were also the first to adopt TCP/IP in parallel to their existing proprietary protocols like Systems Network Architecture (SNA) (Kapoor, 1992) and DECnet.

---

[21]DARPA (Defense Advanced Research Projects Agency), ARPANET and the Origins of the Internet.

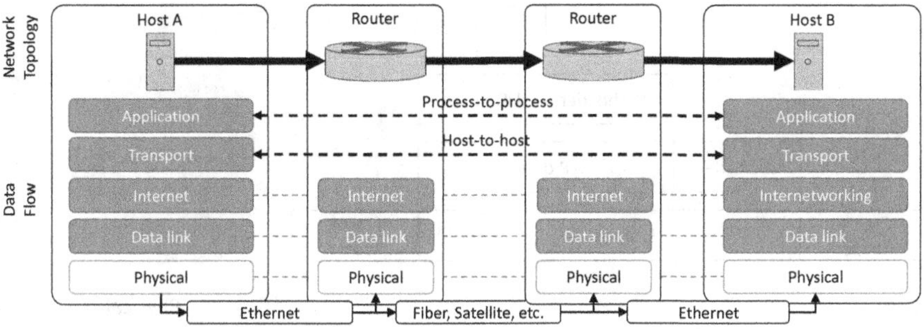

Fig. 8 End-to-end communication principle of the Internet protocol suite

It is important to understand that the first implementations of TCP/IP were successfully deployed in the mid 1980s (Halabi, 1997), while the OSI model and standards created by the ISO and ITU were just being finished in 1984.

Simple Principles of TCP/IP
TCP/IP defines networking up to layer 4 and provides end-to-end connectivity on these four abstraction layers (see Fig. 8). On the physical layer, we see two Ethernet links connecting hosts to routers (the IP network), while the routers are connected via fiber or satellite links or any other commonly used wide area network links. On top of the physical connections, we have the Data Link layer (layer 2), containing communications technologies for the specific network segments (the link). Above that, the Internet layer delivers connectivity across independent networks, which allows internetworking, commonly known as routing.
The layers above layer 3 handle all end-to-end communications; the Transport layer (layer 4) is responsible for host-to-host communications and finally, the Application layer provides process-to-process application communication and exchange of data. The TCP/IP model and the implemented protocol models are maintained by the IETF (Internet Engineering Task Force).[22]

Abstraction to Isolate Lower Layer Details
The encapsulations used to provide abstraction of protocols and services are aligned with the divisions of the protocol suite into layers of general functionality as shown in RFC1122[23] (see Fig. 9). A RFC is a Request for Comments, which is the way the IETF publishes specifications, communication protocols and procedures as official documents.

RFC 1122 defines how an application uses a set of protocols to send its data down the other layers, where they are further encapsulated at each layer.

The described abstraction design principle allows for the isolation of upper layer protocols from the details of lower layers for example, when transmitting bits over

[22]IETF (Internet Engineering Task Force), www.ietf.org

[23]Braden R., IETF, RFC 1122—Requirements for Internet Hosts—Communication Layers, 1989, retrieved 2017-06-02.

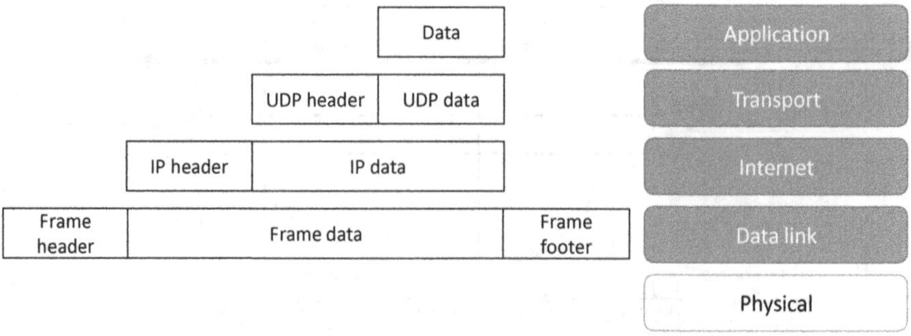

Fig. 9 Encapsulation of application data through layers defined in RFC 1122

Ethernet looking after all the access methods like CSMA/CD or a fiber link etc. The lower layers, in turn, do not need to know the details of each and every application and the protocols it uses.

Internet Protocol Suite Layers Defined

The Application Layer

The Application Layer describes how applications create user data and communicate this data to other applications on either the same or a remote host. Due to the layering approach, the applications and processes use the services provided by the underlying layers, mainly the Transport Layer, which ensures reliable or unreliable pipes to other processes (Stevens, 1994).

Communications partners are described by the application architecture; examples are the client-server model or peer-to-peer networking. The Application Layer is also the layer where all higher layer protocols like SMTP[24] (Simple Mail Transfer Protocol), FTP[25] (File Transfer Protocol), UDP[26] (User Datagram Protocol), SSH[27] (Secure Shell), HTTP[28] (Hyper Text Transfer Protocol), etc. operate (Stevens, 1994). The different processes are addressed via ports, essentially representing services.

A very important protocol operating at the Application Layer is the Simple Network Management Protocol[29] (SNMP), which defines SNMP agents that expose management data on the managed systems as variables, permitting active

[24]Klensin J., IETF, RFC 2821—Simple Mail Transfer Protocol, 2011.

[25]Postel J., Reynolds J., IETF, RFC 959—File Transfer Protocol, 1985.

[26]J. Postel, IETF, RFC 768, User Datagram Protocol, 1980.

[27]Ylönen T., Secure Shell (SSH-1), 1995.

[28]Fielding R., et al., IETF, RFC 2068, Hypertext Transfer Protocol, 1999.

[29]Case J., et al., IETF, RFC 1098, A Simple Network Management Protocol, 1990.

management tasks like modifying and applying new configurations through remote modification of these variables (Rose, 1991).

Transport Layer

The Transport Layer is responsible for host-to-host communications on either the same or different hosts, which can be either located on the same local area network or on remote networks, interconnected by routers, providing a channel of communication needs for applications.

In the connection-oriented implementation, the Transmission Control Protocol[30] (TCP) provides connection establishment, reliable transmission of data and flow control. In the connectionless implementation, the User Datagram Protocol (UDP) provides an unreliable datagram service.

The Internet Layer Establishing Internetworking and the Internet

The main task of the Internet layer is to exchange layer 2 data (datagrams) over network boundaries, using packets on layer 3. One of its main tasks is to provide a uniform networking interface, hiding the actual topology or the layout of the underlying network connections (Stevens, 1994). The Internet layer is, therefore, the layer that establishes internetworking and, moreover, it defines the Internet (Halabi, 1997).

There is a clear definition of addressing and routing structures (Halabi, 1997) that is used by the TCP/IP protocol suite. Of course, the primary protocol in this scope is IP, the Internet Protocol, defining IP addresses. The routing function of the Internet Layer handles the transport of packets to the next IP router that has connectivity to a network closer to the final destination.

Link Layer

The Link Layer has to look after the handling of the networking methods specific to the local area networks used, where the hosts are communicating without routers to run, for example, the necessary media access protocols such as CSMA/CD for Ethernet. The Link Layer covers all necessary local area network protocols and topologies as well as the interfaces for effective transmission of Internet layer datagrams or packets to the next neighbor hosts (Stevens, 1994).

OSI Reference Model

Beginning in the 1970s the definition and standardization for the architecture of networking systems by the ISO (International Standardization Organization) and the CCITT (Comité Consultatif International Téléphonique et Télégraphique) started, now known as ITU-T (Telecommunication Standardization Sector of the International Telecommunications Union). By 1983, the similar networking models

---

[30]DARPA, IETF, RFC 793, Transmission Control Protocol, 1980.

	Layer	Data Unit	Function	Examples
Host Layers	7. Application	Data	High Level APIs including resource sharing, remote file access, directory services and virtual terminals	HTTP, FTP, SMTP, SSH, TELNET
	6. Presentation	Data	Translation of data between a networking service and an application including character encoding, data compression and encryption/decryption	HTML, CSS, GIF
	5.Session	Data	Managing communication sessions, i.e. continuous exchange of information in the form of multiple-back-and-forth transmissions between two nodes	RPC, PAP, SSL, SQL
	4. Transport	Segments / Diagram	Reliable transmission of data segments between points in a network including segmentation, acknowledgement and multiplexing	TCP, UDP, NETBEUI
Media Layers	3. Network	Packet	Structuring and managing a multi-node network including addressing, rooting and traffic control	IPv4, IPv6, IPSec, AppleTalk, ICMP
	2. Data Link	Frame	Reliable transmission of data between two nodes connected by a physical layer	PPP, IEEE 802.2, L2TP, MAC, LLDP
	1. Physical	Bit	Transmission and reception of raw data bit streams over a physical medium	Ethernet Physical Layer, DSL, USB, ISDN, DOCSIS

Fig. 10 OSI reference model

of these two standard bodies were merged into one standard, called "The Basic Reference Model for Open Systems Interconnection", which generally is referred to as the Open Systems Interconnection Reference Model (see Fig. 10), or simply the OSI Model (Stallings, 1990a). The OSI Model was published in 1994 by both the ISO as standard ISO 7498[31] and the CCITT (called today ITU) as standard X.200.[32]

The OSI Model has two major components, the abstract model for networking, which is usually called the Basic Reference Model or Seven Layer Model (because it consists of seven layers) and a set of specific protocols.

Many aspects of the OSI design evolved from practical experiences with networks like the ARPANET and others. The networking system is divided into seven layers and within each layer one or more entities implement their functionality.

Each of these entities interacts directly with the layer immediately beneath it and provides facilities for usage by the layer above it. Protocols enable an entity in one host to communicate and interact with a corresponding entity at the same layer in another host. There are service definitions describing the functionality provided to a (N)-layer by a (N−1)-layer of the seven layers of this model.

At each level N, two entities exchange PDUs (protocol data units) through the specific layer N protocols (for example on the network layer or layer 3, the protocol data units are called "packets").

It is obvious that the OSI model puts an abstract architecture in place for defining communication and networking which, of course, can be implemented in different ways but at least gives us a basic understanding of what functionalities and features should be covered by each layer. The history of networking has shown many implementations of the OSI model, but the one which became most prominent and is still in use today is the Internet protocol suite and we will cover its success in more detail later in this section.

[31]ISO/IEC 7498-1:1994, Information Technology—Open Systems Interconnection—Basic Reference Model: The Basic Model, 1994.

[32]ITU, Recommendation X.200 (07/94), Information Technology—Open systems Interconnection—Basic Reference Model: The Basic Model, 1994.

ATM: Attempt to Integrate Data and Voice

ATM (Asynchronous Transfer Mode) started in the late 1980s as a high speed-networking standard for integrating voice and data communications into one network (Malamud, 1992). ATM describes how data can be carried over a cell based data link layer, Layer 2, of the OSI based reference model (Ginsburg, 1996). ATM is based on a connection-oriented model, which means that connections between end points need to be set up before any data (cells) can be transferred. In the following, we will have a more detailed look at ATM[33] and compare its principles to the IP as well as the OSI models, since this will allow us to get a much deeper understanding of these different networking approaches (Partridge, 1994).

As ATM uses asynchronous time-division multiplexing, data is encoded into small, fixed sized packets, the ATM cells. This is significantly different from packet switched networks and the Internet Protocol, which uses variable length packets (layer 3) and frames (layer 2). Furthermore, ATM uses connection-oriented communications, where VCs (virtual circuits) have to be established between two endpoints before any communication or data transfer can start. Virtual circuits can be either permanent, meaning dedicated connections pre-provided by service providers, or set up and disconnected on demand and as needed using dedicated signaling protocols. The choice of a connection-oriented method is no surprise, as ATM was driven and developed by telephone companies for their long-distance networks since you always have to setup a telephone connection before you can communicate, thus the telephone companies were working with what they knew best.

Connection Oriented Versus Connectionless

This connection oriented approach (Wright, 1993) is also one of the biggest differentiators to connectionless packet based networking, where data are enveloped into a data frame (on the data link layer) and then sent from a source to a destination end point (device, computer, etc.), using either pure layer 2 (bridging and switching) or layer 3 routing technology. Both approaches, connectionless and connection oriented have their advantages and disadvantages. Connection oriented communication is better suited for voice or other similar communications (mainly between human beings) where stringent timing is needed as well as controlled jitter and delay, whereas connectionless communication is generally better suited for sending data between computers or machines.

Over time, the connectionless approach won the battle as the universal networking technology based on packet switching and packet routing. Eventually, even quasi connection oriented approaches were implemented on top of classic routing via pre-assigned end to end labeling of data packets, also known as MPLS[34] (Multi-

[33]ITU-T, I.150: B-ISDN Asynchronous Transfer Mode functional characteristics, 1999.

[34]Rosen E., et al., IETF, RFC 3031, MPLS—Multiprotocol Label Switching Architecture, 2001.

Protocol Label Switching) with additional enhancements like traffic engineering or fast rerouting for packet networks using routed traffic. All this evolution happened during the 1990s, when ATM was a very prominent and promising approach for unified voice and data communications at the beginning of that decade but was gradually replaced by packet communication and routed networks towards the end of the decade.

Who Was Behind the Definition of ATM?
ATM was significantly driven and developed by the ATM Forum, which was founded in 1991 and the founding president and chairman was Fred Sammartino of Sun Microsystems. The ATM Forum described ATM as "a telecommunications concept defined by ANSI and ITU-T (also formerly known as CCITT) standards for carriage of a complete range of user traffic, including voice, data and video signals". It was designed to meet the needs of B-ISDN (Broadband Integrated Services Digital Network), which was defined at the end of the 1980s and offered a common transport network for telecommunication and computer networks.

The goal of this design was to support traditional high flows of traffic like file transfers, as well as real-time, low latency traffic from voice and video. ATM describes the lowest three layers of the OSI reference model, the physical layer, the data link layer and the network layer (DePrycker, 1993). As a core protocol, ATM was built to support SONET/SDH backbones of Public Switched Telephone Networks (PSTNs) as well as Integrated Services Digital Network (ISDN).

ATM Networking Differences
We already mentioned that ATM is different from common data link technologies like Ethernet in several ways. The most important difference is that ATM networking devices, called ATM switches, use point-to-point connections between the endpoints so that data can flow directly from source to destination. This means there is no routing involved during the communication phase, as this routing is only performed for the connection setup and the end-to-end path stays the same for the complete communication (Bay-Networks, 1996). Thus, ATM does not utilize routing during the communication phase like packet based router networks, which forward each packet according to the current best route, which they constantly derive through the background communication of routing protocols and updating of their routing tables.

ATM Cell Structure
ATM cells are 53 bytes in length, consisting of a 5 byte header and 48 bytes of data. There is a slight difference between the User Network Interface (UNI) (ATM-Forum, 1995) and the corresponding UNI ATM Cells and the Network Network Interface (NNI) and the specific NNI ATM Cells between provider networks. The NNI Cells do not have the Generic Flow Control (GFC) field of the end system, which is used by additional virtual path bits instead, but have the same overall length of 53 bytes (see Fig. 11).

The PT field is used to designate various kinds of cells for OAM (operations, administration and management) purposes and for the delineation of packet

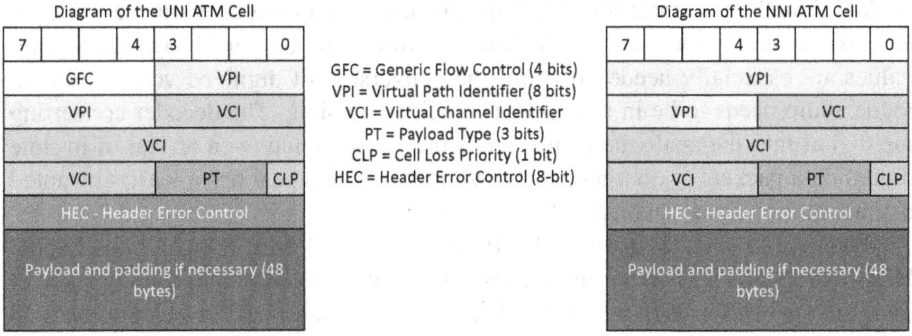

Fig. 11 ATM cells at the UNI and NNI

boundaries in some ATM Adaptation Layers (AAL), which are needed to map different traffic types such as voice, video or packet data into ATM cells.

The HEC field is used by ATM link protocols for driving a CRC based framing algorithm, allowing for the locating of ATM cells without any overhead and the 8-bit CRC is used for correction of single-bit header errors as well as detection of multi-bit header errors. In the case of multi-bit header errors, the current and subsequent cells are dropped until a cell with no header errors is found.

The GFC field is reserved by NNI cells in order to allow local flow control and sub-multiplexing between system users, which allows several terminals to share a single network connection, something very similar can be done between two Integrated Services Digital Network (ISDN) phones in order to share a single basic rate ISDN connection. This shows, once more, that the source of ATM comes from the voice world.

At the NNI, the cell format is exactly the same as at the UNI with the exception that the GFC field is reallocated to the VPI field, allowing for extended 12 bit VPI values. This raises the capability of a single NNI ATM connection to allow from almost 212 VPs to up to almost 216 VCs each (in practice some of the VP and VC numbers are reserved). We should remember that ATM uses connections that are characterized by virtual circuits and multiple virtual circuits can be collected into one virtual path.

Cells Versus Packets

ATM uses small cell sizes (small data packets) resulting from the idea of achieving optimal transmission conditions for different traffic types. Voice can be squeezed into relatively small voice packets, but these can easily interfere with large data traffic packets, if they have to share the same link. In order to avoid unpredictable queuing delays for real time traffic such as voice or video, all ATM packets (called cells) have the same, small cell size. This fixed cell structure guarantees seamless and in time switching by hardware without generating the typical delays introduced by variable length frames and packets, which are forwarded by either switches or routers.

Additionally, the small cell size helps to reduce jitter or delay variance, which is equally important for voice traffic and end-to-end round trip delays. Low jitter values are especially needed because the conversion of digitized voice into analogue audio needs to be in real time as much as possible. The decoder converting the digital into the analogue signal needs the constant and even arrival of in time digital data packets to do a proper job without loss of signal or, at least, unwanted degradation of signal quality.

When ATM was designed in the 1980s and early 1990s, Synchronous Digital Hierarchy (SDH) was one of the highest speed data transmission technologies over wide area networks (Wright, 1993). Discussing data transmission technologies and optical networking in more detail is beyond the scope of this book, but you can find an excellent description of Next Generation Optical Networks in (Tomsu & Schmutzer, 2002).

Typical transmission speeds were up to 155 Mbps or even 622 Mbps. A 155 Mbps SDH connection could easily carry a 135 Mbps payload, but we should be aware that many links in that Plesiochronous Digital Hierarchy (PDH) infrastructure were much slower at these times, ranging from 2 Mbps (E1 link) to 34 Mbps (E3 link) in Europe and 1.544 Mbps (T1 link) to 45 Mbps (T3 link) in the USA. Sending a long data packet consisting of, for example, 1.500 bytes via such a high-speed link takes 77.42 us to transmit, whereas on a slow speed link like a 1.544 Mbps T1 line, transmission of a 1.500 bytes packet would take 7.8 ms, introducing an unacceptable delay for real time traffic. It is obvious that queuing delays introduced by such large data packets in addition to other delays resulting from packet generation easily exceeds the acceptable delay and jitter for voice traffic in order to achieve acceptable voice quality.

Thus, ATM had to support low jitter interfaces and cells were the preferred design choice in order to guarantee short queuing delays while, at the same time, they allowed for the handling of data traffic. This was why 48 byte data units were used and either filled with real-time or non-real time traffic. 5 byte headers were added in order to allow reassembling of the data units later at the destination.

There is an interesting story behind the 48 bytes, as the decision for this number was not a technical one, but rather a political one. During standardization at CCITT (ITU today) there were (as often was the case in these processes) different parties involved representing different nations, and, of course, asking for different approaches. In the case of the ATM payload type, the United States asked for a 64-byte payload, as they felt this number would be a perfect compromise between larger payloads more optimal for data transmission and shorter payloads optimal for real time. In opposition to this was the opinion from Europe, opting for 32-byte payloads in order to simplify voice applications with respect to echo cancellation through the use of small cell sizes. There were long discussions and, while many European countries eventually moved into the US camp, France and a few others held their position for the shorter cell size. Finally, a 48-byte payload was chosen as a political compromise between the two camps.

The Success of the Internet Protocol Suite

What Is So Special About the Internet Protocol Suite?

The Internet protocol suite, commonly known as TCP/IP (Transmission Control Protocol and Internet Protocol) describes a computer networking model and a set of communication protocols. It became the de facto computer networking standard and the basis of the Internet during the 1990s. In this section, we will discuss why TCP/IP finally won the battle to become the basis for modern networking and why the OSI reference model as well as ATM disappeared.

Internet Protocol Suite Versus OSI Model

The Internet protocol suite and its layered protocol stack design were already in use at the beginning of the 1980s, long before the standard for the OSI model was ready in 1994. There are, of course, a number of key differences between the Internet protocol suite and OSI (David M. Piscitello, 1993) that we will discuss now, in order to better understand why TCP/IP finally succeeded.

Although the OSI reference model was more detailed than the Internet protocol suite, there were never many successful implementations and deployments of OSI protocols. Most of these protocols did not gain much popularity, the only exceptions were X.400[35] defining the Message Handling System (MHS), X.500[36] defining Directory Services Protocols and ISIS[37] or the Intermediate System to Intermediate System routing protocol. It should be mentioned that there have also been many, vendor specific implementations mainly throughout the 1980s and until the middle of the 1990s like DECnet, AppleTalk, Banyan Vines, etc. (Dickie, 1994; Malamud, 1992) that gradually all disappeared, as the Internet protocol suite became the dominating standard.

One of the big success factors of TCP/IP is that it uses a much looser layering than the OSI model, which is known for its strict layering. This looser approach allowed for greater flexibility and freedom when designing and implementing.

Another important feature of TCP/IP is that it leaves reliable communication to hosts. We already discussed that abstraction allows higher layers to offer services, which cannot be offered by lower layers. Maybe the most important difference overall is that IP is not designed to be reliable and is a best effort delivery protocol, while OSI initially defined a reliable connection oriented approach, which was later extended to include connectionless services as well.

From the beginning, IP was designed so that all transport layer implementations must choose how to guarantee reliability and leave the final task of reliable communication to the end nodes. This becomes obvious when looking at the

---

[35]ITU-T, X.400: Message Handling System and Service Overview, 1999, retrieved 2016-12-05.

[36]ITU-T, X.500: The Directory: Overview of Concepts, Models and Services, 1988–2016, retrieved 2016-12-05.

[37]Parker J., RFC 3787, Recommendations for Interoperable IP Networks using Intermediate System to Intermediate System (IS-IS), 2004, retrieved 2016-12-05.

mechanisms supported by UDP (User Datagram Protocol), like providing data integrity via checksum but without any guaranteed delivery and TCP providing data integrity and delivery guaranty through retransmission, but still leaving acknowledgements of packet reception to the receiver.

Why ATM Was Not Successful

While ATM stayed popular in certain environments for a long time (and still is on some occasions today), the fixed length short cell approach and the connection oriented nature of ATM finally proved not to be as useful for networking solutions and packet switched networks based on routing became the dominant solution.

Router networks soon began to integrate ATM and the most popular approach to do so was Multi Protocol Label Switching (MPLS),[38,39] which, over the years, was further developed by the IETF (Internet Engineering Task Force) and some other standardization bodies like the ITU-T. MPLS is still a very prominent standard for advanced routing mechanisms, although for a long time most of them are no longer related to ATM but more to Traffic Engineering, Fast Reroute etc. Very often, used MPLS implementations are MPLS VPNs (Virtual Private Networks), allowing enterprises and service providers to dramatically improve scalability, performance, security and efficiency of VPNs by the use of MPLS (Tomsu & Wieser 2002).

The Internet Protocol Suite Was Easy to Implement

In the end, the Internet protocol suite reached the goal of becoming an open-standard protocol suite and this success was and still is mainly driven by the IETF. One of the biggest secrets behind this success was that the specifications were simpler and easier to implement with fewer and less rigid defined layers than the complicated OSI architecture and protocol stack plus they were open and publicly available standards written down in Request for Comments (RFC's)[40] by the IETF. While these RFCs often needed revisions and so did their implementations, this concept was much better suited for a wide acceptance and usage and allowed an easier implementation and deployment for real-world protocols.

RFC 1958[41] shows a snapshot of the architecture available in 1996 and it becomes immediately obvious that there is no suggestion for a stack of layers, as there is only a reference to the existence of an internetworking layer and generally the upper layers, which is the major difference between the IETF and the OSI approaches: *"The Internet and its architecture have grown in evolutionary fashion from modest beginnings, rather than from a Grand Plan. While this process of evolution is one of the main reasons for the technology's success, it nevertheless seems useful to record a snapshot of the current principles of the Internet architecture."*

---

[38]Rosen E. et al., IETF, RFC 3031, Multiprotocol Label Switching Architecture, 2001.

[39]IETF, Document Index, https://datatracker.ietf.org/wg/mpls/documents/, retrieved 2016.

[40]IETF, RFC Index, 2016, https://tools.ietf.org/rfc/

[41]Carpenter B., IETF, RFC 1958, Architectural Principles of the Internet, 1996.

Another example of less emphasizing the rigid layering approach is RFC 1122 and 1123 respectively. RFC 1122[42] describes "Host Requirements" and while its structure is loosely referring to a four layer model, it also covers many other architectural principles above layering: *"RFC 1122 covers the communications protocol layers: link layer, IP layer and transport layer; its companion RFC 1123*[43] *covers the application and support protocols."*

While the model for the Internet protocol suite lacks many formalisms of the OSI model and associated documents, the IETF does not use a formal model, nor does it consider this to be a limitation, which is expressed in a comment by David Clark, one of the Internet pioneers and author of "The Design Philosophy of the DARPA Internet Protocols", Computer Communications Review 18: 4, August 1988, pp. 106–114: *"We reject kings, presidents and voting. We believe in rough consensus and running code."* This nicely shows why the Internet protocol suite, as defined by the IETF, was so successful.

Internet Protocol Next Generation aka IPv6

Only two things are infinite, the universe and human stupidity, and I'm not sure about the former.—Albert Einstein

In the fall of 1991, it became apparent that the capacity of the Internet to support many millions of users was beginning to deplete the existing address space of IPv4 (IP Version 4). Scott Bradner and a group of others recognized this at this time when the IPv4 address space would not be nearly enough to satisfy the coming needs of future generation of the Internet. We need to understand that this was as early as 1991, when we were still far away from the mobile boom of smartphones, intelligent connected devices and sensor networks and additionally, the Internet of Things would appear almost 20 years later. Scott Bradner wrote the following in his book (Scott Bradner, 1996):

> *The Internet's astounding growth has begun to stress the technology that supports it, namely the current Internet protocol suite. IPv4. At first look IPv4 appears as if it should be able to handle an Internet of scale projected for the next decade or two. Its 32-bit address structure can enumerate over 4 billion hosts on as many as 16.7 million networks. However, the actual address assignment efficiency is far less than that, even on a theoretical basis, and this inefficiency is exacerbated by the division of IPv4 addresses into three separate classes, Class A, Class B, and Class C. Address assignments were made assuming that the Internet world consisted of numerous small organizations containing fewer than 250 computers each (Class C networks), a smaller number of larger organizations with up to 64000 computers each (Class B networks), and a few very large companies with up to 15 million computers each (Class A networks).*

[42]Braden R., IETF, RFC 1122, Requirements for Internet Hosts—Communication Layers, 1989.

[43]Braden R., IETF, RFC 1123, Requirements for Internet Hosts—Application and Support, 1989.

Given the address assignment rate in 1991, Class B address space would have been exhausted by March 1994. The solution for surviving with the current IP address space was to assign multiple Class C addresses instead of one Class B address. That caused different problems, since the routing tables of the backbone routers grew by an exponential rate, which would lead to memory exhaustion even when most modern memory was used.

This triggered the Internet community and the IETF to initiate the switch to new technologies and techniques, and the IPng movement was started.

The Way Towards IPng

In November 1991, the IETF formed the Routing and Addressing (ROAD) group, who came up with several recommendations like adopting the Classless Inter-Domain Routing (CIDR)[44] route aggregation, helping to reduce the growth rate of routing tables plus exploring new approaches for bigger addresses needed for the foreseeable growth of the Internet. At the Boston IETF meeting in July 1992, the call for IPng proposals went out in combination with forming a number of working groups and, a year later at the July 1993 Amsterdam IETF meeting, the decision was finally made to design a new Internet Protocol and it soon became apparent that IPng had to solve more than just the address scaling and exhaustion issues of IPv4.

The IPng Directorate was formed soon after and the members came from all around the industry: J. Allard (Microsoft), Steve Bellovin (AT&T), Jim Bound (Digital), Ross Callon (Bay Networks), Brian Carpenter (CERN), Dave Clark (MIT), John Curran (BBN Planet Corporation), Steve Deering (Xerox Corporation), Dino Farinacci (Cisco Systems), Eric Fleischman (Boeing Computer Services), Paul Francis (NTT), Paul Mockapetris (USC/ISI), Rob Ullmann (Lotus Development Corporation) and Lixia Zhang (Xerox Corporation).

In October 1993, a new Address Lifetime Expectations (ALE) Working Group was formed and chaired by Tony Li (Cisco Systems) and Frank Solensky (FTP Software) with the goal of estimating the remaining lifetime of the IPv4 address space and if a more stringent address allocation and utilization would provide more transition time until IPng was ready. It soon turned out that none of the more stringent assignment policies was necessary and that network service providers should help new customers in renumbering their networks to conform to the providers CIDR assignments. A lot of other efforts were taken to keep IPv4 alive as long as possible, one of the most prominent being the deployment of CIDR to shrink routing table entries with initially measurable success in 1994. But address range, address allocation and routing table size were only some of the issues, which would be solved with IPng.[45]

[44]Fuller V., Li T., IETF, RFC 4632, Classless Inter-domain Routing (CIDR): The Internet Address Assignment and Aggregation Plan, 2006, retrieved 2016-12-05.

[45]IPngWG, IETF, IPng Documents, https://datatracker.ietf.org/wg/ipngwg/documents/, 1995–2001, retrieved 2016-12-05.

IPng/IPv6 Advancements

IPv6 brings several essential improvements to the future of networking and the next generation of Internet in general (Hagen, 2014).

Larger Address Space
As already discussed, one of the main advantages of IPv6 over IPv4 is the larger address space, which comes from the IPv6 address length of 128 bits as compared to the 32 bits used in IPv4. This allows for 2^{128} addresses in total or approximately 3.4×10^{38} addresses. It was the intent not only to increase the address space, but also to enable much more efficient route aggregation as well as the implementation of special addressing features without the need to use complex Classless Inter Domain Routing (CIDR) methods necessary for the smaller IPv4 address space. In IPv6, the standard size of a subnet is 2^{64} addresses, which is the square of the size of the entire IPv4 address space. This larger subnet space also improves the network management and routing efficiency above hierarchical route aggregation.

Multicast
The transmission of a packet to multiple destinations in a single send operation is an inherent part of IPv6 and not an optional feature as with IPv4. IPv6 builds on features and protocols from IPv4 multicast, but with changes and improvements. It does not use a special broadcast addresses like IPv4, but sends a packet to the link-local all nodes multicast group address ff02::1. There are also new multicast implementations making life easier like embedded rendezvous point addresses in an IPv6 multicast group address, simplifying deployment of inter-domain solutions.

Stateless Address Auto Configuration (SLAAC)
This means that any IPv6 host can perform automatic self-configuration when connected to an IPv6 network through the use of the Neighbor Discovery Protocol (NDP)[46] via Internet Control Message Protocol version 6 (ICMPv6)[47] to route discovery messages.

When a new host is connected to a network, it sends a link-local router solicitation multicast request for its configuration parameters and a router responds to this request by advertising the Internet Layer configuration parameters.

If the IPv6 stateless auto configuration is not supported by an application, there is also the possibility that the network may use a configuration with states by using Dynamic Host Configuration Protocol version 6 (DHCPv6).[48] Of course, hosts could still be configured manually via static methods. For routers, a stateless configuration can be achieved with a special router renumbering protocol.

---

[46]Narten T. et al., IETF, RFC 4861, Neighbor Discovery for IPv6, 2007.

[47]Conta A. et al., IETF, RFC 4443, Internet Control Message Protocol (ICMPv6), 2006.

[48]Droms R. et al., IETF, RFC 3315, Dynamic Host Configuration Protocol for IPv6, 2003.

Network Layer Security

The very well known and widely established Internet Protocol Security (IPSec)[49] was actually developed for IPv6, but found wide acceptance and deployment long before IPv6 took off in IPv4 installations. IPsec is now an optional feature of IPv6.

Simplified Processing by Routers

IPv6 uses a simplified packet header and packet forwarding process. The IPv6 header is at least twice the size of the IPv4 header but packet processing by routers is more efficient, because less processing is required. This enables even better end-to-end Internet design with most processing happening in the leaf nodes as opposed to in the network nodes (the routers).

For example, IPv6 routers do not care for packet fragmentation instead, the IPv6 hosts need to perform path Maximum Transfer Unit (MTU) discovery, perform the end-to-end fragmentation and send packets, which fit into the MTU of 1280 octets. Also, the IPv6 header has no checksum protection, as integrity protection needs to be handled by the link layer or error detection and correction of higher layer protocols such as TCP or UDP. This requires a checksum in UDP as well. The routers in the data path do not need to recompute a checksum as soon as header fields change, such as Time To Live (TTL) or Hop Count (HC). This is also why the TTL field in IPv6 was renamed to Hop Limit, as the routers do no longer need to compute the time a packet has spent in a queue.

Mobility

Mobile IPv6[50] avoids triangular routing as used in IPv4. This is why it is as efficient as native IPv6, allowing entire subnets to be moved to new router connection points without renumbering.

Options Extensibility

The size of the IPv6 header is 40 octets at its minimum, options are used to implement extensions, opening the opportunity to extend the protocol later without affecting the core packet structure.

Jumbograms

An IPv6 node can optionally handle packets over the limit of IPv4, which are 65,535 octets. This is referred to as Jumbograms,[51] which can have a length of up to 4,294,967,295 ($2^{32}-1$) octets in order to improve performance on high-MTU links and are indicated by the Jumbo Payload Option header.

Privacy

IPv6 supports globally unique IP addresses, same as IPv4. This allows tracking of the network activity of each device attached, reemphasizing the end-to-end network design principle of the early Internet. Network prefix tracking is less of a concern if

---

[49]Kent S. et al., IETF, RFC 4301, Security Architecture for the Internet Protocol, 2005.

[50]Johnson D. et al., IETF, RFC 3775, Mobility Support in IPv6, 2004, retrieved 2016-12-05.

[51]Borman D. et al., IETF, RFC 2675, IPv6 Jumbograms, 1999, retrived 2016-12-05.

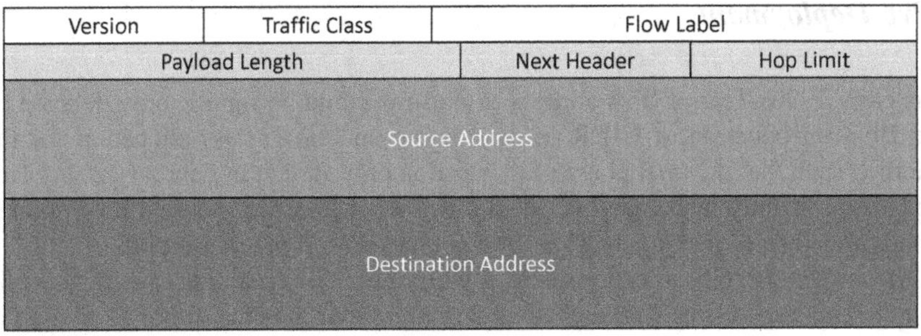

Fig. 12 IPv6 packet header

the user's ISP assigns a dynamic network prefix via DHCP. Also, privacy extensions have little impact on the protection of the user from tracking if the ISP assigns a static network prefix as, in this case, the network prefix is the unique identifier for tracking and the interface identifier is secondary.

In IPv6 with Auto-configuration,[52] the Interface Identifier (the MAC address) of an interface port is used to make its public IP address unique, by exposing the type of hardware used and providing a unique handle for a user's online activity.

If Auto-configuration is not used, the interface address is globally unique as opposed to the Network Address Translation (NAT)[53] masked private networks. Special privacy extensions for IPv6 were defined to address these privacy concerns.

IPv6 Packet Format

Like IPv4, IPv6 packets also have two parts, a header and a payload (see Fig. 12).

The IPv6 header has a fixed length of 40 octets as already mentioned. In front of the source and destination IPv6 address are the fields for Traffic Class Options, a Flow label, and the Payload Length. The Next header field tells the receiver how to interpret the data following the header. If the packet contains options this field contains the option type of the next option. If it is the field of the last option, it points to the upper layer protocol. Finally, the Hop Limit field follows.

[52]Thomson S. et al., IETF, RFC 4862, IPv6 Stateless Address Autoconfiguration, 2007, retrieved 2016-12-05.

[53]Srisuresh P. et al., IETF, RFC 2663, IP Network Address Translator (NAT) Terminology and Considerations, 1999, retrieved 2016-12-05.

IPv6 Deployment

The early 1990s, fear of IPv4 address exhaustion could be significantly delayed by the 1993 introduction of CIDR in the routing and IP address allocation for the Internet, such that the final phase of exhaustion only started in February 2011. This was a significantly later than predicted. Still, as of late 2013, only 4% of domain names and 16% of networks on the Internet had IPv6 protocol support.

Meanwhile IPv6 has been implemented in all major operating systems used in commercial business as well as home consumer environments. It has been used in some major world events like the 2008 Summer Olympic Games and some governments including the federal governments in the USA and China have issued guidelines and requirements for IPv6 capability.

By 2014, IPv4 still carried more than 99% of worldwide Internet traffic and, by February 2016, IPv6 traffic was tracked at about 1.4% with an annual growth rate of 0.8% per year thus far. In December 2015, the number of IPv6 users of Google services reached 10% for the first time with an annual growth rate of 4.3%. By the middle of 2015, deployment of IPv6 on web servers was around 16%, varying widely worldwide.

References

ANSI. (1987). *X3T9.5 (now X3T12) Fiber distributed data interface*. Washington, D.C.: ANSI American National Standards Institute.

ANSI/IEEE. (1985a). *802.3 Carrier sense multiple access with collision detection*. New York: IEEE – Institute of Electrical and Electronics Engineers.

ANSI/IEEE. (1985b). *802.4 Token passing bus access method*. New York: The Institute of Electrical and Electronics Engineers.

ANSI/IEEE. (1985c). *802.5 Token ring access method*. New York: The Institute of Electrical and Electronics Engineers.

ANSI/IEEE. (1990). *802.6 Local and metropolitan area networks: Distributed queue dual bus (DQDB) subnetwork of a metropolitan area network (MAN)*. New York: The Institute of Electrical and Electronics Engineers.

ATM-Forum. (1995). *ATM user network interface (UNI) specification* (Version 3.1 ed.). Upper Saddle River, NJ: Prentice Hall.

Bay-Networks. (1996). In T. V. Mathias Hein (Ed.), *ATM LAN guide*. Köln: Fossil Verlag.

Bernhard Albert, A. P. (1994). *FDDI and FDDI-II architecture, protocols and performance*. Boston, MA: Artech House.

Buckwalter, J. T. (1999). *Frame relay – Technology and practice*. Reading, MA: Addison Wesley.

Burg, U. V. (2001). *The triumph of ethernet*. Stanford, CA: Stanford University Press.

Ceruzzi, P. E. (1998). *A history of moder computing* (2nd ed.). Cambridge MA: MIT Press.

Charles E. Spurgeon, J. Z. (2013). *Ethernet switches*. Newton, MA: O'Reilly.

David M. Piscitello, A. L. (1993). *Open systems networking – TCP/IP and OSI*. Reading, MA: Addison Wesley.

DePrycker, M. (1993). *Asynchronous tranfer mode – Solution for broadband ISDN* (2nd ed.). New York: Ellis Horwood.

Dickie, M. (1994). *Internetworks*. VNR Communications Library.

Elliott, B. (2002). *Designing a structured cabling system to ISO 11801* (2nd ed.). Cambridge: Woodhead.

Ginsburg, D. (1996). *ATM – Solutions for enterprise internetworking*. Harlow: Addison Wesley.

Hagen, S. (2014). *IPv6 Essentials* (3rd ed.). Sebastopol, CA: O'Reilly and Associates.

Halabi, B. (1997). *Internet routing architectures – The definitive reource for internetworking design alternatives and solutions*. Indianapolis, IN: Cisco Press.

Hans-Georg Göhring, F.-J. K. (1990). *Token ring – Grundlagen, Strategien, Perspektiven*. Bergheim: Datacom Verlag.

Held, G. (2003). *Ethernet networks* (4th ed.). Chichester, UK: Wiley.

Kapoor, A. (1992). *SNA – Architecture, protocols and implementation*. New York: McGraw-Hill.

Larry L. Peterson, B. S. (2012). *Computer networks – A systems approach* (5th ed.). Amsterdam: Elsevier.

Madron, T. W. (1988). *Local area networks – The second generation*. New York: Wiley.

Malamud, C. (1992). *Stacks – Interoperability in today's computer networks*. Englewood Cliffs, NJ: Prentice Hall.

McNamara, J. E. (1982). *Technical aspects of data communication*. Maynard, MA: Digital Equipment Corporation.

Minoli, D. (1993). *1st, 2nd & next generation LANs*. New York: McGraw Hill.

Partridge, C. (1994). *Gigabit networking*. Reading, MA: Addison Wesley.

Paul Ferguson, G. H. (1998). *Quality of service – Delivering QoS on the intrnet and in corporate networks*. New York: Wiley.

Perlman, R. (1992). *Interconnections – Bridges and routers*. Reading, MA: Addison Wesley.

Rose, M. T. (1991). *The simple book – An introduction to management of TCP/IP based internets*. Englewood Cliffs, NJ: Prentice Hall.

Scott Bradner, A. M. (1996). *IPng – Internet protocol next generation*. Reading, MA: Addison Wesley.

Stallings, W. (1989). *Handbook of computer communications standards volume 3 – The TCP/IP protocol suite* (2nd ed.). London: Stallings/MacMillan.

Stallings, W. (1990a). *Handbook of computer communications standards volume 1 – The open systems interconnection (OSI) model and OSI-related standards* (2nd ed.). London: Stallings/McMillan.

Stallings, W. (1990b). *Handbook of computer communications standards volume 2 – Local area network standards* (2nd ed.). London: Stallings/MacMillan.

Stevens, R. W. (1994). *TCP/IP illustrated – The protocols* (vol. 1). Reading, MA: Addison Wesley.

Tanenbaum, A. S. (2003). *Computer networks*. Upper Saddle River, NJ: Prentice Hall.

Tomsu, P., & Schmutzer, C. (2002). In D. Cullen-Dolce (Ed.), *Next generation optical networks*. Upper Saddle River, NJ: Prentice Hall.

Tomsu, P., & Wieser, G. (2002). In T. D. Nadau (Ed.), *MPLS-based VPNs, designing advanced virtual networks*. Upper Saddle River: Prentice Hall.

Wright, D. (1993). *Broadband: Business services, technologies and strategic impact*. London: Artech House.

Managing Virtual Storage

As far as the laws of mathematics refer to reality, they are not certain; and as far as they are certain, they do not refer to reality

Albert Einstein

We describe how storage virtualization frees the applications to know about details like specific drives, partitions or storage subsystems in which their data reside, by adding a new layer of software and hardware between storage systems and servers.

This enables administrators to identify, provide and manage distributed storage as if it were a single and consolidated resource and increase availability at the same time. We will deal with the Shared Storage Model and discuss the different types of storage virtualization.

Why Storage Virtualization

Whenever dealing with computers, storage is needed. Storage was usually implemented as a disk drive and, as soon as more storage was required, a bigger disk drive had to be deployed. Soon, even the increase of storage offered by adding multiple disk drives was not enough and the management of multiple disk drives became harder, which was one of the main reasons why RAID (Redundant Array of Independent Disks) network attached storage and storage area networks were developed. These developments usually meant that managing and maintaining thousands of disk drives became an ever more difficult and almost impossible task.

The answer to this dilemma was storage virtualization. By adding a new layer of software and/or hardware between storage systems and servers, the applications no longer had to know on which specific drives, partitions or storage subsystems their data resided. Administrators can identify, provision and manage distributed storage as if it were a single and consolidated resource, while also increasing availability, because applications are no longer restricted to specific storage resources and are, therefore, not affected by failure of one or even multiple storage resources.

High Level Technical Background

The principle of storage virtualization is the introduction of an intermediate layer that becomes the primary interface between servers and storage. From the server side, the

M. Oppitz, P. Tomsu, *Inventing the Cloud Century*,
DOI 10.1007/978-3-319-61161-7_6

virtualization layer looks like a single storage device and, conversely, all the individual storage devices see the virtualization layer as their only server, making it easy to group storage systems even from different vendors into tiers of storage.

This virtualization layer acts as a shield to servers and applications from any changes in the storage environment, allowing users to hot swap a tape or disk drive, for example. Another task of the virtualization layer is the management of data copying services, allowing data replication, whether for a snapshot or disaster recovery, to be handled entirely by the virtualization system in the background via a common management interface.

Storage virtualization can be classified in host based, storage device based and network based approaches.

Host based is where traditional device drivers handle physical devices and, with a software layer above the device driver intercepting I/O requests, look up metadata and redirect the I/O requests.

Storage device based is where virtualization is usually built into the storage fabric, with RAID controllers allowing other storage devices to be attached downstream. This allows for the building of a hierarchical system, where a primary storage controller handles pooling and manages metadata, allowing the attachment of other storage controllers. The primary storage controller can be either a dedicated hardware appliance, although newer systems are using switches. In addition, these systems also allow replication and migration services across different controllers.

Network based is where storage virtualization is viewed as a network based device, making use of Fiber Channel[1] networks connected as a SAN (Storage Area Network), where lately appliance or switch based implementations are most common.

Definition of Storage Virtualization

We have seen, so far, that storage virtualization is used to enable better functionality and advanced features in computer storage systems by pooling of physical storage from multiple network storage devices into a virtual single storage unit, making use of either host, storage device or network based virtualization.

In order to discuss storage virtualization in more detail, we will rely on the work done by SNIA[2] (Storage Networking Industry Association), which was formed in 2001. SNIA defines storage virtualization in their SNIA Technical Tutorial about Storage Virtualization[3] as follows:

> The act of abstracting, hiding, or isolating the internal functions of a storage sub-system or service from applications, host computers, or general network resources, for the purpose of enabling application and network-independent management of storage or data. The application of virtualization to storage services or devices for the purpose of aggregating functions or devices, hiding complexity, or adding new capabilities to lower level storage resources.

[1]Fiber Channel Industry Association, http://fibrechannel.org.

[2]SNIA, Storage Network Industry Association, http://www.snia.org, retrieved 2016-11-24.

[3]Bunn F. et al., SNIA, SNIA Technical Tutorial, Storage Virtualization, 2004.

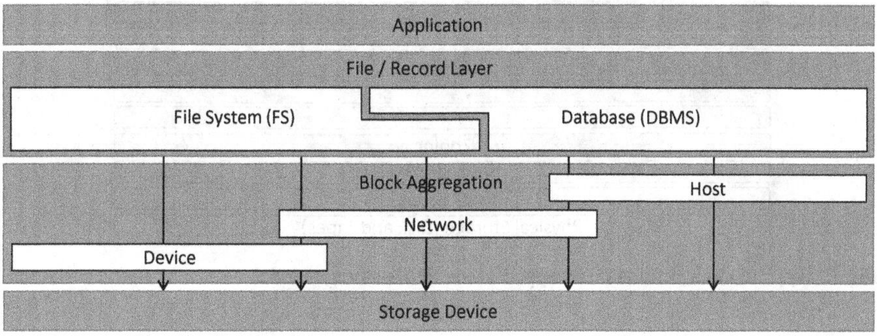

Fig. 1 The SNIA shared storage model

Shared Storage Model

The Shared Storage Model[4] was an early effort by the SNIA Technical Council to show how the introduction of layers in modern storage architectures enables the complete range of storage functions (see Fig. 1).

The main four layers of the shared storage model (from the bottom up) are the storage devices (like tape or disk drives), the block aggregation layer, the file/record layer and on top the application layer. On each layer, virtualized objects may be presented to the layer above.

Virtualization of Different Storage Functions
Several techniques exist for the virtualization of different storage functions (see Fig. 2), including

- Physical storage (layer 1 devices)
- RAID (redundant array of independent disks) groups
- Logical unit numbers (LUNs)
- Storage zones
- LUN subdivision
- LUN masking and mapping
- Host bus adapters
- Logical volumes and volume management
- File systems and database objects (table space, rows, columns)

There are several devices covering these virtualization functions like disk arrays, array controllers, storage switches, storage routers virtualization appliances, host bus adapters, operating systems and finally application layer software.

[4]Yoder A., SNIA, SNIA Shared Storage Model, http://www.snia.org/education/storage_networking_primer/shared_storage_model, 2003.

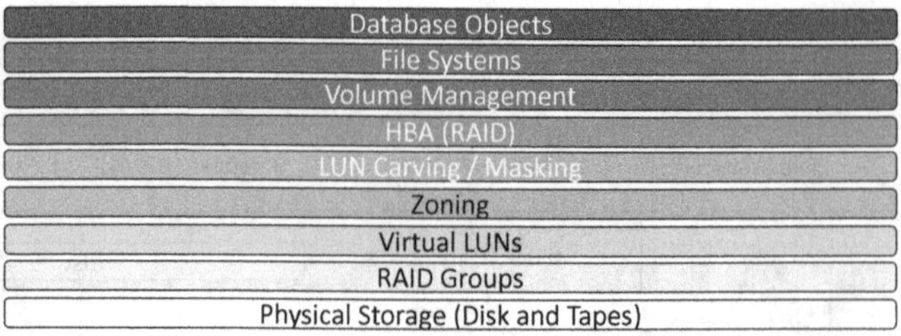

Fig. 2 Typical storage functions mapped to the SNIA storage model

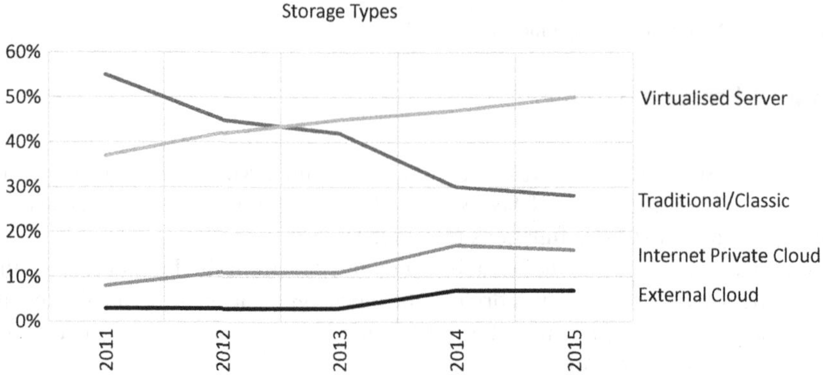

Fig. 3 Data/storage evolution from classic to virtualized and cloud environments

Traditional vs. Virtualized Storage Architectures

Traditional storage architectures use discrete physical drives, with almost no functions handled on layer 2 for detail abstraction. The traditional model does not scale, as almost everything like paths, access, configuration, connections etc. must be managed and handled individually and, furthermore, it does not allow for the necessary control.

There are numerous benefits when virtualized architectures are deployed because it is much easier to manage virtual resources compared to physical resources. This effect is also adopted and accepted by the market (see Fig. 3).

The main goal is to achieve automated storage management, in order to reduce the burden and need of manually handling storage management, especially when providing a predictable quality of service in Storage Area Networks (SANs[5]) (Cunnings, 2004). No longer is the knowledge of vendor specific characteristics of the single components necessary, as virtualization brings the required abstraction of details in order to make the universe of managed entities look simple, ordered and uniform (see Fig. 4).

[5]SNIA, What Is a Storage Area Network?, http://www.snia.org/education/storage_networking_primer/san/what_san, 2016, retrieved 2016-12-06.

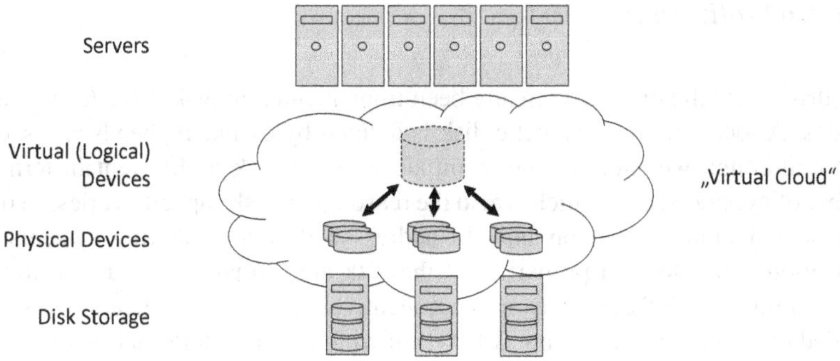

Fig. 4 A virtual storage connection

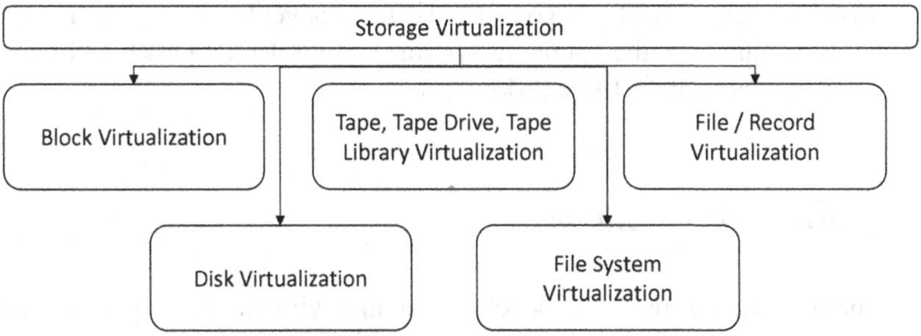

Fig. 5 Five different types of storage virtualization

Why Choose a Virtual Storage Connection Model?
Changing the physical devices (layer 1) into virtual devices (layer 2), describes a simple model of virtual storage. Any server accessing virtual layer 2 devices will not see a difference in accessing physical layer 1 devices and, further to that, the virtualization hides any changing, adding or replacing components of the physical layer 1, thus there will be no disruptions of the layers above.

In short, there are three main improvements resulting from storage virtualization: easier storage management in heterogeneous IT environments, better availability and downtime elimination through automated management and, finally, improved storage utilization (Clark, 2005).

Different Types of Storage Virtualization

We can differentiate between five different types of storage virtualization, which is also nicely described by the SNIA storage virtualization taxonomy.[6] These types are block, disk, tape, file system and file virtualization (see Fig. 5).

[6]Bunn F. et al., SNIA, Storage Virtualization, http://www.snia.org/sites/default/files/sniavirt.pdf, 2004, retrieved 2016-12-06.

Disk Virtualization

Disk drive virtualization had already been implemented in disk drive firmware for decades. A location on a magnetic disk is defined by cylinders, heads and sectors (CHS), although we should keep in mind that each disk is different in terms of number of cylinders, etc., which is also the reason why disk capacity varies. In order to present a unique form on how to address a disk to operating systems and applications, the physical properties of the disk are virtualized by disk firmware, transforming the CHS addresses into consecutively numbered logical blocks. This is called logical block addressing (LBA), allowing for the determination of the disk size simply by the number of LBA.

Another benefit of the disk virtualization firmware is ensuring that the magnetic disks always look defect free to the operating systems and applications even if, during disk lifetime, some blocks go bad and can no longer be reliably used to write and read data. This way, the operating system does not have to track bad records and the host only sees defect free disks.

Tape Storage Virtualization

Tape storage virtualization can be classified into virtualization of tape media (so called cartridges) and virtualization of tape drives.

Tape Media Virtualization

In tape media virtualization, online disk storage is used as a cache. This emulates the reading and writing of data to and from the tape in order to improve backup performance, as well as the service life of tape drives. This performance improvement comes from the disks acting as a buffer to smooth out fluctuations caused either by busy hosts or by the network. This is also a huge improvement because tapes move more with constant speed (the so-called shoe shining effect is lowered), thus lowering wear and tear on the tape media as well as on the recording heads.

Tape media virtualization also helps in improving utilization capacity levels through the emulation of a large number of relatively small tape media volumes. Meanwhile, data is accumulated on disks before being written with constant streaming speed to the tape, and each medium can also be filled to its optimum level. Further compression of files in the disk cache is improved before they are written to the tape, as well as improved restore performance by avoiding time consuming mount and unmount operations through intelligent emulation at the disk cache layer.

Tape Drive Virtualization

Networking of tape drives through SAN technologies like Fiber Channel[7] allow tape drives to be shared between multiple servers, where the access to the drives is

[7]FCIA, Fiber Channel Industry Association, http://fibrechannel.org, 2016, retrieved 2016-12-06.

controlled through the network rather than individual servers. This allows for the sharing of physical tape drives in tape libraries among multiple host systems, which significantly saves hardware resources. Now, with networking tape drives another challenge appears, which is how to best keep different applications or servers from clashing during access and leading to potential data corruption.

In this case, tape virtualization allows for the establishment of tape drive pools with guaranteed data integrity, where single drives can typically appear as several virtual drives assigned to individual servers. The access to the tape drive is handled by a tape drive broker, reserving and mapping a physical tape drive to the hosts virtual drive and, after operation completes, the physical tape drive is returned back to the tape drive pool. Also, defective tape drives can be displaced in a non-disruptive way by other drives out of the pool. Storage consumers are unaware of all this functionality, as it runs in the background.

File System Virtualization

Several forms of file system virtualization exist like networked remote file system, also called network attached storage (NAS), where dedicated file servers manage shared network access to files in the file system. This can be shared by many hosts on the network even those running different operating systems. Another form of file system virtualization is based on simplified database management, which makes use of file system virtualization in database environments. Storing database table spaces and transaction logs on file systems instead of on raw disk drives allows for better and improved management but brings some additional overhead, which degrades in turn performance.

File/Record Virtualization

Hierarchical Storage Management (HSM) is the most widely deployed form of file virtualization (William, 1999). The idea behind HSM is to automate the migration of rarely used data to cheap secondary storage media like optical discs or tapes or low cost to disc storage. HSM needs to be, of course, transparent to the users and applications, offering location transparency. Migrated files must be easily retrievable, which is guaranteed by a combination of pointers in the file system and metadata from the HSM application.

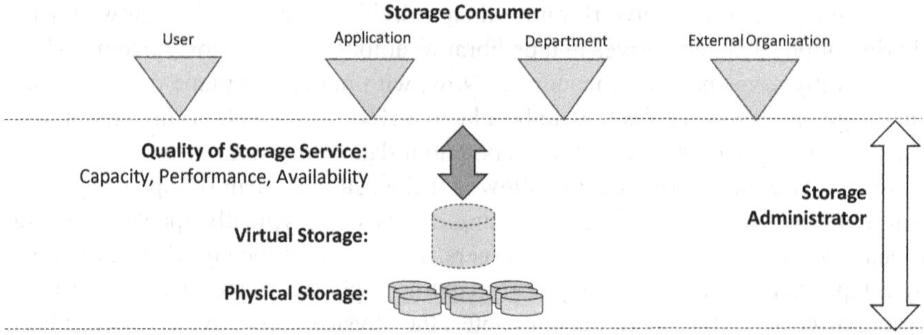

Fig. 6 Logical view of storage

Block Virtualization

Block virtualization deals with block level disk services and is a relatively new method of storage virtualization happening on layer II of the SNIA shared storage model. Instead of manipulating a single magnetic disk representing it as a logical block address, as done in disk virtualization, block virtualization represents the next logical step by virtualizing several physical disks to represent a single logical drive (often also called block aggregation).

By doing so, block virtualization can overcome the physical limits of individual devices without the need for additional intelligence in applications. Applications just see new virtual disks with larger logical block address ranges. There are also other services, which can be covered by the block virtualization layer helping to improve performance, availability, etc. required by the storage consumer (i.e. application) (see Fig. 6).

To the storage consumer, the physical aspects of storage are irrelevant and they usually do not want to deal with technical details and just define the needed storage services. Block virtualization allows the control of physical storage assets by combining them to form logical volumes with sufficient capacity, performance and reliability. Consumers can simply see logical volumes, which cannot be distinguished from physical disks, hence block virtualization generates virtual storage devices from physical disks, which are just as fast, available and have the capacity the storage consumer requires.

References

Clark, T. (2005). *Storage virtualization: Technologies for simplifying data storage and management*. Upper Saddle River, NJ: Addison Wesley.

Cunnings, R. (2004). *SNIA technical tutorial: Storage network management*. Colorado Springs, CO: SNIA.

William, J. (1999). *The holy grail of data storage management* (1st ed.). Upper Saddle River, NJ: Prentice Hall.

From Physical to Virtual Servers

Innovation distinguishes a leader from a follower
Steve Jobs

In this chapter, we show that server virtualization that is the most dominant form of virtualization in use today, helping to partition one physical server into multiple servers, and enabling each of these servers to run its own operating system and applications while performing as if they all were individual servers.

We detail the most common server virtualization models and have a look at the Open Virtualization Format (OVF). We see how this allows packaging and distribution of software, the so-called virtual appliance, in order to run on virtual machines.

Server Virtualization Overview

Server virtualization is the most dominant form of virtualization in use today. Server virtualization[1] is a technology for partitioning one physical server into multiple virtual servers, where each of these servers can run its own operating system (OS) as well as applications and actually perform as if they were individual servers (David Rule, 2007).

There exist three main principles to server virtualization that we will describe first.

Bare Metal Hypervisor: Type 1
In this case, the guest OS runs directly on the hardware, which requires a Hypervisor Type 1, often called a Bare Metal Hypervisor (see Fig. 1). This type

[1]VMware, White Paper, Virtualization Overview, https://www.vmware.com/pdf/virtualization.pdf, 2006, retrieved 2016-12-06.

M. Oppitz, P. Tomsu, *Inventing the Cloud Century*,
DOI 10.1007/978-3-319-61161-7_7

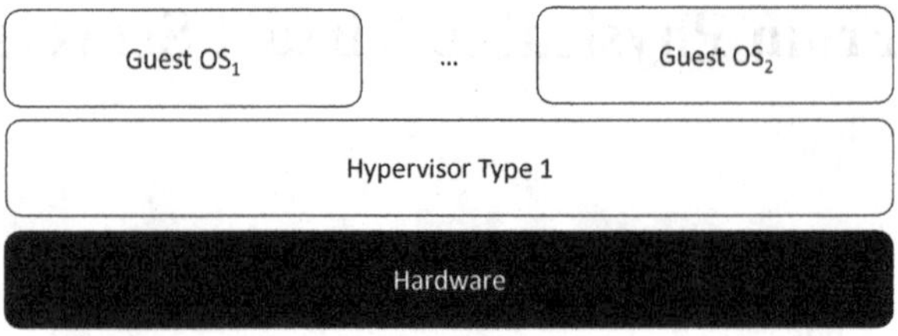

Fig. 1 Hypervisor type 1 for running on bare metal

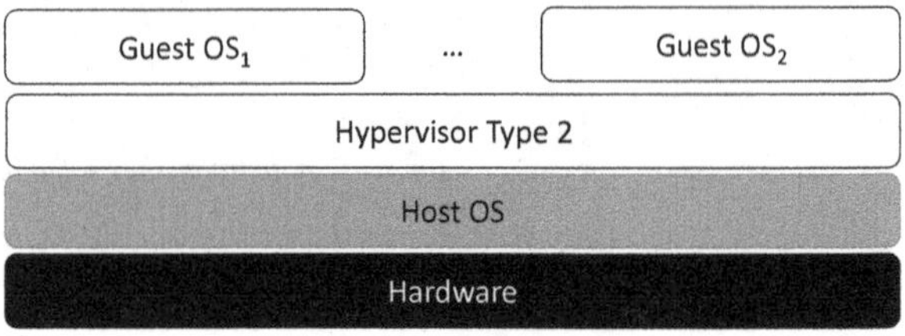

Fig. 2 Hypervisor type 2 for running on host OS

of hypervisor is the software needed to support multiple virtual machines, such as VMware ESXi[2] or XEN[3] in a bare metal environment.

Hosted Hypervisor: Type 2

In this case, the guest OS runs on the host OS, which requires a Hypervisor Type 2 or a Hosted Hypervisor (see Fig. 2). This type of hypervisor is the software needed to support multiple virtual machines, for example Windows Virtual PC[4] in a hosted environment.

Hypervisor Type 0

A third type of Hypervisor exists that can be called Type 0 and it is a mixture of both Type 1 and Type 2 hypervisors, as is the case with Linux Kernel Virtual Machine (KVM).[5]

In general, server virtualization uses server resources that consist of individual physical servers, processors, operating systems and applications, and masks them for server users creating multiple, isolated user spaces on the same operating system, also called containers (see Fig. 3).

This can be achieved by dividing a physical server into multiple isolated environments, sometimes also called virtual private servers, instances, guests, containers or emulations.

Examples of operating system level virtualization are Windows Server 2003, 2008, 2012, etc. Typically, multiple users can remotely login and use the system,

[2]VMware, ESXi, Purpose-Built Bare-Metal Hypervisor, http://www.vmware.com/products/esxi-and-esx.html, 2016, retrieved 2016-12-06.

[3]Linux Foundation, XEN Project, https://www.xenproject.org, 2016, retrieved 2016-12-06.

[4]Microsoft, Windows Virtual PC, https://www.microsoft.com/en-us/download/details.aspx?id=3702, 2016, retrieved 2016-12-06.

[5]Linux, Linux KVM, http://www.linux-kvm.org/page/Main_Page, 2016, retrieved 2016-12-06.

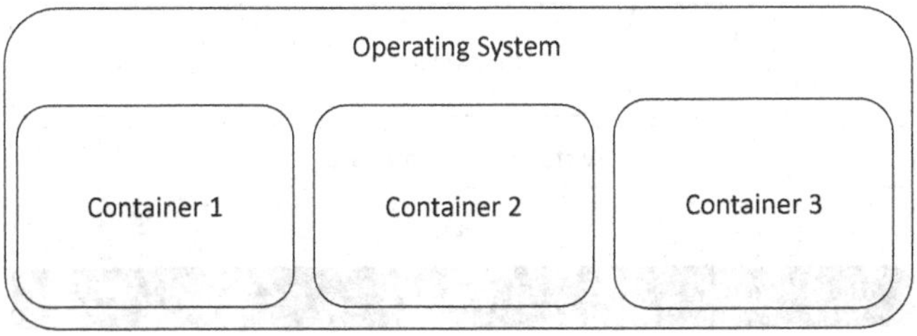

Fig. 3 Containers as isolated user spaces on same OS

although there is only one operating system instance available. The guest operating systems can be similar or even the same as the host, while applications of one user cannot affect other users.

Server Virtualization Methods in Detail

Several server virtualization implementations exist today (David Rule, 2007; Williams and Dittner, 2007).

Native Virtual Machine
The Native Virtual Machine concept uses the host/guest paradigm, where each guest runs on a virtual imitation of the hardware layer and, as such, the guest (host) operating system can stay without any modifications (see Fig. 4). In this virtualization mode, each VM has a separate instance of an operating system that is installed on the hardware.

This allows for different operating systems to be used by different guests while the guests do not even know the host's operating system, as it is not aware that it is not running on real hardware. The operating system on which the virtualization application runs is called Host OS, as it provides the execution environment for the virtualization application.

Hypervisor Based Virtualization
Nevertheless, real computing resources are used from the host and, thus, the hypervisor based model uses a thin software layer referred to as a virtual machine monitor (VMM), which acts as a hypervisor to coordinate instructions on the CPU level, validating all guest issued CPU instructions and managing any executed code requiring additional privileges (see Fig. 5).

The hypervisor also handles all dispatching and queuing and takes care of returning hardware requests made by VMs. The administrative virtual machine console, which acts as the administrative operating system, runs on top of the hypervisor in order to manage and administer the VMs. The VVM holds the role of the hypervisor and runs on physical hardware. This enables administrators to create different VMs with their own separate hardware including CPU, hard drives and memory. It is a big advantage that guest operating systems do not need to be modified for this approach.

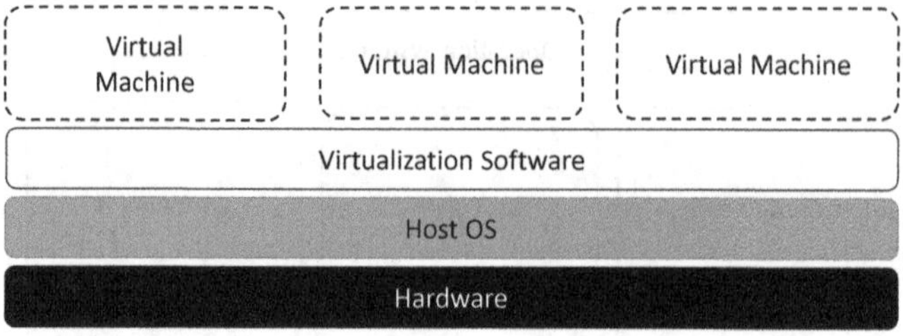

Fig. 4 Native virtual machine

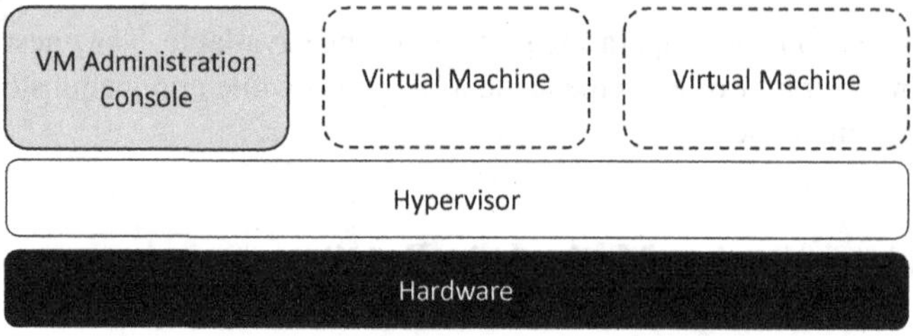

Fig. 5 Hypervisor based virtualization

Paravirtualization

Paravirtualization[6] is based on the previously described hypervisor virtualization and was developed to eliminate most of the emulation overhead resulting from software-based virtualization (see Fig. 6). This means that guest operating systems need to be recompiled and/or modified before they can be installed on a VM, which makes it the most efficient type of virtualization but, on the downside, definitely reduces security and flexibility.

Paravirtualization Using XEN

Paravirtualization using XEN[7] uses modifications of the guest operating systems that enable additional performance enhancements, since the modified OS's can directly communicate with the hypervisor and no longer need emulation services (see Fig. 7). XEN is the typical implementation that uses paravirtualization based on a Linux kernel for support of the administrative part and, additionally, leverages hardware virtualization, thus allowing unmodified OS versions on top of the hypervisor.

---

[6]VMware, Understanding Full Virtualization, Paravirtualization and Hardware Assist, http://www.vmware.com/techpapers/2007/understanding-full-virtualization-paravirtualizat-1008.html, 2008, retrieved 2016-12-06.

[7]XEN, Paravirtualization (PV), https://wiki.xen.org/wiki/Paravirtualization_(PV), 2016, retrieved 2016-12-06.

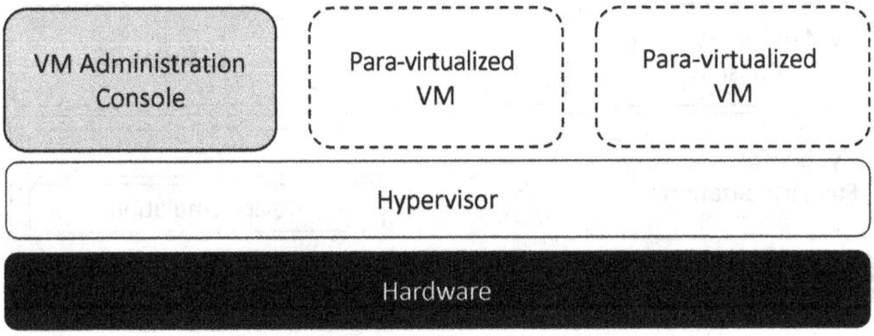

Fig. 6 Paravirtualization

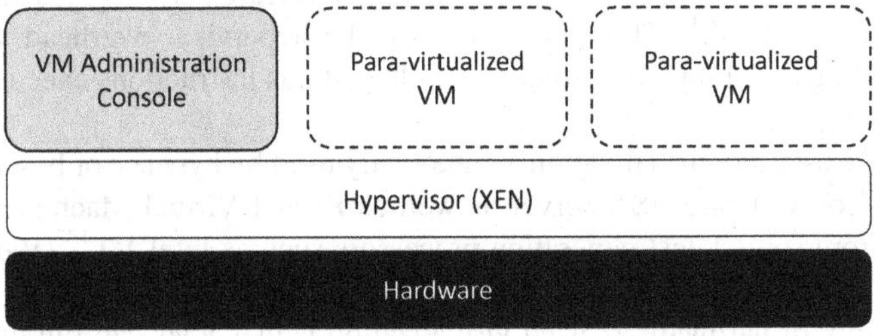

Fig. 7 Paravirtualization with XEN

Full Virtualization

Full virtualization supports the widest range of guest operating systems. It allows complete simulation of the underlying hardware (see Fig. 8). Full virtualization is based on the paravirtualization model and, if needed, contains functions for the emulation of the underlying hardware, where the hypervisor traps machine operations performed by the OS in order to modify or read the system status, as well as for input output operations. The hypervisor then emulates the operations in software, returning status codes as if these codes were delivered by real hardware that allows unmodified operating systems to run seamlessly on top of the full virtualization hypervisor.

This method is, for example, used by the ESX (Elastic Sky X)[8] server from VMware, running a customized version of Linux that is called Service Console and was recently replaced by ESXi as the administrative OS. Nevertheless, the performance of full virtualization is generally not quite as good as paravirtualization.

Hardware Virtualization

Hardware virtualization, often called hardware-assisted virtualization, is a combination of paravirtualization and full virtualization with a hypervisor. The big difference is that it is only appropriate for systems supporting hardware

[8]vmware, ESXi, http://www.vmware.com/products/esxi-and-esx.html, 2016, retrieved 2016-12-06.

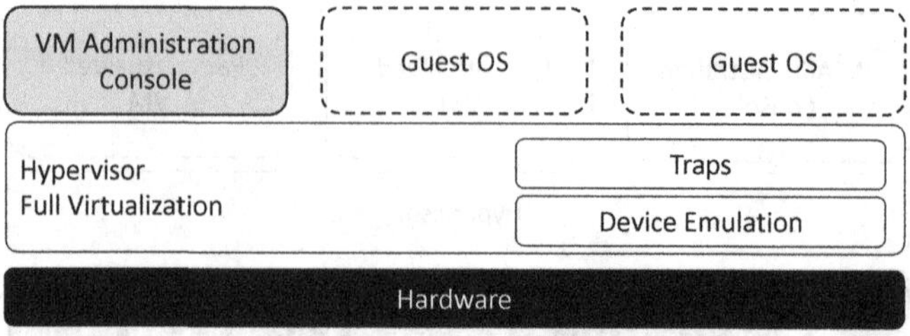

Fig. 8 Full virtualization

virtualization, thus it needs to rely on system architecture hardware extensions like Intel VT[9] or AMD V.[10] This makes most of the hypervisor overhead associated with emulating and trapping obsolete, as well as status instructions executed within the guest OS.

Hardware assisted virtualization can be easily used by hypervisor based systems like XEN[11] or VMware ESX server as well as Kernel Virtual Machine (KVM)[12] available from some latest generation processors such as Intel VT[13] (Vanderpool) and AMD V[14] (Pacifica).

The beauty of hardware assisted virtualization is that VMs can run unmodified operating systems, since the hypervisor is using hardware support for virtualization.

Kernel Virtualization

In kernel virtualization no hypervisor is required, since a separate version of the Linux kernel runs an associated VM as a user process on the physical host thereby enabling multiple, simultaneously running VMs (see Fig. 9). A device driver in the Linux kernel is used to communicate with the VMs and special processors like Intel VT or AMD V Kernel virtualization are used by User Mode Linux (UML)[15] and by Kernel Virtual Machine (KVM).[16] While UML can run without separate

[9]Intel, Intel Virtualization Technology (Intel VT), http://www.intel.com/content/www/us/en/virtualization/virtualization-technology/intel-virtualization-technology.html, 2016, retrieved 2016-12-06.

[10]AMD, AMD Virtualization Solutions (AMD V), http://www.amd.com/en-us/solutions/servers/virtualization, 2016, retrieved 2016-12-06.

[11]Linux Foundation, XEN Project, https://www.xenproject.org, 2016, retrieved 2016-12-06.

[12]Kernel Virtual Machine (KVM), http://www.linux-kvm.org/page/Main_Page, 2016, retrieved 2016,12-06.

[13]Intel, Intel Virtualization Technology (Intel VT), http://www.intel.com/content/www/us/en/virtualization/virtualization-technology/intel-virtualization-technology.html, 2016, retrieved 2016-12-06.

[14]AMD, AMD Virtualization Solutions (AMD V), http://www.amd.com/en-us/solutions/servers/virtualization, 2016, retrieved 2016-12-06.

[15]User Mode Linux (UML), http://user-mode-linux.sourceforge.net, 2016, retrieved 2016-12-06.

[16]Linux, Linux Kernel Virtual Machine (KVM), http://www.linux-kvm.org/page/Main_Page, 2016, retrieved 2016-12-06.

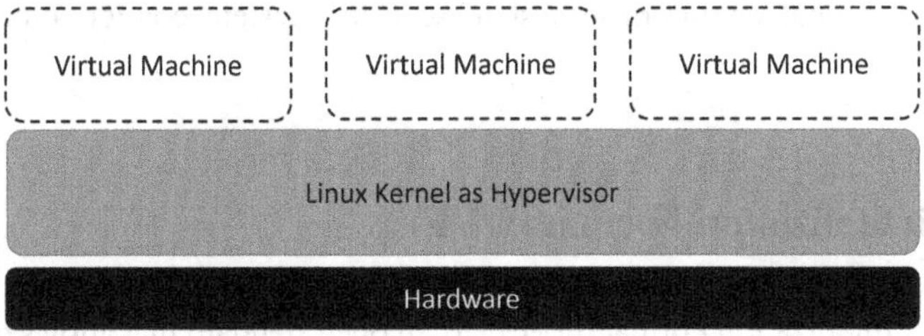

Fig. 9 Kernel virtualization model (KVM)

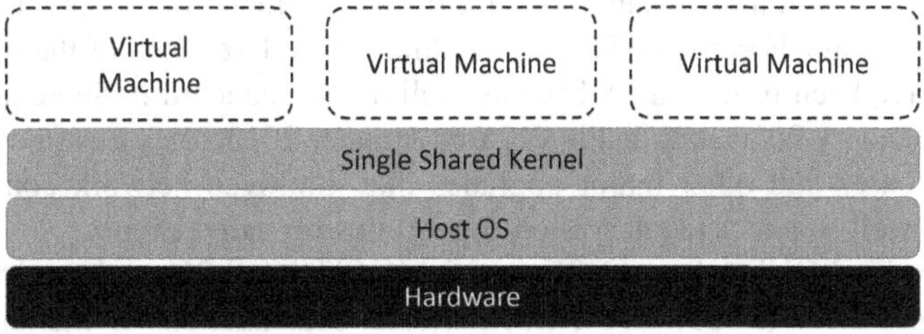

Fig. 10 Shared kernel virtualization

administrative software for execution and management of its VMs, KVM is similar to a full virtualization with the Linux kernel serving as the hypervisor.

System Virtualization aka Shared Kernel Virtualization

System Virtualization, often called Shared Kernel Virtualization (Hess and Amy, 2009), usually runs multiple and logically distinct systems on a single OS kernel instance (see Fig. 10). This is made possible by the "change root" concept (chroot) that we find on modern UNIX based systems. With chroot, the kernel uses root file systems (e.g. initial RAM disks or RAM file systems) for loading drivers and performing system initialization.

With the chroot command, the kernel can switch to other root file systems anytime and mount any other on-disk file system as the final root file system, which is an elegant way to continue system initialization or configuration from that new file system. This results in either system virtualization, as long as all virtual severs run a single copy of a specific OS, or server virtualization in case different virtual servers are running different OSs (or different versions of the same OS.)

By sharing a single instance of the OS kernel, system virtualization is usually more lightweight than using complete virtual machines including the kernel of the server virtualization approach. Thus, a single host can run more virtual servers than the number of VMs it could support.

Prominent examples of system virtualization are Linux VServer,[17] FreeVPS, FreeBSD,[18] and OpenVZ Virtuozzo Containers.[19] Using these types of system level virtualization guarantees high scalability for as many users as necessary, while

[17]Linux, VServer, http://www.linux-vserver.org/Welcome_to_Linux-VServer.org.

[18]FreeBSD, https://www.freebsd.org.

[19]OpenVZ Virtuozzo Containers, https://openvz.org/Main_Page.

single users cannot compromise system security, system configuration and file systems.

Open Virtualization Format (OVF)

The Open Virtualization Format (OVF)[20] is an open standard that allows packaging and distributing software, referred to as virtual appliance, to run on virtual machines. It dates back to September 2007, when VMware, Dell, HP, IBM, Microsoft and XenSource submitted the proposal for OVF to the Distributed Management Task Force (DMTF). Meanwhile, several versions of the OVF specification have been rolled out, V1.0.0 as preliminary standard in September 2008, V1.1.0 in January 2010 and, finally, V2.0 in January 2013. They covered emerging cloud use cases and other major upgrades like improved network configuration support as well as package encryption capabilities for safe delivery.

ANSI ratified OVF 1.1 as ANSI standard INCITS 469-210 and the Joint Technical Committee 1 (JTC1) of the International Organization for Standardization (ISO) and the International Electrotechnical Commission (IEC) adopted OVF 1.1 as International Standards in August 2011.

We will give a brief overview of OVF and its merits for virtualization in this section.

From Virtual Machines to Virtual Appliances

OVF describes an open, secure, portable, efficient and extensible packaging format for distribution of software to run on virtual machines. This is done in a way that is open to any hypervisor and processor architecture and can even carry one or more virtual systems deployable to virtual machines in one, so-called OVF package.

The main driving factor for developing a portable meta-data model for VM distribution to and between virtualization platforms is the increasing popularity of the virtual infrastructure. OVF allows for the packaging of applications together with certified operating systems into virtual machines. This allows independent software vendors to transfer these preconfigured units through test, development and production. Now, we can deal with pre-deployed and tested applications packaged as virtual machines, which we call virtual appliances. Virtual appliances are enabled by a vendor neutral standard for the packaging of VMs and metadata required for automatic and secure installation, configuration, running the appliance on any virtual platform and tearing it down again, if necessary.

[20]Distributed Management Task Force (DMTF), Open Virtualization Format (OVF), https://www.dmtf.org/standards/ovf, 2009–2015.

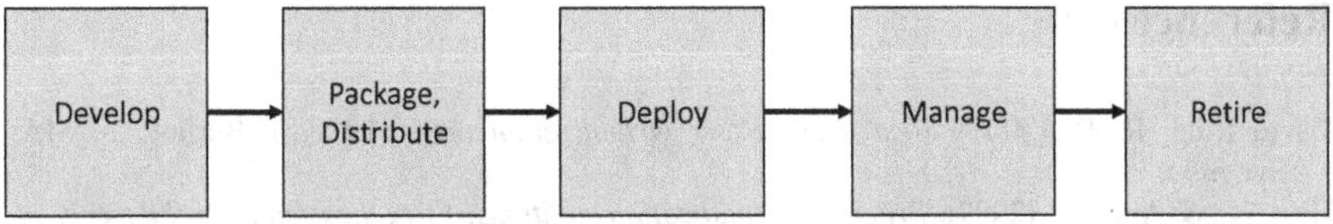

Fig. 11 Virtual appliance life cycle

Virtual Appliance Life-Cycle

The software life cycle of virtual appliances is clearly defined in by OVF as the virtual appliance life cycle (see Fig. 11).

Typically, a service consists of one or more VMs including the relevant configuration and deployment data and is packaged in the OVF format. After the development phase, the OVF package is distributed and deployed, and, during the management phase, the virtualization product is in use with ongoing maintenance and upgrades to the appliance, which is highly dependent on the contents of the VM in the OVF. Finally, when the virtual appliance is no longer needed, it is retired.

Advantage of Virtual Appliances

The use of virtual appliances changes software distribution, allowing for the application builder's optimization of their products and delivery of turnkey solutions and software services to the end user.

Solution providers can build virtual appliances more cost effectively than hardware appliances. Prepackaging the operating system that the virtual appliance uses reduces the typical application/OS compatibility testing and certification. This allows the independent software vendor to reinstall the application in the specific OS environment it is intended for.

End users get a more simplified way to do software management for their appliances because they can leverage a standardized, automated as well as efficient set of processes, replacing current, conventional OS and application specific management duties.

Further modern enterprise applications require a Service Oriented Architecture (SOA)[21] usually with multiple tiers, where each tier contains at least one or, in many cases, multiple virtual machines, rendering a single VM model no longer suited for distribution of a multi-tier service. Complex applications require customization during install time for their network or other custom specific properties.

Virtual appliances can be pre-packaged in run time format to include hard disk images and configuration data suitable for particular hypervisors and optimized for execution. For an efficient software distribution, numerous additional features like portability, platform independence, verification, signing, versioning and licensing terms can be supported.

[21]The Open Group, Service Oriented Architecture (SOA), https://www.opengroup.org/soa/source-book/soa/soa.htm.

References

David Rule, R. D. (2007). *Best darn server virtualization book* (1st ed.). Burlington, MA: Syngress.

Hess, K., & Amy, N. (2009). *Practical virtualization solutions: Virtualization from the trenches.* Upper Saddle River, NJ: Prentice Hall.

Williams, D. E., & Dittner, R. (2007). *Best Damn server virtualization book period.* Burlingston, MA: Syngpress Publishing/Elsevier.

Software Defined Virtual Networks

If you can't explain it to a six-year-old, you don't understand it yourself

Albert Einstein

We describe why network virtualization, also known as Software Defined Networking (SDN), already has a pretty long history going back to the middle of the 1990s. We discuss what became the limiting factors of legacy networking technologies that led to the invention of programmable networking and what we understand when we talk about future networking architectures.

We dive into details about the concept and premises of SDN and describe its architectural components, before we discuss the different control plane concepts and outline why the economic impact of SDN means a real disruption for legacy networking.

We have a look at a number of popular open SDN concepts like OpenFlow, OpenDaylight and Open Compute Project and, finally, give examples of the three typical and most influential SDN implementations from VMware, Cisco and Big Switch Networks.

Some SDN History and Evolution

Over the 1990s and 2000s, the core network elements for IP networks were switches (forwarding data on layer 2) and routers (forwarding data on layer 3). Switches and routers started as pure packet forwarding engines and they became smarter over time, but with a very big downside, as all these advances were mainly proprietary and controlled by the switch and router vendors. Only a select number of players (vendors) had the ability to control and add new functionalities, highly restricting the possibility for customers and users to make even the simplest changes to their networks.

Why did this existing, old IP architecture, which had mainly been designed to deliver packets between stationary hosts with globally, distributed routing protocols

M. Oppitz, P. Tomsu, *Inventing the Cloud Century*,
DOI 10.1007/978-3-319-61161-7_8

along with manual configuration become suddenly insufficient? It was mainly the evolution of virtualization and cloud computing that required networking to change significantly (Victor Moreno, 2007). It was time for Software Defined Networking.

SDN has a pretty long history and one of the first SDN projects was AT&T's GeoPlex in the mid 1990s that leveraged the network APIs and dynamic aspects of the Java[1] language as a means to implement middleware networks. This networking middleware used one or even more operating systems running on computers connected to the Internet and offered a service platform to manage networks and online services. The main goal behind this approach was to have a Soft Switch that could reconfigure physical switches in the network and load them with new services from an Operational Support System (OSS).[2]

1998 The Soft Switch Concept

It was in 1998 when Mark Medovich from Sun Microsystems[3] and Javasoft founded the Soft Switch startup WebSprocket in Sunnyvale California. This soft switch was based on a new network operating system and on an object-oriented runtime model that could be modified by a network compiler and class loader in real time. This allowed for the writing of applications with Java threads that contained a WebSprocket kernel, network, and device classes. The applications could later be modified by a networked compiler/class-loader.

In July 2000, WebSprocket released VMFoundry, a Java to bare metal runtime compiler, and VMServer, a networked device compiler/classloader application server. This allowed preloading custom networked devices with images created by VMFoundry and deploying them on the network with the connection to VMServer via UDP or TCP. VMServer could now proactively or reactively load and extend network protocol methods and classes on the target system. This allowed an early implementation of SDN that was not confined to a set of limited actions managed by an SDN controller because the control plane contained code that could change, override, extend or enhance network protocols on operating networked systems.

2001 Early Network Programmability Ideas

In 2000, the Gartner Group saw the emergence of programmable networks based on this original SDN concept as the next big thing for the Next Generation Internet. Gartner introduced Supranet, which was the fusion of the physical and virtual worlds as the "Internet of Things" and WebSprocket was selected as one of the top emerging technologies in the world. In early 2001, Ericsson and WebSprocket entered a license agreement to build the first commercial soft switch and it became possible to port Ericsson's complete call control software in a matter of days.

---

[1]Oracle, Java Software, https://www.oracle.com/java/index.html, 2017, retrieved 2017-02-06.

[2]OSS Line, The Definition of OSS and BSS, http://www.ossline.com/2010/12/definition-oss-bss.html, 2010, retrieved 2017-02-06.

[3]Oracle, Oracle and Sun Microsystems, https://www.oracle.com/sun/index.html, 2017, retrieved 2017-02-06.

The Supranet Consortium was founded and Ericsson announced WebSprocket as the enabling technology of the Supranet Transaction Server (STS), which was the first comprehensive framework to deliver any networked service. As telecom markets deflated in 2001, Ericsson's soft switch development was ended early, marking the end of the first commercial SDN soft switch.

2007 The Invention of OpenFlow

One of the innovators in the area of network virtualization was Martin Casado, who began his career auditing networks at a government intelligence agency. He was trying to secure networks against terrorist attacks but became increasingly frustrated with legacy networking technologies that prevented him from solving mission critical problems. What he needed was a better platform that was radically different, so he decided to build this platform himself. In 2007, Casado connected with Stanford Professor Nick McKeown and UC Berkley Professor Scott Shenker, who were also frustrated with the current state of networking because it was almost impossible to innovate on the network. Real network innovation requires being able to work with real production networks but, unfortunately, these networks were too fragile to allow for this.

Thus, they thought of a platform that completely virtualized the network making it as flexible as a virtualized server. They completely decoupled networking software from proprietary hardware and developed OpenFlow[4] (Azodolmolky, 2013), which is the protocol running between the control plane and the data plane and will be discussed later in this chapter.

Suddenly, anybody or any company could add any kind of functionality to the network, not just the major networking vendors. Finally, this new platform could be infinitely scalable by using commodity hardware. In order to fund this big project, they started a new company, Nicira Networks, which was later acquired by VMware[5] and their first product, the Network Virtualization Platform (NVP),[6] became a prominent representative for SDN.

Legacy Networking Limitations Driving SDN

Locked by Old Networking Architecture

Requirements for traditional networking architecture changed significantly throughout the 2010s (Underdahl & Kinghorn, 2015). This marked a very important driver for SDN. Enterprises were forced to squeeze the most out of their networks, which they achieved by device level management tools and increasingly more

---

[4]Open Networking Foundation (ONF), OpenFlow, https://www.opennetworking.org/sdn-resources/openflow, retrieved 2017-05-29.

[5]VMware, https://www.vmware.com/support/acquisitions/nicira.html, retrieved 2017-05-29.

[6]VMware and Nicira, Network Virtualization Platform (NVP), http://www.vmware.com/company/acquisitions/nicira.html, retrieved 2017-05-29.

manual processes. The same was true for carriers, mainly driven by the increasing demands for mobility and higher bandwidth, while, at the same time, profits declined by escalating capital equipment costs on one side and declining revenue on the other. At the same time, existing, legacy network architectures could no longer meet the requirements of modern users, and enterprises and carriers and also network architects began to be constrained by a number of legacy networks limitations, mainly the complexity of these network architectures, modern policy requirements, constantly changing traffic patterns and increased demand for highly agile networks.

For example, Amazon's Amazon Web Services (AWS)[7] (Michael Wittig, 2015), which is one of the largest cloud offerings in the world today, had to run a feature poor network in the middle of the 2010s in order to allow enough flexibility to support their necessary services. This was unfortunately also true for most enterprises, as moving to cloud computing and datacenters would have meant a reversal of the last 20 years of networking features from their applications. The old IP networking architecture could simply no longer handle these requirements because their extremely smart but proprietary switches and routers were unfortunately closed to almost all of the new world networking requirements.

Complexity of Legacy Network Architectures

One of the most limiting factors in legacy networks has become complexity. Network architecture needs changed significantly over the past few decades, which led to advanced networking protocols. These promise to deliver the required higher performance, reliability, and support ubiquitous connectivity while, at the same time, allow for more stringent security. Most of these protocols were developed as solutions to solve specific problems, but fundamental abstractions were simply not inherent, which resulted in the complexity of legacy networks.

Examples showing the complexity become obvious as soon as devices need to be moved or added, as this requires touching multiple switches, routers, firewalls or other network appliances, including Web authentication etc. In addition, access control lists (ACLs), virtual LANs (VLANs), quality of service (QoS) and a number of other protocol based mechanisms need to be configured, which can only be achieved using device level management tools in these legacy network architectures. In the end, this complexity rules legacy networks and forces them to remain rather static, especially as potential service interruptions need to be kept as low as possible.

Requirements of Modern Network Architectures

Today's applications and user behavior require highly dynamic networks. This comes, in a great part, from server virtualization, which significantly increases the number of hosts requiring network connectivity and means that hosts may no longer have a fixed physical location as this was the case in legacy network times a

---

[7]Amazon, Amazon Web Services (AWS), https://aws.amazon.com, retrieved 2017-05-29.

decade or longer ago. Before we used virtualization, applications mainly resided on a single server, primarily exchanging traffic with connected clients, while, in modern networks, applications are usually spread via multiple virtual machines (VMs) exchanging the necessary traffic flows with each other. To make things even more complex, these VMs are migrated in order to optimize and rebalance server loads. This, in turn, causes physical end points of existing traffic flows to change again and again, in many cases even very rapidly. This migration behavior of VMs in modern networks challenges legacy networking addressing schemes and namespaces, but also the notion of segmented, router based legacy network design.

The nature of many legacy enterprise networks carrying voice, data and video traffic simultaneously, often called voice, video and data network convergence became a trend introduced at the end of the 1990s (Ellis, Pursell, & Rahman, 2003; Lee, 1999). These networks were also known as multi-service networks and needed a very sophisticated, differentiated implementation (Armitage, Carlberg, & Gleeson, 2000) for the different applications, which were mostly provisioned manually. As soon as these networks consisted of multiple vendor equipment, the network parts coming from different vendors needed separate configuration and adjustment of parameters such as network bandwidth and QoS on a session and per application basis. That, again, made the network pretty static, a fact that highly contradicted the requirements for dynamic adaptation to fast and continuously changing traffic, application and user demands on modern networks.

Modern Policy Requirements

Another huge limitation was the fact that network wide policies had to be configured manually on thousands of devices (Stallings, Foundations of Modern Networking – SDN, NFV, QoE, IoT, and Cloud, 2015). Imagine bringing up a VM in a legacy network and then needing hours or days to configure ACLs across the whole network. Today's networks became very complex because of the constantly increasing support for mobile users, requiring the application of a consistent set of policies for access, security, QoS and many more features that leaves enterprises vulnerable to security breaches, non-compliance with regulations and many more negative issues.

Rapid Demand for Growth

In today's (and future) networks, the rapid demand for growth is inherent and we have seen how the network became vastly more complex with the ever-increasing number of network devices that need to be configured and managed continuously because, in modern virtualized data centers, traffic patterns change very dynamically and unpredictably. This is true for most data centers used by enterprises and modern business applications but, of course, also mega operators such as Google, Yahoo, Facebook or Amazon even more seriously face these scalability changes (Khan & Zomaya, 2015).

These challenges originate from the necessary large scale processing algorithms and the resulting associated datasets across the entire computing pools in use. When end user applications are searching the entire World Wide Web for instant results it becomes pretty obvious that the number of computing elements explodes and, in many cases, data set exchanges between compute nodes can easily reach petabyte scales (Hooda, Kapadia, & Krishnan, 2014). In order to handle these requirements, new hyper-scale, high performance network architectures are necessary—see also "hyper-scale data center" later in this chapter, offering low cost connectivity for up to millions of servers, making it easy to understand why the configuration of these networks can no longer be done manually.

Requirement for Agile Networks

The need for carriers to offer higher value and competitively differentiated services to their customers requires new, sophisticated and dynamic multi-tenant networks in order to serve all the groups of users with their different applications and performance needs (Kale, 2016). The idea of agile networking is nothing new. In 1998, agility was seen as the next step in business reengineering (Metes, Gundry, & Bradish, 1998).

We have already discussed how key operations such as steering different customers traffic flows throughout the network in order to provide the required customized performance controls, policies and on-demand delivery became too complex to implement with legacy carrier scale networks, and required specialized network nodes at the edges, which increased not only CapEx, but also time to market new services.

As carriers and enterprises are looking to introduce new networking capabilities and services (Stallings, Foundations of Modern Networking – SDN, NFV, QoE, IoT, and Cloud, 2015) in order to answer the highly dynamic requests and needs of their customers and users, legacy vendor equipment product cycles of usually 3–5 years cannot keep up with these new demands any longer. Further to that, many networking vendor implementations are closed and very often lack standard implementations opposed to the promises of the vendors. This is because many networking vendors follow their old and successful pattern of locking in customers by offering semi-standard and highly proprietary solutions. In the end, this limited the ability of network operators to adapt the network to the ever-increasing demand of modern applications and use cases. The problem grew to such an extent that finally the industry created new architectures and standards for open network programmability that we generally refer to as "software defined networking," in order to overcome the mismatch between modern market requirements and legacy networking capabilities.

Unfortunately, some of the legacy vendors were still trying to offer halfhearted SDN implementations in 2016, in order to keep the existing revenue stream from their legacy networks as long as possible, thus binding many customers. Fortunately, this trend will end as soon as companies realize the need to stay competitive in their offers to customers who need the full potential of programmatic networking capabilities.

SDN Disrupting Legacy Networking

Networks are at the heart of information technology (IT), as they enable new architectures that create new business opportunities. This gives a whole new meaning to IT as the principal enabler of new businesses, making the network the central point of these movements as long as the architecture is capable and intelligent enough and can be adapted in a timely enough fashion.

There was already a movement back in the middle of the 1990s called Service Oriented Architecture (SOA),[8] which facilitated the cooperation of a large number of networked computers in order to exchange information via services without any human interaction. Changes to programs and applications were introduced to the complete infrastructure of servers, which enabled better flexibility, scalability of applications and services. SOA was the vehicle that introduced web services, which became the new way for accessing functional building blocks with standard Internet protocols. SOA was platform and language independent and, by offering faster development, testing, deployment and management of IT infrastructures, changed how development, management and operation looked at technology.

Network Virtualization or Software Defined Networking (SDN) is very similar to SOA, as SDN and network APIs allow for the easy programming of the network, receiving state from the network and services. It opens up the control of the network and services to everyone who is capable of doing so or any respective application able to take control. There are many views of SDNs but at its core it defines an architecture allowing for the simplification of networks while making them far more reactive to new business' needs and requirements resulting from new and changing services supported by the network. This allows IT to be up to speed as soon as new business opportunities show up through programming access by operators, enabling automated orchestration and management approaches and applying different policies across multiple network elements, like servers, switches and routers. The key advantage of this new flexibility is a result of the decoupling of the applications performing the operations mentioned above from the operating systems running in the network elements.

Through SDN, control of the network can be achieved without the need to change any parts of the physical networking infrastructure. Thus, the new architecture can be introduced without displacing old platforms and, additionally, allow better support for the needs of dynamic VM migration required by modern applications. This also applies to cases like mobility and sensor networks, also referred to as Internet of Things.

As soon as the concepts of SDN are applied to carrier networks, this is referred to as Network Functions Virtualization (NFV)[9] (Pierre Lynch, 2014), which

---

[8]Service Oriented Architecture, SOA, http://www.service-architecture.com/articles/web-services/service-oriented_architecture_soa_definition.html, retrieved 2017-05-29.

[9]ETSI, Network Function Virtualization, NVF, http://www.etsi.org/technologies-clusters/technologies/nfv, retrieved 2017-05-29.

introduces one of the most profound paradigm shifts the networking industry has faced to date. NFV started being driven by ETSI in 2012 by moving proven functions such as routing, policy, firewall, DPI, and many others from running on dedicated hardware appliances to running on virtualized server platforms with the aim of achieving massive efficiencies.

Carriers generally agree that NFV is necessary, but it still remains hard for them to identify the benefits and find an applicable migration towards NFV enabled networks. This is due to the need for applying a completely new network architecture based on virtualization and software control.

White Boxes and SDN

Today, in the middle of the 2010s, we are at a major turning point for networking and the whole networking industry. As we have shown, there is a fundamental change happening (or has maybe has already happened) in the way networks need to be designed and run in order to cope with the multitude of modern dynamic business' needs and applications. The legacy routers and switches that, for many years, dominated carrier and enterprise networks and built the backbone for the Internet are being gradually replaced by smaller, so called "white boxes," which are based on off-the-shelf merchant silicon and software defined networking (Göransson & Black, 2014).

So far, this revolution has happened quietly behind the scenes but it has, meanwhile, increased by a considerable number with around a million white box network elements installed in 2015 alone. Actually, in the biggest data centers today run by Microsoft, Amazon, Google and Facebook you can find many more white boxes than switches and routers from legacy networking gear vendors, exceeding more than $1 billion worth of network equipment purchases each year. Plus, all of them have moved away from traditional networking methods and deploy SDN instead.

We have talked about many trends as to why this transformation in networking is happening. One of the most important reasons for enterprises owning and operating large networks is to customize and differentiate their network and service offers to best serve their own and also their customer's needs. Carriers like AT&T, BT and DT, on the other side, want and need to take control of the software controlling their networks. Another driving reason is the growing availability and acceptance of merchant silicon for switching and routing, which we can see is analogous to the arrival of the cheap microprocessor thereby enabling the growth and popularity of the personal computer. The merchant switching and routing chips combined with open, third-party software enable operators and owners of networks to offer more innovative and competitive services in a much more dynamic timeframe compared to legacy networking technologies.

SDN to Mainframe Analogy

A similar pattern of changing and maturing technology could already be seen in the evolution of mainframe computers. Mainframes were built on proprietary microprocessors that were controlled by the few vendors dominating the mainframe market, developing their own operating systems (OSs) and running their own proprietary applications. This vertically integrated, closed model of computing lasted until the first off-the-shelf microprocessors were available in the middle of the 1980s and started the revolution of the personal computer. They allowed development of newer and smaller operating systems, where Microsoft and Apple succeeded in turn to deliver them. This, again, allowed a multitude of new applications based on the "open interface" between commoditized hardware and software and finally turned the complete mainframe industry upside down.

When we compare this situation to today's networking industry, we find a lot of similarities. The networking industry and its major vendors dominated the market for almost three decades by selling networking equipment in the form of closed boxes. The secret of this success was that these closed boxes enabled simple, plug and play options, so that all over the world people with some experience could easily build the parts and pieces of their networks and the Internet as a whole.

Also, this model proved to be a great growth engine for more than 25 years. The success of Cisco Systems,[10] undeniably the most prominent player in the networking arena since its origins back in the 1980s, demonstrates this point. Over time, these networks became more and more sophisticated and the network elements became much more complex, such that they can have taken the shape, size and power consumption of the mainframes of the 1980s. Additionally, the technologies they are based on come from a few selected vendors and are like mainframes but vertically integrated. Given the many hundreds of billions of dollars invested in these networks, there are understandably strong interests by these vendors to keep it that way as long as possible.

SDN Disruption

SDN started to disrupt the myth that software and hardware components in network elements cannot be separated without a disaster. In fact, modern SDN implementations demonstrate the opposite very well and that, in turn, is already turning the whole networking industry as of today upside down and forcing the introduction of an open horizontal model. It is pretty easy to predict what could very likely happen to the networking industry based on the experiences we have from mainframes back in the 1980s. In the middle of the 1980s the old IBM model was no longer sustainable, as it allowed moving but far too slow. Competition from personal computers and open applications became too big and there was no chance for IBM to have sustained the necessary level of innovation by themselves.

Legacy networking gear vendors built and sold boxes integrating hardware and software and their real differentiators were in the excellence and know-how to design and run layer 2 and layer 3 protocols that decided where data packets should be forwarded in the network. In this closed market, it was pretty easy to always add some nice bells and whistles on top of standardized solutions in order to have better sales pitches for quasi-standard but, in many cases, de-facto proprietary solutions. This had changed drastically by the middle of 2015 as merchant silicon started to

[10]Cisco Systems, http://www.cisco.com, 2016, retrieved 2016-12-09.

make the legacy networking gear hardware manufacturers irrelevant, same as low cost microprocessors did in the middle of the 1980s for the computing industry. As a result, the servers in data centers were increasingly replaced with white boxes based on merchant silicon from Intel and others using Microsoft or Linux operating systems and no longer server technology from the big vendors. For the networking equipment industry, the same started to happen and SDN, plus merchant silicon renders control over the network back to the network operators of carriers and enterprises.

Prominent examples of this movement into SDN are carriers like NTT, AT&T or DT, followed by companies like Amazon, Facebook, Google or Microsoft. The deployment of SDN running on merchant silicon allows these companies to design and deploy networks in a much cheaper, faster and more dynamic way, making it easier to better fit individual needs. For example, the Google WAN backbone interconnecting its data centers is completely based on SDN, allowing Google to test and roll out their new applications at an unprecedented speed, while, at the same time, effectively guaranteeing bandwidth for different traffic flows and working quickly around link failures. Google also based its network on OpenFlow[11] and has run a vertically integrated network since the middle of the 2000s, only deploying white boxes and merchant silicon since then. Also, AT&T builds its new network of the future based on open equipment and technologies like NFV[12] and SDN. This, in turn, will completely change the way the AT&T network is built and what software it is running, changing its culture and operations because SDN is simply the best fit and economically the way to go.

All these facts make SDN a real revolution for the networking industry and it will be interesting to see how this will change and even disrupt many things of the past by turning the user and vendor landscape upside down.

Concept and Promise of SDN

The evolution we have described so far in this chapter paves the way for SDN as the solution for enabling these modern requirements. Today's concept of SDN did not actually change much from the original in 1998. It still defines an architecture, which shall be dynamically manageable, cost-effective and adaptable, while being suitable for the high bandwidth and dynamic nature of modern applications. The core principle is still that this SDN architecture decouples network control and forwarding functions, which allows network control to become directly programmable and the underlying infrastructure to be abstracted from applications and network services.

[11]ONF, Open Networking Foundation, OpenFlow, https://www.opennetworking.org/sdn-resources/openflow, 2016, retrieved 2016-12-09.

[12]ETSI, Network Function Virtualization (NFV), 2016, retrieved 2016-12-09.

Separation of Control and Forwarding Plane
Abstracting the control from the forwarding functions allows agile and dynamic adjustment of network-wide traffic flows to meet the fast and dynamically changing needs of today's (and the future's) demanding applications. SDN has a centralized management concept, as network intelligence is logically centralized in software based SDN controllers, which maintain a global view of the network, and look to applications and policy engines as a single, logical switch.

Advanced Network Programmability
Programmatic configuration of SDN means that network managers can configure, manage, secure and optimize network resources quickly via dynamic, automated SDN programs. It is easy for network managers to write these programs now because they no longer depend on proprietary software. Finally, a vendor neutral implementation through open standards as discussed in the sections "Open SDN Implementations".

OpenFlow and OpenDaylight (ODL) allows SDN to support simplified network design and operation due to the fact that instructions are provided by SDN controllers instead of multiple, vendor specific devices and protocols (Stallings, Foundations of Modern Networking – SDN, NFV, QoE, IoT, and Cloud, 2015).

High Level View of SDN

When looking a bit deeper into the overall SDN concept, we immediately recognize that there is much more than just that basic separation of control and forwarding planes.

The basic components that are always present in SDN are the application plane, the control plane and the data plane. But, in addition, there needs to be the management and administration planes, communicating the SLAs and Contracts with the SDN Applications, the Policy Configuration and the Performance Monitoring with the SDN Controller and the Element Setup with the network elements (see Fig. 1).

SDN Control to Data-Plane Interface (CDPI)
The SDN CDPI is the interface between the SDN Controller and the SDN Datapaths. This interface provides the programming control of all forwarding operations, statistics reporting, capabilities advertisements, event notification, etc. This interface should be implemented in an open, vendor-neutral and interoperable way to enable the full power of SDN.

SDN Northbound Interfaces (NBI)
SDN NBIs are the interfaces between SDN Applications and SDN Controllers that offer abstracted network views to the applications and enable direct expression of network behavior and network requirements. This can happen at any level of abstraction (latitude) and across different sets of functionalities (longitude). These

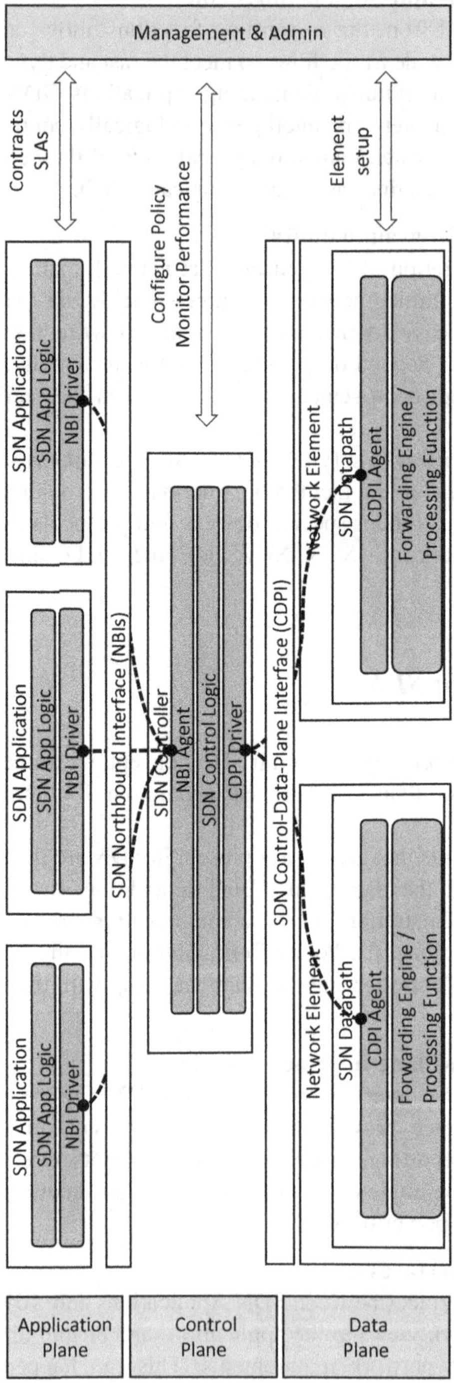

Fig. 1 A high-level overview of the software-defined networking architecture

interfaces should be implemented in an open, vendor-neutral and interoperable way as well in order to enable the full power of SDN.

SDN Applications
SDN applications communicate their network requirements and their desired network behavior via the northbound interface (NBI) to the SDN Controller. They can use the abstracted view of the network for their decision-making purposes as well. SDN Applications contain the SDN Application Logic and the necessary NBI Drivers. In case the SDN Applications themselves expose another layer of abstracted network control, they will offer one or multiple higher-level NBIs through NBI agents.

SDN Controller
The SDN Controller sits at the heart of the whole SDN concept and, as such, forms a centralized entity that translates requirements from the SDN Applications into network configurations and down to the SDN data paths, while, at the same time, providing the abstracted network view to the SDN Applications. The SDN Controller needs one or more NBI Agents to communicate with the SDN Applications via the NBIs and the SDN Control Logic and the SDN Control to Data Plane (CDPI) driver to communicate with the network elements via the CDPI.

The logically centralized definition of the SDN Controller does not give any implementation details like controller hierarchies, federation of multiple controllers, virtualization and slicing of network resources, nor does it provide communication interfaces between controllers.

SDN Datapath
Finally, the SDN Datapath represents a logical network device, offering visibility and uncontended control over the advertised forwarding and data processing capabilities of that specific network device, while the logical representation may contain all or only a subset of the physical resources. The SDN Datapath needs a CDPI agent, a set of traffic forwarding engines (minimum of one), as well as a set of traffic processing functions (zero or more). The forwarding engines and processing functions may simply do the forwarding between the external interfaces of the Datapath, as well as perform the internal traffic processing or traffic termination functions.

A single, physical network element may contain one or more SDN Datapaths, which results in an integrated physical combination of communication resources managed as a unit, while a typical SDN Datapath can also be defined across multiple physical network elements. The above logical definition does not describe any implementation details, like the logical to physical mapping, the management of shared physical resources, interoperability with non-SDN networks, or data processing functionality including any layer 4 to layer 7 functions.

Centralized Versus Distributed Control and Data Planes

So far, we have seen that one of the fundamental benefits of SDN comes from the separation of control and data planes in network devices and we have already discussed that this is not a new concept. We will give a brief overview of control and data plane architectures in this section and recommend some further reading (Nadeau & Gray, 2013), or (Paul Göransson, 2014).

With the separation of control and data planes a lot of other questions need to be answered, like how far away from the data plane could the control plane be placed and how many instances of the control plane need to be active in order to guarantee high availability requirements as well as resiliency. In fact, the classic set up in legacy network elements with switches and routers was to have a control plane in each networking device, representing the distributed control plane approach, as each L2 switch or L3 router has its own control plane. With SDN and the clear separation of the control plane and the data plane also a semi-centralized or a fully centralized control plane become possible now.

Classic Control Plane Approach

In the classic control plane approach, the control plane is fully distributed, which is what we have described so far as the basis of classic computer networking and describes how the Internet grew up and still works today. In this approach, every network device runs its completely separate control plane and, depending on the sophistication and level of the device, one or several data planes. In this model, all the separate and distributed control planes need to cooperate with all the other control planes in the remote devices of the network, with typical examples being layer 2 protocols for bridging and switching or layer 3 protocols for routing in order to guarantee the proper network operation.

Semi Centralized Control Plane Approach

In the semi-centralized control plane approach, the centralization allows for new capabilities without completely replacing every device capabilities and leaves a sort of remaining control plane with the device. This results in a kind of hybrid approach that allows for the combining of distributed control plane functions like ARP[13] or MAC processing in the device as an underlay part, whereas the centralized control plane takes care of the operational functionality in an overlay part.

Fully Centralized Control Pane Approach

In the fully centralized control plane approach, no more control plane functions exist at the device level, which means that a network element becomes "dumb" while possibly being a fast switching device. This "dumb" device is fully controlled remotely by a centralized intelligence, which yields to a very extreme type of

[13]Plummer, D., IETF, An Ethernet Address Resolution Protocol (ARP), 1982, updated by RFC 5227, 5494, https://tools.ietf.org/html/rfc826, retrieved 2016-12-09.

implementation, especially when high availability and scalability are taken into account.

Control Plane

The control plane creates the data sets for the forwarding table entries, which are then used by the data plane for traffic forwarding from the ingress ports to the egress ports on the network devices and, by so doing, determines the actual network topology. This network topology consisting of all the data sets is stored in the routing information base (RIB) that is usually exchanged with control planes of other routers within a network either via routing protocols or manual configuration in the case of static routes. This process guarantees a proper operation of the network particularly avoiding loops or taking care of certain policies like preferred path, quality of a path, delay, traffic priorities, etc. Once the RIB is stable and consistent, the Forwarding Information Base (FIB) is populated, which is, in turn, used by the data plane to forward traffic.

Considering switching and routing as networking functions, we are dealing with layer 2 and layer 3 control planes respectively. In a layer 2 network element called bridge or switch, the control plane has to handle hardware or physical addresses such as IEEE MAC addresses, while, in a layer 3, the network element called the router uses the control plane network addresses such as IP addresses for packet forwarding.

Layer 2 and Layer 3 Forwarding

Layer 2 networks forward based on MAC addresses and need to ensure loop free forwarding through the Spanning Tree Protocol (STP)[14] and the flooding of broadcast, unicast unknown and multicast traffic, which can easily result in scalability limitations and challenges. Layer 2 control protocols like IEEE SPB/802.1aq[15] and IETF TRILL[16] were developed to address some of the limitations of layer 2 networking.

Layer 3 networks forward on behalf of network addresses, and the most commonly used became IP addresses as outlined in the chapter "The Success of the Internet Protocol Suite." Actually, forwarding in IP networks is based on the destination IP prefix, which includes network prefixes from a number of address families for unicast and multicast addresses. This allows the connection of layer

[14]IEEE, IEEE 802.1D, Standard for Local and Metropolitan Area Networks, Media Access Control (MAC) Bridges, http://standards.ieee.org/getieee802/download/802.1D-2004.pdf, 2004, Revision of IEEE Std 802.1D 1998, retrieved 2016-12-05.

[15]IEEE 802.1aq, Shortest Path Bridging, http://www.ieee802.org/1/pages/802.1aq.html, 2005,-2012, retrieved 2016-12-05.

[16]Eastlake D., et al., IETF, RFC 6326, Transparent Interconnection of Lots of Links (Trill), https://tools.ietf.org/html/rfc6326, 2011, retrieved 2017-05-29.

2 domains and helps significantly reduce and overcome the afore mentioned layer 2 scaling problems. Layer 3 networking easily allows building hierarchical network structures, connecting together layer 2 networks with layer 3 routers and, by connecting more of the layer 3 routers, builds up even larger networks, which are, in turn, interconnected with gateway routers. Packet routing is performed on layer 3 between the different networks, while only the frame or datagram part of the packet is forwarded to layer 2, as soon as the destination sub-network is reached before finally delivering it to the destination host.

Hybrid Layer 2 and Layer 3 Forwarding

In the attempt to combine layer 2 switching and layer 3 IP routing, a few new protocols were invented, all with their own benefits and major application areas and these hybrids are all a mixture of the previously described and clearly separated mechanisms for layer 2 and layer 3 forwarding respectively.

One of the most prominent is Multi-Protocol Label Switching (MPLS),[17] actually some more than 20 years after its introduction a complete protocol suite, with the goal of sharing the advantages of fast packet switching on ATM with the complex but very capable path signaling techniques from IP. MPLS was invented in the late 1990s and up to today is widely spread in many core and enterprise networks (Harnedy, 2007; Tomsu & Wieser, 2002).

Another approach is Ethernet Virtual Private Network (EVPN),[18] which tries to solve layer 2 scaling issues by tunneling remote layer 2 bridges (switches) together over tunnels in IP infrastructures (Krattiger, Kapadia, & Jansen, 2016). It uses either MPLS or Generic Route Encapsulation (GRE)[19] as tunneling mechanisms, in order to not effect the IP networking by layer 2 scaling issues. This works by exchanging reachability information between the bridges (switches) as data in a newly added BGP[20] address family, together with some other optimizations finally limiting the amount of layer 2 addresses that need to be exchanged.

The last approach is the Locator/ID Separation Protocol (LISP)[21] described in RFC 6830, which allows for the solving of some shortcomings of the distributed control plane model for multi-homing. This is achieved by adding new addressing domains and, at the same time, separating site addresses from providers with a new map and encapsulation control and forwarding protocol.

---

[17]Rosen E., et al., IETF, RFC 3031, Multiprotocol Label Switching (MPLS), https://tools.ietf.org/html/rfc3031, 2001, retrieved 2017-05-29.

[18]Sajassi A., et al., IETF, RFC 7209, Requirements for Ethernet VPN (EVPN), https://tools.ietf.org/html/rfc7209, 2014, retrieved 2017-05-29.

[19]Farinacci D., et al., IETF, RFC 2784, Generic Route Encapsulation (GRE), https://www.ietf.org/rfc/rfc2784.txt, 2000, retrieved 2017-05-29.

[20]Rekhter Y., et al., IETF, RFC 4271, A border Gateway Protocol 4 (BGP-4), https://www.ietf.org/rfc/rfc2784.txt, 2006, retrieved 2017-05-29.

[21]Farinacce D., et al., IETF, RFC 6830, https://tools.ietf.org/html/rfc6830, 2013, retrieved 2017-05-29.

Forwarding and Data Plane

The task of the forwarding or data plane is to handle incoming datagrams or frames on any type of media, either on wire or wireless and, if the sanity check of the frame is correct, a lookup is performed in the FIB[22] containing the forwarding information that was pre-programmed by the control plane. As long as datagrams can be directly forwarded based on the FIB entries, this process is often referred to as fast forwarding or fast path because no further processing has to be performed by the control plane.

In case the packets destination address is unknown, it has to be sent to the route processor in order to allow the control plane to process this packet consulting the RIB, forward the packet and populate the FIB with the new destination address, thus allowing subsequent packets with the same destination address to be forwarded directly by the FIB lookup. This software lookup or software path is usually more time consuming than pure, fast path forwarding based on the FIB tables. Also, the FIB tables can reside in several different forwarding entities like software, hardware accelerated software, commodity silicon, field programmable gate arrays (FPGAs),[23] or specialized silicon like ASICs used by router vendors like Juniper, Cisco or others.

Usually performing lookups in hardware tables (fast path) allows significantly higher packet forwarding performance. It is no wonder that many of the legacy network equipment vendors have deployed and cultivated this design in their network elements (routers or switches). It is important to mention that recent evolutions in I/O processing of cloud based computing have significantly increased even software based forwarding, thus it became possible to use these techniques for low to mid-range performance beginning in the middle of the 2010s.

Separation of Control and Data Planes

The separation of control and data planes is a well-known approach and has been widely implemented over the last two decades. Usually any modern switch or router with a multiple slot chassis has a control plane running on a dedicated processing card (route processor) or two if redundancy is required. The data plane is spread via multiple line cards, which all have their own forwarding processors (see Fig. 2).

The line cards and the route processor are connected in the network element via a bus or kind of high speed internal network, so that packets that cannot be directly forwarded based on the FIB entries can be sent as fast as possible to the route processor in order to determine the correct destination egress port and then populate

[22]Trotter G., IETF, RFC 3222, Terminology for Forwarding Information Base (FIB) based Router Performance, https://tools.ietf.org/html/rfc3222, 2001, retrieved 2017-05-29.

[23]Intel, Intel FPGAs, https://www.altera.com/products/fpga/overview.html, retrieved 2016-12-05.

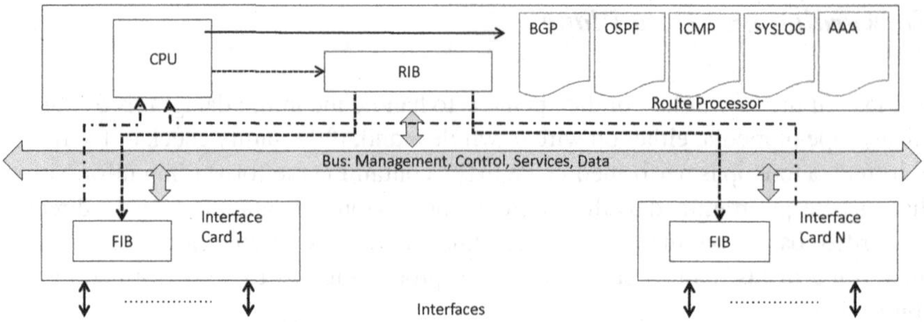

Fig. 2 Control and data plane implementation in modern network device

the RIB as well as the FIB. There are even some protocols optimized for this type of architecture like MPLS, which carries control traffic based on the IP protocol suite. For such a protocol, there can easily exist a separate route processor with the fixed label based switching performed by dedicated and specialized packet processors on different line cards.

It is important to understand that there are also other flows throughout the network elements (routers) besides data packets, since management, services etc. have to be supported as well.

Advantages of Separation

This separated control and forwarding plane architecture is typical for a distributed control plane implementation, where each network element has its own control plane (route processor) and separated forwarding or data plane (interface cards). The necessary management and services flows show that much more tasks have to be performed by each network element than just the forwarding of packets, which we will discuss in the chapter "Different Functional Planes of Network Elements." Through the separation of the control and data planes, network operators can finally scale the forwarding and processing cards independently and on demand during operation of the network, thus allowing the separate evolution of the control and data planes.

This identifies one of the big advantages of SDN, as the move from distributed control plane to centralized control pane (fully centralized or semi centralized) achieves upgrades in the control of the service plane more easily, as the upgrade no longer happens in a vendor locked in the environment, but in a Commercial Off The Shelf (COTS) compute environment. Further separation of the control plane from other functional planes allows even more scalability as these processes can also run on COTS hardware and this can be either within the switch or router or remotely in a datacenter.

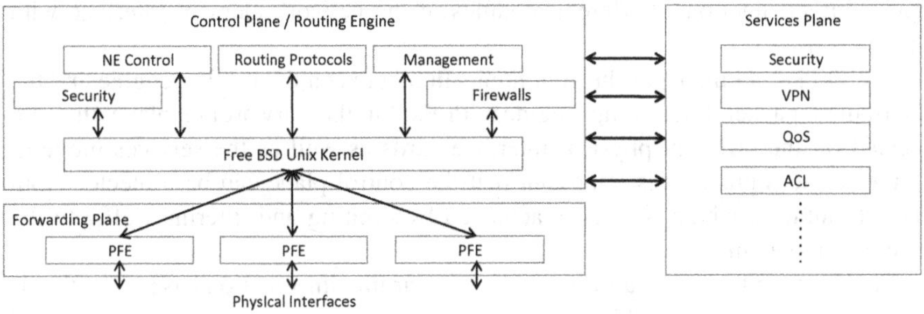

Fig. 3 Different functional planes of network elements

Different Functional Planes of Network Elements

As we have already mentioned, several functional planes exist in network elements like service planes and management planes besides the control plane. This additional logical separation allows for cleaner interworking of the different functions but, at the same time, is also implemented in multiple different ways depending on the vendor or the specific product of a vendor. When looking at switches and routers, we have products from a number of vendors today. Over the past two decades, the most prominent representatives became Cisco Systems[24] and Juniper Networks.[25] By the middle of the 2010s, both finally offer not only the classic distributed control plane implementations, but also SDN based solutions for their networking products. We will now look into some general ways of implementation of the different network element planes, describing the specific functionality and interworking.

At the heart of the control plane resides the operating system of the network element that is often based on an open source like FreeBSD[26] UNIX. Of course, different vendors modify and adapt this kernel individually not only at the time when this was implemented but even more so to add their proprietary functionality that determines and differentiates the final product.

In many cases, the complete network element functionality is separated into three planes: the control plane, the forwarding plane and the services plane (see Fig. 3). Typically, the control plane runs on a central card (hardware) of the network element that is called the routing engine and forms the central intelligence of the specific network element like the router, switch or different network appliance. This central intelligence is responsible for control and communication with the interfaces, and the chassis runs the routing protocols, firewalls, security and

[24]Cisco Systems, http://www.cisco.com, 2016, retrieved 2106-12-09.

[25]Juniper Networks, https://www.juniper.net/us/en/company/, 2016, retrieved 2016-12-09.

[26]FreeBSD, The FreeBSD Project, https://www.freebsd.org, retrieved 2016-12-05.

management functions. Different modules of the network element interface with the OS Kernel.

The OS kernel manages the communication between the routing engine and the distributed packet forwarding engines (PFEs) of the forwarding plane that are located on the different physical interface cards as well as the services modules of the services plane. It is important that the control plane can be protected from security attacks, which is usually achieved by limiting and filtering traffic to the central route engine.

Based on the local forwarding table or forwarding information base (FIB) that is stored on the distributed PFEs, the forwarding plane will forward packets to other network devices connected via the physical interfaces. The FIB is populated and always updated by the central routing engine depending on the results achieved from the routing protocols with the applied firewall, security and other policies. This local FIB also allows the PFE to work autonomously without the need to communicate back to the central routing engine for standard packet forwarding and even allows continuous forwarding when the central routing engine is updated, exchanged or simply fails.

This separate services plane concept enables different service types depending on the platforms (switch, router, network appliance). Typical service types supported are packet prioritization in case of delay and time sensitive traffic as is the case for voice and video, address translation between networks or user access control to proprietary network sections or applications.

Evolution of Control Plane Concepts

Distributed Control Planes

The Internet, as it evolved over the past decades, is based on the distributed control plane model. It is worth mentioning that this model is based on eventual consensus, as it works by all the network elements (routers) collecting reachability information via their proxy routers in order to develop a localized view of a network. This model can only build an eventual consensus view of the network because the propagation delays of the reachability updates have to be taken into account, which is typical for this distributed control plane model. Inherently, this model shows repeated non-synchronization that can potentially lead to suboptimal forwarding till synchronization is back again.

In both IP and MPLS forwarding, this distributed control plane model is based on the routing and reachability information, which finally allows the data plane to setup the paths throughout the network. This is in general based on routing paradigms and according to routing protocols, specifically the Interior Gateway Protocols (IGPs) that are freely available as IETF drafts and standards, thus we will not dig any deeper in this book. Nevertheless, there is a great introduction in (Nadeau & Gray, 2013) about distributed control plane models as well as creating

IP undelays, convergence times, load balancing, high availability, MPLS overlays and replication, which is highly recommended for further reading.

Centralized Control Planes

The main advantages of centralized control planes are the simplification of programming control and the view of the network being provided to an application. An application no longer needs to contact individual network elements in order to get information about the network or to make end-to-end changes. Now it can interact with some control points that ensure that this information is accurate. The route server for IP networks and the ATM switch controller mark some early historical implementations of centralized control planes and even products in the middle of the 1990s existed like a solution from Ipsilon Networks using deterministic routing based on combining IP and ATM, which later became obsolete through TAG switching from Cisco and finally MPLS.

Centralized Control Plane Example ATM/LANE

ATM/LANE (Asynchronous Transfer Mode/LAN Emulation) was already a popular implementation of the centralized control plane approach dating back to the middle of the 1990s (Black, 1997). ATM uses either permanent (static) manually configured virtual circuits (PVCs) or Switched Virtual Circuits (SVCs). For SVCs, ATM used a subnet based NSAP addressing scheme for the endpoints of ATM connections over an ATM network, which could be either Variable Bit Rate (VBR) or Committed Bit Rate (CBR) based on the Quality of Service (QOS) requirements. PNNI (Private Network to Network Interface—an ATM Forum standard) was used as a dynamic distributed routing protocol of the IP environment to distribute the NSAP to VCI/VPI (Virtual Circuit Identifier/Virtual Path Identifier) mappings.

LANE, which was also defined by the ATM Forum, allowed ATM networks to appear to higher layers like a LAN (Local Area Network) with the same MAC layer service interface. As such, the LANE protocol allowed for the creation of an overlay network on the ATM switched undelay network in a way that all this operation was transparent to the switches.

In order to achieve this functionality of transparently connecting ATM hosts and LAN hosts via an ATM cloud, LANE defined a LAN Emulation Server (LES), a Broadcast and Unknown Server (BUS) as well as LAN Emulation Client Server (LECS). The LES had to handle MAC registration and control, and acted essentially as the ARP server for the emulated LAN (ELAN). The LEC worked as the protocol interface on the host's or network's elements between the MAC layer (layer 2) and the higher layers. The LES and the BUS worked in tandem to handle broadcast and multicast traffic for a specific ELAN. The LECS ensured that the database contained the domain wide LEC/ELAN mappings and provided the ATM address of the LES for a specific ELAN (see Fig. 4).

The LANE protocol allowed the operation of a software defined network on a mixed ATM and IP network and hence shows a typical example of logical overlay supported by multiple servers in a controller-like functionality, although without programmatic interfaces and only rudimentarily high availability.

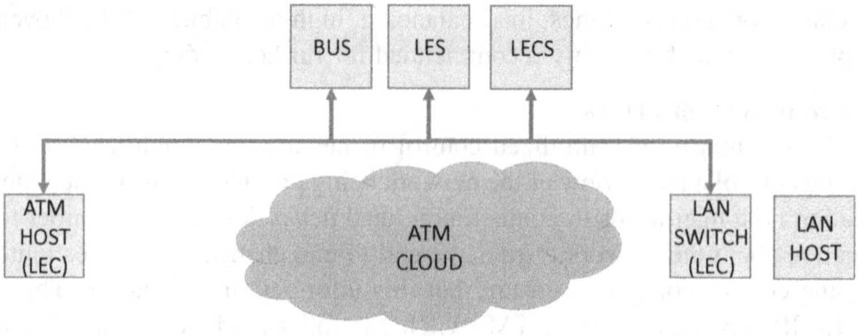

Fig. 4 ATM LANE with clients and servers

Centralized Control Plane Example Route Server

Other prominent and already long term deployed examples of centralized control are the Route Server,[27] offering Internet service providers a centralized control point for handling the scale of external peering points, or the Route Reflector,[28] which was developed and standardized for scaling of internal peering information (Halabi, 1997). Simply speaking, route servers are control points used by Border Gateway Protocol (BGP)[29] that receive BGP control state updates, the so-called Network Layer Reachability Information (NLRI)[30] from all the participants within an Autonomous System (AS).[31,32] After applying policies and filters to the updates, they calculate best paths before creating a separate RIB per participant, which in turn is returned to the participant. Route servers can also interface to routing registries, holding route objects such as policies, prefixes, authentication info and autonomous system numbers that are used for automatic provisioning of stand-alone peers (edge routers) as well as route servers. Route reflectors provide similar tasks to internal peers.

While both route processors and route reflectors represent the central control plane concept, there still remains the lack of standardized programmability and there is also no standard API defined, which should not be the case with a full SDN solution.

---

[27]Jasinska E., et al., IETF, RFC 7947, Internet Exchange BGP Route Server, https://datatracker.ietf.org/doc/rfc7947/, 2016, retrieved 2017-05-29.

[28]Bates T., et al., IETF, RFC 4456, BGP Route Reflection: An Alternative to Full Mesh Internal BGP (IBGP), https://tools.ietf.org/html/rfc4456, 2006, retrieved 2017-05-29.

[29]Rekhter Y., et al., IETF, RFC 4271, A Border Gateway Protocol 4 (BGP-4), https://tools.ietf.org/html/rfc4271, 2006, retrieved 2017-05-29.

[30]Bates T., et al., IETF, RFC 4760, Multiprotocol Extensins for BGP-4, https://tools.ietf.org/html/rfc4760, 2007, retrieved 2017-05-29.

[31]Hawkinson J., et al., IETF, RFC 1930, Guidelines for creation, selection and registration of an Autonomous System (AS), https://tools.ietf.org/html/rfc1930, 1996, retrieved 2017-05-29.

[32]Huston G., IETF, RFC 5398, Autonomous System (AS) Number Reservations for Documentation Use, https://tools.ietf.org/html/rfc5398, 2008, retrieved 2017-05-29.

Open SDN Implementations

OpenFlow

So far, we have mainly discussed control and data plane concepts and will now concentrate on OpenFlow architectures and protocols, as OpenFlow was one of the first standardization attempts of SDN until today.

OpenFlow[33] was created as part of network research from Stanford University in the 2010s in order to enable the use of campus networks for research and experimentation. One of the main goals was to completely replace the layer 2 and layer 3 functionalities of commercial switches and routers. In 2011, the Open Networking Foundation (ONF) was launched by Deutsche Telekom, Facebook, Google, Microsoft, Verizon and Yahoo, with the aim of standardizing and commercializing SDN solutions and OpenFlow. The goal of the founding members was to use SDN and OpenFlow in their production networks and do active marketing and promotion of related topics.

As such, ONF has played an important role in increasing awareness and attention to SDN from the very beginning and is a great repository for according documents and specifications, including a whitepaper about SDN.[34] OpenFlow was first defined in the OpenFlow Switch Specification,[35] which was published by the ONF in 2011 with the most current version being Version 1.5.1.[36]

A schematic diagram of the OpenFlow architecture is shown in Fig. 5, where we can see the data plane and the control plane being connected with the OpenFlow protocol and the application plane on top connected via APIs. The key components of the OpenFlow model became part of the common SDN definition with the clear separation of the control and data plane. For ONF, the control plane is managed by a logically centralized controller system. OpenFlow is the first standard protocol used between the controller and the agents on the network elements in order to instantiate states. The necessary network programmability is provided from a centralized view via an extensible API. As such, OpenFlow defines a set of protocols and an API, and is neither a product nor a product feature. Without application programs defining the flows that have to go on the network element, the controller would not do anything.

---

[33]Open Networking Foundation (ONF), OpenFlow, https://www.opennetworking.org/sdn-resources/openflow, retrieved 2017-05-29.

[34]ONF, ONF White Paper, Software Defined Networking: The New Norm for Networks, https://www.opennetworking.org/images/stories/downloads/sdn-resources/white-papers/wp-sdn-newnorm.pdf, 2012, retrieved 2017-05-29.

[35]ONF, OpenFlow, OpenFlow Switch Specification Version 1.1.0, http://archive.openflow.org/documents/openflow-spec-v1.1.0.pdf, 2011, retrieved 2017-05-29.

[36]ONF, Openflow, OpenFlow Switch Specification Version 1.5.1, https://www.opennetworking.org/images/stories/downloads/sdn-resources/onf-specifications/openflow/openflow-switch-v1.5.1.pdf, 2015, retrieved 2017-05-29.

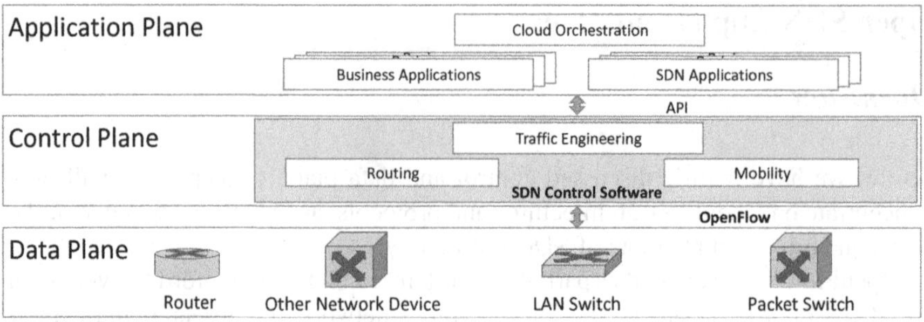

Fig. 5 OpenFlow architecture

SDN and OpenFlow Functionality

Like we already discussed in this chapter, the task of the OpenFlow SDN controller is to define data flows that are forwarded on the data plane. In order to allow a flow to be forwarded through the network, it must first get permission from the controller, which checks if the flow is permitted based on the network policy. As soon as the controller allows the flow, it computes the route for this flow through the network elements and stores an entry for this flow in each of the network elements in its domain (we will discuss the domain concept later in this section). This concept of performing all the complex functions in the controller leaves switches and routers to simply manage flow tables with the entries in the flow tables being populated by the controller. The standardized API being used for communication between the controller and the switches is defined in the OpenFlow specification.

This concept enables a remarkably flexible SDN architecture,[37] as it can operate with different types of switches and at different protocol layers, allowing the implementation of Ethernet switches (Layer 2), Internet routers (Layer 3), transport (Layer 4) switching, or application layer switching and routing and as such SDN still relies on the common forwarding functions on networking devices, like packet forwarding based on flow definitions.

SDN switches encapsulate and forward the first packet of a flow to the according SDN controller, leaving the controller to decide if the flow should be added to the switch flow table. If the flow is added to the flow table, the switch from now on forwards incoming packets out the selected port based on the flow table entry. Of course, the flow table also can hold priority information calculated by the controller. Also, the switch can either temporarily or permanently drop packets of a particular flow based on how it is programmed by the controller, allowing the use of packet dropping for security purposes in order to end Denial of Service (DoS) attacks or fulfill traffic management requirements.

Thus, the SDN controller manages the forwarding state of the switches in the SDN through a vendor neutral API defined and standardized by the ONF as the OpenFlow protocol. This allows the controller to address a broad variety of operator

[37]ONF, SDN Architecture, https://www.opennetworking.org/images/stories/downloads/sdn-resources/technical-reports/TR_SDN_ARCH_1.0_06062014.pdf, 2014, retrieved 2017-05-29.

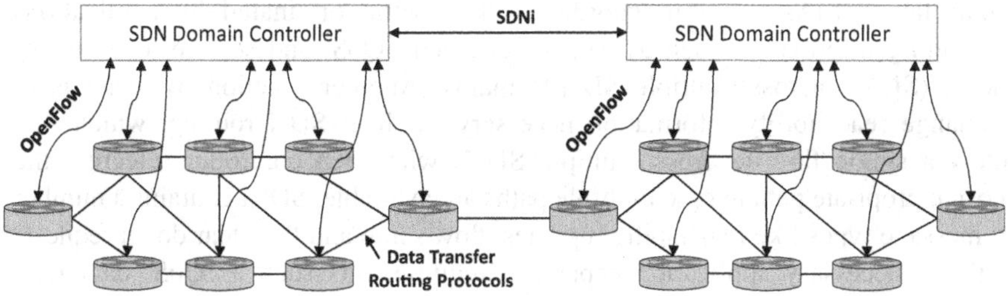

Fig. 6 SDN domains

requirements without any changes needed at the lower level network aspects including topology. The decoupling of the control and data plane enables applications to deal with abstracted network elements without the need to know details about how a specific device operates. This means that network applications residing in the applications plane see a single API to the controller, which allows fast creation and deployment of new applications to orchestrate network traffic flow in order to meet specific performance or security requirements for whatever customer.

SDN Domains

When networks become larger, such as in large enterprise or carrier networks, using only one controller for managing the complete network soon becomes a scaling and redundancy problem. In this case of large SDN networks, the deployment of multiple non-overlapping SDN domains[38] is the better solution with a separate controller per domain (see Fig. 6).

There are many reasons for splitting the network into multiple domains. The most obvious one is scalability, as the number of devices that can be managed by a single controller is limited. Another reason can be privacy as, for example, a carrier could choose or is forced to run different privacy policies for different SDN domains. This can become the case if sets of customers require their own privacy policies so that certain networking information like network topology in that specific domain must not be disclosed to an external entity. In case a carrier network consists of different types of infrastructure, like legacy and newer parts, it can often make sense to divide the network into multiple, individually manageable SDN domains in order to allow incremental deployment.

As all the SDN controllers in the different SDN domains need to communicate in order to exchange routing information etc., there is currently a protocol developed by the IETF called SDNI,[39] which stands for "interfacing SDN Domain

[38]ONF, SDN Architecture, pages 28-32, https://www.opennetworking.org/images/stories/downloads/sdn-resources/technical-reports/TR_SDN_ARCH_1.0_06062014.pdf, 2014, retrieved 2017-05-29.

[39]Yin H., et al., IETF, Internet Draft, SDNi: A Message Exchange Protocol for Software Defined Networks (SDNS) across Multiple Domains, https://tools.ietf.org/html/draft-yin-sdn-sdni-00, 2012, retrieved 2017-05-29.

Controllers." SDNi need to coordinate flow setup originated by applications containing information such as path requirement, QoS, and service level agreements (SLAs) across multiple SDN domains. Another function of SDNi is to exchange reachability information necessary for inter-SDN routing, which will allow a single flow to cross multiple SDNs with each controller selecting the most appropriate path in case multiple paths are available. SDNi contains a number of message types like reachability updates, flow setup and flow tear down requests with all necessary application capability requirements such as QoS, data rate, latency etc. It also includes capability updates; both network related capabilities like data rate and QoS and system and software capabilities available in the domain.

OpenFlow Concept and Requirements

All the talk about SDN is nice but in order to be able to realize all the advantages and benefits we have discussed so far, it is necessary to meet two basic requirements. First, there needs to be a common logical architecture in all switches, routers and all network elements to enable seamless management by an SDN controller. While this logical architecture can be implemented in different ways with different vendors, equipment and different network elements, it must be guaranteed that the SDN controller sees a uniform logical switch function. Second, the protocol between the SDN controller and the network element must be a standard and secure protocol.

OpenFlow addresses both of these requirements with the specification of the logical structure of the network switch and element functions[40] and the protocol between the SDN controllers and the network elements.

OpenFlow Standardization Status

We are currently looking at the OpenFlow specification 1.3.0. While the original and widely implemented specification 1.0 was developed at Stanford University, ONF released OpenFlow 1.2 after inheriting the project from Stanford. Meanwhile, Openflow 1.3[41] not only significantly expands the functions of the original specification, but will likely become the stable base for future commercial specifications, as ONF intends for this version to stay the stable target for chip and software vendors. There is, in fact, not much change planned for the foreseeable future.

Logical Switch Architecture

The OpenFlow switch architecture and the basic OpenFlow environment is illustrated in Fig. 7: OpenFlow switch The SDN controller communicates with OpenFlow's aware and compatible switches using the OpenFlow protocol via the Secure Sockets Layer (SSL). OpenFlow switches connect to each other as well as to

---

[40]ONF, Openflow, OpenFlow Switch Specification Version 1.5.1, https://www.opennetworking.org/images/stories/downloads/sdn-resources/onf-specifications/openflow/openflow-switch-v1.5.1.pdf, 2015, retrieved 2017-05-29.

[41]ONF, OpenFlow v1.3, Open Networking Foundation Announces OpenFlow v1.3 Test Specification, https://www.opennetworking.org/news-and-events/press-releases/2495-open-networking-foundation-announces-openflow-v1-3-test-specification, 2015, retrieved 2017-05-29.

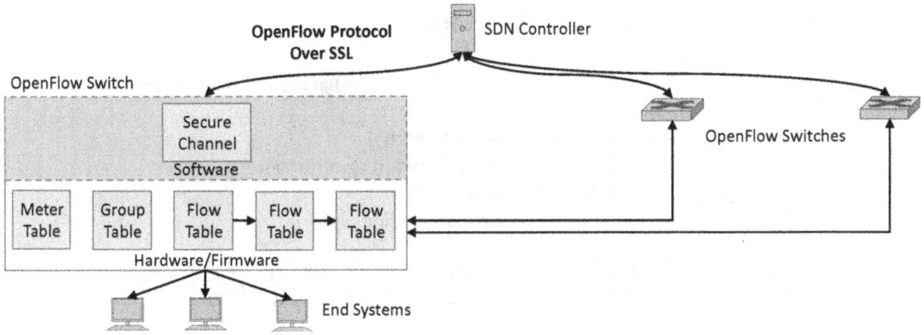

Fig. 7 OpenFlow switch

end systems, which are the sources and destinations of the packet flows. Inside each OpenFlow switch a series of tables, either implemented in hardware or firmware, build the basis for managing the flows of the packets through the switch (see Fig. 7).

There are three table types defined by the OpenFlow specification in the logical switch architecture. The most well known, due to its name, is the Flow Table, mapping incoming packets to a particular flow and also specifying the functions that need to be performed on the packets. There can also be multiple flow tables operated in a pipeline fashion. Any flow table may direct a flow to a Group Table, which, in turn, may trigger a variety of actions that affect one or more flows. A Meter Table can trigger several performance related actions on a flow.

A flow is not defined by the OpenFlow specification and it is also not defined in any of the OpenFlow literature. What it basically means is a sequence of packets traversing the network, sharing a set of header field values. This could either be the set of packets between a certain source and destination IP address, or all packets identified by the same VLAN identifier.

To work in an OpenFlow environment, any device wanting to communicate to an SDN Controller must support the OpenFlow protocol. Through this interface, the SDN Controller pushes down changes to the switch or router flow-table, giving network administrators several innovations to partition traffic, control flows for optimal performance, and start testing new configurations and applications.

Match, Action Statistics

In order to make it easier to understand what is behind a flow, we will start with a very high level view of OpenFlow. It is important to understand that OpenFlow introduces the concept that flow entries are no longer stored permanently on the network elements, enabling protocol configuration beyond the unstandardized and often vendor limited protocol configurations. These flow table entries are based on a Match rule, an Action part and Statistics part.

The Match rule determines the flow table entry based on the Ingress Port, layer 2 information like the Ethernet field and VLAN ID as well as layer 3 information like IP source and destination address and port. This provides additional masking capabilities in order to allow the network emulation of IP destination forwarding behavior, layer 2 and layer 3 source and destination routing behavior and enables

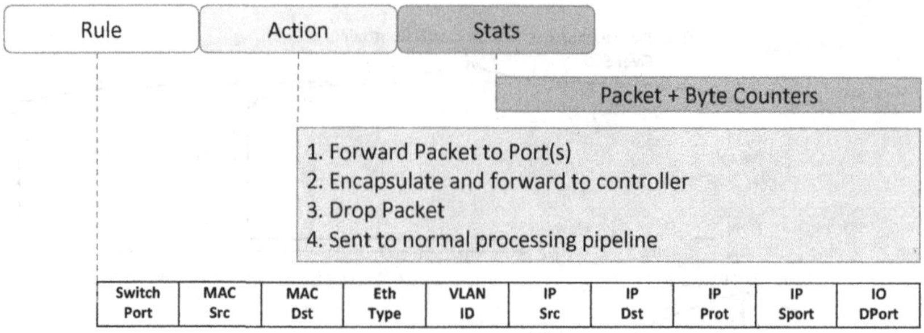

Fig. 8 Wire protocol and flow table entries for manipulation

very strong policy based routing or other advanced match/forwarding mechanisms, much more than in a classical distributed control environment.

The Action part defines forwarding of packets to specific ports, encapsulation and forwarding to controllers, dropping of packets or sending packets to the normal processing pipeline. Plus, there is the Modify Action, which could make the network element behave like a service appliance performing services like Network Address Translation (NAT), firewall, etc. (see Fig. 8).

How the Flow-Table Is Built

We have already identified that the flow table is the basic building block of the logical switch architecture in OpenFlow. This means that each packet entering a switch has to pass through one or more flow tables and each flow table is made of entries that consist of six components:

- **Match Field:** used to select the packets matching the values in the field
- **Priority:** this defines the relative priority of table entries
- **Counters:** they are updated for matching packets. Also, OpenFlow specifies multiple timers, including the number of received bytes and packets per port or per flow table, as well as per flow table entry, the number of dropped packets and the duration of a flow
- **Instructions:** these are the actions that should be taken in case the flow matches
- **Timeouts:** this defines the maximum amount of idle time before a flow is expired by a switch
- **Cookie:** this is an opaque data value chosen by the controller and can be used to filter flow statistics, flow modification or flow deletion, but is not used when processing packets

Flow tables can include a so-called table miss entry, where every field is a match and this has the lowest possible priority of 0.

The Match Fields component of a table entry consists of an ingress port (either physical or virtual), Ethernet source and destination address with the possibility of checking only some parts of the address or match any value. IPv4 or IPv6 protocol number indicating the next header in the packet, IPv4 or IPv6 source and destination address either an exact address, a subnet mask value, a bitmask value or a wildcard for any match. TCP source and destination ports again exact match or wildcard value, User Datagram Protocol (UDP) source and destination ports with again exact match or wildcard value. Any switch that is OpenFlow compliant must support the preceding match fields.

There are a number of fields, which are optional, ranging from the Physical Port designating the underlying physical port when a packet is received on a logical port to the Ethernet Type field, VLAN ID, protocol type and code fields like for ICMP, ARP, etc., IPv6 Flow Labels as well as Multiprotocol Label Switching (MPLS) Label Values and many more. This allows OpenFlow to be used with multiple types of network traffic covering a variety of protocols as well as network services.

Flow versus Path

Now we are able to define the term flow much more precisely. An individual switch sees a flow as a sequence of packets that match a specific flow table entry. Thus, the definition is packet oriented and, as such, a function of the values of header fields of the packets that constitute a flow as opposed to being a function of the path the packets follow through the network. A specific path is hence defined by and mapped onto a combination of flow entries on multiple switches.

Action and Action Set

In the instruction component of a table entry, we find the set of instructions that need to be executed for all packets matching that entry. We have Actions and Action Sets.

An **Action** describes the packet forwarding, packet modification as well as group table processing operations like forwarding to a specific output port. This can include setting the queue ID for a packet in order to determine which queue is attached to a specific output port, processing of packets through a specified group, push or pop tag fields for VLAN or MPLS packets, set field actions to modify the values of respective header fields in the packet, various change Time To Live (TTL) actions for IPv4 TTL or IPv6 Hop Limit and MPLS TTL.

The **Action Set** describes the list of actions that are associated and accumulated with a packet while the packet is processed by each table and executed as soon as the packet exits the processing pipeline. There exist four instruction types. These include how to direct a packet through the pipeline with the Goto-Table instruction, how to perform an action on a packet when it is matched to a table entry, how to update actions by merging specified actions into a current action set or clear all the actions in the action set for a specific packet and finally, how to update metadata where a metadata value can be associated with a packet in order to carry information from one table to the next.

OpenFlow Protocol

The OpenFlow protocol defines the messages that are exchanged between the OpenFlow controller and the OpenFlow switch and is implemented on top of SSL or TLS (Transport Layer Security), in order to guarantee a secure OpenFlow channel. The OpenFlow protocols consist of two parts, a wire protocol and a configuration management protocol.

Wire Protocol: The wire protocol is responsible for establishing a control session, defines the message structure for the exchange of flow modifications (flowmods), collects statistics and finally defines the basic structure of a switch with ports and tables.

Configuration Management Protocol: The configuration management protocol is based on NETCONF[42] and uses Yang data models to allocate physical switch ports to a specific controller, defines active and standby high availability and what to do in case of controller connection failure. OpenFlow can configure the basic operation of OpenFlow command and control, while booting or maintaining network elements is still left to future versions.

We already discussed that the OpenFlow protocol allows the controller to add, update or delete actions to the flow entries in the flow tables of the switch. There are three message types supported:

Controller-to-Switch messages are initiated by the controller and may require a response from the switch, allowing management of logical switch states by the controller. There is also the possibility of a packet-out message, which allows a switch to send packets to the controller in order to allow the controller to direct the packet to a switch output port.

Asynchronous messages are sent without any solicitation from the controller, like the status messages to the controller and the packet-in message, allowing the switch to send a packet to the controller in case there is no flow table match.

Symmetric messages are also sent without solicitation from controller or switch and include hello messages during connection establishment between controller and switch, echo request and reply messages for measuring latency or bandwidth of the controller switch connection as well as verifying device operation and, finally, experimenter messages which are reserved for future features in future OpenFlow versions.

OpenFlow Conclusions

The OpenFlow protocol enables the controller to manage the logical structure of a switch, but it does not cover the details on how the switch implements the OpenFlow logical architecture. This demonstrates one of the biggest benefits of OpenFlow SDN implementations by allowing a vendor neutral and independent approach for dynamic management of complex networks. It allows OpenFlow based SDNs to still use many already installed legacy network technologies like VLANs or MPLS infrastructures and become the foundation for large carrier and enterprise networks, cloud infrastructures and also networks supporting big data and analytics. Especially OpenDaylight (see the chapter on OpenDaylight) will demonstrate an Open Source approach that very well demonstrates the future of Open SDN.

[42]Enns E., et al., IETF, RFC 6241, NETCONF, Network Configuration Protocol, https://tools.ietf.org/html/rfc6241, 2011.

OpenDaylight (ODL)

The OpenDaylight (ODL)[43] project is a collaborative Open Source project hosted by the Linux Foundation. The main goal is to accelerate the adoption of SDN and create a foundation for Network Function Virtualization (NVF). All software is written in Java.

The OpenDaylight project was founded in April 2013 and the founding members were Arista Networks, Big Switch Networks, Brocade, Cisco, Citrix, Ericsson, HP, IBM, Juniper Networks, Microsoft, NEC, Nuage Networks, PLUMgrid, Red Hat and VMware, who pledged to donate software and engineering resources to the OpenDaylight open source framework in order to help support of an open source SDN platform for consumers, partners and developers.

The first code from the OpenDaylight project was named Hydrogen[44] and was released in February 2014, including an open controller, a virtual overlay network, protocol plug-ins as well as switch device enhancements.

Hydrogen was the first ODL code released in February 2014 and consisted of 15 projects that were separated into three different editions with the base edition being the ODL Controller (actually the SDN controller) (see Fig. 9).

Helium[45] was the second code release for ODL controllers and came out in September 2014. The main features were a new user interface as well as a simplified and customizable installation process based on the use of Apache Karaf containers and offered deeper integration with OpenStack based on improvements in the Open vSwitch database integration project as well as new features like Security Groups, Distributed Virtual Router and Load Balancing as a Service.

Lithium[46] was the third ODL controller software released in June 2015. With the third Open SDN release, the OpenDaylight platform broadened programmability of intelligent networks as well as support of virtualized and cloud environments. Major improvements were in the areas of security, automation, new and enhanced APIs and support of six new protocols and network services for cloud data center platforms. In 2015, Lithium demonstrated that the OpenDaylight project offered immense capabilities for building advanced SDN and NVF solutions. OpenDaylight had experienced tremendous growth since its inception, having already become the largest and most successful open source networking project with 466 people contributing over 2.3 million lines of code. This tremendous speed could very well be a great indicator for where the future of Open Source based Software Defined Networking might be headed.

---

[43]OpenDaylight, Linux Foundation, https://www.opendaylight.org, 2016, retrieved 2017-05-29.

[44]OpenDaylight, Linux Foundation, Hydrogen, https://www.opendaylight.org/software/downloads/hydrogen-base-10, 2014, retrieved 2017-05-29.

[45]OpenDaylight, Linux Foundation, Helium, https://www.opendaylight.org/software/downloads/helium, 2014, retrieved 2017-05-29.

[46]OpenDaylight, Linux Foundation, Lithium, https://www.opendaylight.org/software/downloads/lithium, 2015, retrieved 2017-05-29.

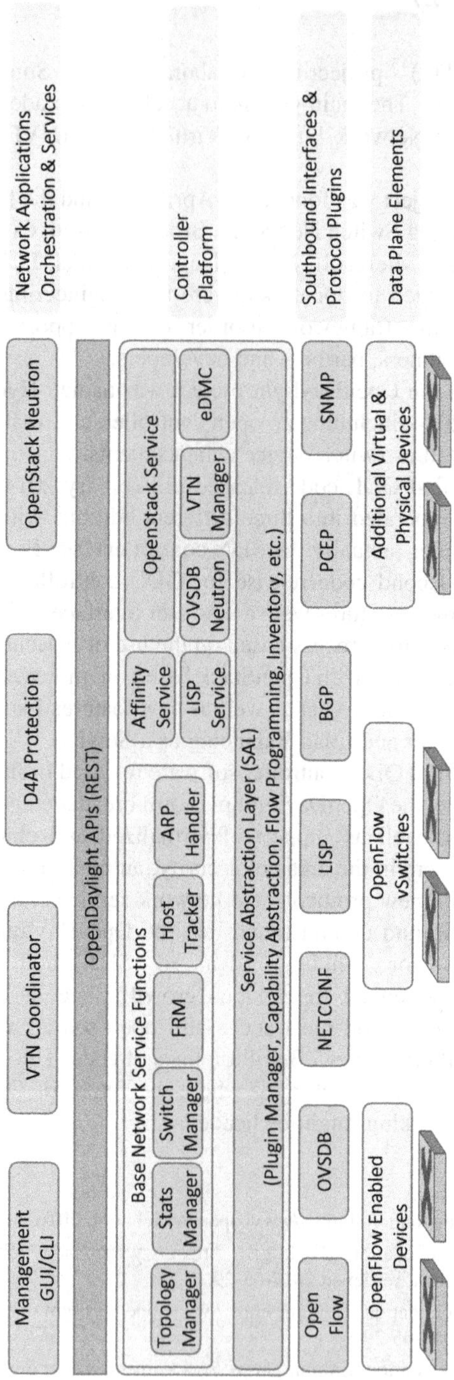

Fig. 9 ODL controller hydrogen

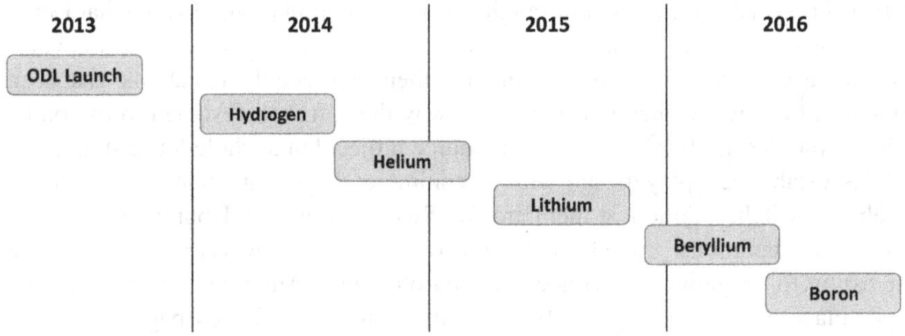

Fig. 10 ODL controller roadmap

The Beryllium[47] controller release followed in February 2016 and, in late 2016, the Boron[48] release showed that there is a solidly planned roadmap for ODL, which is a great indicator for the successful future of ODL based SDN (see Fig. 10).

Open Compute Project (OCP)

In 2011, the Open Compute Project[49] originated from Facebook with the intention of building and customizing design servers, software and data center components in general for its own infrastructure and openly share all these designs. Meanwhile, OCP has found many supporting and active members[50] like Goldman Sachs, Arista Networks, Rackspace and many more.[51] This can be seen by the illustrative list of Incubation Committee (IC)[52] members that includes a lot of leading people from the whole industry like Andy Bechtoldsheim who is also the chair of the IC.

Meanwhile, it became cult-like to follow and support OCP and Internet companies. Enterprises started monitoring their activities in order not to be left out. In 2015, Apple, Cisco, Juniper Networks and Nokia joined OCP, followed in 2016 by Google and Lenovo to mention just a few of the well-known players.

[47]OpenDaylight, Linux Foundation, Beryllium, https://www.opendaylight.org/software/downloads/beryllium, 2016, retrieved 2017-05-29.

[48]OpenDaylight, Linux Foundation, Boron, https://www.opendaylight.org/odlboron, 2016, retrieved 2017-05-29.

[49]OCP, Open Compute Project, http://www.opencompute.org, 2016, retrieved 2016-12-09.

[50]OCP, Organization and Board, http://www.opencompute.org/about/organization-and-board/, 2016, retrieved 2016-12-09.

[51]OCP, OpenComputer Foundation: Member Directory, http://www.opencompute.org/about/member-directory/, 2016, retrieved 2016-12-09.

[52]OCP, OCP Incubation Committee, http://www.opencompute.org/assets/assets/IC-Slide-for-the-website-as-of-16-April-16.pdf, 2016, retrieved 2016-12-09.

In a simplified way, you could say the OCP is to computer hardware what Linux was to software, which means it should become open source, which allows everybody to use the designs and also modify them. Especially Facebook has been leading in this area as they not only give away the software designed to run on its switch—the Wedge 100[53]—as an open-source project but a whole lot of startups as well as established players are selling commercial software that works on the Facebook switches. Amongst them are Big Switch Networks, Ubuntu and Apstra.

OCP has definitely gained momentum over the past few years and it will be interesting to see how these concepts based on "Open-Anything" will change the marketplace over the next years. For sure, the change has already begun.

SDN Market and Implementations

We have seen that the most important and disrupting characteristics of SDN are programmability, separation of control and data planes, and the centralized management of network state. The SDN controller is at the heart of the complete SDN concept, regardless of whether this is based on a proprietary or semi proprietary vendor implementation, or an available standard like OpenFlow, OpenDaylight. In this section, we will have a brief look at some of the most popular SDN controller services from vendors as well as open source solutions now.

Up to today, the most popular, commercially available products include VMware (vCloud/vSphere) and Nicira (NVP), Big Switch Networks (BCF), Juniper/Contrail, Cisco (ACI), Brocade, HP and Nuage, to mention only the most popular ones. Some of these commercial products provide both proprietary solutions as well as open source controllers.

The Ever-Expanding Functionality of Controllers

Expanding on the original task of data center orchestration including the management of compute, storage, virtual machines and finally network states in the data centers, more recent incarnations of SDN controllers allowed for the management of network abstraction in combination with the resource management in data centers with Open Source APIs like OpenStack[54] and CloudStack.[55] This evolution was driven by the expansion of SDN applications out of the data center network and into other network areas that do not require the tight coupling of virtual resources like processing and storage.

This also triggered the requirements for new network elements like switches, routers and bridges to get virtualized and be controlled by a hypervisor. This is

[53]Facebook, Facebook Code, Wedge 100, https://code.facebook.com/posts/1802489260027439/wedge-100-more-open-and-versatile-than-ever/, 2016, retrieved 2016-12-09.

[54]OpenStack, https://www.openstack.org, 2016, retrieved 2017-05-29.

[55]Apache CloudStack, CloudStack, Open Source Cloud Computing, https://cloudstack.apache.org, 2016, retrieved 2017-05-29.

known as network services virtualization or, more popularly called, Network Functions Virtualization (NFV),[56] (Stallings, Foundations of Modern Networking – SDN, NFV, QoE, IoT, and Cloud, 2015) which is adding an ever increasing number of such elements to future generation network architectures, all requiring even more controllers to operate the things building these architectures.

Virtual switches and virtual routers are the most apparent elements and lowest common denominators for networking and have certain advantages as well as disadvantages compared to their hardware based brothers. This is due to the often much simpler hypervisor based implementation compared to purpose built routers and switches.

We will now look in more detail at three implementations of VMware, Cisco and Big Switch Networks. This choice out of the myriad of available SDN implementations in 2016 was made because these three examples cover a wide range of how SDNs can be realized based on what we have discussed so far.

VMware and Nicira

VMware[57] was founded in 1998 and is very well known for its data center centric application suite based on a hypervisor ESX, called ESXi[58] in a later release, and a hypervisor switch called vSphere Distributed Switch (VDS).[59] This data center orchestration solution was based on a proprietary SDN controller and agent and became a de facto standard during the 2010s. The ESXi hypervisor used in vSphere provides a web interface for management in addition to command line language (CLI) management options as well as a client API and vCenter visualization.

VDS is the abstraction of a single, logical switch and represents what was originally a collection of individual virtual switches, the so-called vSphere Standard Switches (VSSs).[60] This allows vCenter to act as the management and control point for all VDS instances and separates management and data planes of individual VSSs. VMware allows for abstraction of physical cards within the VDS, load balancing, failover as well as traffic shaping, VLAN assignments and security. The operator can reuse all these abstractions as configuration templates, in order to further ease and speed up deployment. Once provisioned, the single components

[56]ETSI, Network Functions Virtualization (NVF), 2016, retrieved 2017-05-29.

[57]VMware, http://www.vmware.com, retrieved 2017-05-29.

[58]VMware, ESXi, Purpose-Built Bare-Metal Hypervisor, http://www.vmware.com/products/esxi-and-esx.html, retrieved 2017-05-29.

[59]VMware, vSphere Distributed Switch, http://www.vmware.com/content/dam/digitalmarketing/vmware/en/pdf/techpaper/vsphere-distributed-switch-best-practices-white-paper.pdf, 2012, retrieved 2017-05-29.

[60]VMware, vSphere Standards Switches, https://pubs.vmware.com/vsphere-60/index.jsp?topic=%2Fcom.vmware.vsphere.networking.doc%2FGUID-350344DE-483A-42ED-B0E2-C811EE927D59.html, retrieved 2017-05-29.

that form the network operation (called the ESXi vswitch) can continue operation even if the controller, the vCenter Server[61] fails or is partitioned from the network. High Availability (HA) is achieved by a master slave approach, where a single agent is elected as master of a fault domain and all the other agents are slaves, creating a highly scalable VM health monitoring system.

Nicira Network Virtualization Platform (NVP)

Nicira, founded in 2007, released its Network Virtualization Platform (NVP)[62] in 2011 and, in contrast to VMware, it is more of a classic network controller with the network being the resource managed. VMware acquired Nicira[63] in 2012 for $1.26 billion. By the beginning of the 2010s, cloud computing had disrupted VMware as many other legacy technology companies and this acquisition helped to alleviate the pressure that VMware faced in building a comprehensive strategy to compete with Amazon Web Services (AWS)[64] and OpenStack.[65] This also put VMware suddenly in direct competition with Cisco, although these two companies have had a deep relationship in some areas in the past.

The interesting fact is that Nicira was a significant contributor to OpenStack, as the network interface for OpenStack uses the Nicira API. This suddenly made VMware part of the OpenStack community and, as such, VMware now became part of a longer-term play for the networking market. Needless to say, this acquisition shook up the market, especially competitors like Cisco who were now forced to introduce their own open SDN solution, the Application Centric Infrastructure (ACI),[66] which will be discussed later.

The clear goal for VMware was that they had understood the need for software defined networking and orchestration for future clouds hosting big data as well as service environments and the need to flexibly program the network to fit these needs.

VMware SDN Market Share by 2016

By 2016, VMware was the revenue market leader for SDN Overlays based on the 2012 Nicira acquisition and their NSX platform with more than 1200 paying NSX customers. The secret is that VMware brings computer science techniques of hardware abstraction that were successfully deployed on servers to the network,

---

[61]VMware, vCenter Server, https://www.vmware.com/products/vcenter-server.html, 2016, retrieved 2017-05-29.

[62]OpenStack, VMware/Nicira NVP Deep Dive, https://www.openstack.org/summit/portland-2013/session-videos/presentation/vmware-nicira-nvp-deep-dive, retrieved 2017-05-29.

[63]VMware, VMware and Nicira, http://www.vmware.com/company/acquisitions/nicira.html, 2012, retrieved 2017-05-29.

[64]Amazon, Amazon Web Services (AWS), https://aws.amazon.com, 2016, retrieved 2017-05-29.

[65]OpenStack, https://www.openstack.org, 2016, retrieved 2017-05-29.

[66]Cisco, Cisco Application Centric Infrastructure (ACI), http://www.cisco.com/c/en/us/solutions/data-center-virtualization/application-centric-infrastructure/index.html, 2016, retrieved 2017-05-29.

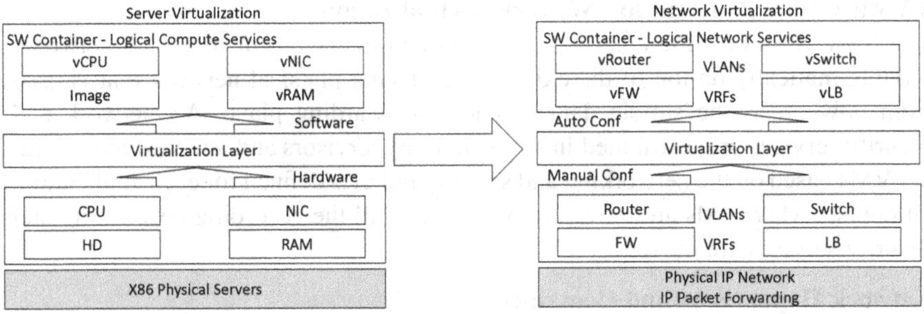

Fig. 11 VMware—from server virtualization to network virtualization

which allows them to create a fully abstracted network residing on top of a physical infrastructure as we will see later in this section.

They have the opposite issue as Cisco since the physical networks in the data center still remain important, where legacy network vendors like Cisco have dominated in the past.

From Server Virtualization to Network Virtualization

Soon after VMware acquired Nicira they came up with their NSX[67] platform that introduced the network hypervisor as the functional equivalent to a server hypervisor, already well known from their server virtualization products. We remember that for server virtualization a software abstraction layer is introduced in order to reproduce familiar attributes of x86 physical servers like CPU, RAM, Disk or NIC (Network Interface Card) in software allowing then to be programmatically assembled in any arbitrary combination to produce unique VMs in seconds. The same concept sits behind the network hypervisor that reproduces the complete set of Layer-2 to Layer-7 networking services like switching, routing, access control, firewalls, QoS and load balancing, but now in software.

This allows for a similar programmatic assembly in any arbitrary combination like we already know from server virtualization, but this time to produce virtual networks in seconds. Similarly, as VMs are independent from the underlying x86 platform, virtual networks are now independent from the underlying IP network infrastructure, allowing the treatment of the network as a simple pool of transport capacity consumable on demand (see Fig. 11). This network virtualization also opens the path to Software Defined Data Center (SDDC),[68,69] which we will discuss later.

[67]VMware, VMware NSX, Network Virtualization Platform for Software Defined Data Center, http://www.vmware.com/products/nsx.html, 2016, retrieved 2017-05-29.

[68]VMware, Software Defined Data Center (SDDC), http://www.vmware.com/solutions/software-defined-datacenter/in-depth.html, 2016, retrieved 2017-05-29.

[69]VMware, Technical White Paper SDDC, http://www.vmware.com/content/dam/digitalmarketing/vmware/en/pdf/techpaper/technical-whitepaper-sddc-capabilities-itoutcomes-white-paper.pdf, 2015, retrieved 2017-05-29.

VMware NSX Platform for Network Virtualization

The NSX allows for the creation, provision and management of virtual networks programmatically on top of the existing underlying physical network that is used from now on as the simple data or packet-forwarding plane. All network and security services are distributed in software to hypervisors and attached to individual VMs based on the networking and security policies defined for each application. Of course, when VMs are moved to other hosts, all the according networking and security services are moved as well.

Network Hypervisors and Containers

NSX defines a multi-hypervisor solution, leveraging the already well-known server hypervisors present in data centers. NSX has the task of coordinating the vSwitches and network services that are pushed to them from connected VMs, in order to finally build the network hypervisor for the creation of virtual networks. The idea of software containers presenting logical compute services to applications as used for virtual machines is very similar to the concept that a virtual network is now a software container presenting logical network services[70] like virtual switches, virtual routers, virtual firewalls, virtual load balancers, virtual private networks (VPNs), etc. to connected workloads. All these network and security services are delivered in software and leave only the IP packet forwarding to the underlying physical network.

The NSX controller is connected via a RESTful API[71] to a cloud management platform (CMP)[72] in order to receive the virtual network and security service requests for a specific workload, before it distributes the necessary services to the according switches and finally finishes setting up the logical connection between the workloads. This allows combining different virtual networks with different workloads on the same hypervisor as well as the creation from virtual networks between two nodes to advanced complex network topologies for multi-tier applications.

VMware Conclusion

By 2013, the NSX has been deployed in full production by a number of large cloud service providers as well as global enterprise and financial data centers worldwide, the most prominent customers being AT&T, NTT, Rackspace, eBay or PayPal using this elegant solution. The beauty and success of the NSX approach lies in the usage of well-known and widely accepted virtualization strategies for server virtualization that are now easily and effectively extended to the virtual network

---

[70]VMware, Technical White Paper, The VMware NSX Network Virtualization Platform, http://www.vmware.com/content/dam/digitalmarketing/vmware/en/pdf/whitepaper/products/nsx/vmware-nsx-network-virtualization-platform-white-paper.pdf, 2013, retrieved 2017-05-29.

[71]VMware, Network Virtualization Blog, Automation Leveraging NSX REST API Guide, https://blogs.vmware.com/networkvirtualization/2016/06/new-automation-leveraging-nsx-rest-api-guide.html#.WD7ia3eZNXI, 2016, retrieved 2017-05-29.

[72]VMware, Cloud Management Platform (CNP), http://www.vmware.com/solutions/virtualization/cloud-management.html, 2016, retrieved 2017-05-29.

as well. As soon as customers have a VMware server virtualization deployed, this extension to virtualize their network with whatever underlying hardware becomes, obviously, very attractive.

The NSX solution easily allows for data center automation, with significant configuration simplification for VLANs, firewalls, etc., isolated development of test and production environments on the same physical infrastructure, data center enhancement through distributed network and security services where administration is still centralized and easy provisioning of multi-tenant clouds. Given all these benefits, the Software Defined Data Center (SDDC) became a reality.

Cisco ACI: Application Centric Infrastructure

Cisco[73] introduced what they call Application Centric Infrastructure (ACI)[74] as their SDN solution in 2014 and they consider ACI just another methodology of SDN, like OpenFlow or Network Virtualization. Fair enough to say that numerous features make ACI a perfect fit on top of a legacy grown infrastructure like Cisco's routers and switches, but, of course, it also provides a pretty advanced and future proof feature set for network programmability, which allows making networks work in a more modern way.

Cisco ACI Market Share by 2016

In late 2015, ACI hit the 1000 customers mark with more than 5000 Nexus switches installed, the switching platform that is necessary for the ACI solution. One of the major differentiators of the ACI SDN solution is based on custom ASICs that gather application traffic flow data that allow the building of some interesting SDN applications around analytics of applications and network traffic. Nevertheless, Cisco at the same time faces the threat from alternative network solutions based on white-boxes and bare–metal switches often based on open source code. See the section on Big Switch Networks. IT and Dev Ops teams are adopting these new, open environments with remarkable speed that is cutting down the legacy Cisco market, so it will be interesting to see how well they can compete with ACI.

Drivers for ACI

The argument for why ACI is necessary is pretty similar to what we have already seen as the driving points from the rest of the industry, but it is interesting to see how the biggest incumbent networking gear vendor argues this trend. Due to Cisco, there are a number of major technical shifts happening today leading to why SDN and ACI became necessary.

[73]Cisco, Cisco Systems, http://www.cisco.com, 2016, retrieved 2017-05-29.

[74]Cisco, Application Centric Infrastructure, http://www.cisco.com/c/en/us/solutions/data-center-virtualization/application-centric-infrastructure/index.html, 2016, retrieved 2017-05-29.

1. Most importantly, compute power increased constantly over the past years and meanwhile this increase accelerated in even shorter time intervals, requiring infrastructures supporting much higher bandwidth. New server platforms became available enabling higher I/O throughput that increases server access speed from 1 Gbps to 10 Gbps, which goes hand in hand with multi-core architectures like 12 cores per socket. This means a huge amount of data that needs to be handled since this data has to get in and out of the servers and as soon as server performance and I/O speeds grow to 10 Gbps, also the infrastructure needs to support these 10 Gbps access speeds in the access switches, while the aggregation switches needing to run at 40 Gbps at least.
2. Looking at the virtualization landscape, this is very different from what it was originally considered. Around 2010, virtualization was already around 70% and many industries were looking forward to a 100% virtualization by 2013. The fact is that even in 2016 the majority of environments are still around over 70% virtualized. There are many reasons for this including traditional applications not working very well with virtualization, and modern applications for this purpose just being under design and able to span across servers in a bare metal fashion.
3. There are new architecture requirements resulting from new applications like big data that require increasingly east west traffic between virtual machines. Most old infrastructures were designed for north south traffic in a typical 3-tier design consisting of core, aggregation and access, with the core in the north, connecting the campus, the WAN and the users, while south being the servers. All network, hardware and applications were designed for this architecture.

From 3-Tier North-South to 2-Tier East-West Leaf Spine Designs

Today traffic patterns are primarily east west, which requires a huge change in architecture and generally this can be best supported by leaf-spine designs. Furthermore, all hardware for the old north south architectures was designed for oversubscription, toning down speed when hitting the core because WAN links or campus links were slower than the server links, so the hardware had to be oversubscribed to save cost (see Fig. 12).

The two basic assumptions were to aggregate down to some link that is slower at some point and that not every port on a switch is lit up by a server at the same time at full rate. These assumptions are no longer valid in today's data centers, as big data applications, high performance computing, etc. will light up every switch port at the same time. There is the need to support the actual bandwidth in the back of the switch in order to support what is offered up to the demanding modern applications.

Thus, topologies today have moved from 3-tier north south architectures to spine leaf designs, which are 2-tier architectures optimized for east west traffic,[75] based on non-blocking, non-oversubscribed hardware, which are also the best design

[75]Cisco, White Paper, Cisco Data Center Spine-and-Leaf Architecture; Design Overview White Paper, http://www.cisco.com/c/en/us/products/collateral/switches/nexus-7000-series-switches/white-paper-c11-737022.html, 2016, retrieved 2017-05-29.

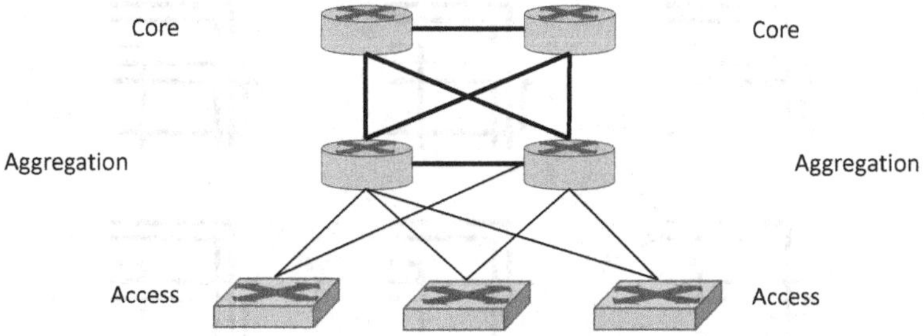

Fig. 12 Classic 3-Tier access-aggregation-core datacenter topology

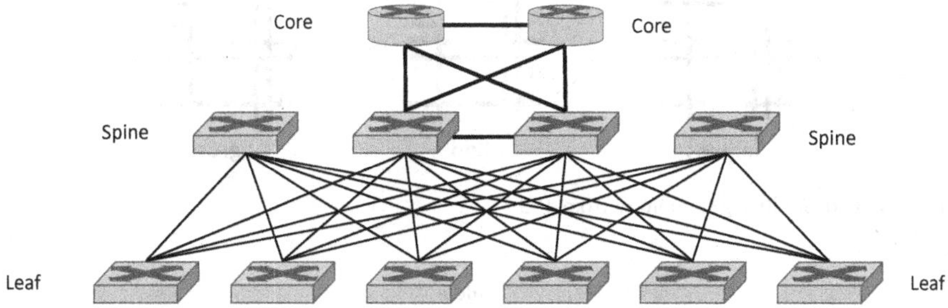

Fig. 13 2-Tier leaf-spine datacenter topology

practices required by all software and virtualization vendors selling SDN solutions (see Fig. 13).

Leaf spine designs are often mentioned to be Clos topology. This topology is named after Charles Clos, who, in 1953, solved a very similar problem in the telecommunications industry. He invented a method of creating a multistage network based on the so-called Clos topology (see Fig. 14) that can grow beyond the largest switch in the network.

The Clos topology offers a non-blocking design allowing predictable performance as well as scaling characteristics. Actually, what we find today, often referred to as leaf-spine topologies or designs, is identical to three-stage Clos topologies. It can also be referred to as folded three-stage Clos topology because the ingress and egress points are folded back on top of each other with the spine switches being simple layer 3 switches, whereas the leaves are the top-of-rack (TOR) switches that provide connectivity to the servers (see Fig. 15).

Application and Networking Silos

Next, what happened over the years until 2015 was that applications and networks usually developed very much. While the application people usually talk application language like application tier policy and dependencies, security requirements, service level agreements, application performance, compliance, geo-dependencies and tenants; the networking people usually talk networking language like VLAN, IP addresses, subnets, firewalls, quality of service, load balancers and access lists. This means we had to deal with two different languages and it is often hard for an

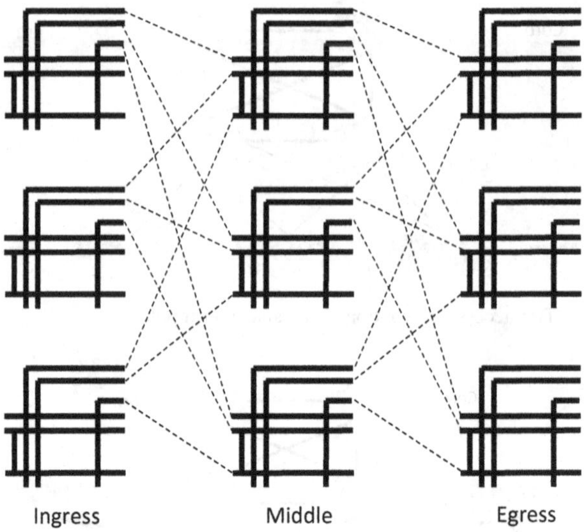

Fig. 14 Clos multistage topology from 1953

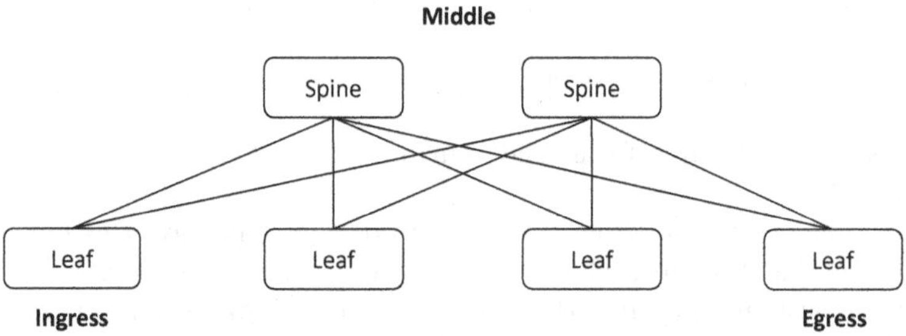

Fig. 15 Leaf-spine topology

application designer to understand networking terms, as it is at the same time hard for a networking architect to understand application language terms.

What ACI helps to do is an automatic translation between these two silos. We can think of an example of Google translating between English and German. As we all know, the results from Google translate are far from being perfect, but, if two people are really willing to understand, they will be able to do so even if the results are only 70% correct. The same is true for ACI and the translation of application requirements to the infrastructure requirements and vice versa (see Fig. 16).

The main goal of ACI is to offer a common operational model for simplified operation, better performance and scalability, overcoming the drawbacks of the traditional silo approach with separate operational models for applications, networks, security and cloud teams. This will guarantee the application agility required by modern businesses and allow for the necessary quick and seamless changes. ACI

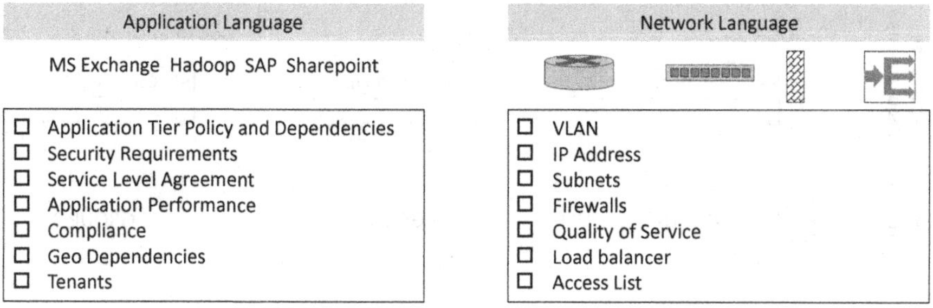

Application Language
MS Exchange Hadoop SAP Sharepoint
Application Tier Policy and Dependencies
Security Requirements
Service Level Agreement
Application Performance
Compliance
Geo Dependencies
Tenants
Network Language
VLAN
IP Address
Subnets
Firewalls
Quality of Service
Load balancer
Access List

Fig. 16 Applications vs. network—two languages

resides in the data center and is built with centralized automation and policy-driven application profiles[76] and is positioned to offer the flexibility of software with the scalability of hardware performance.

Advantages of ACI Policy Model: Endpoint Groups

ACI uses a simple policy model as groups of things connect to other groups of things while the necessary policy is applied. This leaves the designer to identify the groups of things that should connect, which could be routed ports, servers, virtual machines, containers etc., tell the system which other groups they should connect to, and what is the policy in between.

This results in an application network profile that is a logical description of the end-to-end policy requirements for an application.[77] Each of the elements of an application network profile are re-useable objects, which means you can build a policy contract that can be redeployed as soon as a similar application is used. Policy in this model is very flexible, it can be any to any connectivity in a very simple way or it can be very strict like service chains connecting groups to load balancers, firewalls etc., in essence this takes groups of things and maps them to applications adding a lot of flexibility for different application needs by profiles (see Fig. 17).

Group Policy Model of ACI and Flexibility of Profiles

ACI allows forming groups for different policies[78] like different security zones, application development, VLANs or when different profiles per application are needed and easily scales up to several thousands of profiles.

---

[76] Cisco, Cisco ACI Fundamentals, Chapter: ACI Policy Model, http://www.cisco.com/c/en/us/td/docs/switches/datacenter/aci/apic/sw/1-x/aci-fundamentals/b_ACI-Fundamentals/b_ACI-Fundamentals_chapter_010001.html, 2016, retrieved 2017-05-29.

[77] Cisco, Cisco ACI, Endpoint Groups (EPG) Usage and Design, http://www.cisco.com/c/en/us/solutions/collateral/data-center-virtualization/application-centric-infrastructure/white-paper-c11-731630.html, 2016, retrieved 2017-05-29.

[78] Cisco, Cisco ACI Fundamentals, Chapter: ACI Policy Model, http://www.cisco.com/c/en/us/td/docs/switches/datacenter/aci/apic/sw/1-x/aci-fundamentals/b_ACI-Fundamentals/b_ACI-Fundamentals_chapter_010001.html, 2016, retrieved 2017-05-29.

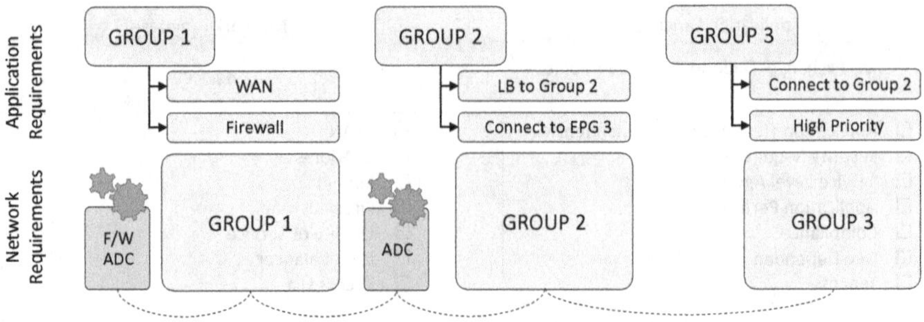

Fig. 17 How ACI understands and speaks application needs

In case of different security zones like a DMZ for untrusted traffic and a production network separated by a firewall allows sharing of services like DHCP, DNS or LDAP between the zones by using groupings to map the security zones.

In case of application development, groups can be used to help transit through the stages like developing applications, testing applications and finally bringing the applications to production. What this means is that applications can be easily tested on a production-like infrastructure or on an infrastructure that is already very close to the final production network.

In case of VLANs that are often used as policy boundaries with a subnet per VLAN, policies can be easily implemented between VLANs by using the layer 3 boundaries and mapping the VLANs into the group model. Mapping VLANs into groups and routing traffic to firewalls or controling elements enables the definition of what is allowed between the groups. This makes it easy to limit traffic from specific applications to certain groups and forbid traffic that is not allowed by certain policies.

Finally, profiles can be used per application like in a three-tier web app that consists of web-host, application and database in order to get visibility on a case by case basis, allowing the application to program the network through best practices coming from the application vendor. This also allows taking care of usually more complicated applications that need to contain additional elements like DHCP, DNS or monitoring services to support specific ecosystems. The usage of the group model brings a clear advantage over conventional CLI or GUI programming for easier and faster application deployment (see Fig. 18).

Application Policy Infrastructure Controller

The Application Policy Infrastructure Controller (APIC)[79] is the unifying automation and management point for the Application Centric Infrastructure (ACI) data center fabric. The APIC provides centralized access to all fabric information, handles and optimizes the security and application lifecycle for scale and performance and supports flexible infrastructure provisioning across physical containers

[79]Cisco, Application Policy Infrastructure Controller (APIC), http://www.cisco.com/c/en/us/products/cloud-systems-management/application-policy-infrastructure-controller-apic/index.html, 2016, retrieved 2017-05-29.

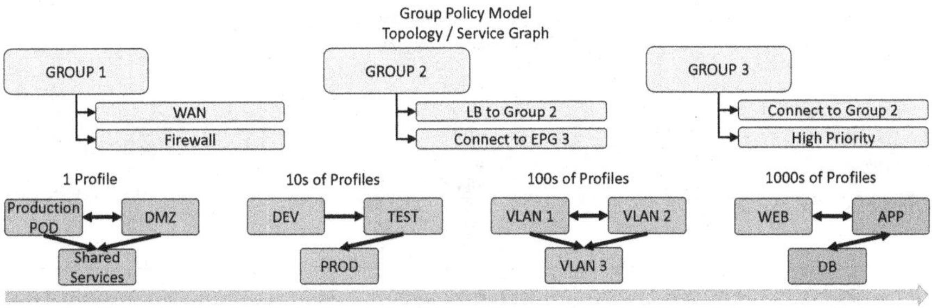

Fig. 18 Flexibility of profiles

and virtual resources. APIC forms a common policy platform for physical, virtual and cloud computing. It enables building and enforcing application-centric network policies, is based on open standards framework and supports southbound as well as northbound APIs. It further allows the integration of third-party layer 4–7 services as well as virtualization and management and offers scalable security and multitenant environments.

APIC uses Cisco OpFlex as a southbound protocol that allows enabling policies to be applied across physical and virtual switches.

OpFlex, OpenFlow and Group Policy API

The southbound protocol used by APIC is OpFlex[80] that is pushed by Cisco as the protocol for policy enablement across physical and virtual switches and is standardized by the IETF.[81] It is important not to confuse OpFlex with OpenFlow,[82,83] that was already discussed, but for the sake of simplicity here are again the most important differences.

OpenFlow is one of the first and most widely deployed communication protocols used for SDN. The main goal of OpenFlow is to ensure consistent policy enforcement across the underlying infrastructure looking to centralize all functions on the SDN controller, while OpFlex centralizes policies allowing the controller to offer greater resiliency, availability and scalability, by moving some of the intelligence to hardware devices that use their legacy network protocols. This approach is, of course, typical for a legacy networking gear vendor like Cisco as it supports the argument to stay longer with a network vendor to lock in and sell more proprietary

[80]Cisco, White Paper, OpFlex: An Open Policy Protocol White Paper, http://www.cisco.com/c/en/us/solutions/collateral/data-center-virtualization/application-centric-infrastructure/white-paper-c11-731302.html, 2014, retrieved 2016-12-02.

[81]Smith M., et al., IETF, Internet Draft, OpFlex Control Protocol, https://tools.ietf.org/html/draft-smith-opflex-00, 2014, retrieved 2016-12-02.

[82]ONF, OpenFlow, https://www.opennetworking.org/sdn-resources/openflow, 2016, retrieved 2016-12-02.

[83]ONF, ONF SDN Evolution, https://www.opennetworking.org/images/stories/downloads/sdn-resources/technical-reports/TR-535_ONF_SDN_Evolution.pdf, 2016, retrieved 2016-12-02.

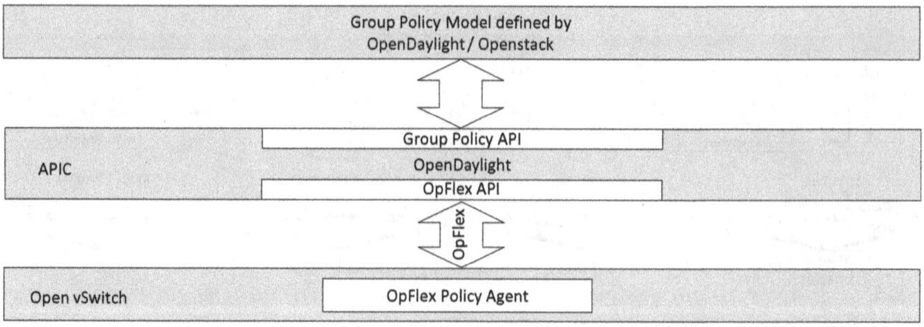

Fig. 19 APIC and protocols

and usually more expensive hardware compared to a fully open SDN approach that totally relies on open and generally cheaper white box hardware instead.

To make this work, an OpFlex agent must be embedded in the switches and routers to support the Cisco OpFlex protocol. As a result, Cisco is working on an open source OpFlex agent that can be used across platforms. Microsoft, IBM, F5, Citrix, Red Hat, Canonical, Embrane and AVI Networks have already committed to embed this agent in their solutions (see Fig. 19).

The northbound APIs of APIC is optimized for integration with existing orchestration and management frameworks and is compatible with OpenStack.[84] This is important as it allows for the full integration with the open cloud operating system developed by OpenStack that should allow for seamless control of networking, storage and compute resources in different environments via the group policy ACI. Vice versa, these ACI policies are communicated southbound via OpFlex to the networking and virtualization elements.

Big Switch Networks

Big Switch Networks[85] has its roots in a Stanford research team that had pioneered SDN and was founded in 2010. In 2013, they started to package technology components into bare metal SDN fabric solutions that allows for the combination of SDN software with the very competitively priced bare metal switches or also known as white box hardware. This allows building advanced global datacenters with a high degree of network automation at very competitive price points. At the

[84]OpenStack, Open Source Software for Creating Private and Public Clouds, https://www.openstack.org, 2016, retrieved 2016-12-02.

[85]Big Switch Networks, http://www.bigswitch.com, 2016, retrieved 2016-12-12.

heart of this implementation is the Big Cloud Fabric (BCF),[86] which is in fact a Leaf-Spine Clos Fabric and optimally suited for VMware and OpenStack based data centers.

The Hyperscale Data Center

Hyperscale networking is a new design philosophy for data centers that promises significant CapEx reduction through the use of bare metal switches, improvements in automation as well as operational simplification based on latest SDN evolutions, high resiliency and availability and easily replicable building blocks called "PODs."[87] Big Switch Networks is pioneering this new design approach with the Big Cloud Fabric (BCF),[88,89] solution that is using these new design principles for building hyperscale data centers and will be discussed in more detail throughout this section.

Bare Metal Hardware (Switches) for Cost Reduction

We speak of bare metal switches[90] when the hardware is operating system independent and the applications software running on this hardware. Bare metal hardware is either available from branded vendors like Dell,[91] HP[92] or Lenovo[93] or white-box vendors like Quanta Computers,[94] Super Micro, etc. This separation of software (OS and applications) and hardware is paving the way for tremendous innovations in all separate areas like CPU, hardware, OS and applications.

We have already discussed the big obstacles of the networking industry, where networking software, hardware and ASICs were offered by vertically integrated vendors that resulted not only in high prices but even worse, in an usually slow pace of innovation. Since merchant silicon from companies like Intel, Marvell or

---

[86]Big Switch Networks, Hyperscale Networking for All with Big Cloud Fabric, http://go.bigswitch.com/rs/974-WXR-561/images/BigSwitch_BigCloudFabric_WP_FINAL.pdf, 2014, retrieved 2106-12-02.

[87]Big Switch Networks, Core and Pod Data Center Design, http://go.bigswitch.com/rs/974-WXR-561/images/Core-and-Pod%20Overview.pdf, 2016, retrieved 2016-12-21.

[88]Big Switch Networks, Big Cloud Fabric, http://www.bigswitch.com/products/big-cloud-fabrictm/big-cloud-fabric, 2016, retrieved 2016-12-02.

[89]Big Switch Networks, Secure and Resilient SDN with Big Cloud Fabric, http://go.bigswitch.com/rs/974-WXR-561/images/BCF-White-Paper-Secure%20and%20Resilient%20SDN-2.pdf?_ga=1.81813843.492494998.1463156306, 2016, retrieved 2016-12-02.

[90]Big Switch Networks, Tutorial: White Box/Bare Metal Switches, https://www.bigswitch.com/sites/default/files/presentations/onug-baremetal-2014-final.pdf, 2014, retrieved 2016-12-02.

[91]Dell, Network Devices, http://www.dell.com/us/business/p/networking-products?stp_redir=false&~ck=mn, 2016, retrieved 2016-12-02.

[92]HP, HP Enterprise, https://www.hpe.com/us/en/products.html, 2016, retrieved 2016-12-02.

[93]Lenovo, Lenovo XClarity, https://www.lenovo.com/images/products/system-x/pdfs/datasheets/xclarity_ds.pdf, 2016, retrieved 2016-12-02.

[94]Quanta Computers, Server, http://www.quantatw.com/Quanta/english/product/qci_es.aspx, 2016, retrieved 2016-12-02.

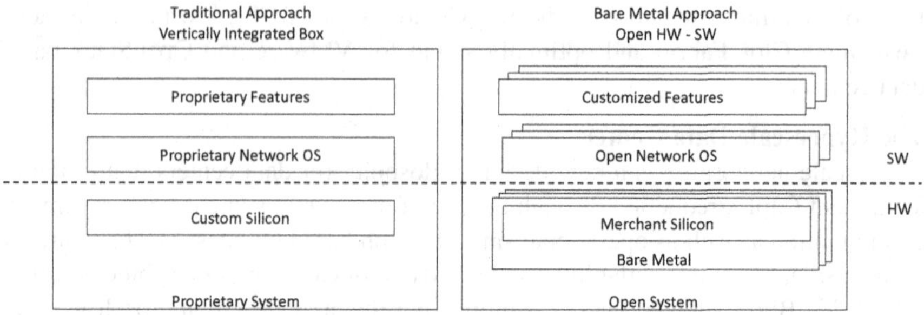

Fig. 20 Proprietary system versus open system

Broadcom[95] became available throughout the last decade, this disaggregation started to grow in the area of networking hardware and ASICs and meanwhile merchant silicon based Ethernet switches are gaining increasingly market share over the previous proprietary solutions. This disaggregation based on bare metal switching hardware and in house developed software stacks that are often open source based are poised for dramatic cost reductions, especially when deployed in large data centers (see Fig. 20).

The Bare Metal Approach Disruption
This disruption happening through the bare metal approach was significantly supported by initiatives like the Open Compute Networking Project[96] as well as Open Network Linux (ONL),[97] raising further awareness and expanding the acceptance of the bare metal switching approach. It allows datacenter operators to build their infrastructures from a wide range of suppliers, both the Top-of-Rack (TOR) and Spine Switches. This allows for never before known flexibility and free choice of vendors and healthy competition, while at the same time drastically reducing CapEx and speeding up innovation, because each layer of the stack can evolve now independently of all the other layers.

Big Switch Networks Bare Metal SDN Implementation
Big Switch Networks bare metal SDN fabric's share the same codebase and operational model across their physical hardware platforms as well as hypervisors, allowing the use of open source as well as commercial parts in combination. As such they are either running an Open Network Linux (ONL) SDN operating system called Switch Light OS, that is based on an open source effort within the Open

---

[95]Broadcom, 100-Gigabit Ethernet (GbE) Capable Switching Silicon, https://www.broadcom.com/products/Switching/Carrier-and-Service-Provider/BCM88600-Series, 2016, retrieved 2016-12-02.

[96]Opencompute, Open Compute Projects, Networking, http://www.opencompute.org/projects/networking/, 2016, retrieved 2016-12-02.

[97]Opennetlinux, Open Network Linux, https://opennetlinux.org, 2016, retrieved 2016-12-02.

Compute Project on the bare metal switches and Switch Light VX,[98] that is a user space software agent for KVM[99] based virtual switches for their hypervisors.

Switch Light OS that is Linux based allows consistent data plane programming on merchant silicon based physical switches as well as hypervisor vSwitches. Physical platforms can be chosen from a wide variety including high performance switching silicon from Broadcom, but also a number of Linux distributions utilizing KVM hypervisor on Intel x86 platforms.

The Switch Light software uses a common interface for programming layer 2 and layer 3 data plane forwarding rules and policies on both bare metal switches as well as virtual switches from the centralized SDN controller. The Big Switch OpenFlow agent with open extensions on the switches enables programming rules and policies to the forwarding tables in the physical switches, or in case of virtual switches on the Open vSwitch forwarding tables.

OS Based on Open Source components ONIE and ONL

Switch Light OS is based on the Open Network Linux (ONL) distribution that is available from the Open Compute Project for bare metal Ethernet switch hardware, allowing separation of software and hardware procurement. ONL uses the Open Network Install Environment (ONIE)[100] for installations onto on board switch flash memory. The standard ONL distribution includes the Debian[101] Linux kernel, installation scripts, device drivers as well as networking bootloader allowing network boot functionality for many available bare metal switches.

This implementation based on emerging industry standards like Open Network Install Environment (ONIE) and ONL for bare metal switch hardware is a perfect example for what is already achievable in 2016 with open source based components. This offers the possibilities to either run traditional networking operating systems from vendors, Switch Light OS or hardware diagnostic tools as well, guaranteeing simple and low risk migration paths to SDN.

Core and Pod Designs for Faster Innovation

We already discussed that the traditional datacenter design approach was based on North-South traffic and a tree shaped network topology with a core, aggregation and access switching and routing layers (see Fig. 21). This design requires well-known protocols like Spanning Tree in order to avoid loops in the tree topology and in addition link aggregation technologies often based on proprietary solutions like[102]

---

[98]Big Switch Networks, Switch Light, http://www.bigswitch.com/products/switch-light, 2016, retrieved 2016-12-02.

[99]Linux-kvm, Kernel Virtual Machine (KVM), http://www.linux-kvm.org/page/Main_Page, 2016, retrieved 2016-12-02.

[100]ONIE, Open Network Install Environment, http://www.onie.org, 2016, retrieved 2012-12-02.

[101]Debian, The Universal Operating System, https://www.debian.org, 2016, retrieved 2016-12-02.

[102]IETF, RFC 6326, Transparent Interconnection of Lots of Links (TRILL) Use of IS-IS, https://tools.ietf.org/html/rfc6326, 2011, retrieved 2016-12-02.

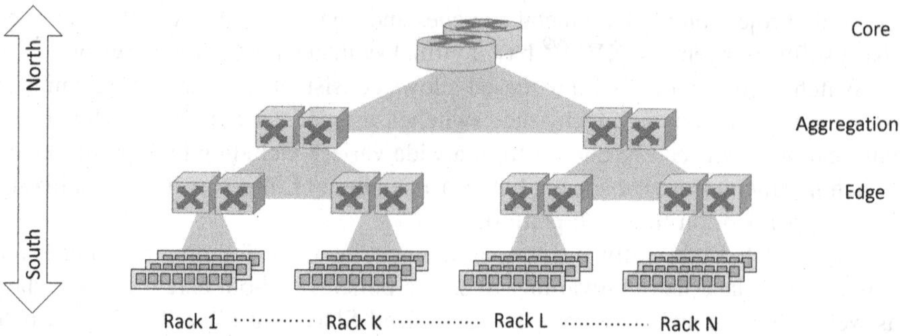

Fig. 21 Traditional 3-tier datacenter design

or Fabric Path[103] in order to overcome various Spanning Tree protocol shortcomings.

But there are many more shortcomings and problems present in modern data center requirements. One of the most serious is that, as already discussed previous in this section, the original tree shaped design approach is no longer optimal for East-West traffic between the server nodes within the data center that contributes the most dominant portion of modern data center traffic. To make it even worse, the classic north-south designs cannot hold up for the modern requirements of scale-out data centers that require infrastructures to be added and removed fast, allow zero impact upgrades and efficient failure management. Failures in a single zone using a traditional data center design can easily impact adjacent zones and in worst case the whole data center.

Another big issue of traditional tree shaped designs became the layer 2 to layer 3 boundaries at the aggregation layer in terms of bandwidth, CPU utilization as well as forwarding table sizes that usually require expensive upgrades with bigger boxes. Finally, the classic data center designs do not allow easy innovation as they typically need to be upgraded by newer boxes from the same networking vendor that we already discussed as the often-unfortunate vendor "lock in."

The core and pod datacenter design should overcome these limitations as it is built around a router core that connects a number of independently designed Pods (see Fig. 22). In this context, a Pod is an atomic unit within a large data center, typically designed as one entity containing network, compute and storage and interfaces to the core via layer 3 devices. Pods can consist of either a few racks or a few hundred racks and easily be home for several thousand VMs.

The biggest advantage resulting from this design is that automation at the POD level is generally much simpler than at the data center level. This allows easier adoption of newer technologies with close look at price/performance improvements

[103]Cisco, White Paper, Cisco FabricPath Best Practices, http://www.cisco.com/c/dam/en/us/products/collateral/switches/nexus-7000-series-switches/white_paper_c07-728188.pdf, 2016, retrieved 2016-12-02.

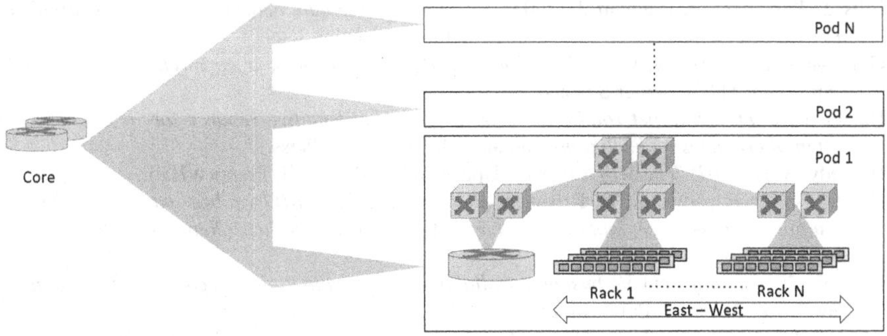

Fig. 22 Core and pod datacenter design

as new instances of pod designs can be simply attached to the data center core using either static routes or traditional routing protocols. This results in an additional separation of innovation cycles from capacity planning decisions and as such defines a key element of hyper-scaling a data center.

Big Switch Networks Big Cloud Fabric (BCF) Solution
This optimized and flexible core and pod datacenter design allows for very scalable implementations and hence is called hyperscale data center. The central element of BCF is the BCF Controller. This is a logically centralized SDN controller implemented as a cluster of VMs or hardware appliances for high availability (HA). The Bare Metal Leaf and Spine Switches are based on 10G or 40G Ethernet switch hardware from different vendors, or the Switch Light vSwitch that is a high performance SDN vSwitch optimized for unified physical and virtual Clos design and run the purpose-built Switch Light operating.

The combination of APIs and orchestration system integrations with OpenStack and CloudStack offer network provisioning, security as well as workflow audits. One of the key advantages of this BCF architecture is the possible unified P+V solution that should allow significantly better economic results compared to legacy implementations and storage workloads, while at the same time fully ensure the support for open industry standards, making these standards applicable to new hypervisor designs.

We will look closer into the future of network virtualization in the chapter "Network Virtualization."

References

Armitage, G., Carlberg, K., & Gleeson, B. (2000). *Quality of service in IP networks*. Indianapolis, IN: Sams Publishing.
Azodolmolky, S. (2013). *Software defined networking with openflow*. Birmingham: Packt.
Black, U. D. (1997). *ATM, volume III: Interworking with ATM*. Toronto: Prentice Hall.

Ellis, J., Pursell, C., & Rahman, J. (2003). *Voice, video and data network convergence, architecture and design, from IP to wireless*. Amsterdam: Elsevier Science.
Göransson, P., & Black, C. (2014). *Software defined networks, a comprehensive approach*. Cambridge, MA: Elsevier Science.
Halabi, B. (1997). *Internet routing architectures – The definitive reource for internetworking design alternatives and solutions*. Indianapolis, IN: Cisco Press.
Harnedy, S. J. (2007). *The MPLS primer*. Upper Saddle River, NJ: Prentice Hall.
Hooda, S. K., Kapadia, S., & Krishnan, P. (2014). *Using TRILL, fabric path, and VXLAN: Designing massively scalable data centers (MSDC) with overlays (Networking technology)*. Indianapolis, IN: Cisco Press.
Kale, V. (2016). *Agile network business: Collaboration, coordination, and competitive advantage*. New York: Auerbach Publications.
Khan, S. U., & Zomaya, A. Y. (2015). *Handbook on data centers*. New York: Springer.
Krattiger, L., Kapadia, S., & Jansen, D. (2016). *Building data centers with VXLAN EVPN (Networking technology)*. Indianapolis, IN: Cisco Press.
Lee, D. C. (1999). *Enhanced IP services for cisco networks*. Indianapolis, IN: Cisco Press.
Metes, G., Gundry, J., & Bradish, P. (1998). *Agile networking*. Englewood Cliffs, NJ: Prentice Hall.
Michael Wittig, A. W. (2015). *Amazon web services in action*. Shelter Island, NY: Manning Publications.
Nadeau, T., & Gray, K. (2013). *SDN: Software defined networks*. Sebastopol, CA: O'Reilly.
Paul Göransson, C. B. (2014). *Software defined networks – A comprehensive approach*. Burlington, MA: Morgan Kaufmann.
Pierre Lynch, M. H. (2014). *Demystifying NFV in carrier networks: A definitive guide to successful migrations* (1st ed.). North Charleston, SC: CreateSpace Independent Publishing Platform.
Stallings, W. (2015). *Foundations of modern networking – SDN, NFV, QoE, IoT, and Cloud*. Indianapolis, IN: Pearson Education.
Tomsu, P., & Wieser, G. (2002). In T. D. Nadau (Ed.), *MPLS-based VPNs, designing advanced virtual networks*. Upper Saddle River, NJ: Prentice Hall.
Underdahl, B., & Kinghorn, G. (2015). *Software defined networking for dummies, Cisco special edition*. Hoboken, NJ: Wiley.
Victor Moreno, K. R. (2007). *Network virtualization*. Indianapolis, IN: Cisco Press.

Building the Internet

We reject: kings, presidents, and voting. We believe in: rough consensus and running code
MIT professor Dave Clark, at the
24th annual July 1992 IETF conference?

What we know today as the Internet has its roots in the 1960s and, until today, has expanded to the largest technical structure ever build. With three billion users, more than one billion websites, millions of servers and routers and millions of applications and services, it spans the globe as a unique global structure providing services for individuals, organizations and corporations. The structure is based on a stack of technologies and standards developed, in three phases, in the last two decades of the twentieth century:

- The fundamentals for wide area computer networks were laid in the 1960s leading to several competing implementations of computer networks
- In the 1970s and 1980s, the agreement on standards for networking protocols, operating systems, and programming languages prepared the way for the global acceptance and economic foundation for a commercially successful use of a global computer network
- Finally, the introduction of the World Wide Web in the 1990s led to the rapid grow of web applications and to the development of cloud computing and cloud services.

Following the path of technical developments and the decisions based on architectural assumptions and agreed standards, we find a set of major elements that propelled the acceptance of the Internet as the carrier for the World Wide Web and cloud services. These elements are:

- A consensus about a global coding system for numerical and character based data
- A family of operating systems, based on UNIX and the open software paradigm (LINUX and distributions)
- A set of high level programming languages (C, C++, Java)

M. Oppitz, P. Tomsu, *Inventing the Cloud Century*,
DOI 10.1007/978-3-319-61161-7_9

- A widely-used standard for the interaction between human users and content provided, consisting of HTML, HTTP as protocols and browsers and web servers as applications
- A set of accepted standard formats for the interchange of graphical, audio and video data
- A set of application-to-application interfacing architectures (SOAP, REST)
- The extension of the network to wireless services (wireless LAN, Bluetooth and various mobile phone protocols)
- The introduction of a global positioning system (GPS) as a major trigger for location based services.

What all these technologies and standards have in common is that they contributed greatly to the development of the Internet, the World Wide Web and cloud services. Each Internet service or cloud service we use today is based on some of these technologies. These technologies also had a major impact on the IT market and the rapid growth of the Internet and the adoption of the Web. What is more, all these technologies and standards are publicly available and based on open software licenses or public standards.

The making of the Internet and the World Wide Web was also effected by a major change in how technology is developed and distributed to the user. Today, this change is known as the open software movement. Today's Internet and web services are based on a stack of open software services and the paradigm of open communities using the net and cloud based services. This disruptive evolution is one of the creative factors in cloud based services and effects the social impact and the ecosystems of cloud services. The most important elements in this process are:

- The idea of using the net for building cross border communities
- The commercialization of the Internet around 1988
- The open software movement, creating the nexus of a new system of software development and distribution
- The creation of new businesses using the Internet and the World Wide Web

We will drill down to the development and major effects of each of these milestones in the following chapters.

Preparations

The Internet, the World Wide Web and the cloud services offered today are composed of several basic concepts. One—and maybe the most important—is the connection of machines (computers) and the integration of human users to a complex structure of processes, networks and computing devices. The substantial idea behind service ecosystems is the combination of one central network using provider technology, standards and inexpensive end-devices to enable a successful business. Behind that concept, several major challenges were waiting for a solution.

The creators of the Internet and the World Wide Web had to answer the question of how to connect machines with each other and also how to make that integration visible and usable for human users. Starting with the beginning of the computer area in the 1950s, it took more than 60 years to find and implement solutions for these questions.

Connecting Machines and People

1960–1990 The Code Wars

One of the major challenges to computer connection is the standard used for data exchange. The transition from simple streams of bits to machine-readable data known as "code" took a long time. In the 1960s, computers from different vendors used their own method of coding raw data from bits (0 and 1) to digits, numbers or characters.

Many coding methods were used. IBM introduced EBCDIC as an 8-bit coding method together with the release of their System 360 in 1963. At the same time, the ASA (American Standard Association, later American Standards Association or ANSI) introduced ASCII (American Standard Code for Information Interchange), which was, at that time, a 7-bit coding system. Both systems were not compatible. The fact that all the code points were different was less of a problem for interoperating with ASCII than the fact that sorting EBCDIC put the lowercase letters before the uppercase letters and those before the numbers, exactly the opposite order of sorting ASCII used.

The coding wars lasted for more than one decade and cost a lot of money to those who wanted to exchange data between computers from different vendors. Finally, ASCII was accepted as the standard coding method and, triggered by the booming PC business, and it was the most common character encoding at the beginning of the World Wide Web.

ASCII still had its disadvantages especially when it came to the coding and transport of language specific characters. In December 2007, it was finally surpassed by UTF-8, which includes ASCII as a subset and allows for the coding of all language specific characters including all European languages, as well as simple Chinese and Japanese. UTF-8 was first, officially presented at the USENIX conference in San Diego in 1993. It was proposed and defined by the X/Open committee and the Unix System Laboratories. Today, most of the applications and data base systems use UTF-8 thus enabling the storing and exchanging of text in most of the different local character sets.

1969 ARPANET

The summer of 1968 became known for its social revolutions and accompanying political unrest from Prague to Paris to Chicago. The world seemed to be teaming with new political and social changes. At the same time, another revolution started. Larry Roberts, program manager at the ARPA IPTO (Information Processing

Techniques Office), sent out an offer to 140 computer companies requesting them to build a special purpose minicomputer to serve as the routers for the ARPANET project. Only a dozen companies decided to bid (IBM not among them) and, after an internal discussion, the US company BBN (Bolt Beranek and Newman) received the contract. BBN was founded 1948 by the two MIT professors, Richard Bolt und Leo Beranek, and had two brilliant MIT assistant professors, Robert Kahn and Frank Heart, on their engineering team who worked out the basic design for the new Interface Message Processor. Robert Kahn later became, together with Vincent Cerf, one of the fathers of the Internet. The first Interface Message Processor was shipped to UCLA (University of California Los Angeles) in August 1969, and the second one to SRI (Stanford Research Institute). On October 29th, 1969,[1] the first host-to-host connection was established between Los Angeles and Palo Alto. The Internet was born in the year of Woodstock, Vietnam War protests, Altamont and the first landing of a man on the moon.

The project was financed through a collaboration between the US Department of Defense and a group of research facilities assembled under the Advanced Research Project Agency (ARPA). ARPA was founded in 1958 as reaction to the Sputnik-shock. At its beginning, the agency concentrated on ballistic missile and nuclear weapon test projects. After the transfer of the civilian space programs to NASA and the military programs to the military service organization, ARPA started to focus on information processing and, in 1962, formed the ARPA IPTO (Information Processing Techniques Office), led by J. C. R. Licklider, a computer scientist and pioneer in the field of interactive computing, time sharing systems and networking. Licklider left ARPA in 1964 for IBM and later went back to MIT, but was the forerunner of the ARPANET concept.

As already described in detail in the chapter "Networking Technologies and Standards," the concept was based on the idea of packet switching, which simply means organizing the data transmission between hosts through the transfer of small "data packages" each of them having the destination and source addresses on its digital envelope and being "routed" along the network from the sending host through network connections and Interface Message Processors to the receiving host. Deciding in favor of packet switching instead of circuit switching (setting up a dedicated connection between the communicating hosts like in a telephone call) or message switching (setting up a reserved connection for the complete message) was one of the most important design decisions for digital networks. It was made because of the expected performance increase and reliability compared to circuit switching or message switching. All types of digital communication protocols used today are based on packet switching. The foundations for packet switching procedures were laid by Paul Baran at RAND in 1960, Leonard Kleinrock at MIT in 1961, Larry Roberts at MIT Lincoln Labaratories in 1963, and British scientist, Donald Davies, at the National Physical Laboratory (NPL) in 1966.

[1]Note of the author: this is also my 12th birthday.

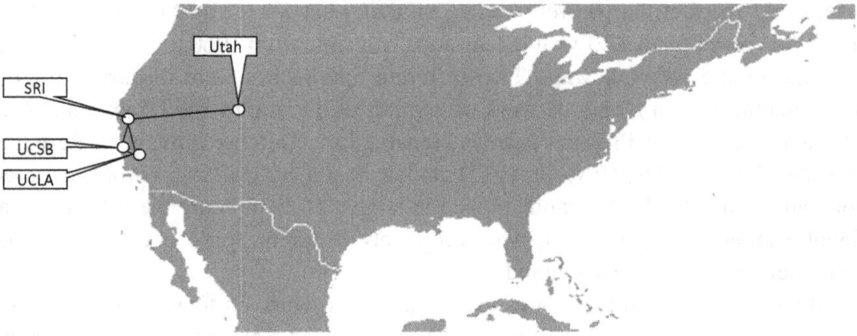

Fig. 1 Arpanet in December 1969

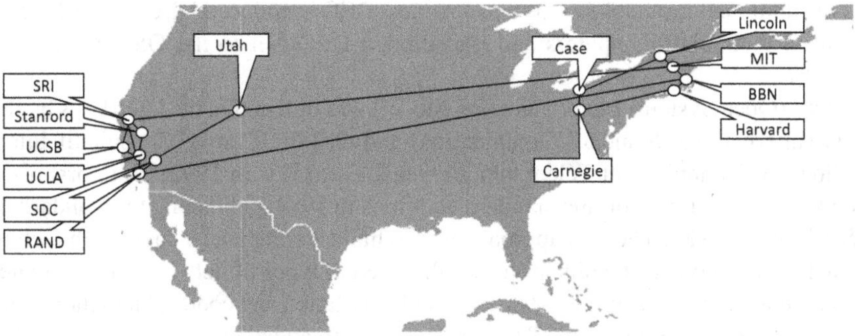

Fig. 2 Arpanet in December 1970

By end of 1969, the ARPANET included four nodes: UCLA, Stanford SRI, UCSB and University of Utah (see Fig. 1). In March 1970, the ARPANET reached the east coast connecting BBN in Cambridge, Massachusetts and a dozen more nodes by the end of that year (see Fig. 2).

In the early years of ARPANET, between 1969 and 1982, the Network Control Program (NCP) was used to organize the transport of data packages between ARPANET hosts and Interface Message Processors. From the beginning, ARPANET was based on packet switching, but so were other networks like the ALOHA Net in the Islands of Hawaii, the PARC LAN in Palo Alto or the PRNET in San Francisco. The problem was a lack of a common standard, all those networks used different types of protocols rendering them not compatible or interoperable. In 1973, the ARPA IPTO engineer, Robert Kahn, was selected to solve that problem. Together with Vincent Cerf he began to develop a network standard protocol that was later known by the name of TCP/IP.

1970 The Protocol Competition

The internet today is based on two major protocol elements: IP and TCP. All other services are based on these two layers. Back in the beginning of Arpanet and the first international computer networking concepts, many competing ideas were discussed. The fact that we rely, today, on that what we see as basic layers of the Internet evolved from a 20-year discussion. This time from 1960 to 1980 was called the time of the protocol wars. Two different approaches were discussed for the implementation of a global network of computers. Driven by the US-based ARPA (Advanced Research Projects Agency, renamed the Defense Advanced Research Projects Agency (DARPA) in 1972) and a nexus of US and European based companies and standard organizations, the largest of them being the ISO (International Standard Organization), two completely different approaches for a global computer network were discussed.[2]

At the same time, the IT industry had started to work on their own proprietary concepts and networking protocols. IBM introduced SNA (System Networking Architecture) in 1974 to connect their mainframes to peripheral devices and, later, to other computers. Using its market power, IBM achieved rolling out SNA as the dominant protocol for large companies. DEC introduced DEC Net in 1975 to connect their PDP-computers and transitioned DEC Net to the OSI architecture in 1987.

OSI (Open Systems Interconnection Model) was initiated in the late 1970s by the ISO (International Standard Organization) and CCITT (Comité Consultatif International Téléphonique et Télégraphique, renamed to ITU in 1992). Both organizations joined their networking standard activities in 1983, publishing the standard as ISO 7498[3] in 1984. The OSI approach was to build a complete, 7-layer architecture which could serve as the blue print for all different types of network and computer interconnection thus leaving the task of implementation and code open to those who would accept the architecture. This strategy lead to an extensive but also complex set of definitions discussed and created by many working groups and committees. The result was a "standard elephant," which was large but also slow and influenced by the idea of building a comprehensive but more theoretical model of networking.

On the other side of the Atlantic, the Arpanet group around Vincent Cerf used a completely different strategy. Starting a little bit earlier in 1973, the team concentrated on using the best existing ideas around packet switching networks and focused on building running code and operable connections between the Arpanet computers. The used style was completely different from the organizations of ISO and CCITT. The Arpanet team applied a bottom-up process style of unstructured working groups. The team members worked on specific questions and problems and contributed their results as so-called RFC (Request for Comments) to the group. Having its origins more in scientific research, this process became the model for later projects like the World Wide Web and the open source movements. The

[2]The Protocol Wars: http://www.computerhistory.org/revolution/networking/19/376, retrieved 2016-11-16.

[3]ITU: "Open Systems Interconnection Model", www.itu.int/rec/dologin_pub.asp?id=T-REC-X.200 retrieved 2016-10-12.

```
Network Working Group                                    Vinton Cerf
Request for Comments: 675                                Yogen Dalal
NIC: 2                                                   Carl Sunshine
INWG: 72                                                 December 1974

          SPECIFICATION OF INTERNET TRANSMISSION CONTROL PROGRAM

                         December 1974 Version

1.  INTRODUCTION

   This document describes the functions to be performed by the
   internetwork Transmission Control Program [TCP] and its interface to
   programs or users that require its services. Several basic
   assumptions are made about process to process communication and these
   are listed here without further justification. The interested reader
   is referred to [CEKA74, TOML74, BELS74, DALA74, SUNS74] for further
   discussion.
```

Fig. 3 Networking Working Group RFC 675

method was called "Rough Consensus and Running Code," the result was TCP/IP and the Internet.

In 1973, Vincent Cerf started a networking research group at Stanford resulting in the first TCP/IP specification in 1974. The term "Internet" originates from that discussion, it was first used in one of the RFC papers, probably RFC675,[4] published by Vincent Cerf, Yogen Dalal and Carl Sunshine in December 1974 (see Fig. 3).

In the following years, many RFCs were submitted and tests were performed between Stanford, Massachusetts and the University College London. In 1979, the Internet Configuration Control Board[5] (ICCB) headed by Dave Clark from MIT was founded by Vincent Cerf to advise on technical issues. The ICCB was transformed into Internet Architecture Board[6] in 1981 to supervise all activities regarding the standardization and definition of Internet related technologies. In 1982, the US Department of Defense declared TCP/IP as the standard for all military connections and, on January 1st, 1983, the migration of Arpanet to TCP/IP was officially completed. From August 25th to 27th, 1986, the Internet Architecture Board organized a 3-day workshop on TCP/IP in Monterey, California for the computer industry, promoting the protocol and leading to its increasing commercial use. The event was attended by 250 vendor representatives. Beginning in 1988, the US Federal Network Council allowed commercial use of the network connections for US based service providers. Consequently, Arpanet was decommissioned in 1990. In 1995, the Internet was fully commercialized.

Today, the Internet's infrastructure (connections, lines and routers) is provided by a small group of approximately 20 tier-1 providers and hundreds of tier-2 providers around the world, also known as Internet Service Providers. This infrastructure is financed by the users paying monthly fees or having Internet access

---

[4]RFC675: https://tools.ietf.org/html/rfc675 retrieved 2016-10-12.

[5]Internet Architecture Board History: https://www.iab.org/about/history/ retrieved 2016-10-12.

[6]Internet Architecture Board (IAB): https://www.iab.org/ retrieved 2016-10-12.

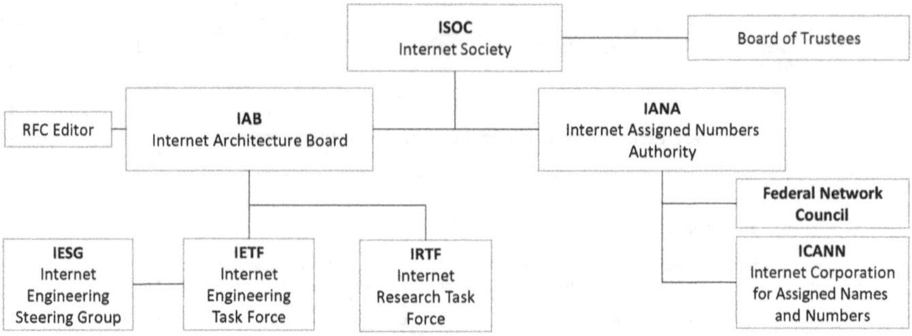

Fig. 4 Internet Society and connected organizations

included in their phone rates. The Internet policies (not to mix up with World Wide Web policies), its architecture and technology is supported and supervised by a group of non-profit organizations and initiatives. Those organizations share a policy of open development and consensus, they are based on the idea of developing and maintaining standards as an open community. As the umbrella organization, the Internet Society (ISOC) was founded in 1992 to provide leadership in Internet-related standards, education, access, and policy (see Fig. 4) *"...to promote the open development, evolution and use of the Internet for the benefit of all people throughout the world"*[7]

All work on technical specifications is concentrated in the IETF (Internet Engineering Task Force): *The mission of the IETF is to make the Internet work better by producing high quality, relevant technical documents that influence the way people design, use, and manage the Internet.*[8] The IETF is accompanied by the IESG (Internet Engineering Steering Group) and the IRTF (Internet Research Task Force). The IAB (Internet Architecture Board) serves as a coordination and supervising platform for all architecture related missions. The IANA (Internet Assigned Numbers Authority) and ICANN (Internet Corporation for Assigned Names and Numbers) serve as authorities for the registration of Internet addresses and domain-names. As the organizer of the yearly IETF conferences, the IETF is one of the major proponents of an open discussion within all Internet architecture related topics. The IETF conferences are open to everybody. At the 24th annual IETF conference, Dave Clark declared the motto of IETF in his "A Cloudy Crystal Ball/ Apocalypse Now"[9] presentation: *"We reject: kings, presidents, and voting. We believe in: rough consensus and running code"*.[10] Today, this motto is used in the

[7]Internet Society Mission: http://www.internetsociety.org/who-we-are retrieved 2016-10-12.

[8]IETF Mission Statement: https://www.ietf.org/ retrieved 2016-10-12.

[9]Dave Clark 1992: "A Cloudy Crystal Ball / Apocalypse Now": https://groups.csail.mit.edu/ana/People/DDC/future_ietf_92.pdf, retrieved 2016-11-16.

[10]Dave Clark: "A Cloudy Crystal Ball-Visions of the Future" in "Proceedings of the Twenty-Fourth Internet Engineering Task Force, *1992*" https://www.ietf.org/proceedings/24.pdf retrieved 2016-11-29.

"IETF TAO[11]" and was also used as a t-shirt logo at the 83rd-annual conference in Paris.

1971 Getting Data Files from One Computer to Another

Back in the 1980s, the Internet was a great concept and enabled, for the first time in history, the connection of different types of computers using TCP and IP. Packet switching turned out to be the most efficient way to build a global network of computers, most of them being hosts or servers. The users of the internet at that time were, besides the members of ARPA, universities and other research facilities in most cases. By 1981, the number of hosts had grown to 550 nearly doubling to 940 in 1984. At that point in time interfacing between computers connected via TCP and IP was restricted to applications like FTP (file transfer protocol). FTP (File Transfer Protocol) was defined by Abhay Bhushan in April 1971. It should provide a standard method of moving and copying files from one computer system to another. In the beginning, FTP was implemented on top of the predecessor of TCP/IP, the NCP protocol (Network Control Program). After the success of TCP/IP, it was replaced by a TCP/IP version and serves as the standard file transfer methodology today.

1971 The First Email

Exchanging data between machines is a nice thing to have, but in the end, it is about the question of how to integrate the human user into that value chain. The idea was very simple: why not also use computers connected to each other to send messages between the users of those computers. It was the programmer Raymond Samuel Tomlinson who implemented the first email system in 1971 on the Arpanet. At that time, Tomlinson was working for BBN Technologies and was asked to change a program called SNDMSG, which sent messages to other users of a time-sharing computer, which ran on TENEX, an operating system for DEC-computers. He added code to SNDMSG so messages could also be sent to users on other computers via Arpanet. Tomlinson's implementation was the first system able to send mail between users on different hosts connected to the Arpanet. Tomlinson also decided to use the "@" sign to separate the user from their machine, which has been used in email addresses ever since. The first email ever sent was not preserved and Tomlinson describes it as insignificant, something like "QWERTYUIOP." When Tomlinson showed his implementation to a colleague, the colleague said, "Don't tell anyone! This isn't what we're supposed to be working on." What a tremendous misjudgment! In 2015, more than 190 billion emails were sent and received per day.

There was still a long way to go in finding a standard to make personal messaging available for users of different computer systems. In the early years, sending and receiving emails was based on transferring files using FTP, which was inefficient and complex. In 1980, Jon Postels, a computer scientist at USC (University of Southern California), proposed a new protocol that began to remove the

[11] IETF Tao: https://www.ietf.org/tao.html, retrieved 2016-11-16.

mail's reliance on FTP. The later called SMTP (Simple Mail Transfer Protocol) became widely used in the early 1980s.

Building the Basement: Unix and C

In parallel to the development of connecting computers via a standardized network, the idea of a standardized workbench for programmers and operators of computers began to evolve. In the 1960s and 1970s, computers of various vendors ran their proprietary operating systems. The new approach was based on the idea of building a portable, multi-tasking and multi-user operating system. It was Ken Thompson and Dennis Ritchie, two engineers at Bell Labs, who decided to implement the first version of a new operating system on a PDP-7 in 1969. Their idea was to create an operating system much smaller than the existing Multics system, based on a hierarchical file system and using a concept of computer processes, device files and a command line interpreter. Thompson and Richie implemented the new operating system in about a month's time. This was the software that would later become known as Unix. For the time being, Bell Labs decided to implement and use that new operating system on other machines used by Bell Labs.

The Bell Labs was a research institution that had its roots in Graham Bells' telephone company of 1898. Later renamed AT&T Bell Labs, it became the research branch of the Bell Group for nearly 100 years and founded the tradition of creating technology standards in the Bell System. Bell Labs is one of the most successful research institutions and was home to many innovators and scientists among them Nobel-prize winners like John Bardeen, Walter H. Brattain, and William Shockley (innovation of the transistor) and Arno A. Penzias and Robert W. Wilson (discovery of the cosmic microwave radiation). In 1996, AT&T renamed and reformed Bell Labs into the new company Lucent Technologies. In 2015, Lucent was acquired by Nokia for $16.6 billion.[12]

Thompson and Ritchie received the Turing Award in 1983 for their work on operating system theory, and for developing Unix. The Turing Award selection committee wrote: "*The success of the UNIX system stems from its tasteful selection of a few key ideas and their elegant implementation. The model of the Unix system has led a generation of software designers to new ways of thinking about programming. The genius of the Unix system is its framework, which enables programmers to stand on the work of others.*"

1972 The C Programming Languages

The first version of UNIX was written by Thompson and Ritchie in assembler for the PDP-7 computer. Assembler is a type of programming language that enables

[12]Trevis Team: Nokias 16.6 billion acquisition of Lucent, Forbes 2015: http://www.forbes.com/sites/greatspeculations/2015/04/16/nokias-16-6-billion-acquisition-of-alcatel-lucent-explained/#5306ba3c4d6d retrieved 2016-10-12.

programmers to write a sequence of instructions for a specific computer. Software written in assembler can only be executed on the hardware it was designed for. Porting assembler programs to another computer is impossible, the program must be rewritten for the instruction set of the target computer. The solution to this problem is found in the so-called high-level programing languages. Those languages are hardware-independent. Programs written in a high-level-language must be translated by a compiler-program which converts the program into the computer specific instructions.

This approach facilitates the design of portable software: the program itself is reusable on different machines if there is a compiler-program available which translates the software into the computer specific, so-called machine-code. The first high-level programing languages were developed in the late 1950s and early 1960s (FORTRAN and COBOL).

During the 1960s, it became evident that those early programming languages had their limitations in supporting more complex data structures and algorithms. A new family of high-level programming languages was proposed by computer scientists in Europe and the US. Those languages followed the approach of structured programming, which requests the extensive use of sub routines, block structures and loops instead of the old methods of GOTO-instructions leading to something that was called "spaghetti-code." Structured programming aimed at improving the clarity, quality, and development time of software. Early examples of structured programming languages are ALGOL (Edsger W. Dijkstra, Netherlands, 1960), BCPL (Martin Richards GB, 1966), PASCAL (Niklaus Wirth 1970, Switzerland), MODULA (Niklaus Wirth, 1975). PASCAL became a widely-used programming language in the 1970s and 1980s for education, but also for scientific and commercial software development. Pascal was also the primary high-level language for Apple's Lisa computer (1983) and the early Apple Macintosh (1984).

In the Bell Labs, Thompson and Ritchie, having implemented the first version of UNIX in 1969, concluded that a highly standardized operating system has to be written in a high-level programming language to facilitate portability to different computers. Unsatisfied with the existing high-level programming languages, they simply decided to build their own. Their aim was to design a language which follows the ideas of structured programming, but also provides constructs that allow for accessing and mapping efficiently to typical machine instructions, which was not the case in PASCAL or MODULA. Their initial attempt was a language called "B" as an extension of Martin Richards BCPL. Around 1972, they came out with the design and first implementation of the "C programming language." In 1972, Unix was rewritten by Ritchie and Thompson in the C-programming language, contrary to the general notion at the time "that something as complex as an operating system, which must deal with time-critical events, has to be written exclusively in assembly language." This implementation became famous as the impressive, proof-of-concept for the C programming language and the portability of operating systems and system software. In 1978, Dennis Ritchie and Brian Kerninghan published the book "The C Programming Language" which became the informal specification of C for many years.

After the successful implementation of UNIX in the C programming language in 1972, UNIX and C became first choice and de-facto standard for operating systems and systems software from the middle of the 1970s until today, though both key technologies went through many transitions in the following decades. From a distant point of view, the introduction of UNIX and C in 1972 defines the point in time when engineers could put away their screw drivers and welding irons and concentrate on the design of software. Based on Kernighan and Ritchie's book during the late 1970s and 1980s, many C-compilers were implemented for a wide variety of computers including mainframes, minicomputers and PCs including the IBM PC and its followers. C became the "work horse" for system software and applications.

Developing a new software product using C enabled the developer to offer that software to different computers if a C-compiler for the machine was available. This was the time when software developed into a new industry segment. Software companies like SAP, Microsoft and others were founded and highly standardized operating systems and software applications were developed as hardware independent products. The statistics of international analysts started to recognize software as a category of its own and software engineering and software development became professions of their own. The C programming language was extended to C++ by the Danish computer scientist Bjarne Stroustrup at Bell labs in 1998. C++ added the concept of object-orientation to C allowing a much higher level of abstraction using classes and objects to describe the data and processes implemented and executed by the software.

1974–1984 The UNIX Wars

The UNIX evolution took a slightly different but comparable path. In the beginning, UNIX was offered as a license to educational organizations and commercial users by AT&T, being the owner of Bell Labs. In 1973, AT&T released Version 5 Unix (not to be mixed up with AT&T's Version V from 1983) for a license price of $20,000 (which is valued atmore than $80,000 today). This version of UNIX ran on DEC machines only (PDP and VAX).

As more of the source code was rewritten in the C language, several universities started their own UNIX projects porting UNIX to other computer architectures. One of the first projects was the Berkley Software Distribution (BSD). The Bell Labs (AT&T) UNIX code arrived at Berkley in 1974, and computer science professor Bob Fabry immediately decided to start a UNIX project. As a result, the first Berkeley Software Distribution (1BSD) was released in 1978. A major trigger for the final success of BSD was the decision by DARPA in 1978 to fund Berkeley's Computer Systems Research Group (CSRG), which would develop a standard Unix platform for future DARPA research in the VLSI (Very Large Scale Integration) Project. The team at Berkley concentrated on improved implementation of nearly all operating system functions and added network support implementing to the OSI stack as well as TCP/IP.

In early 1980, BSD was the most popular non-commercial UNIX distribution. Nevertheless, Berkley soon faced legal discussions with AT&T regarding intellectual property and licensing rights between AT&T's UNIX V and Berkley's BSD. At the same time, AT&T was involved in an anti-trust case with the US Department of Justice, preventing AT&T from turning UNIX into a product. The case was settled in 1983 leading to the "breakup of the Bell system". Thus, AT&T immediately started to commercialize Unix System V, claiming intellectual property rights for the complete source code. The move nearly killed Unix. At Berkley, the team decided to rewrite the complete source code to avoid copyright violations and to enable the independent licensing of BSD. This slowed down the development of BSD and was one of the reasons the open source community started several open source UNIX projects and additionally it triggered the foundation of the GNU project (**G**NU's **n**ot **U**nix) in the same year by Richard Stallman. It took until 1994 and several lawsuits between Berkley and AT&T before it could be settled.

Due to AT&T's decisions, 1983 was the starting point for the wide spread development of a huge number of different Unix implementations and distribution. At that point of time, Berkley's BSD and AT&T'2 UNIX V were the only available UNIX implementations. Unix V was used at AT&T and elsewhere, BSD was popular in the university community. Computer science students moving from university to enterprises wanted to continue using Unix and observers began to see potential in the Unix universal operating system. In 1983, the Unix kernel consisted of only 20,000 lines of C-code and was more than 75% machine independent. Soon, the Unix kernel was ported to more different computers and processors and the large computer manufactures soon recognized that UNIX is major threat for their proprietary operating systems. At that point of time, the big manufacturers maintained and sold their own operating systems for their computers, but rapidly started to support Unix or to create their own distribution (see Fig. 5).

Microsoft released the first Xenix-Version using a AT&T license and providing an implementation for Intel architectures. The Santa Cruz Operation (SCO) released SCO Xenix in 1983, followed by SCO Unix in 1989. Sun released their Sun OS based on Motorola processors in 1982, followed by the Sun Solaris operating system in 1991. Hewlett-Packard released HP-UX for their HP 9000 servers in 1982. IBM introduced AIX Version 1 in 1986 for the IBM 6150 RT

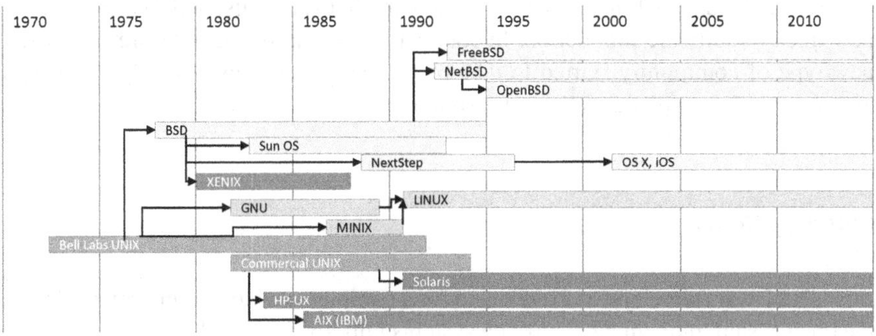

Fig. 5 UNIX version history

workstation. Steve Jobs used a BSD based Unix as the operating system, NeXT Step, for his Next computer in 1989. The Next computer and NeXT Step was also the platform on which Tim Berners-Lee created the first Web browser. After the acquisition of Next by Apple in 1997 NeXT Step became the foundation of Apples IOS and OS X operating systems. In 1992, AT&T sold its UNIX rights to Novell, which sold them again to the UNIX company, Santa Cruz Operation (SCO), in 1995.

By the late 1980s, the situation on the UNIX market was chaotic and far from reaching a standardized operating system platform. In 1988, the IEEE Computer Society took over responsibility in creating a commonly accepted standard for compatibility between operating systems. Richard Stallman suggested "POSIX" as name of the new family of standards including the application programming interface, the command line shell and utilities interfaces. Based on POSIX it should be possible to certify different implementations of UNIX as well as other compliancy standards in operating systems thus enabling compatibility between operating systems. Today, most of the UNIX distributions as well as some other operating systems including Windows are either POSIX certified or at least POSIX compliant.

In the early 1990s, the UNIX market was still dealing with lawsuits between the different UNIX providers, diverse views on standards and definitions and different license models. It seems that the successful introduction of UNIX as a standardized foundation for operating computers and networks was being blocked by many unsolved issues.

The problem was recognized by the major UNIX vendors, who were also receiving competitive pressure from Microsoft. Between 1993 and 1996, several initiatives were taken to create an operable and working community for the standardization of UNIX. As final step, in 1996, several of these platforms merged into the Open Group, which took over the responsibility and representation of the UNIX community. Today, the Open Group is a vendor- and technology-neutral industry consortium with over five hundred member organizations. It is the certification body for the UNIX trademark and publishes the "Single UNIX Specification." The Open Group also uses POSIX as the standard definition for UNIX distributions.

Despite the unsatisfying situation in the late 1980s, this also led to a completely new track of software production. In contrast to the traditional software vendors, a new type of community started to take over the software business. Open systems and open source evolved as the new paradigm.

Open Systems

The UNIX-wars were one of the triggers for a complete new paradigm referred to as the open system movement. The open systems initiatives created a parallel universe to the traditional forms of distribution and monetizing software and applications.

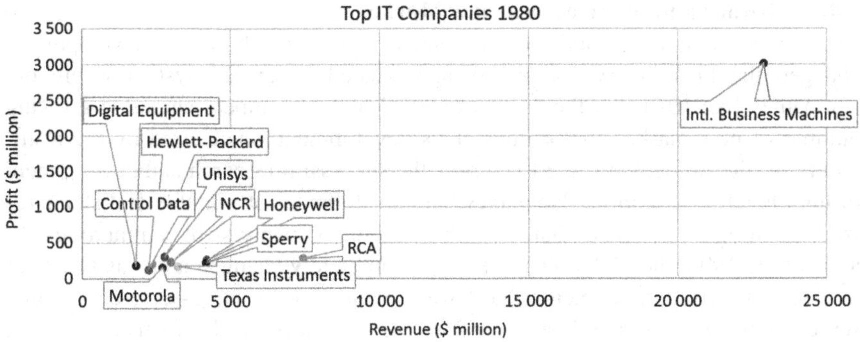

Fig. 6 IBM and other top IT companies in 1980

Open systems and open software began to impact the ecosystem of information technology and became one of the major drivers behind internet applications and, later, the development of the cloud service market. The open systems movement had its origins in the idea of the "Open Society" first suggested in 1932 by the French philosopher Henri Bergson, and later developed by Sir Karl Popper in his book "*The Open Society and Its Enemies*" (Popper, 1945). Popper defines an open society as a system "in which individuals are confronted with personal decisions" in opposition to a "magical or tribal or collectivist society." In opposition to open systems, which allow free interaction, closed systems are isolated from their environment.

In information technology, open systems are defined as a class of computer systems that provide interoperability, portability and follow open software standards. Today, a major part of Internet and cloud architecture is based on open systems like Linux, Java and many so-called open source products.

Back in the 1960s, all computers were mainly closed systems, designed and produced by a comparably small number of information technology manufacturers led by IBM, with around 60% market share, and followed by a group of seven other more specialized suppliers called "IBM and the seven dwarfs" (UNISYS, DEC, Burroughs, Honeywell, NCR, Control Data, and Digital) (see Fig. 6).

By 1980, many new enterprises had entered the market (DEC, Data General, Hewlett Packard, Wang—Minicomputers, Apple—Microcomputers). AT&T, General Electric and Philco had left the market. The market structure was still shaped by the product segments mainframe, minicomputer, microcomputer and peripherals. In a Datamation-report from 1986[13] analyzing revenue numbers between 1982 and 1985, the product segment "Software" or "Computer Services" does not appear before 1983. Software and services were not a product category, they were still a commodity provided by the manufacturers of the hardware.

[13]Datamation, June, 1986.

1980 Software and Services as a New Market

Between the mid-1970s and the mid-1980s, several revolutionary developments changed the IT industry completely and created a new ecosystem where old companies like IBM and the "seven dwarfs" had to compete against a growing number of new market companions. This development was caused by the availability of microprocessors at a low price, the successful research and development of high level programming languages, the development of portable and standardized operating systems and, finally, by the creation of open system communities. As a result, by 1996, the global IT market had completely changed and was also split into several different segments. IBM was still leading in large-scale-computing, servers, desktop, peripherals and services. HP had developed a focus on servers, PC's and on its successful printer business. The Japan-based enterprises, Hitachi, Fujitsu and NEC, developed to become competitors in the large scale main frame market. The newcomer, Cisco, entered the new network market with its routing and switching products. The PC market reached 10 years after the release of the IBM PC a significant volume and was populated by traditional IT companies (IBM, HP) together with new brands (Compaq, Toshiba, Dell, Apple and Acer). Software and services had developed into a market segment of its own with companies concentrating on these new segments like Microsoft and EDS, Anderson Consulting and Cap Gemini.

The major reason for the rapid development of a software and service market was the availability of high level programming languages and development tools, which enabled companies like EDS, Anderson, Cap Gemini and many others to design and implement software for different hardware architectures as reusable application packages. In parallel, software became a product of its own and, due to the reusability of software on different machines, the license agreements became more important. Software could be copied much easier from one computer to another and copyright, intellectual property, license fees and terms of usage were handled more strictly by software producers to protect their business.

Outside of the software companies and the community of computer engineers, students and researchers at universities had a completely other approach to software. They saw software as a common good, finding their pride in improving existing code and creating new programs and applications, which was seen as a kind of contribution to a better world of information technology. The movement was soon coined "Free Software." License fees for new types of portable operating systems like UNIX and strict copyright rules were, in their opinion, exactly the wrong direction to move towards. The UNIX wars between 1973 and 1984, despite all their negative impacts, seemed to confirm the academic's beliefs.

1983 Richard Stallman and the Free Software Foundation

Frustrated by the long-lasting discussions about software licensing, intellectual property rights and impressed by the idea of open standards, the US programmer and activist, Richard Stallman, started the "GNU-Project" in 1983 to create a free, UNIX-like operating system. Stallman was convinced that software should be distributed in such a way that its users have the freedoms to use, study, distribute and modify it. Stallman called it "free software" and, in 1984, founded the Free Software Foundation[14] as a platform to support the free software movement. The Free Software Foundation also became the platform and financial supporter of the GNU-project, hiring developers and selling manuals and tapes with the code. The GNU "UNIX" project never succeeded completely, but the basic idea was taken over by others like Andrew Tannenbaum and Linus Thorwald in Europe.

Richard Stallman's ground breaking idea was the introduction of the GNU GPL (General Public License). In 1989, Stallman worked together with a lawyer on a new type of license agreement for free software. The GNU (General Public License) enabled developers or teams to protect their software to a certain extent, but further guarantees users of software the right to use, study, modify and copy the software. The major disruption in the GPL idea is the so-called "copyleft" principle. It simple says that if somebody modifies software and distributes it to others, this modified version must be distributed under the GPL license as well. GPL was one of the major success factors for Linux, giving the programmer of Linux code the guarantee that his work would benefit the whole world and remain free.

1987 Towards an Open Operating System Standard
In 1987, Andrew Tannenbaum from the Freien Universität Amsterdam introduced Minix (Minimal Unix) as Unix implementation on Intel processors for students. Minix was based on UNIX V and, later, allowed Linus Thorwald to implement his own free Unix version. In the US, FreeBSD was released by a group of Berkley students in 1993. They took the original BSD code from 1991 called Net-1, which still contained six files of AT&T code, and replaced those files in an 18-month reengineering project to avoid copyright violations. They placed the new version under the name "FreeBSD" on an FTP server. Today, FreeBSD is used by many companies as the starting point for their Unix implementation. A concurrent group at Berkley released another version of Unix under the name of NetBSD in 1994. The father of most open source Unix distributions today is Linus Thorwald, a Finnish computer scientist.

1991 Linus Thorwald and Linux
Thorwald attended the University of Helsinki between 1988 and 1996 working on operating systems, one of them being Tannenbaum's Minix. Not satisfied with the limited licensing options that Minix restricted to educational organizations, Thorwald decided to work on his own operating system kernel based on Tannenbaum's MINIX. In a first posting, Thorwald introduced his project in a Usenet use group as a personal hobby: *"Hello everybody out there using minix, I'm doing a (free) operating system (just a hobby, won't be big and professional like gnu) for 386(486) AT clones. This has been brewing since April, and is starting to get ready."*[15]

[14]Free Software Foundation: https://www.fsf.org, retrieved 2016-11-30.

[15]Linus Thorwald: "Hello everybody out there using minix" https://groups.google.com/forum/#!msg/comp.os.minix/dlNtH7RRrGA/SwRavCzVE7gJ retrieved 2016-11-29.

The first source code pieces were uploaded to an FTP server in September 1991. Thorwald himself selected the name "Freax" as a combination of "Free," "Freak" and "x". His coworker at the University of Helsinki, Ari Lemmke, convinced him to use "Linux" as the project name, and today Linux is possibly the most successful product with its creator's name as part of the product name. Two years later, a community of 100 developers had joined the project and, in 1994, the Linux version 1.0 was released under the supervision and judgment of Linus Thorwald. In the same year, the first two commercial LINUX distributers, Red Hat and SUSE, published version 1.0 of their Linux distribution. The IT industry began to understand that Linux could be a game changer. In 1998, major companies, such as IBM, Compaq and Oracle announced their support for Linux. In 2005, Google acquired Android, a company that provides a smart phone operating system based on Linux.

Today, Linux based operating systems have a market share of around 30% for Web servers[16] and around 80% in smart phones and tablets. Of the top 500 fastest supercomputers in the world, more than 98% of them run on Linux. Linux, with its more than 280 releases, became one of the tree major operating systems together with UNIX and Windows. Only in desktop computing does Linux has a market share of only 3%.

The Linux development is coordinated and supervised by the Linux Foundation.[17] The major project is still the Linux kernel with more than 4000 developers contributing to new releases. Linus Torvalds remains the ultimate authority on what new code is incorporated into the standard Linux kernel. The Linux Foundation also hosts up to 50 Linux related open source projects, thus being one of the largest open source platforms worldwide and providing a model for open source development. The foundation itself is financed by partners and sponsors among them all the large IT companies. The distribution of Linux to the customers is done by the Linux distributors as commercial companies delivering Linux and offering services and additional products to their customers. The largest Linux distributers today are Redhat, SUSE, Debian, Ubuntu, Linux Mint.

Open Source Today

Since the introduction of Linux, open source moved down a road to success by creating a new paradigm: software can be successfully developed, distributed and maintained by an open source community. Soon after Linux was released, many other open source projects had been started, delivering successful tools and application as open software using the GNU license. In 1999, Eric Raymond published his book "The Cathedral and the Bazaar" on software engineering in an open source or free software environment. In his book, Raymond described 19 lessons to learn as good practice for open software development. Most of these lessons deal with motivation and quality as basic drivers for good software. The fundamental claim of the open software movement is that a much higher level of quality can be achieved

[16]Linux for Web servers: https://w3techs.com/technologies/details/os-linux/all/all

[17]Linux Foundation: https://www.linuxfoundation.org, retrieved 2016-11-30.

when software is designed and implemented by a highly-motivated community that follows the principles of nonhierarchical working teams and makes decisions based on a consensus instead of an appeal to authorities. As a prerequisite, the open source movement also claims that software and its source code must be open to everybody for learning, usage and improvement.

During the last decade, open source became a vital contributor to the IT economy and the cloud services nexus. Projects like Linux or Apache provide major software technologies and architectures. It also became evident that open source projects hosted by large foundations are much more efficient in developing new software than old school R&D departments in corporations. New business models were created consisting of open source platforms (mostly non-profit foundations) inviting corporate and private members to be sponsors.

Rollout of the Internet

In the late 1970s, the Internet connected around 180 hosts, most of them in research facilities and universities as a part of the Arpanet. Usage of ARPANET was restricted to research organizations involved in ARPA projects (see Fig. 7).

Connecting to the Internet

1979 Dial-Up Modems
Connecting to the Internet was not so easy at that time. You had to have access to a workstation or PC that was connected to a computer acting as a host and connected to an Internet Exchange Point (IXP). That was only the case for researchers or students working in a federal or private research facility or university. With the rise in personal and home computers, many private users wished to access new media from their home office. The device that made it possible to build a connection

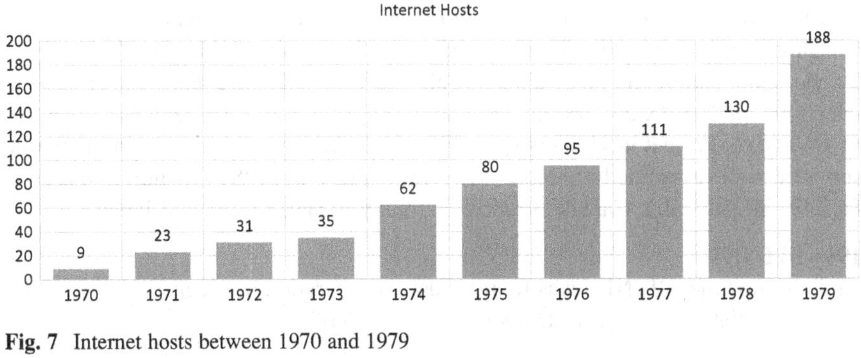

Fig. 7 Internet hosts between 1970 and 1979

between a home computer and the global network was the modem. A modem (modulator/demodulator) is a small device that translates the analog signal delivered by a phone line into the digital signal for a computer and vice versa. Using a modem enables the connection of a home computer via a phone line to a remote host.

The problem was that the phone companies like AT&T in the US had a monopoly and were blocking each attempt to use their lines for other purposes other than telephone calls. The way out was to use a so-called acoustical coupler, which transformed the digital signal into tones that could be transmitted via a phone line at a rate of 300 bits per second. In 1981, the Hayes Smart-modem was released, which could be connected directly to the phone line and was compatible to the Bell standard uses from AT&T. The Hayes Smart-modem was sold for $299 in the 300-baud version and, later, for $699 in the 1200 baud version. This was exactly the missing link needed to ignite the online revolution. AT&T and other telecommunication providers were forced by federal communication controls to step back and accept the usage of phone lines for data traffic.

The First Communities

Researchers at that time already followed a long tradition in exchanging research results and information. The academic community already followed the idea of belonging to a global community, not blocked by national borders. Connecting with researchers in other locations or countries was previously conducted through a few, occasional personal meetings at conferences or on phone—very expensive—or sending letters—very slow. The existence of a network based communication technology soon attracted researchers in the US and later also in Europe. Putting together the Arpanet architecture based on TCP/IP, the file transfer method FTP and the electronic mail implementation based on SMTP led to the simple idea of creating a community of researchers using a set of functions to take their communication to the next level.

1981 CSNET

In 1981, the CSNET was the first computer network that provided information exchange services for the scientific community. It was founded by the US National Science Foundation and offered a mail relay service and name service over dial-up telephone, or X.29/X.25 terminal emulation. Starting with three nodes (University of Delaware, Princeton University, and Purdue University) in 1981, the network services were expanded to more than 180 nodes by 1984 including research facilities in Australia, Canada, France, Germany, Israel, Korea, and Japan.

1981 BITNET

At the same time, BITNET was founded as a co-operation between US universities and was initiated by the City University of New York.

1985 NSFNET
In 1985, the National Science Foundation (NSF) started the NSFNet program to link research facilities involved in the NSF-funded supercomputing program. The connections built for the NSFNet became the first part of a nationwide internet backbone. The Internet of that time was used for scientific projects and still far away from commercial use. All of these networks and services were still funded either by ARPA project funds or by research facilities or universities. In 1995, the NSFNET was decommissioned and the so-called "New Internet Architecture" was introduced by defining four network access points (NAP) where commercial Internet Service Providers (ISP) could connect with one another based on peering agreements. The open NAPs were a major technical step for the commercialization of the Internet. Today, this task is taken over by the modern internet exchange points (IXP).

1986 A Complex Landscape
Around 1986, the landscape of computer networks was colorful and dominated by networks and backbones between research facilities, universities and contractors. DEC, XEROX and IBM had all started to implement their own versions of packet switching networks. Each of those networks run their own backbones and some of them had started to build gateways into other networks. A map of the Arpanet gateways dated June 18th, 1985, shows an impressive collection of different networks and their gateways to each other [18] (see Table 1).

The result was a rather complex architecture consisting of leased lines between different organizations—most of them labs or universities—and interfaces allowing users of one network to access certain services provided by befriended networks. The move to the common Internet standard TCP/IP was in progress, but not yet completed.

In the mid-1980s, the number of Internet hosts began to grow from around 2000 in 1985 to more than 300,000 in 1990. The boost for that growth was the commercialization of the Internet (see Fig. 8).

The Commercialization of the Internet

In the late 1980s, many organizations and commercial corporations began to develop interest in the global standardized network infrastructure. Outside of the Internet, some telecommunication companies had already started to build and offer electronic communication services to their customers. These first steps also brought up discussions about payment, business models and the question of whether commercialization and privatization of the Internet is a good thing. In the late 1980s, the first internet service providers (ISPs) emerged to take over the connection services

[18] Primary Internet Gateways—1985 June 18: http://www.livinginternet.com/i/ii_arpanet_gateways.htm

Table 1 Networks in 1985

Network	Country/Organization	Focus	Start	End
ARPANET	US	Research	1969	1990
DEC Net	DEC	Commercial	1975	2010
IBM VNET	IBM	Internal	1975	Existing
XEROX	XEROX	First Ethernet Implementation	1977	1990
USENET	US	Research	1979	Existing
SUNET	Sweden	Research	1979	1993
NETNORTH	Canada	Research	1980	1981
CSNET	US	Research	1981	1986
BITNET	US	Research	1981	2007
CCNET	DEC	Research	1982	1990
MILNET	US	Defense research	1983	Existing
JANET	UK	Research and education network	1983	Existing
NSFNET	US	Research	1986	1995

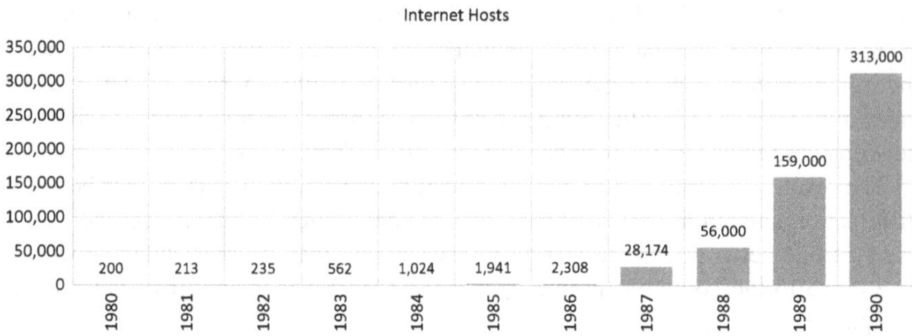

Fig. 8 Internet hosts between 1980 and 1990

and the operating of lines between the Internet nodes. It seemed that the research project of the Internet could be moved to an internationally available infrastructure operated by private companies as Internet Service Providers offering and billing connections and services to their customers. Therefore, the Arpanet was decommissioned in 1990. The Internet was fully commercialized by 1995, the same year that the NSFNet, a backbone provider, was decommissioned.

1979 Usenet: The First Internet Community

It was a group of researchers (Tom Truscott, Steve Bellovin and Jim Ellis) at the University of North Carolina at Chapel Hill and the Duke University that put together the puzzle in 1979 and created Usenet based on UUCP (Unix-to-Unix-Protocol). UUCP is a protocol suite integrated into the operating system UNIX and allows the exchange of email and files between Unix machines. Thus, Usenet was restricted to Unix machines but not to Arpanet connections. It could also be implemented using simple telephone lines and was also called the "poor man's Arpanet." Usenet was designed to be user community based on the idea of so-called "newsgroups" for specific topics, organized in a hierarchical order. For example, "sci.physics" and "sci.math" are within the science ("sci") branch, "comp.software" is a node within the software-branch. As a registered user of Usenet, you could register to one or many newsgroups thereby receiving news or contributing to the discussion. Thus, Usenet was the first Internet community creating an information exchange based on concepts, which are known today as blogs and forums. Interestingly enough, many terms we use today like "post", "thread", "FAQ" (frequently asked question) or "spam" have its origins in the first Usenet community. Within the next decade, UseNet became one of the most important information exchange communities for research institutions and universities around the world.

By 1984, the Usenet community had expanded to 1000 Usenet terminals at universities and research facilities and Usenet had evolved to become the first global network based communication media. Today, UseNet still exists but has lost importance due to the many other services available like blogs, social networks and different news and mail services. UseNet is still used for the exchange of binary data in the group "alt.binary." Being restricted to UNIX-users and the research community, UseNet had no economic influence, but had a major impact on the way people connect and exchange information. It was one of the first types of social networks, being a forerunner to the World Wide Web (1990) and the social networks starting in the mid-1990s.

1979 The Source

One of the competitors of CompuServe was The Source (Source Telecomputing Corporation) providing consumer connection services, news sources, weather, stock quotations, a shopping service, electronic mail, various databases, online text of magazines, and airline schedules. The Source's value proposition was one of the first directed to what cloud services would become a little bit later: *"It's not hardware. It's not software. But it can take your personal computer anywhere in the world."* The services were in operation between 1979 and 1989 after which time The Source was acquired by CompuServe.

By 1987, CompuServe had 380,000 subscribers and was the largest online service provider, compared to 320,000 at the Dow Jones News/Retrieval, 80,000 at The Source, and 70,000 at Genie.

1980 CompuServe and MicroNet

Providing Internet access for private households was a logical step as soon as the Internet was available for commercial and private use. One of the first to move on this was CompuServe, which was founded in 1969 as a subsidiary of Golden United Life Insurance and provided data center services for corporate customers based on time sharing systems and also offered its own packet switching network. In the early 1980s, CompuServe started to provider consumer services under the name of CIS (Consumer Information Service) later renamed to MicroNet and marketed through Radio Shack. The services contained dial-up connections, file transfer, one of the first chat programs, electronic mail, computer games and text versions of newspapers which collectively became known as the "CompuServe Experiment." Reading newspapers was not very practical, a download of a $0.20 newspaper would take up to 6 hours to download at costs of $5 per hour. Through the 1980s, CompuServe extended their services and aggressively recruited members building up a user community of more than 300,000 subscribers.

1983 MCI Mail
The first commercial email service was launched by MCI Communications as MCI Mail, allowing the sending and receiving of emails between MCI customers. It was the first commercial email service in the US. Access to the service was provided using a 110-, 300-, 1200-, 2400- or 5600-baud modem connected to a standard telephone landline. MCI Communications was acquired by WorldCom in 1998, the MCI Mail service was officially decommissioned in June 2003.

1984 Cisco
The rapid growth of the network infrastructure also became a major business for providers of routers and switches. The Silicon Valley based company Cisco was founded in 1984 by Leonard Bosnack, who was in charge of the Stanford University computer science department's computers, and his wife Sandy Lerner. Bosack had developed multi-protocol router software at Stanford, which became Cisco's first product. The company went public in 1990 with a market capitalization of $224 million and became one of the most successful corporations in the Internet infrastructure business.

1985 The Well
In the mid-1980s, the supporters of alternative social models began to discover the Internet as a new platform for communication and collaboration. This group was already well organized and welcomed the opportunity to use the new technology for the digital exchange of information and as a discussion forum. One of the main proponents was Stewart Brand, editor of the legendary "Whole Earth Catalog,"[19] a publication and American counterculture magazine published by Brand between 1968 and 1972. Together with Larry Brilliant, he founded "The Well" a digital version of the "Whole Earth Catalog" in 1985, which became the precursor of every online business and blog. The Well still exists today and "*...provide a cherished watering hole for articulate and playful thinkers from all walks of life*"[20] It also offered one of the first dial-up services and followed the transition to Web technology up to today. Building and maintaining a virtual community was one of the major visions of Brand and Brilliant.

1985 AOL Quantum Link
Beginning in the late 1980s, private companies took over. AOL (America On Line) was founded by Bill von Meister in 1983 under the name of CVC (Control Video Corporation) and became one of the most recognized brands on the Web in the US and a pioneer of the Internet for consumers. In 1985, an online service for home computers (Commodore 65, Apple II) was introduced under the name of Quantum

[19]The Whole Earth Catalog: http://www.wholeearth.com/ retrieved 2016-12-16.
[20]The Well: http://www.well.com/ retrieved 2016-12-16.

Link and Apple Link, later followed by PC Link for IBM PCs. In the beginning, the services provided primarily computer games and were later followed by email and instant messaging. Entering the Web age in the early 1990s, AOL became one of the most successful online service providers with more than 25 million users in 2002. Following an aggressive acquisition strategy, AOL purchased Netscape in 1999 and merged with Time Warner in 2000. In the following years, AOL rapidly declined not being able to follow the transition from dial-up connections to broadband. In 2015, AOL was acquired by Verizon for $4.4 Billion, and is a subsidiary of Verizon today.

1987 UUNET

One of the first Internet service providers and tier-1 provider was UUNET founded in 1987. UUNET still exists as a brand of Verizon. Others followed, creating an ecosystem of Tier-1 providers operating the backbones and many Tier-2 or ISP and offering a connection to the Internet for private persons and corporations.

1988 Commercial Use for Corporations

The coordination of networking in the US was the responsibility of the Federal Networking Council (FNC), a sub-organization of the US National Science and Technology Council. In 1988, the FNC decided to allow the development of a gateway between the Internet and MCI Mail. In the same year, many other commercial e-mail services got permission to operate similar connections. This was the starting point of the commercialization of the Internet. Three years later, the non-profit organization, Advanced Network Services (ANS), established by IBM, Merit and MCI was allowed to carry commercial data over the Internet backbone. NSF required ANS to charge the cost to its customers, to set aside any profits in a pool to enhance network infrastructure and finally—to ensure that commercial traffic did not diminish NSFNet traffic.

1990 The Last Mile: Dial Up Versus Broadband

At the beginning of the Internet and the first years of the World Wide Web, the connection of a private home to the global net was organized via phone lines and dial-up modems. While a typical LAN at that time operated at a speed of 10 Mbit/s, the data rates with modems were much more limited and slowly grew from 300 bit/s to 1200 bit/s in the early 1980s and to 56 Kbit/s by the late 1990s. It soon became clear that the restricted access to the Internet would block the distribution. In the US, the National Information Infrastructure initiative made broadband Internet access a public policy issue. Having broadband access to the internet provider offers a completely different type of user experience.

Broadband access not only speeds up the download and upload of data but also allows an "always-on"-modus, so there is no need to dial-up to the ISP exchange point since the connection is always active. At the end of the 1990s, dial-up connections using modems could provide a download speed of 4800 bit/s and 2400 bit/s for upload. In 1984, the CCITT (now ITU) proposed a new technology called ISDN (Integrated Services Digital Network) based on the idea of a simultaneous transport of digital data, video and voice over a telephone connection. The

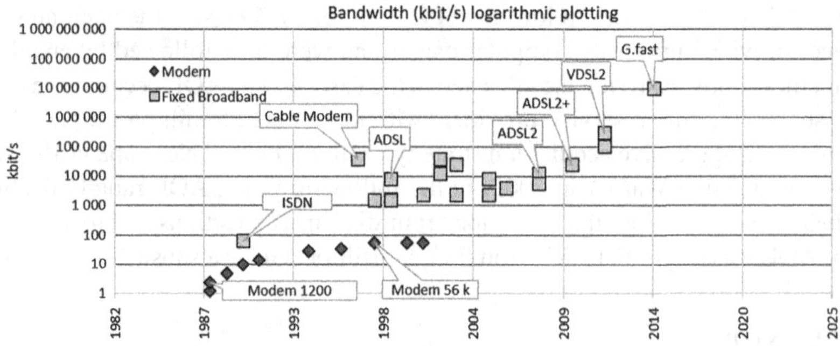

Fig. 9 Bandwidth for the last mile: Dial-up modem versus ISDN

standard was approved and led to the digitalization of the telephone network. The first ISDN routers became available around 1990 speeding up internet connections to 64 Kbit/s which was more than 10 times the bandwidth compared to dial-up modems (see Fig. 9).

For the next decade, ISDN became a work horse for digital connections to the internet for companies and private households.

The IT Market at the End of the 1980s

The IT market had not changed very much since the 1970s. IBM was leading with a market share of more than 50% followed by Digital and Hewlett Packard as the major producers of mid-range servers and PCs. The rest of the seven dwarfs (Unisys, Burroughs, Control Data, Honeywell, Data General and NCR) were far behind and loosing territory (see Fig. 10).

Microsoft and Oracle began to grow into software businesses, Apple was on the radar screen for the first time and the processor manufacturers Intel, National Semiconductor, Texas Instruments and Motorola competed against each other in the growing micro-processor market (see Fig. 11).

The Internet Before the Web

At the end of the 1990s, the Internet had been expanded to 300,000 hosts and was operated by backbone providers (tier-1) and a growing number of Internet service providers offering connectivity to companies as well as a growing number of private households. The Internet infrastructure was based on connections operated by the telephone providers and backbone operators running the network access points and internet exchange points (see Fig. 12).

Three years later, in 1993, the number of hosts would break the 1 million mark. In these 3 years, a disruptive engineering achievement changed the IT market and the application of the Internet forever.

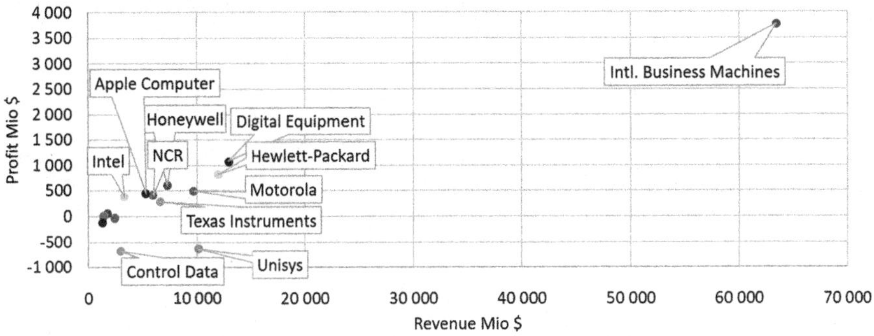

Fig. 10 Top IT companies 1990

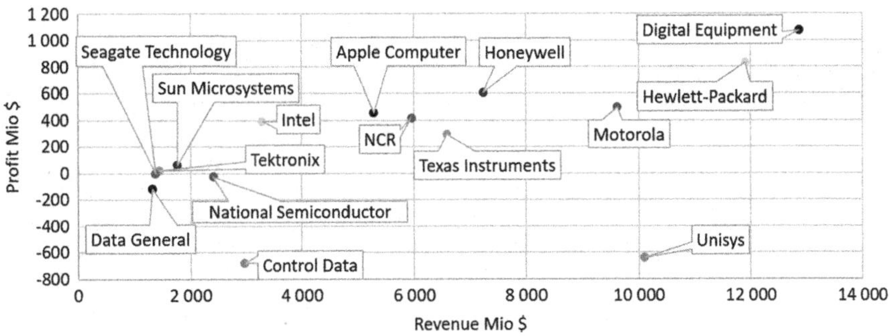

Fig. 11 Top IT companies 1990 without IBM

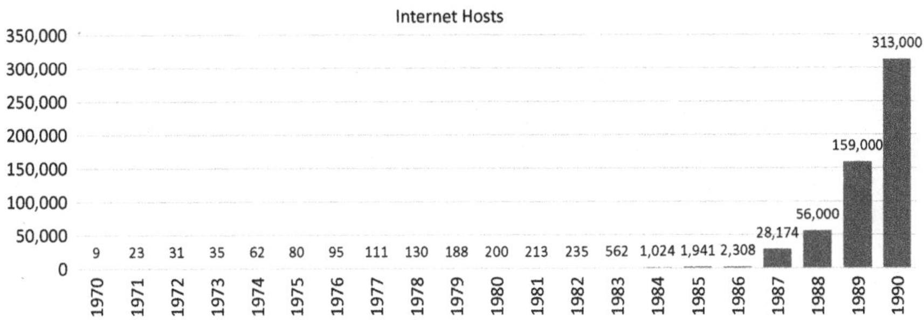

Fig. 12 Internet hosts between 1970 and 1990

Reference

Popper, S. K. (1945). *The open society and its enemies*. London: Routledge.

World Wide Web

I just had to take the hypertext idea and connect it to the Transmission Control Protocol and domain name system ideas and—ta-da!—the World Wide Web ...

Tim Berners-Lee

The World Wide Web Is Born

1990 The First Webpage

The World Wide Web was born just before Christmas in a lab in a European research institute for nuclear physics. On December 20, 1990, in Switzerland the first website went live, 1 year later on December 12, 1991, the Web jumped over the Atlantic. At Stanford University, the first website outside of Europe was launched increasing the total number of worldwide websites to ten. Today, the total number of websites online is more than 900 million and the number of World Wide Web users is just below 3.5 billion[1] (see Fig. 1).

It was a team of engineers and scientists at CERN that put together what was already there and connected the dots with a simple but ground breaking idea. Back in 1990, the world of computing had already reached the individual user. Personal computers were introduced around 1980 and the user interface development went through many rapid development steps. The first "Windows"-based user interface was developed in the early 1970s by XEROX Parc with the XEROX Alto workstation, Windows 1.0 was introduced by Microsoft in 1982, and the Apple Macintosh was released in 1984. Providing a graphical user interface to consumers was the hype of the decade. But connecting the local PC to other sources of data and information was difficult and far from easy to use.

In 1980, Tim Berners-Lee was a 25 years old contractor at CERN, the European Organization for Nuclear Research in Geneva, Switzerland for no more than half a year. During this time, he recognized that sharing research papers within the academic community was time consuming and painful. The work of the

[1]http://www.internetlivestats.com/ retrieved 2016-12-16.

M. Oppitz, P. Tomsu, *Inventing the Cloud Century*,
DOI 10.1007/978-3-319-61161-7_10

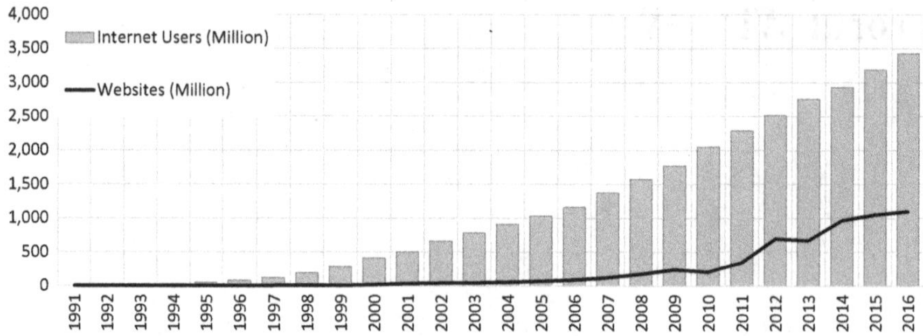

Fig. 1 Internet users and websites between 1991 and 2016

international research community is mainly based on sharing and exchanging information structured in research papers that are stored in the local computers and hosts of worldwide distributed research organizations. Getting easy and quick access to this information in a globally connected world is vital. So, Tim Berners-Lee started to think about an easy method to combine local workstation computing with the opportunities offered by a global network like the Internet. At the same time he was convinced that this method should be based on the idea of a simple and highly standardized protocol for information retrieval and should also follow the idea of the thin client. That would allow the implementation or migration of the software used to retrieve and display information from remote computers on different types of workstations independent from the hardware or operating system used. Berners-Lee started to think about a standardized information format, which could be used to store documents on computers and retrieve them from other workstations.

As a preliminary concept, he proposed using hypertext and built a prototype named ENQUIRE. Hypertext is a concept that allows for the display of text on a computer screen including so-called "hyperlinks," which refer to other parts of the same text or other documents and allow the user access to these text parts by pointing on the link—today we would "click on it." The principles of hypertext have many parents going back to the 1940s. A legend says that the short story, "The Garden of Forking Paths" by Jorge Luis Borges, served as the inspiration for the hypertext concept. Several conceptual works and some implementations were introduced by the 1980s, none of them successfully adopted by the community. Berners-Lee left CERN in December 1980 to work with Image Computer Systems in the UK and returned to CERN again in 1984.

In 1989, CERN had become the largest Internet node in Europe and Berners-Lee again returned to his idea of using hypertext, but this time together with the capabilities of the Internet to provide a simple and effect full way to create and distribute information in the research community: *"I just had to take the hypertext idea and connect it to the Transmission Control Protocol and domain name system ideas and—ta-da!—the World Wide Web . . . Creating the Web was really an act of desperation, because the situation without it was very difficult when I was working at CERN later. Most of the technology involved in the Web, like the hypertext, like the Internet, multi-font text objects, had all been designed already. I just had to put them together. It was a step of generalizing, going to a higher level of abstraction, thinking about all the documentation systems out there as being possibly part of a larger imaginary documentation system."*[2]

Berners-Lee wrote the first proposal in 1989 and a revised version in 1990,[3] which was accepted by his manager, Mike Sendall Fig. 2.

The idea for this concept originated in a new protocol on top of TCP/IP named HTTP (hypertext transfer protocol) and a simple but effective language for structuring information named HTML (hypertext markup language). Based on these 2 elements, it was possible to develop the two missing links: the so-called HTTP server, a comparably simple piece of software, is installed on a host or server enabling the retrieval of HTML documents stored on that server. The HTML browser is the software needed on the remote workstation to request a view of a HTML document from the server. The underlying transport protocol is TCP/IP. Mike Sendall bought a Next computer, which was the machine used by Tim Berners-Lee to build his first version of what was later called the World Wide Web. The first webpage was set to life by Tim Berners-Lee on the Next computer on December 20, 1990. The page still exists[4] and is hosted by CERN as a historical milestone in Internet history (see Fig. 3):

1994 The W3C: World Wide Web Consortium

Berners-Lee was convinced by the idea that software, standards and information are resources that should be available to everybody and thus made his idea available and free. There is no patent on the World Wide Web or the underlying concepts and implementations, the usage is royalty free. To provide an international platform for further development of the "WWW," Berners-Lee founded the W3C (World Wide Web Consortium) in 1994 at MIT. The W3C is the main international standards organization for the World Wide Web and also decided that its standards should be royalty free and easily adopted by everyone. Since its start, the W3C has created and maintains many standards shaping the Web, among them HTML itself, but also standards like XML, SOAP and CSS. The W3C has more than 400 members including large IT enterprises like Apple, Google, Microsoft, Oracle, Cisco, AT&T, IBM and others. In 2004, Tim Berners-Lee was knighted and promoted to Knight of the British Empire. Sir Timothy Berners-Lee is still a professor at MIT and director of the World Wide Web Consortium. He still promotes the free access and usage of the Web and is one of the pioneers in favor of Net Neutrality.

---

[2]Berners-Lee, T. (2007, June 22). *Interview with Tim Berners-Lee*: http://www.achievement.org/autodoc/page/ber1int-1, retrieved 2016-11-16.

[3]Berners-Lee: "Information Management: a Proposal", http://info.cern.ch/Proposal.html, retrieved 2016-11-16.

[4]CERN: "The first Webpage" http://info.cern.ch/hypertext/WWW/TheProject.html, retrieved 2016-11-16.

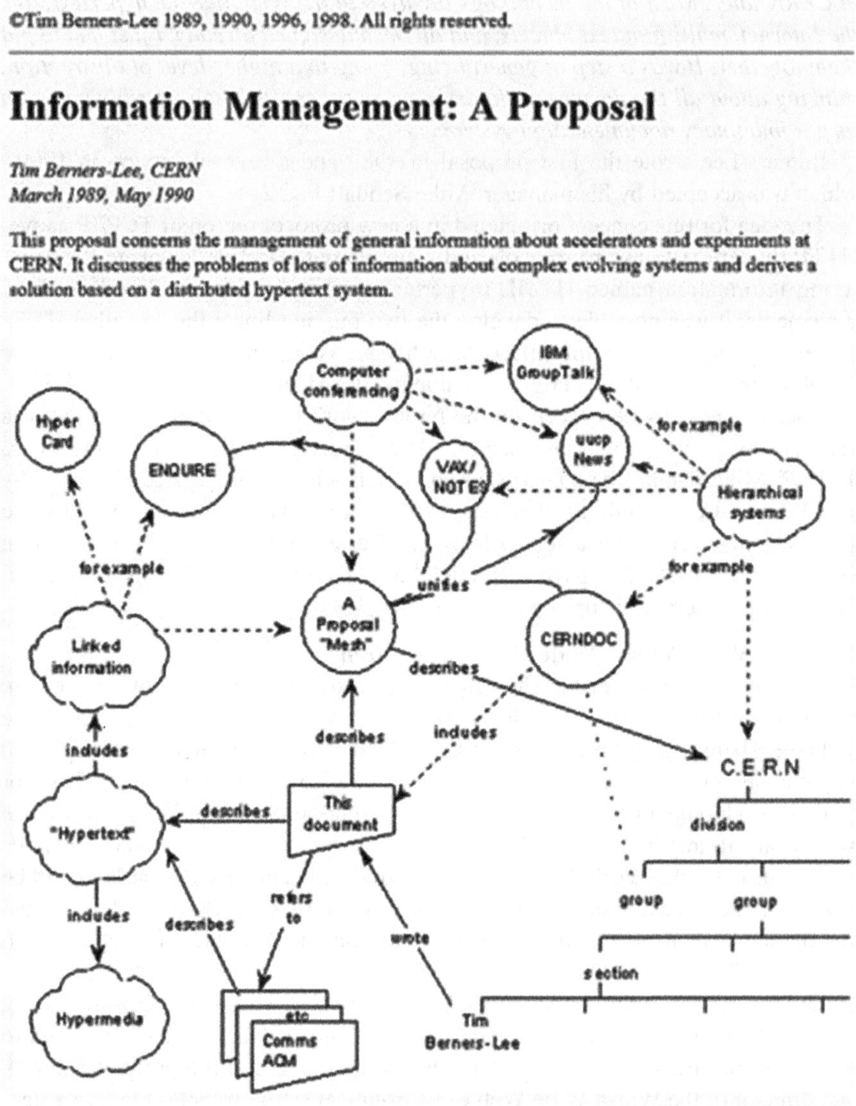

Information Management: A Proposal

Tim Berners-Lee, CERN
March 1989, May 1990

This proposal concerns the management of general information about accelerators and experiments at CERN. It discusses the problems of loss of information about complex evolving systems and derives a solution based on a distributed hypertext system.

Fig. 2 Berners-Lee proposal from May 1990

Berners-Lee is also the one who is responsible for the two slashes ("//") in a Web address, it's his design. In a Times article in October 2009, he admitted that they were actually unnecessary: "*He told the Times newspaper that he could easily have designed URLs not to have the forward slashes. "There you go, it seemed like a good idea at the time," he said. He admitted that when he devised the Web, almost 20 years ago, he had no idea that the forward slashes in every Web address would*

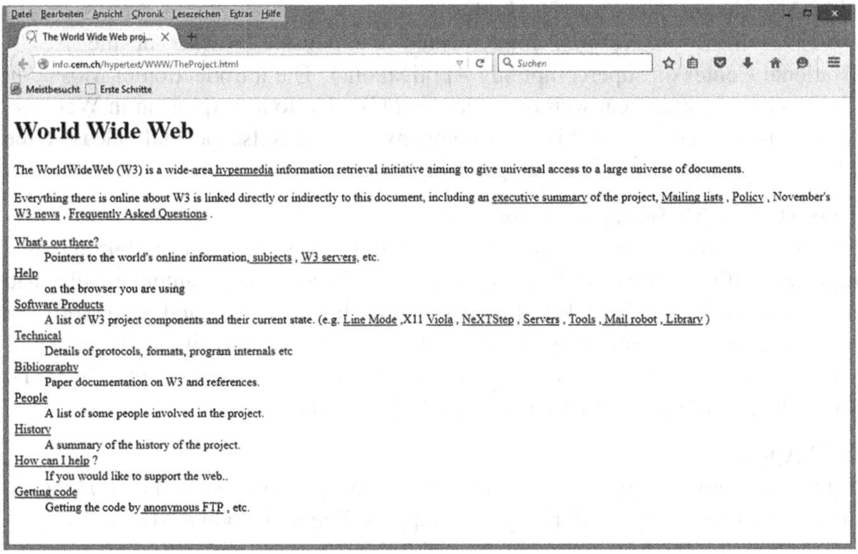

Fig. 3 The first web page

cause "so much hassle". His light-hearted apology even had a green angle as he accepted that having to add // to every address had wasted time, printing and paper"[5]

Browsers

The concept was refined during the next couple of years. In 1994, the final specification of URL (unique resource locator) was published, enabling the reference of websites by names in addition to the IP-address.

1991 Berners-Lee Worldwide Web Browser
The first browser was developed by Tim Berners-Lee and the CERN team and named "WorldWideWeb," the name was later changed to Nexus. The "WorldWideWeb"-browser was implemented on a Next computer and not portable to other machines. This restricted the usage during the first years to Next-users. At the same time, other teams started to implement text-only browsers for UNIX machines. Still, the lack of widely available browsers was a blocking issue for the acceptance of the World Wide Web.

[5]BBC. (2009). Berners-Lee 'sorry' for slashes http://news.bbc.co.uk/2/hi/technology/8306631.stm retrieved 2016-12-10.

1993 Mosaic and Netscape Navigator

In 1993, Marc Andreessen implemented the Mosaic-browser at the NSCA (National Center of Supercomputing Applications). The introduction of Mosaic in 1993—the first graphical web browsers for PCs—led to an explosion in Web use. Marc Andreessen started his own company named Netscape and released the Mosaic-influenced Netscape Navigator in 1994.

1995 Microsoft's Internet Explorer

Microsoft followed the idea and introduced its Internet Explorer in 1995 bundled together with Windows 95, so that every Windows user automatically had Microsoft's browser. That lead to the so-called browser wars and Microsoft was charged of using its monopoly to the disadvantage of its competitors. The case was settled in 2001 and Microsoft had to open its software programming interfaces and allow PC manufacturers to bundle its product with third party software.

2003 Apple Safari

Safari was introduced by Apple at the MacWorld conference in 2003 as an alternative to Netscape. Safari is part of Apples iPhone, iPad and MacOS.

2004 Mozilla and Firefox

In 1998, Netscape created the Mozilla Organization to co-ordinate the development of the Mozilla Application Suite. Netscape was acquired by America Online in 1999 for $10 billion. Because of the acquisition, AOL scaled back its involvement with the Mozilla Organization and, in July 2003, the Mozilla Foundation was launched to ensure Mozilla could still survive without Netscape. The Mozilla browser finally led to the Firefox browser in 2004, one of the most popular products today. Today, browsers belong to the large group of free software. Providing the basic building blocks of the World Wide Web as open source components became a hallmark approach of the web community.

Marc Andreesen, the creator of Mozilla, left AOL in 2000 to form Loudcloud together with Ben Horowitz and others. Andreesen and Horowitz started a new career as investors and incorporated as Andreessen Horowitz in 2009, still one of the most successful venture capital firms in Silicon Valley.

2008 Google Chrome

Google started its own browser project in 2008 releasing the first version of Chrome for Microsoft Windows. The Android version followed in 2012. The focus of Chrome was on security and performance. Today, Goggle's Chrome is the most widely used browser with more than 70% of all browser usage.

Figure 4 shows how the browser usage evolved between 2007 and 2016.

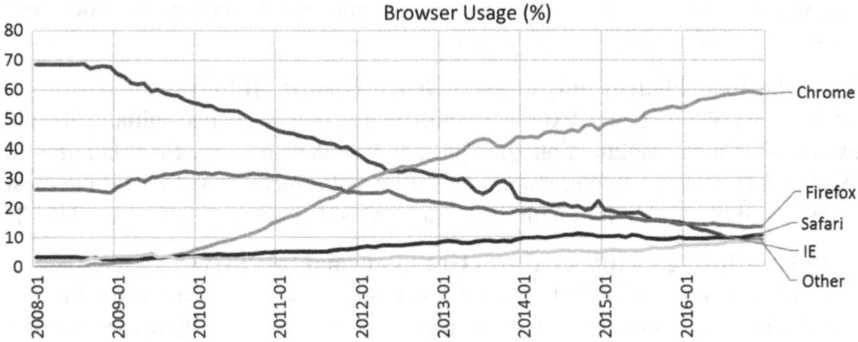

Fig. 4 Browser usage between 2007 and 2016

Sharing Pictures, Music and Video

In the 1980s, it became clear that computers are not only number crunchers but text processing machines. Today, more than 70% of network traffic or 38 petabytes per month is audio or video data. Forecasts are saying that this share of data volume will have a CAGR (Compound Annual Growth Rate) of 31% until 2020.[6]

With the rise of graphical user interfaces (Xerox Work Station 1973, Apple Macintosh 1982 and Microsoft Windows 1982), it became evident that graphic representation of data would be an important demand in the future. The problem was that there was no standard for graphical data and it was also clear that—compared to text—images needed much more storage space and bandwidth. The method of storing and processing images at that point of time was simple: images were translated into a matrix of dots called pixels each of them either taking 1 bit of storage (black and white) or 1 byte of storage (equal to 256 colors or shades of grey). Storing, processing or exchanging graphical data using that structure was costly, so the challenge was to find methods to standardize graphic and media formats and develop methods to reduce their size by applying compression algorithms.

1987 Graphical User Interface Format (GIF)

One of the first attempts was made by CompuServe, one of the first network service companies in the US. CompuServe introduced the GIF (Graphical User Interface Format) in 1987 using the LZW (Lempel–Ziv–Welch) data compression, which was one of the most efficient compression algorithms at that point of time. The format supported 8-bit per pixel thus allowing 256 colors per pixel. The LZW compression was patented in 1994 by Unisys and led to controversy between

[6]CISCO VNI Forecast and Methodology, 2015-2020 http://www.cisco.com/c/en/us/solutions/collateral/service-provider/visual-networking-index-vni/complete-white-paper-c11-481360.html, retrieved 2016-11-30.

Unisys and CompuServe. As a consequence, CompuServe focused on the development of PNG (Portable Network Graphics).

1992 The Joint Photography Experts Group Format (JPEG)
JPEG (Joint Photography Experts Group) is a standard created initially for the exchange of high resolution photographs and was developed by Eric Hamilton in 1992. JPEG defines a compression method for color pictures and is still the most widely used file format for pictures. Its standard is defined by ISO/IEC and ITU-T.

1993 The Moving Pictures Experts Group Format (MPEG)
With the introduction of CDs and DVDs, it soon became apparent that moving pictures or video would be one of the important data types distributed via the new media. The International Standard Organization (ISO) together with the International Electrical Commission decided to establish a working group to develop standards for audio and video compression- the "Moving Picture Experts Group" (MPEG). MPEG-1 was released as the standard in 1993 and defined the coding of moving pictures and the associated audio stream.

1993 Page Description Format (PDF)
It was a Silicon Valley based company that filled in one of the last gaps in file format standards. Adobe was founded by John Warnock and Charles Geschke after leaving XEROX PARC in 1982 and with the plan of developing and selling the PostScript page description language. Adobe soon became a successful company accelerated by the Apple Computer decision to license PostScript for use in its LaserWriter printers. In 1991, John Warnock created the first page description format with the name "Camelot" as the forerunner for PDF. The PDF specification was released by Adobe free of charge in 1993 but still controlled by Adobe. In 2008, Adobe decided to release PDF as open standard and passed the control to ISO. Since then, PDF is the standard format for document exchange between different applications.

1994 MP3
Storing and transport of audio data leads to a classical dilemma: if the size of the data file is small so that storing and transport is easy, the quality of the recording is poor. Enhancing the quality leads to huge file sizes and timely and expensive transport. The challenge was to find a compression method that reduces the data size without any quality loss. MP3 is part of the MPEG-1 standard and was developed by a group of scientists at the Fraunhofer Institute in Erlangen, Germany in cooperation with Bell Labs and Thomson. The method the team used is referred to as "perceptual coding" and uses psychoacoustic models to reduce those parts from the audio data that are less audible to humans. The result is a compression rate between 75 and 90% with a still high level of sound quality. The first MP3 encoder was released in 1994 and soon found its application in not only MP3-players but, more importantly, in the growing numbers of audio exchange platforms on the Web.

1995 Portable Network Graphics (PNG)
Finally, it was time for an improved version of the graphics interchange format GIF. GIF is limited to 256 colors and was also still patented by Unisys. A group of graphics enthusiasts lead by Thomas Boutell started to work on an improved standard for graphics files in 1996. The standard was approved by the W3C in the same year and became an international ISO standard in 2003.

Starting with Web Portals and Search

Once browsers like Mosaic were available for PC's, the next logical step was to create business in providing services for the rapidly growing community of web users. New founded companies began creating offers to build the first web portals. The idea was to collect content that was available on the Web and built a portal page with links to that content's pages. The race was on and portals became the first entry point for web users. Those early portals were organized like catalogues of web links ordered in categories and focused on themes like news, weather reports, financial information and mailing lists.

The concept of providing portals as lists of web links was not exactly what Berners-Lee was trying to initiate because it structured the Web in a hierarchical order, which was defined by the creators of the portal as opposed to the network structure of linked documents where users follow their own paths to seek information on the World Wide Web. The missing link was something that allowed the user to explore that network of links easily and free of defined categories and hierarchies. With the growing number of pages and URL's available on the World Wide Web, the idea was raised to provide search functions, which would allow the user to growl through the Web without using predefined lists of links. Beginning in the early 1990s, several projects were started, mostly at universities, implementing different types of search engines for either FTP-servers or later web pages examples being the W3Catalog (University of Geneva), the World Wide Web Wanderer (MIT), Lycos (Carnegie Mellon University) (see Table 1).

1993 IMDb
One of the first services launched on the Internet and later the Web was IMDb (Internet Movie Data Base). IMDb has its origins in a UseNet mailing list about movies from the 1980s. It was moved to the Web under the name of "Cardiff Internet Movie Database" in 1993. In 1996, the "Internet Movie Database" was incorporated in the U.K. and 2 years later acquired by Amazon for approximately $56 million.[7]

[7] Amazon sec-filing 1998: http://www.secinfo.com/dr643.7kp.c.htm#2ndPage retrieved 2016-12-16.

Table 1 Timeline of early portals and search engines

Year	Service	Type	Country
1990	IMDb Internet Movie Database	Media	UK
1993	W3 Catalog	Portal and Search	USA
1993	World Wide Web Wanderer	Search	USA
1994	Lycos Web portal and search engine	Portal and search	USA
1994	BBC Networking Club	Portal	UK
1994	Microsoft MSN	Portal	USA
1995	AltaVista search engine	Portal and search	USA
1995	Microsoft Hotmail web-based email	Mail	USA
1995	The Globe portal and social network	Social network	USA
1995	Yahoo! Search	Portal and search	USA
1996	Netscape browser	Browser	USA

1994 Yahoo!
One of the first companies that adopted the idea of offering information and content to a wide target group via the Internet was Yahoo!. Yahoo! was founded by Jerry Yang and David Filo in Sunnyvale, California in January 1994. Both were students at Stanford University and had created a web page under the name "Jerry and David's Guide to the World Wide Web." They renamed their page to Yahoo! and provided it as portal to the World Wide Web. Within a short time, Yahoo! received one million hits and Filo and Lang realized that this type of service had a huge business potential. Yahoo! was incorporated in 1994 and had its initial public offering in 1996, raising $33.8 million. In the following years, Yahoo introduced additional services including mail service, financial information services and one of the first Web search engines: Yahoo! Search. Yahoo! followed the strategy of acquiring other companies to diversify their services offered. During the dot-com bubble, Yahoo! was shaken like many other Internet companies but in the end survived. Today, Yahoo! is still one of the major service providers on the Web generating a revenue of $4968 billion (2015).

1994 Microsoft MSN
Microsoft, under its founder Bill Gates, was very reluctant about the Internet and the World Wide Web in the years between 1990 and 1995. The philosophy of open communication and open standards and software was not exactly in line with Bill Gates and Microsoft's understanding of software and information distribution. As a first step, Microsoft decided to create a closed shop platform consisting of Microsoft's own web browser and the "Microsoft Network" as the web service for Microsoft customers including news services, mail services (Hotmail) and a dial-up internet connection. It soon became evident that this approach would not lead to success. MSN was a direct competitor of AOL and Yahoo! and not seen as a strategic product within Microsoft. On the other side, the consumers began to understand that access and services in the new World Wide Web are basically open and free, which did not fit into Microsoft's business strategy at that time. It would take many years for Microsoft to gain traction again in the World Wide Web

and cloud services market. Today, Microsoft is one of the major players in platform as a service for business solutions including its office application platform and its Azure platform.

1994 BBC Networking Club
On the other side of the Atlantic, the BBC began to think about embracing the new technology from a media company's point-of-view. The BBC Networking Club was launched on April 11, 1994 and offered access to the Internet and one of the first bulletin boards called "Auntie," providing information and asking for feedback about BBC programs. The service was rather expensive compared to today: Joining fee was £25 and the monthly subscription was £12 and included access to an early type of social networking and a dialup Internet connection service.

1994 Lycos
Lycos Inc. was established in 1994 as a spin-off from Carnegie-Mellon and soon became a successful business that offered a web portal including search features. Lycos had its first public offering in 1996 and was one of the first profitable Internet businesses in the 1990s.

1995 Hotmail
Providing mail as a service was one of the first applications delivered to the growing community of web users. Having an email account in the early 1990s was restricted to employees of large companies or members of universities and research facilities running their own mail servers. Running private email accounts required setting up a private email server and was complicated and costly. The idea of using the Web as the platform for a public email service open for members of web portals appeared in the early 1990s when email as a service was introduced by several providers. One of the first was Hotmail, founded in 1995 by Sabeer Bhatia und Jack Smith, using "HTML" as the alternative acronym for the company name HoTMaiL. The service grew rapidly, collecting 12 Million users within the first 2 years. Hotmail was acquired by Microsoft for $400 Million in 1997 and became known as Microsoft Hotmail, a major element of Microsoft's MSN. At the same time—in 1997—Yahoo launched Yahoo! Mail, an email service of its own. Together with Google Mail—launched in 2009—Microsoft Hotmail and Yahoo! Mail are the major email providers in the US today.

1995 AltaVista
AltaVista was created by Paul Flaherty, Louis Monier and Michael Burrows at Digital Equipment Corporation's Network Systems Laboratory and was launched in 1995 as one of the first search engines on the Web. One year later, AltaVista became the exclusive search engine for Yahoo!. After the acquisition of Digital by Compaq, it became clear that AltaVista's strategy as a part of Compaq was no longer successful, it also lost its market share especially to Google and was finally shut down in 2011.

1996 Netscape
Providing efficient searches soon turned out to be a deal maker for portal providers. Users began to embrace the idea of free search and browsing through the growing number of pages on the Web. In 1996, Netscape was looking for an effective search engine to provide in their portal services and decided to close deals with five of the major search engines: Yahoo!, Magellan, Lycos, Infoseek, and Excite. Netscape payed $5 million a year and each search engine would be in rotation on the Netscape search engine page.

Improving Efficiency: Java, PHP and Web Services

In the same decade, a group of computer scientists at SUN Microsystems started to think about software for media and network based applications like interactive television. In this case, the application needed to be distributed between a centralized server and local devices like TV sets which are not primarily computers.

1995 Java
The basic idea behind Java was that an application can be executed by a so-called "Java Virtual Machine," a piece of software that can run not only on computers but also on TV sets or—maybe—coffee machines. The story goes that this was the origin of the name "Java." James Gosling, Mike Sheridan, and Patrick Naughton initiated the Java language project in June 1991, and the first public implementation was released as Java 1.0 in 1995. The promise of the Java design was "Write Once, Run Anywhere" (WORA). In the beginning, Sun approached ISO to formalize a standard for Java, but drew back from this approach and decided to create the Java Community Process (JCP) in 1998 as the open standardization platform for everybody who wants to contribute to standard technical specifications for Java. Sun Microsystems saw its role as an "evangelist" for that new technology (Sun VP Rich Green, 2002).

Today there are more than 1000 voting members in JCP. Soon Java became the de facto standard for web applications. Therefore, in 2006 and 2007, Sun released its Java Virtual Machine (JVM) as free and open-source software, under the terms of the GNU General Public License thereby finally moving Java to the open source community. After the acquisition of Sun Microsystems by ORACLE in 2010, ORACLE announced that it would follow an open strategy but soon critical statements about the openness of ORACLE were heard, leading to the resignation of the Apache Software Foundation from the JCP-board. Nevertheless, Java is, together with C, the most widely used programming language today. Much of the Internet and web applications are based on Java and its additions and classes like Java Servlets and Java Server Pages. Java also has a strong relationship with HTML and browser architecture thus forming a complete but also expandable framework for web applications.

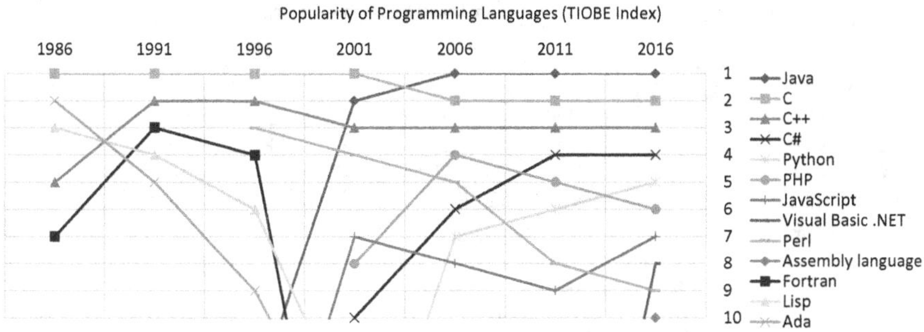

Fig. 5 Programming community Index

The TIOBE Index[8] measures the popularity of programming languages using ratings that are based on the number of skilled engineers world-wide, courses and third party vendors as well as search engines such as Google, Bing, Yahoo!, Wikipedia, Amazon, YouTube and Baidu (see Fig. 5).

Since the introduction of Java and the release of the Java virtual machine as open-source software, Java became the most popular programming language followed by the C-programming language including C++ and C-sharp (C#).

1998 PHP

In 1994, the Danish software engineer, Rasmus Lerdorf, started to create a set of functions to track the visits on his personal webpage coining them as "Personal Home Page Tools" or PHP Tools.[9] Over time, more functions were added including database interaction and a framework supporting the development of dynamic web applications. In 1995, Rasmusen released the source code for PHP, which soon found the interest of a large community of developers around the world. The two Israeli software engineers, Andi Gutmans and Zeev Suraski, joined the PHP team in 1997. The original PHP code was rewritten and expanded leading to the release of PHP 3 in 1998. Within a short time, PHP became one of the most successful script languages for web applications. It is licensed as free software under the so-called "PHP-License,"[10] the development and version policy is coordinated by the PHP Group.[11]

1998 More Bandwidth: DSL, ADSL, VDSL and Cable Connections

ISDN became the major technology for Internet access in 1990. ISDN was based on 2 channels each of them with a bandwidth of 64 Kbit/s. With DSL (Digital Subscriber Loop), an alternative concept was introduced in the early 1990s using higher frequency bands to transmit data via existing copper wires. The concept was

[8]TIOBE Index: http://www.tiobe.com/tiobe-index//, retrieved 2016-11-30.

[9]The History of PHP: http://php.net/manual/en/history.php.php, retrieved 2016-12-16.

[10]PHP License: http://www.php.net/license/ retrieved 2016-12-16.

[11]PHP Group: http://php.net/ retrieved 2016-12-16.

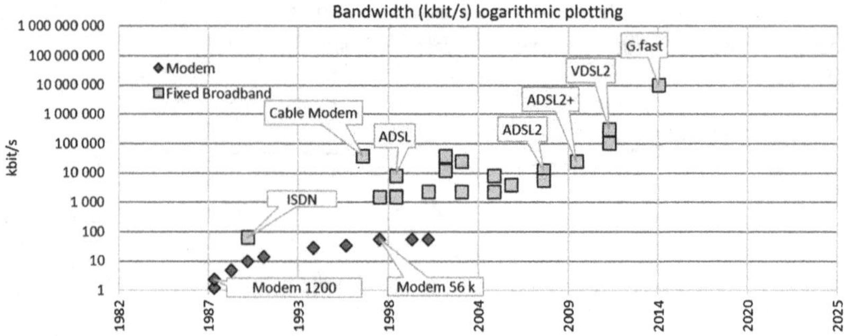

Fig. 6 Bandwidth for the last mile: ISDN versus ADSL and cable modems

refined by introducing ADSL DSL (Asymmetric Digital Subscriber Loop), which split the connection into two modes of transport: one for the fast streaming of data, the other one as the "interleaved channel" for the error-free transport of files and critical data. This allowed Internet service providers to offer more bandwidth for downloading data by reducing the rarely needed upload of data. The first publicly available ADSL routers provided 8 Mbit/s download rate, which was more than ten times the access speed of ISDN routers (see Fig. 6).

At the same time, the TV cable providers started to offer Internet connections using their TCV-cable network instead of the telephone lines. TV cable networks initially had a much higher bandwidth than twisted-pair telephone lines and cable providers could offer that bandwidth as an Internet service provider to their existing customers. The first cable modems were introduced in 1997 and offered a bandwidth of up to 38 Mbit/s thus beating ISDN by a factor of 800 and ADSL by a factor of 4. ADSL and cable modems were improved in the following years achieving access speeds of 100 Mbit/s or more today.

1998 Web Services

Once the Web was introduced and accepted by a rapidly growing number of users, the next question arrived. Services provided on the Web for human users may be based on other Internet services. Each of these services may be responsible for delivering a specific piece of information or cover a specific part of a process. A web page providing information about a city could integrate information about the current weather conditions. A website providing information about a company could integrate the current stock price. The methods available like HTML are not easy to apply to that, they are designed for machine-to-human communication. What was needed was a new definition of a "web service" which simply means a dedicated service, designed to deliver specific information or data that can be easily consumed or integrated into another web service or web page (see Fig. 7).

The introduction of standardized methods for machine-to-machine interfaces like SOAP or REST led to a new concept of web service interaction.

Fig. 7 Web Services

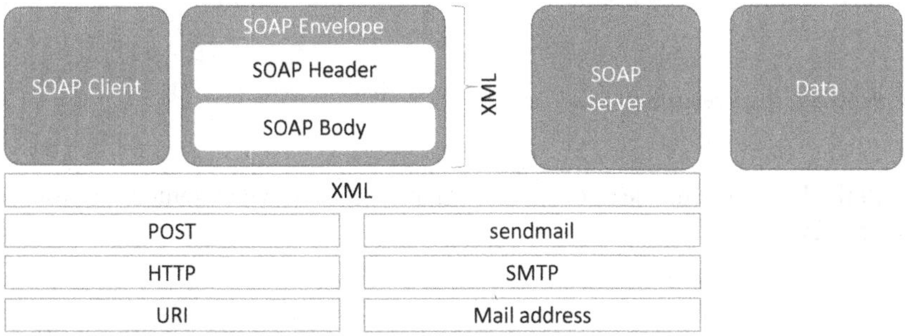

Fig. 8 SOAP: Simple Object Access Protocol

1998 SOAP: Simple Object Access Protocol

In 1998, Dave Winer, Don Box, Bob Atkinson, and Mohsen Al-Ghosein—at that time working for Microsoft—came out with a completely new approach named SOAP (Simple Object Access Protocol). SOAP is based on the idea that a service (SOAP receiver) may accept standardized messages from a requestor (SOAP sender) and answer these messages with a result set consisting of data (e.g. the stock price or weather report data) (see Fig. 8).

The message is structured in an envelope, header and content section. The message data is formatted as an XML-file. XML (Extended Markup Language) was defined at the same time, and W3C (World Wide Web Consortium) recommended XML 1.0 on February 10, 1998. At the same time, several open source SOAP-client libraries and SOAP-server packages were developed by the web community to power the acceptance of the new machine-to-machine communication method. SOAP became the first type of standardized web service.

2000 REST: Representational State Transfer

At the same time, another architecture was proposed for machine-to-machine interaction also with the strange name REST (Representational State Transfer). The term was first introduced in 2000 by Roy Fielding in his doctoral dissertation. REST is also based on HTTP, but uses the so-called CRUD (Create, Read, Update, Delete) functions (see Fig. 9).

The basic idea of REST is that each request is one of these basic operations and thus can easily be communicated via HTTP POST, GET, PUT or DELETE. REST claims to be much easier to scale and to change than SOAP. Today RESTful API's

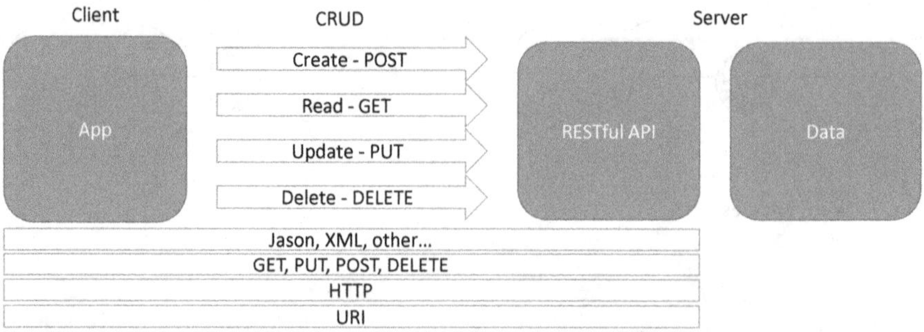

Fig. 9 REST: Representational State Transfer

are all the hype, and the developer's community sees SOAP as something from the last century.

Building First Businesses

In the mid-1990s, the basic elements of the Internet and the Web were available. The Internet was commercial and had started its geographical expansion. Networking and routing technologies had reached a certain level of maturity. The Web, including browsers and web servers, was widely accepted by the users.

Developers had an impressive selection of software tools including UNIX, Linux, C++ and Java. Network infrastructure providers like Cisco developed and delivered network gear and telecom providers started to connect their customer to the net. So, it was time for business! But the road to economic success turned out to be bumpy (see Table 2).

Starting with e-Commerce

In the mid-1990s, it became clear that the World Wide Web would develop into a global service network, which would be a splendid platform for doing digital business with consumers. Within no more than 5 years, several new and disruptive business models were created by new companies that became known as "born on the web." Those business models were based on offering ordering, buying and payment services to private consumers for primarily books, CDs, electronics, games and were soon expanded to all other types of consumer goods. Combining e-commerce portals with focused advertisements and social networking features became the second element of e-commerce, leading to the rapidly growing market

Table 2 Time line of web business services between 1995 and 2000

1995	Amazon e-commerce portal	e-commerce	USA
1995	eBay online auctioning	e-commerce	USA
1996	Craigslist advertisement	Advertisements	USA
1997	Rakuten e-commerce	e-commerce	Japan
1997	Babel Fish translation	Translation	USA
1998	Microsoft MSN Search, Live Search, Bing	Portal and search	USA
1998	PayPal internet payment service	Payment	USA
1998	JD.com	e-commerce	China
1998	Google search	Portal and search	USA
1998	Tencent social network	Social network	China
1999	Alibaba e-commerce	e-commerce	China
1999	Napster peer-to-peer file sharing	File Sharing	USA
2000	Baidu search engine, portal and social network	Portal and search	China
2000	Yandex Search	Portal and search	Russia

of web advertisement. The new portals and platforms started in the US but the idea was quickly adopted by new companies in Japan, China and Russia. By the end of the century, e-commerce platforms had defined and conquered a completely new and global market.

1995 Amazon

In 1994, the Harvard graduate Jeff Bezos read a report about the future of the Internet that projected the annual growth of e-commerce at 2300%. Bezos started to design a business plan by listing products that could be offered and sold via a web portal, concluding that the most promising articles would be books, compact discs and computer hardware and software. Amazon was founded in a garage and incorporated in 1995, starting as an online book store. Its initial value proposition was *"A powerful, yet easy-to-use search engine makes it possible for customers to find titles after a few minutes of searching, without ever leaving their desks. Instead of hours chasing down phone numbers and driving all over town to specialty bookstores, customers order online and books are delivered directly to their doors via UPS or Airborne Express."*[12] Soon, Amazon was known as the largest search engine for books and received a warm welcome from Yahoo! and Netscape as the perfect addition to their portal services. Today, Amazon is one of the largest players in e-commerce creating revenue of more than $100,000 million (2015) including Amazon Web Services.

1995 eBay

Amazon.com is based on the idea of the easy connection of private buyers to professional sellers. At the same time, the French-born Iranian-American computer programmer Pierre Omidyar had the idea to use the World Wide Web to connect private sellers to private buyers by introducing an auctioning platform using the

[12]http://www.urlwire.com/news/100495.html retrieved 2016-12-14.

World Wide Web. EBay was founded by Omidyar in San Jose, CA, in September 1995 and immediately started getting business. The growth factor was impressive: the platform hosted 250,000 auctions in 1996 and expanded to 2,000,000 in January 1997. Because of that success, eBay received $6.7 million in funding from the venture capital firm Benchmark Capital and went public in September 1998. EBay survived the dot-com bubble and today belongs to the major players in e-commerce generating revenue of $8.59 billion (2015).

1996 Craigslist

E-Commerce began to become a huge driver for the new World Wide Web. It soon became evident that advertising would be the logical element to create a huge, new business nexus. One of the first advertising platforms was founded by Craig Newmark under the name "Craigslist." It started as a simple mailing list promoting events in the San Francisco Bay area and soon expanded to other US cities. Today, Craigslist is available in more than 70 countries and creates $381 million (2015) revenue from advertisements.

1997 Rakuten

Rakuten is a Japanese e-commerce company founded by Hiroshi Mikitani in 1997. The company expanded beyond e-commerce into banking, travel, sports and entertainment. Today, Rakuten operates worldwide generating a revenue of JPY 713 billion ($118 billion). Today, it belongs to the group of largest e-commerce companies.

1998 JD.com

The company was founded by "Richard" Liu Qiangdong in July 1998, and its B2C platform went online in 2004. It started as an online store, but soon diversified, selling electronics, mobile phones, computers, etc. The domain name was changed to "360buy.com" in June 2007, and to "JD.com" in 2013. At the same time, JD.com announced its new logo and mascot. Today, JD.com is one of the five largest e-commerce platforms.

1998 PayPal

Organizing e-commerce using the Web needs one crucial element: money transfer or payment. Buying goods like books in the Web or consuming services via the Internet needs a simple but safe method to pay the agreed price. Compared to Europe, money exchange in the US at that time was much more complex and still based on checks. The option of transferring money per wire was restricted and made the payment of online services and purchases clumsy. In 1998, Max Levchin, Peter Thiel, Luke Nosek and Ken Howery decided to establish a company named Confinity to develop payment services for users of Palm Pilots. The company started to extend the service under the name PayPal into payment transactions using email and developed the idea of an Internet-currency that could be used for each purchase or service payment between web-users and providers. PayPal had an ambitious mission but a difficult start, convincing users to join was not so easy and Palm Pilots were very exotic. On the other hand, the business model was clear and the need to turn a profit led to a small fee for customer transactions. PayPal received a $100 million investment in early 2000. In March of that same year, the company merged with X.com, another payment service founded by Elon Musk in 1999. E-commerce platforms like Amazon and eBay were observing the market of payment services carefully, knowing that simple and safe payment was a critical success factor for e-commerce services. PayPal went public in 2002 generating over $61 million and was acquired by eBay the same year.

1999 Alibaba
Alibaba.com was founded as an online business-to-business platform to connect Chinese manufacturers with overseas buyers by Jack Ma and 17 co-founders in 1999. The services were soon expanded into business-to-consumer and consumer-to-consumer platforms. Alibaba filed its IPO in 2014, raising $21.8 billion.[13] Today, Alibaba's revenue is $14 billion (2016) and the Alibaba Group is the world's largest retailer.[14]

Smart Search: Google and Followers

In 1997, the number of websites on the Internet was approaching 120 million and quickly grew to 180 million in 1998 and 280 million in 1999. Navigating through that enormous volume of information and services became a huge challenge. The rapidly growing number of new websites, portals and platforms in different languages began to evolve into a digital and virtual universe of new opportunities for access of data, news, media and products or services. It soon became evident that tools for search, navigation and translation would be key for the future acceptance and growth of the Web.

1997 Babel Fish
Using the web for content retrieval in different languages lead to the idea of building translation services for web users. One of the first applications was developed by a team of researchers at Digital Equipment and launched in 1997 as AltaVista Translation Service at the Web address "babelfish.altavista.com." The name was taken from Douglas Adams' "Hitchhikers Guide to the Galaxy." After the acquisition of AltaVista by Overture Services and the later take over by Yahoo in 2003, the service became known as Yahoo! Babel fish. Babel Fish was the forerunner of many translation services like Microsoft Bing Translator,[15] Google

---

[13]Stephen Grocer: "Alibabas IPO", Wall Street Journal 2014: http://blogs.wsj.com/digits/2014/05/06/alibabas-ipo-filing-everything-you-need-to-know/ retrieved 2016-12-16.

[14]Alibaba passes Walmart as world's largest retailer: https://www.rt.com/business/338621-alibaba-overtakes-walmart-volume/ retrieved 2016-12-16.

[15]Microsoft Bing Translator: http://www.bing.com/translator retrieved 2016-12-16.

Translator[16] and Leo.[17] Today, translation services are a commodity within many Web portals and services.

1998 Google Search

In 1996, two students at Stanford University, Sergey Brin and Larry Page, were looking for a dissertation theme. Both were fascinated not only by computers but also by mathematics and the new ideas of human-computer interaction. With the help of a Stanford professor, Terry Winograd, they began casting for a research topic, considering several ideas, including the design of a self-driving car. Soon, they started to focus on data mining using the growing structure of the World Wide Web. They soon found that there was a missing link: although Tim-Berners-Lee's initial idea was a complete network structure between the pages of the Web, the bi-directional link between pages (the so called "reverse-link") got lost between the initial idea and the implementation. Hyperlinks within Web-pages point to other documents, but there is no reference back. This information is especially valuable if someone wants to rank the importance of a document. Like with a book or research paper, importance grows with the number of references or pages pointing to that document.

Unfortunately, in the Web, the information of which hypertext pages point to a certain document is not stored. Larry Page tried to find a solution to generate this missing element and create something that would map the structure and interconnection between all Web-pages in a bi-directional graph. Page created a Web crawler that scanned all the existing Web pages—approximately 100,000 at that time containing more than 10 million documents– storing the links, the title of the page and the page where the link comes from. The project was called BackRub and was the first prototype of what later would become the Google search engine. The first version was not very efficient and after having collected more than 100 million links, Larry Page ran out of disk space and caused at least one outage of the Stanford University campus network.[18] In a revision of BackRub which was renamed "PageRank," Page and Brin defined an additional goal for the algorithm. On top of counting the links pointing to a certain page, the new ranking algorithm should also evaluate the importance of the page that points to a document, this adding a value or weight to each of the links. A reference from a Web page like the "Financial Times" or "Scientific American" has more weight. This approach led to several more mathematical challenges because of the recursive nature of such an algorithm.

This was exactly the right challenge for Sergey Brin. In 1997, PageRank was moved from the Stanford domain "google.stanford.edu" to the registered domain "google.com." In 1998, Brin and Page published their first paper at a conference in Australia, describing all the principles of PageRank, but without revealing too many secrets to competitors. In this paper, they also introduced the name of their new product: "*In this paper we present Google, a prototype of a large-scale search*

[16]Google Translate: https://translate.google.com retrieved 2016-12-16.

[17]Leo Dictionary: https://dict.leo.org retrieved 2016-12-16.

[18]Walter Isaacson: "The Innovators", Simon&Schuster, 2014.

engine which makes heavy use of the structure present in hypertext." [19] By the end of 1998, Google had a search index of 60 million pages. The page was still marked "Beta" but soon proved much better search results than its competitors Yahoo, Excite and Lycos, Netscape's Netcenter, AOL.com and MSN.com.[20] Raising money for the new company was the next step. With the help of one of their Stanford professors, Brin and Page got into contact with Andy Bechtholdsheim, one of the founders of SUN Microsystems. After the first presentation, Bechtoldsheim immediately understood that Google is not only a disruptive technical approach for users of the Web, but can also be combined with a powerful advertising business model. Bechtoldsheim went to his car to get his check book and wrote a check to Google for $100,000.

Google Inc. was formally incorporated by Brin and Page in September 1998. The new startup moved into a friend's garage in Menlo Park and, in 1999, into an office location in Palo Alto. Following that first investment, others joined soon including Jeff Bezos from Amazon and some of the major Silicon Valley venture capital firms like Sequoia and Kleiner Perkins. Goggle was born and would become one of the most influential Web companies. In 2003 Google moved to the current Google Plex headquarter location at Amphitheatre Parkway, Mountain View

1998 Microsoft MSN Search, Live Search and Bing

In 1998, Microsoft introduced its own search engine for the MSN community under the name of MSN search. The service was later renamed to Windows Live Search and became Bing in May 2008. It was also used by Yahoo! Search. Today, Bing is still the second largest search engine in the US behind Google.

2000 Baidu

Baidu was founded by Robin Li in Beijing in 2000. Li was a former employee of IDD Information Services, a New Jersey division of Dow Jones, and a graduate of Peking University. During his time at IDD, he developed the RankDex site-scoring algorithm for search engines' results page ranking receiving a US patent for the technology.[21] Today, Baidu is China's number one search engine having a market share of more than 60%. Baidu offers many additional services including maps, social media, music, encyclopedias, TV and collaboration. With these services, Baidu competes with Google, Wikipedia, Apple Music and other US companies, which face intermittent blockage by the Chinese authorities. Baidu's revenue was $9.5 billion in 2015 making it to one of the largest web service providers in Asia.

---

[19]Larry Page, Sergey Brin: "The Anatomy of a Large-Scale Hypertextual Web Search Engine", http://infolab.stanford.edu/~backrub/google.html retrieved 2016-12-19

[20]Scott Rosenberg: "Let's Get This Straight: Yes, there is a better sea IDD Information Services, a New Jersey division of Dow Jones and Companyrch engine", http://www.salon.com/1998/12/21/straight_44/ retrieved 2016-12-19.

[21]US Patent: https://www.google.com/patents/US5920859?dq=yanhong+li#v=onepage&q=yanhong%20li&f=false

2000 Yandex Search

Yandex refers to a Russian Internet technology company and search engine launched in 1997 by and Arkady Volozh and Arkady Borkovsky. The company Yandex was founded in 2000, today it's the largest search engine in Russia having a market share of more than 50%. The services were expanded to email, cloud services and payment services. Revenue is generated from online advertisement and reached $0.98 billion in 2015.

Free Content

1999 Napster

One of the basic ideas of the Web was the interchange of information and content between the users. From the beginning, the inventors of the World Wide Web had it in mind to provide a network of users that could collaborate and share documents of any kind in some open and free modus. It soon became evident that documents could also mean audio, video and pictures and that a peer-to-peer file sharing service can also be used to share documents that are beyond the limits of copyright rules and regulations. MP3 as a format for storing digital audio and video files was a perfect foundation for the free exchange of music and movies.

In June 1999, Shawn Fanning, John Fanning and Sean Parker founded Napster as an exchange platform for MP3 files between private users. The platform became tremendously successful very quickly. Users loved to download songs for free that they had already purchased as a CD or sometimes had not. For a short time, it seemed that free music was a human right and, at its peak, Napster had 80 million registered users. Legal rules and regulations were unclear and downloading material under copyright from a platform that was uploaded by other users seemed not to be within the responsibility of the platform. The supporters of an open society approach applauded the initiative and perceived P2P (peer-to-peer) data exchange as a human right. Finally, the music industry acted. In March 2000, the heavy metal band Metallica filed a lawsuit against Napster followed by others. Napster was faced with allegations from the music industry that its users were directly violating copyright law and Napster itself was also responsible for contributory infringement of the copyrights.

Napster was one of the first lawsuits that made it obvious that the platform itself is also responsible for the content that is shared or communicated via its services. Napster lost the case and closed its service in July 2001 announcing its bankruptcy in 2002. Napster's brand and logos were acquired at bankruptcy and auction by Roxio. The brand went through several acquisitions and still exists today as an online music store part of Rhapsody Online Music Services.[22]

[22]http://at.napster.com/

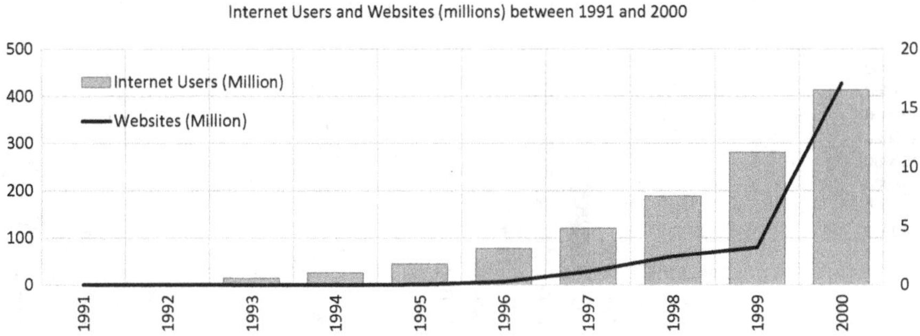

Fig. 10 Internet users and websites Between 1991 and 2003

The Dot-com Bubble

1998 Acceleration

In the mid-1990s, major parts of the economic world where convinced that a new age had begun: the so called "New Economy" powered by Internet and Web technology and creating new business models based on the delivery of information services to corporations and private consumers. Many new companies were founded within a short time to build and deliver those services (see Fig. 10).

The number of Web users grew rapidly from 14 million in 1993 to 280 million in 1999. The number of websites reached 3 million in 1999 compared to 1 in 1990 and around 20,000 in 1995. New companies "born on the web" like Amazon, eBay, Yahoo, Google and many others had started to create new services and new business models and collected hundreds of thousands of users. Most of these companies were startups built by computer science students, hackers or engineers following ideas and challenges they had created at universities, research facilities and computer clubs and communities. Funding of the new companies was soon taken over by VC (venture capital) firms that expected high profits from the new business field created by the Internet and the Web. Investing in Internet startups became a major goal for the financial industry. The company valuations were estimated by the projected number of users that would accept the new services and, given the roaring growth of Internet users and web pages, the sky seems to be the only limit. Raising money for a new idea was easy and the investors were in more than a hurry. Peter Thiel was on his way to the A-round for PayPal in January 2000: "*A South Korean firm wired us $5 million without negotiation or signing any documents*"[23] Once the first companies went public on the stock exchange, the tsunami spilled over to the stock market (see Fig. 11).

Institutional investors as well as private investors started to buy shares from the technology markets, the prices reached levels never seen before. The Internet infrastructure provider, CISCO, became the most valuable company with a market

[23]Peter Thiel "Zero to One", Virgin Books, 2014.

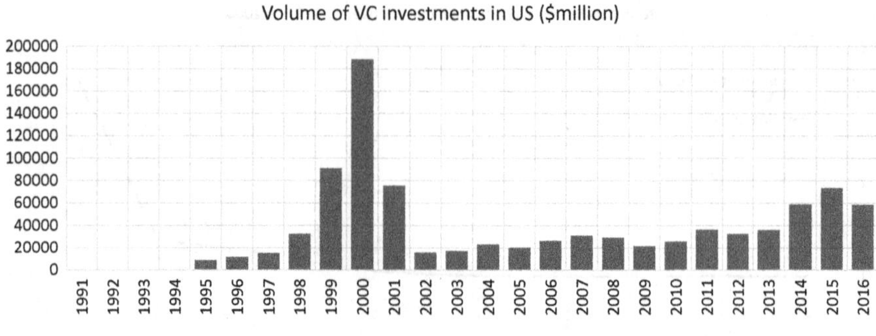

Fig. 11 Volume of venture capital investments in US between 1995 and 2016

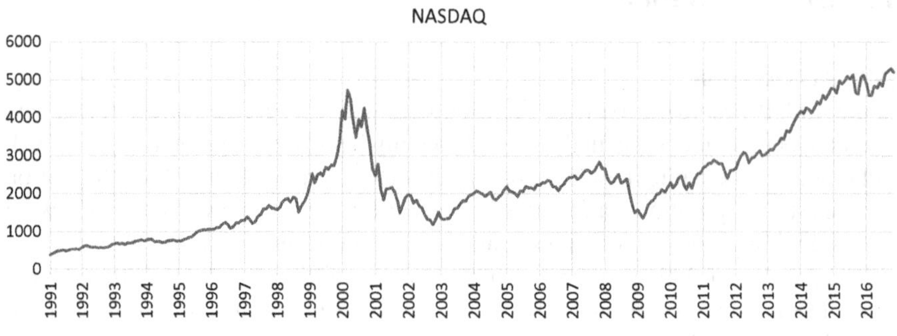

Fig. 12 NASDAQ between 1991 and 2003

capitalization of more than $500 billion. Pet.com became famous as the shortest-lived company: After its IPO in August 1998 it went into liquidation in 268 days. Boo.com was launched in the autumn of 1999 and burned $135 million in just 18 months before it was liquidated in May 2000.

2000 The Big Bang

On March 10, 2000, the party was over. The NASDAQ peaked at 5132.52 intraday, before closing at 5048.62. Afterwards, the NASDAQ fell as much as 78%. The markets collapsed. It would take more than 15 years before the tech market reached that level again (see Fig. 12).

The bubble was supported by the low interest rates between 1998 and 1999 but the major reason was the so-called dot-com theory, which stated that a company should grow its customer base as rapidly as possible even if it produces large losses. Amazon and Google seemed to be the proof for that theory. Both companies did not see any profit in their first years. The claim of the day was "Get large or lose," raising money from VCs or the stock market did not require profit or even a track of revenues. Companies were operated through the capital they had raised and the monthly burn rate defined the life span. Though there had been some startups that seemed to prove that theory as valid, the pattern could not be copied by each type of

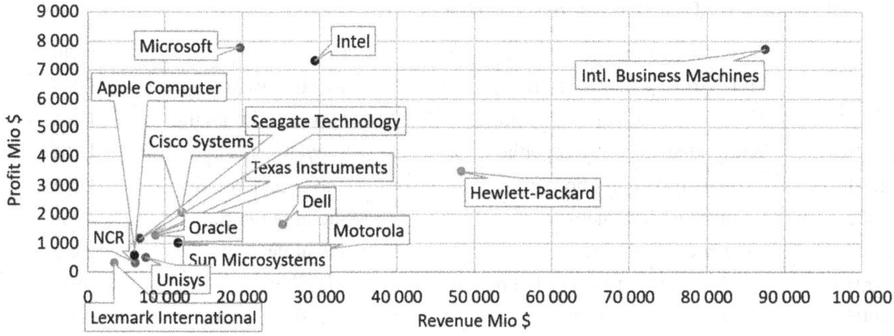

Fig. 13 Top IT companies 2000

Internet enterprise. A huge number of companies did not survive, were acquired by others or, worse, went into bankruptcy taking with them their Investors' money.

The IT Market in the Early 2000s

By the end of 2000, the IT market was still dominated by the major "old business players" led by IBM and followed by Hewlett-Packard, Motorola, Dell and Intel. Software companies like Microsoft and Oracle had started to grow a sustainable business. The only newcomer was Cisco, as one of the major providers of Internet technology (see Fig. 13).

The new market companies that survived the bubble, like Amazon, PayPal, eBay, Google or Salesforce were still under the radar screen. That would change in the first decade of the new millennium. Some of those startups would rise to the top ten of the IT industry.

Web 2.0 and the Social Networks

After the dust had settled, it turned out that the creative power of building disruptive business models based on the Web was still alive. The new movement was called social networks and would lead to something that claimed to be the Web 2.0. Looking at what was created between 1995 and 2000, it was evident that the Web was a great network for the distribution of information and selling stuff to people. The original idea of Tim-Berners Lee to build a huge network for collaboration and sharing ideas and content was still far away. Email services and mailing lists seemed to be the only collaborative tool with content sharing experiencing its first defeat with the closing of free MP3 exchange platforms like Napster (see Table 3).

Table 3 Time line of web 2.0 before smart phones

Year	Company/Service	Type	Country
1997	SixDegrees social network	Social network	USA
2000	TripAdvisor travel and booking	Travel and booking	USA
2001	Wikipedia free encyclopedia	Encyclopedia	USA
2001	Expedia travel and booking	Travel and booking	USA
2002	Friendster social network	Social network	USA
2003	Xing social network	Social network	USA
2003	The Pirate Bay torrent file hosting	File Sharing	Sweden
2003	Myspace social network	Social network	USA
2003	LinkedIn social network	Social network	USA
2003	Apple iTunes music and media store	Media	USA
2003	4Chan anonymous imageboard	Social network	USA
2004	Flickr image hosting	File Sharing	USA
2004	Facebook social network	Social network	USA
2005	YouTube video sharing	Video Sharing	USA
2005	Trivago travel and booking portal	Travel and booking	Germany
2005	Reddit social network	Social Network	USA
2005	Google earth	Data	USA
2007	Google street view	Data	USA

2000 Wikipedia and Wikis

In the aftermath of the dot-com bubble, a new attempt for collaboration and sharing was made by Jimmy Wales, a former financial trader and early internet entrepreneur. After an unsuccessful first enterprise—a web portal named Bomis—Wales decided to concentrate on the idea of building an Internet encyclopedia. Fascinated by the "World Book Encyclopedia" of his childhood, he dreamed of creating something like that on the Web for free use for everybody: *"Imagine a world in which every single person on the planet is given free access to the sum of all human knowledge. That's what we're doing."*[24] The project started in early 2000 as Nupedia, which claimed to be the start of the world's largest online encyclopedia. Wales hired Larry Sanger as editor-in-chief and they started to create articles using a 7-step review process and integrating volunteers as authors and reviewers. It soon turned out that the approach was too clumsy. In November 2000, only two articles on Nupedia were reviewed and published. Wales and Sanger started to search for a better solution and were introduced to Ward Cunningham's "Wiki"-concept and software by the computer programmer Ben Kovitz.

The principle of a "Wiki" is to use a web page editor as a tool for the collaborative creation of web pages following Tim-Berners-Lee idea of a Web where users not only consume pages, but are also able to add remarks or propose changes. Ward Cunningham developed the Wiki-concept back in 1994 and named

[24]Jimmy Wales Interview https://slashdot.org/story/04/07/28/1351230/wikipedia-founder-jimmy-wales-responds retrieved 2016-12-16.

the first implementation "WikiWikiWeb" using the Polynesian word "Wiki" which stands for "quick." Cunningham's first implementation of his Wiki was released in March 1995 on the Internet domain c2.com.[25] "Wikis" are a specialized type of hypertext pages including the typical elements of internal and external links and—more important—a simple editor to create and change those pages. The pages are stored on a wiki-server together with registered users that may create new pages or change existing pages. The change history is completely logged and allows a collaborative design of content. Today, several Wiki implementations exist and are available as free software like MediaWiki[26] or DokuWiki.[27] Wikis are used by many corporations, organizations and private persons to run and organize content like product documentation, project information and collections of content for specific themes.

Wales and Sanger soon understood that a Wiki could be used to speed up the process of editing and review of articles and started Wikipedia as a supporting project for Nupedia. It soon appeared that editing and reviewing articles on Wikipedia was more motivating for the contributors than the Nupedia approach. Wikipedia gained its 1000th article around February 2001, and reached 10,000 articles around September of the same year. In the first year of its existence, over 20,000 encyclopedia entries were created—a rate of over 1500 articles per month. On Friday August 30, 2002, the article count reached 40,000.[28] In 2003, the Wikimedia Foundation[29] was founded by Jimmy Wales to support the future development and financing of Wikipedia. Today, Wikipedia has more than 286,000 active users and more than 65 million registered users. In 2016, the number of Wikipedia articles exceeded 39 million in more than 250 languages. Wikipedia is ranked in the top ten of the most popular web pages (see Fig. 14).

Travel

With the growing use of web portals, chatrooms and blogs, it turned out that people used the new media to also share their travel and holiday experiences with others. Collecting and bundling that content and making it accessible for communities was a logical step and led to portals focused on travel. In the beginning, those portals offered information about places comparable to printed travel guides. Soon they used the community features of the Web to invite users to share their personal stories. In a BBC interview, Tripadvisor's Stephen Kaufer said: "*We started as a*

[25]WikiWikiWeb: http://wiki.c2.com/?WelcomeVisitors retrieved 2016-12-16.

[26]MediaWiki: https://www.mediawiki.org/ retrieved 2016-12-16.

[27]DokuWiki: https://www.dokuwiki.org retrieved 2016-12-16.

[28]History of Wikipedia: https://en.wikipedia.org/wiki/History_of_Wikipedia#Historical_overview retrieved 2016-12-16.

[29]Wikimedia Foundation: https://wikimediafoundation.org/wiki/Home retrieved 2016-12-16.

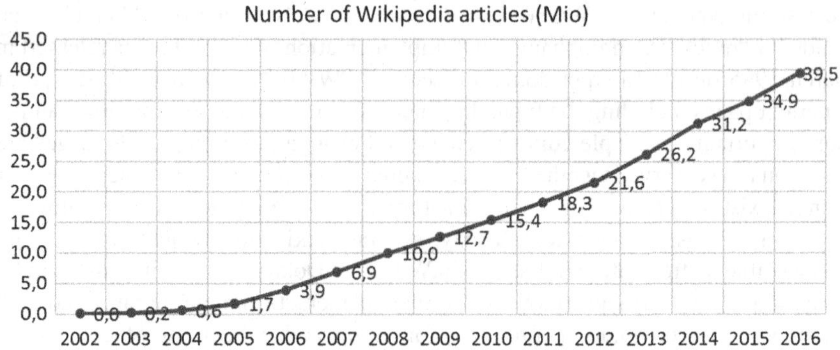

Fig. 14 Number of Wikipedia articles

site where we were focused more on those official words from guidebooks or newspapers or magazines. We also had a button in the very beginning that said, "Visitors add your own review", and boy, did that just take off."[30] As a next step the travel portals started to offer booking features by connecting to existing airline booking systems and hotel room booking systems. Today, hotel and travel booking is one of the largest markets on the Web, covered by hundreds of portals and creating revenues of $489 billion in 2015.[31] The percentage of online travel bookings is estimated by up by 50% in the US.

2000 Tripadvisor

TripAdvisor was founded as a travel website by Stephen Kaufer and Langley Steinert in 2000, acquired by InterActiveCorp in 2004 and sold off in December 2011 in a public offering. In 2015, TripAdvisor generated revenue of $192 billion.

2001 Expedia

The travel platform Expedia was founded as a division of Microsoft back in 1996 and was sold in 1999 to be acquired by InterActiveCorp in 2001. In the following years, many travel platforms were acquired by InterActiveCorp and again sold in 2005 under the name of Expedia Inc. This group included brands like TripAdvisor, Classic Vacations, eLong, Hotels.com, and Hotwire.com. In 2012, Expedia bought a majority stake ($630 million) in Trivago. Today, Expedia is one of the largest online travel platform groups generating revenue of $ 5763 billion in 2014.

2005 Trivago

Trivago was launched as a hotel price comparison website in Germany by Rolf Schrömgens, Peter Vinnemeier, Stephan Stubner and Malte Siewert in 2005. In

[30]BBC Interview with Stephen Kaufer, 2014: http://www.bbc.co.uk/programmes/b04l3cmq retrieved 2016-10-12.

[31]Revenue of online travel bookings, Statista: https://www.statista.com/statistics/238852/online-travel-bookings-worldwide/ retrieved 2016-10-12.

2015 Trivago created a revenue of $500 million being one of the cornerstones of Expedia's business in Europe.

Social Media: Facebook and Others

The years from 2002 to 2006 were influenced by numerous, new companies offering a type of service that became known as social media and would grow to become a tremendous influencer on what we perceive as the Web today. The idea to use the Web as the platform to connect and communicate with friends evolved soon after the Web was born in the mid-1990s. Platforms like GeoCities (1994), The Globe (1995), SixDegrees (1997), Classmates (1997) and Tencent QQ (1999) were typical forerunners of the social media hype in the first decade of this century. The first "modern" social media site was Friendster followed by LinkdIn (2002), MySpace, Xing and 4Chan (2003), and Reddit (2005). Facebook and Twitter were both released in 2006 and are today's market leaders.

2002 Friendster
Friendster was founded by the Canadian programmer Jonathan Abrams as one of the first "modern" social media sites in 2002. Compared to its forerunners, Friendster was based on the approach of building virtual communities of connected people, introducing one of the most important effects in social networking and the later creation of web based platforms. Friendster.com was released in 2002 and collected 3 million users within the first few months.[32] It was funded by Kleiner, Perkins, Caufield & Byers and Benchmark Capital in October 2003 with a reported valuation of $53 million but rejected an offer for acquisition by Google in the same year. Until 2008, Friendster expanded its reach to 115 million registered users before the decline started mainly influenced by the rise of Facebook. In 2009, Friendster was acquired by the large Asian Internet company MOL Global for $26.4 million.[33] The site was moved through several transitions before the final shut down in June 2015.

2002 LinkedIn
LinkedIn was one of the first movers and shakers founded in 2002 by Reid Hoffmann from socialnet.com, a very early dating platform. From the beginning, LinkedIn had its focus on professional contacts addressing business people and offering them a platform to organize their contacts and connect to each other. The approach was successful and LinkedIn reached 1 Million users in 2004 and achieved profitability in 2006. After the first investment round managed by Sequoia

[32]Gary Rivlin, New York Time, 2006: http://www.nytimes.com/2006/10/15/business/yourmoney/15friend.html?_r=1 retrived-2016-12-16.

[33]Michael Arrington, TechCrunch 2009: https://techcrunch.com/2009/12/15/friendster-valued-at-just-26-4-million-in-sale/ retrieved 2016-12-16.

Capital in 2003, the company went public in 2011 at $45 per share. After the first trading day, the closing price was at $94.25. In 2015, LinkedIn had 400 Million members worldwide, generating most of its revenue from advertisements. In 2016, Microsoft announced its acquisition of LinkedIn for $26 billion.

2003 Myspace

One year after LinkedIn's release, another startup began to push into the social media market. A group of employees at eUniverse, a marketing company, began to work on a social media platform based on Friendster's social network features, a platform that was released by Canadian computer programmer Jonathan Abrams in 2002. MySpace soon gained momentum in winning users, reaching the 100 Million users in 2006, but was finally overtaken by Facebook in 2008. After several mergers and reorganizations, MySpace was acquired by Time Inc. in 2016.

2003 Xing

In August 2003, the social media spark jumped over the Atlantic and triggered the foundation of the OPEN Business Club, a platform that concentrated on business users. The OpenBC was renamed to XING in 2006 and gained a lot of attention in German-speaking countries. It reached 1.5 Million members in 2006 and went public the same year becoming the first European Web 2.0 company to go public.

2003 4Chan

4Chan was started in 2003 as the web based "Imagebord" by Christopher Poole, a student from New York City University. The name 4Chan is derived from the Japanese anonymous message board "2channel." 4Chan is linked to the Internet subculture and became famous for provoking media attention and for its anonymous posting culture. In 2015, the site was purchased by Hiroyuki Nishimura, the former administrator of 2channel between 1999 and 2014.

2004 Facebook

Facebook started its operation under the name of "TheFacebook" in 2004 and would become the most successful social media platform in the Western world. At that point in time, Facebook founder, Mark Zuckerberg, already had a track of social media experiences. In 2003, as a Harvard sophomore Zuckerberg wrote a program called Facemash, which collected data including pictures from the web sites of Harvard's private dormitories. The site was extremely successful and attracted 450 visitors and 22,000 photo-views in its first four hours online but also led to some serious discussions about data privacy. It turned out that Zuckerberg had hacked into the private ID's of Harvard student residents.[34]

In February 2004, Zuckerberg launched a new web-site called "TheFacebook." A couple of days after the site was launched, the Harvard seniors Cameron Winklevoss, Tyler Winklevoss, and Divya Narendra accused Zuckerberg of using their ideas to create a competitor to their product, "The Harvard Connection." The

[34]Marc Zuckerberg: Hacker, Dropout, CEO https://www.fastcompany.com/59634/hacker-dropout-ceo retrieved 2016-12-16.

Winklevoss brothers later filed a lawsuit against Zuckerberg, that was settled in 2008 to the tune of $300 million in Facebook shares. In the beginning, access to Facebook was restricted to Harvard students, later expanded to other universities.

In September 2006, Facebook was opened to everyone and started to offer so called group pages for businesses gaining 100,000 business users by 2007. In May 2012, Facebook went public and raised $16 billion based on a company value of $106 billion.[35] Facebook's revenue today is more than $ 27.6 billion (2016) coming mostly from advertising. The company generates a net income of more than $10 billion (2016), and its market cap is more than $340 billion. The number of daily, active Facebook users exceeds 1 billion.

2005 Reddit

Reddit was designed as a discussion forum in 2005 in Medford, Massachusetts by the University of Virginia graduates Steve Huffman and Alexis Ohanian. The owner of "Wired," Condé Nast Publications, acquired Reddit on October 31st, 2006 for an undisclosed price.[36] In October 2014, Reddit raised $50 million in a funding drive including investors Marc Andreessen, Peter Thiel, Ron Conway, Snoop Dogg, and Jared Leto at a valuation of $500 million.[37] Today, Reddit has 234 million users, ranked as the 11th most visited Web-site in US and 25th in the world.

Pictures, Music and Video

In the mid-2000s, digital photography and digital music had reached wide acceptance. The move to digital formats was supported by the formation of the first standards like JPEG (1992), MPEG (1993) and MP3 (1994). Around 2004, digital camera equipment had overtaken film cameras and MP3 had become the de-facto standard for digital music players. Providing services for the growing community of photographers was a logical step in moving personal media content into the cloud, and providing and sharing pictures, music and video out of the cloud.

2003 The Pirate Bay

Though the existence of music download platforms like iTunes seemed to have sorted out the legal situation, the free music community was still active. In 2003, The Pirate Bay was founded as a free download and music exchange platform by a Swedish anti-copyright organization named Piratbyrån (The Piracy Bureau). The

---

[35]Facebook IPO: http://dealbook.nytimes.com/2012/05/17/facebook-raises-16-billion-in-i-p-o/?hp&_r=0

[36]Michael Arrington: TechCrunch, 2006: https://techcrunch.com/2006/10/31/breaking-news-conde-nastwired-acquires-reddit/, retrieved 2016-12-16.

[37]William Alden, NYT Deal Book, 2014: https://dealbook.nytimes.com/2014/10/01/with-reddit-deal-snoop-dogg-moonlights-as-a-tech-investor/, retrieved 2016-12-16.

Pirate Bay was first run by Gottfrid Svartholm and Fredrik Neij. The Pirate Bay is a BitTorrent platform facilitating peer-to-peer file sharing among users of the Bit Torrent protocol. The files itself are not stored on the Pirate Bay's site which would protect the operators of Pirate Bay from any legal accusations: "*No torrent files are saved at the server. That means no copyrighted and/or illegal material are stored by us. It is therefore not possible to hold the people behind The Pirate Bay responsible for the material that is being spread using the site*[38]". Soon after the start of The Pirate Bay, both founders were accused of "assisting in making copyrighted content available" by the Motion Picture Association of America. On May 31st, 2006, the website's servers in Stockholm were raided and taken away by Swedish police for the first time. In the following years, The Pirate Bay went through many lawsuits, shut downs and restarts of operation. Today, The Pirate Bay still exists and offers an index of links to music, video, books, applications and games.

2003 iTunes Store

After the Napster experiment in 1999, music sharing was beyond the borders of legality. In 2002, Steve Jobs started to build agreements with five major record labels to create legal, alternative to peer-to-peer music file sharing and to build a new business model that would change the music industry. The Apple iTunes Store was introduced at the World-Wide Developers Conference in 2002 and opened in April 2003.[39] At its start, it offered 200,000 songs from music companies like BMG, EMI, Sony Music Entertainment, Universal and Warner. Pricing levels per song were between 99 cents and $1.29 and $9.99 per music album. The monthly download rate grew to 200 million songs by end of 2004 and more than 35 billion in 2014.

2004 Flickr

One of the first web sites for sharing pictures and video was introduced under the name of "Flickr" by Ludicorp, a Vancouver based company founded by Stewart Butterfield and Caterina Fake in 2004. The idea behind Flickr was to provide an upload platform for sharing pictures and videos, which could be easily used by photographers and bloggers to host and share images. In March 2005, Flickr was acquired by Yahoo for a price of around $25 million.[40] In 2016, Flickr had 122 million users in more than 60 countries, sharing 10 billion images.[41]

2005 YouTube

In early 2005, Chad Hurley and Steve Chang from PayPal tried to exchange videos they had shot during a dinner party in San Francisco. Finding the technology much

[38]The Pirate Bay: Legal Statement: https://thepiratebay.org/about retrieved 2016-12-16.

[39]Apple Press Release 2003: http://www.apple.com/pr/library/2003/04/28Apple-Launches-the-iTunes-Music-Store.html

[40]Kevin Delaney, Wall Street Journal, 2005: http://www.wsj.com/articles/SB111136815551984786, retrieved 2016-12-16.

[41]Flickr statistics: http://expandedramblings.com/index.php/flickr-stats/ retrieved 2016-12-16.

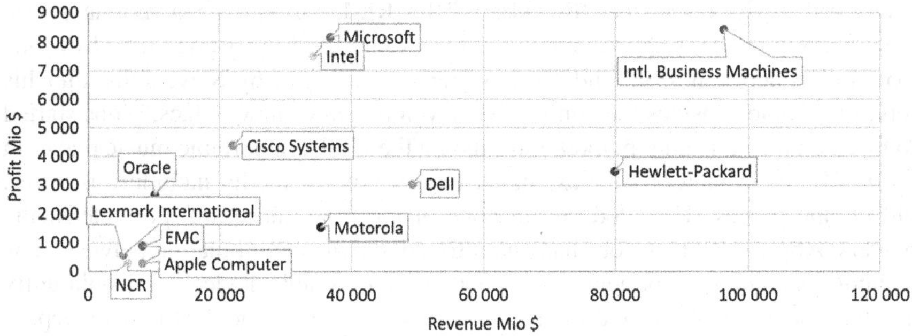

Fig. 15 Top IT companies 2005

too complex, they started to think about an exchange platform for private videos.[42] Within a short time and together with their PayPal colleague Jawed Karim, they developed the YouTube platform within a couple of weeks. The first video was upload to YouTube in April 2005. Receiving an initial $3.5 million investment from Sequoia Capital, the YouTube site was launched officially in December 2005. YouTube become extremely popular within a short time. In July 2006, the company announced that more than 65,000 new videos were being uploaded every day, and that the site received 100 million video views per day.[43] YouTube was acquired by Google in November 2006 for $1.65 billion in Google stock. Today, YouTube is the largest video sharing platform with 7 billion views per day.

Mobile and Smart

The IT Market in 2005

In 2005, the IT market was still dictated by old-style companies. IBM still led in revenue and profit, followed by Hewlett-Packard focusing on printers, servers and PCs, Intel as the major provider of processors and Microsoft leading the software market (see Fig. 15).

The typical end user device was a PC or notebook either as an Intel based machine running Microsoft Windows or one of Apple's Macs. Cell phones were already widely used and the network was based on the GSM and UMTS-protocols. Some attempts had been made to join these two worlds. Nokia released its Nokia Communicator in 1996, Palm introduced the Palm Pilot in 1997. One of the most

[42]John Cloud, Time Magazine 2006: "The YouTube Gurus" http://content.time.com/time/magazine/article/0,9171,1570795,00.html retrieved 2016-12-16.

[43]USA Today 2006: "YouTube serves up 100 million videos a day online" http://usatoday30.usatoday.com/tech/news/2006-07-16-youtube-views_x.htm

successful devices was the "Blackberry" by RIM, introducing a mobile phone version including texting and email in 2003. Apple itself had started the "Newton" experiment between 1993 and 1999, which was stopped by Steve Jobs after his return to Apple. Besides the consequent development of new "Macs," Jobs started to focus on the consumer market, introducing the iPod as a portable music player in 2001. The concept was later expanded to any type of media including pictures, videos and games. The iPod became one of the most successful portable media players. Apple and Steve Jobs had again found a perfect combination between new technology and user experience. The creation of a portable device that would unify entertainment, telephone and Internet access seemed to be the next logical step.

Leaving the Desk

2007 Apples iPhone

On January 9, 2007 Steve Jobs announced the iPhone at the Macworld convention in San Francisco. With this announcement, the smart phone was introduced and would forever change the way people communicate and interact with their friends and business partners.

"Today, we're introducing three revolutionary products. The first one is a widescreen iPod with touch controls. The second is a revolutionary mobile phone. And the third is a breakthrough Internet communications device. So, three things: a widescreen iPod with touch controls, a revolutionary mobile phone, and a breakthrough Internet communications device. An iPod, a phone, and an Internet communicator. An iPod, a phone. . .are you getting it? These are not three separate devices. This is one device. And we are calling it iPhone. Today, Apple is going to reinvent the phone." Steve Jobs Keynote at the MacWorld Expo 2007

The iPhone design went along clearly with Apple's strategy of a "walled garden." Steve Jobs followed a principle that was quoted by Alan Kay from Xerox PARC back in 1982: "*People who are really serious about software should make their own hardware.*"[44] The iPhone combined leading edge hardware including a high-resolution touch screen, camera and the high-performance processor APL0098 operating at 412 MHz and produced by Samsung. The iPhone is operated by Apple's own operating system, iOS, and, from the beginning, delivered a rich set of applications including mail, calendar, contacts, music, video and Apple's Internet browser, Safari. Apple's iTunes was an important part of the new concept. Having access to downloadable music at comparably low prices had been the winning concept of the iPod. The combination of a portable music player with a phone and Internet access created a completely new type of user experience. The concept was new and the target group clear: the iPhone addressed everybody who

[44]Alan Kay talk at Creative Think seminar, July 20, 1982: http://www.folklore.org/StoryView.py?project=Macintosh&story=Creative_Think.txt retrieved 2016-12-16.

wanted to stay connected everywhere and all the time. Not everybody was convinced by the new product. Microsoft's CEO Steve Ballmer said: *"There's no chance that the iPhone is going to get any significant market share."* [45]

The iPhone had a tremendous start. Within the first 2 days Apple sold 270,000 iPhones. The sales numbers climbed to 1.5 million in the fourth quarter of 2007 and to 13.7 million sold units in 2008. With the iPhone, Apple created a new market for smart devices and disrupted business models for the distribution of software, media and information.

2007 Google Maps, Google Earth and Street View

In 2005, Google launched a service, which seemed to be a great feature for playing around on a desktop. It was called Google Maps and provided a complete map of the earth's surface including satellite images, which were very low resolution at that time. Google Maps origins lie in Sydney, Australia, where Lars Rasmussen and his brother Jens Eilstrup Rasmussen implemented the first version of a map software for their company, Where 2 Technologies. The Rasmussen brother pitched the idea to Google and Google decided to acquire Where 2 in 2004. Two other acquisitions led to the release of Goggle Earth in 2005. The Earth Viewer used in Google Earth had its origins in the company Keyhole which was acquired by Google in 2004. The real-time traffic analysis features in Google maps were originally developed by ZipDash, also acquired in the same year.

Combined with the ability of geo-positioning in the iPhone 3G, the map and view services formed the perfect foundation for location based services. Producers of apps for smart phones immediately started to use position and maps to improve their applications providing much more accurate search results based on the geo-location of the user and using maps to enhance usability and data presentation. The location based service feature led to a long-lasting discussion about data privacy. Users of smartphones could now be tracked by their phone or, more precisely, by the apps they run on their smart device.

2008 Location and Navigation: GPS

Having access to the Web using a mobile device is great experience for the user and soon led to the idea of integrating navigation into smart phones based on existing GPS (Global Positioning System) technology. GPS was launched as a project in the US in 1973 to overcome the limitations of previous navigation systems. It is based on 31 satellites with well-known exact positions and carrying a highly precise atomic clock. The satellites continuously send signals with their position and time. The GPS receiver must monitor multiple satellites and can calculate its own position using the data transmitted from the other satellites. After an experimental phase, the first modern Navstar-satellites were launched in 1989 and GPS began operation for military purposes. In 1996, GPS was opened for civilian purposes offering a restricted accuracy, known as SA (selective availability) with a failure

[45]Forbes, 2017: http://www.forbes.com/sites/bensin/2017/01/09/these-are-the-people-who-thought-the-iphone-would-fail/#d4d0def69f0b

rate of 100m. SA was switched off in 2000, and today the accuracy of the GPS system is below 10m with positioning and navigation systems available in civil planes, ships and cars. The GPS receivers were clumsy boxes weighing initially more than 20 kg. The first handheld pocket GPS receiver was introduced in 1988 by Magellan. In the following years, the size, weight and energy consumption dropped year by year. Companies like Garmin and TomTom started to market GPS-based navigation systems for cars and private use. The miniaturization of electronics led to small and lightweight GPS chips in the first years of the 2000s. Today, GPS chip sets have a size of a couple of millimeters and a retail price of around $50.

Apple was the first to use this technology and released its iPhone 3G in 2008 with an integrated GPS module and a map and navigation app. Since then, location based services like maps, navigation and location based searches became standard features of smart mobile devices.

2008 Android and Google

The iPhone operating system, iOS, is based on the Linux-Kernel. iOS is also based on earlier operating system developments by Apple and has its origins in Steve Jobs' NeXT Step, which was brought back to Apple to become the foundation of the MacOS X operating system. In 2003, Andy Rubin, a former developer with Apple, started to build a Linux based operating system for mobile devices called Android. Android was acquired by Google in 2003 to create the foundation of an open operating system for different types of mobile computing devices. Android became the foundation for several smart phones including the Samsung Galaxy, HTC Dream, Motorola Moto and Google Nexus. Today Android's source code is released by Google under open source licenses and used as operating systems in many smart phones. Although most Android devices ultimately ship with a combination of open source and proprietary software, including proprietary software required for accessing Google services. In 2010, Google started to deliver its own series of smartphones under the brand Nexus managing the design, development, marketing, and support of these devices. The production was outsourced to different manufacturers like Samsung, HTC. LG and Huawei.

In tandem, several electronics producers decided to enter the market and develop their own versions of Android smart phones. Sony announced its Sony Xperia in 2008, Samsung announced its Samsung Galaxy in 2009, LG followed with the LG Optima in 2011, Motorola announced its Motorola Moto in 2013, HTC, its HTC One in 2014.

2008 Windows Phone and Nokia

In 2008, Microsoft decided to develop its own mobile operating system. The project was delayed and went through several interim versions. Finally, at the 2011 Mobile World Congress, Steve Ballmer announced a partnership with Nokia to merge Nokia's experience in mobile phone hardware with Microsoft's operating system. Nokia's Lumia Windows Phone was announced later in 2011 and was produced in different versions until 2014. In the meantime, Microsoft developed its own mobile phone version introducing Microsoft Lumia in December 2014.

2010 Tablets: The Apple iPad

In April 2010, Apple released something that claimed to be a mix of a smartphone and a notebook-style PC. It was called iPad and so started the success story of the so-called tablet computer. It is also true that there had been a number of predecessors to the iPad starting with Alan Kay's vision of the Dynabook, described in his 1968 proposal, *A personal computer for children of all ages.*[46] The idea was developed during the following decades leading to devices like the Atari Stylus (1992), Apple's Newton (1993), Palm's Palm Pilot (1996) and the Ericsson's Delphi Pad (2001). After the successful market entry of smartphones led by Apple, it was clear that the technology and software was ready to design a device that had the same high level of user experience as found in a touch screen based smartphone, but simply offered a larger screen. The iPad and the category of tablet computers was born and would become a strong competitor to desktops and notebooks in the following years. With the rise of tablets and mobile operating systems like iOS and Android, the personal end devices took large steps towards Tim-Berners Lee's vision formulated at the outset of the Web: users should have continuing access to the Web whenever and wherever they are, without booting up a machine and waiting for the dial-up connection.[47]

In 2016, the percentage of desktop users was, for the first time, below the percentage of smart phone and tablet users.[48] The number is still growing, and it is expected that web-users especially in non-Western countries start their web experience with mobile devices and will never use a desktop or notebook.

The move to the mobile, smart phone and, later, smart tablet world had tremendous impacts on the acceptance of the Web, the creation of new business models and was a trigger for major shifts in the IT market. Within only a few years, Apple became the most successful IT company in revenue and profit, generated by its iPhone and iPad hardware business and by its App Store business model. Today, roughly 2/3rds of Apple's revenue of $215 billion (2016) is generated by iPhone and iPad sales, more than 10% comes from the App Store and the rest is Apple's Mac business.

A Disruptive Business Model for Software

2008 Apps and App Stores

Though the design of the iPhone was "Apple only," it was also clear that the success of a mobile end device would depend on the options for third parties to add value to

[46]Alan Kay, 1968: "A personal computer for children of all ages" http://www.mprove.de/diplom/gui/kay72.html retrieved 2016-12-16.

[47]Tim Berners-Lee 1999: "Weaving the Web".

[48]http://gs.statcounter.com/press/mobile-and-tablet-internet-usage-exceeds-desktop-for-first-time-worldwide

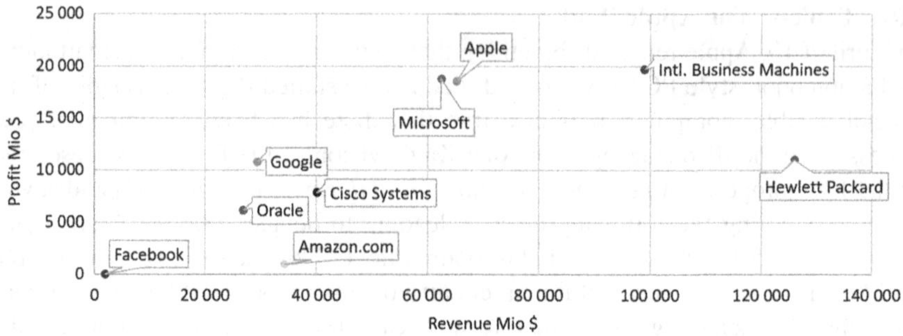

Fig. 16 Top IT companies 2010

the new product. In March 2008, Apple introduced the iOS SDK (Software Development Kit) inviting Software companies to create their own apps for the iPhone and sell it via Apple's App Store. The business plan was simple: a third-party company could deliver its app to the Apple App Store platform, Apple would review the software and offer it in its store. Users may buy and download apps, paying via their Apple account. The revenue is split between Apple and the app producer 70/30. The App Store was launched in July 2008 with 500 downloadable apps. In the following years, the number of app providers and apps grew dramatically reaching 2 million apps and a total of 130 billion downloads by 2016. All other manufacturers of mobile operating systems followed the Apple approach such that today there are app stores for Android mobile devices and Microsoft based mobile devices.

The IT Market in 2010

By the end of the first decade of this century, the IT market started to change. Still in the lead were Hewlett-Packard and IBM, but followed closely by Microsoft's growing software business and Apple, enjoying three of its most successful years after the iPhone launch (see Fig. 16).

Google, Cisco and Oracle were on their way to very sustainable business, while newbies like Amazon and Facebook began to appear on the radar screen.

Reference

Lee, D. C. (1999). *Enhanced IP services for cisco networks*. Indianapolis, IN: Cisco Press.

Cloud Computing

Author and consultant Jerry Weinberg claims to have encountered the question "Why does software cost so much?" more than any other in his long career. The correct answer, he says, is "Compared to what?"
DeMarco (1995) and Weinberg (1992)

What Is Cloud Computing?

A New Value Proposition

What we call cloud services and cloud computing today follows a simple approach. Ownership and responsibility for production and delivery of IT related services is taken over by central organizations either as private businesses or public organizations. These organizations operate the cloud architecture and deliver cloud services to consumers. They take over responsibility for the proper management of the cloud infrastructure and offer the usage of the infrastructure or the services based on that infrastructure to cloud consumers.

Consumers of cloud services like email, storage, applications or other types of services accept these terms because they expect lower costs and a higher level of flexibility compared to the option of owning and running the infrastructure themselves. This simply follows the principle of tool sharing to reduce costs and the idea of accessing technologies and IT solutions which are too complex to implement and run by the consumer alone.

Services Delivered from Remote

Cloud computing is describing the layers of technologies used today to deliver cloud services to consumers. The term "cloud" is simply used because consumers of cloud services use applications or resources they don't see and they don't have to understand completely: they are "hidden in a cloud." If you use a mail service like Gmail, GMX or Hotmail you don't have to operate a mail server or understand how it works, from the user's perspective it is hidden "*...somewhere in the cloud.*" Although many publications talk about "the cloud," there is nothing like a single or

M. Oppitz, P. Tomsu, *Inventing the Cloud Century*,
DOI 10.1007/978-3-319-61161-7_11

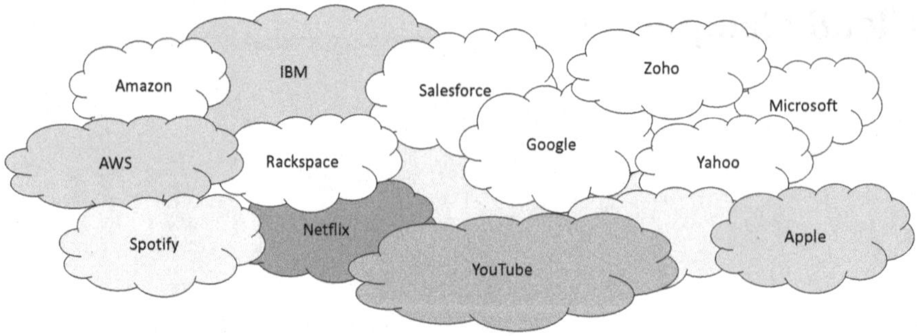

Fig. 1 Many clouds

global cloud. Cloud computing and cloud services are delivered by many different clouds—some of them are public and some of them are private clouds (see Fig. 1).

These clouds—consisting of different cloud computing architectures—are owned and operated by many different companies like Google, Microsoft, Apple, IBM, Amazon and many others.

Coining the Term "Cloud"

The term "cloud" was used long before the IT industry started to create cloud computing services. For engineers, especially in the telecommunication and network business, a "cloud" was used as symbol for a remote part of a blueprint, where the external functionality was clear, but the internal technology used was not fully understood and was not necessary to understand. A "cloud" was something remote like the telephone network that fulfilled its purpose while being something like a black box. Cloud computing points exactly in that direction. The idea is to provide computer or software infrastructure as a "black box:" the internal functionality unknown to the user, who is primarily concerned with the external, provided functions, combined with a guaranteed level of service. It is the same idea as witnessed in electric power, TV or telephone. As the consumer, we are not interested in the internal functionality and how those systems are built and operated, we want to use them at a fair price and at a high level of availability and quality.

Since the beginning of the computer era, the idea of sharing the new technology within a group of people or a larger community was on the mind of the scientists and engineers creating computers. The idea of a worldwide computer network where people could collaborate through sharing computing resources and information was already brought up by computer pioneers like J.C.R Licklider and John McCarthy in the 1960s. Their ideas and initiatives triggered the first concepts, which led to time sharing systems and the beginning of the Arpanet project. After the successful launch of the Internet in the late 1980s, the idea of building a network that allows easy and global access of information was again picked up by the CERN team around Timothy Berners-Lee and led to the successful launch of the World Wide Web.

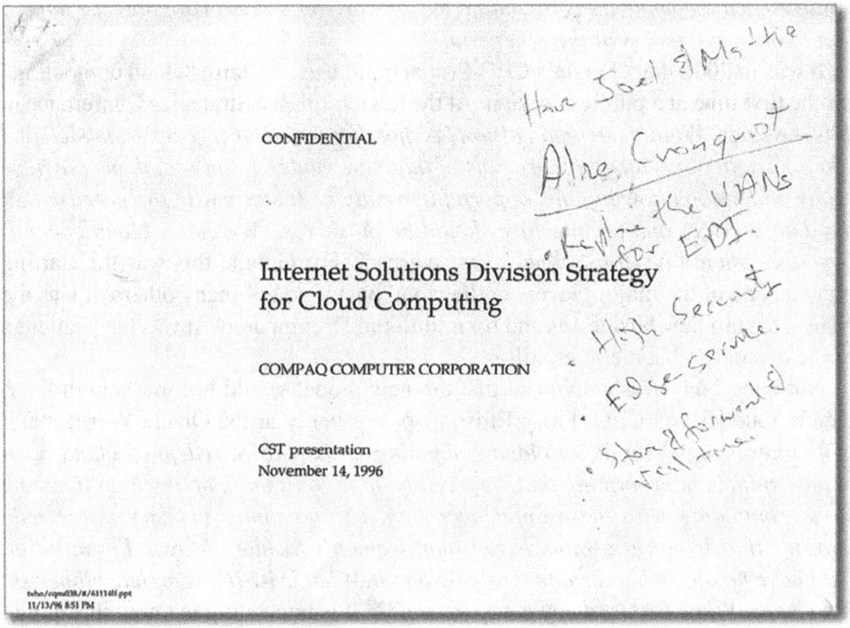

Fig. 2 The Compaq paper from 1996

The concept of networks as a "cloud" were first mentioned in 1996 in an MIT white paper about "The Self-Governing Internet."[1] In the same year on November 14th, 1996, an internal paper describing a business plan was published and discussed within Compaq at an office park outside of Houston (see Fig. 2).

The title of the paper was "Internet Solutions Division Strategy for Cloud Computing."[2] The group consisted of Compaq executives and engineers. For the marketing-executive George Favaloro and the young engineer Sean O'Sullivan, it was clear that this concept will have dramatic outcomes in the future. The story goes that this was the first time the term "cloud computing" was used.[3] (MIT Technology Review, 2011).

Picking Up the Concept

After his return to Apple in 1997, Steve Jobs shared his thoughts about the future of computing with attendees at the Worldwide Developers Conference. One of his statements seemed to point to the direction of cloud computing: "*I don't need a hard disk in my computer if I can get to the server faster... carrying around these*

[1]MIT 1996: "The self-governing Internet" http://ccs.mit.edu/papers/CCSWP197/CCSWP197.html, retrieved 2016-12-16.

[2]MIT Technology Review (2011): http://www.technologyreview.com/files/74481/compaq_CST_1996.pdf, retrieved 2016-12-16.

[3]Antonio Regaldo: "Who coined Cloud Computing" http://www.technologyreview.com/news/425970/who-coined-cloud-computing/, retrieved 2016-12-16.

non-connected computers is byzantine by comparison. I don't care how it's done. I don't care what box is at the other end."[4]

It was in 2006 that Google's CEO Eric Schmid used the term "cloud computing" for the first time at a public occasion. At the Search Engine Strategies Conference in 2006 he said: "*What's interesting [now] is that there is an emergent new model, and you all are here because you are part of that new model. I don't think people have really understood how big this opportunity really is. It starts with the premise that the data services and architecture should be on servers. We call it cloud computing—they should be in a "cloud" somewhere.*"[5] For Google this was the starting point as one of the major players in the cloud business. For many others, it was the chance to start new businesses and for traditional IT companies, it was the challenge of the decade and sometimes still is.

Not everybody was convinced that the new model would become a high flyer. Oracle founder and CEO, Larry Ellison, spoke openly at the Oracle World 2008: "*The interesting thing about cloud computing is that we've redefined cloud computing to include everything that we already do. I can't think of anything that isn't cloud computing with these announcements. The computer industry is the only industry that is more fashion-driven than women's fashion. Maybe I'm an idiot, but I have no idea what anyone is talking about. What is it? It's complete gibberish. It's insane. When is this idiocy going to stop?*"[6] It is interesting to know that at that time Ellison was already invested in one of the most successful Software-as-a-Service businesses named Salesforce, founded in 1999 by the former Oracle executive Marc Benioff. Only four years later, Ellison launched a brand-new Oracle cloud computing service at the Oracle World 2012.

The Hype Before the Cloud

Around 2006, the term "cloud" was widely used by software engineers and business developers. Everybody seemed to be fascinated by the approach of building cloud applications and cloud services. When a Compaq paper mentioned the term cloud computing for the first time—in the mid 90s—the World Wide Web was already the hype of the decade. Introduced by Tim Berners-Lee in 1992, the World Wide Web (WWW), together with the growing number of PCs and the first successful implementation of browsers, offered for the first time an easy way for everybody to become a potential consumer of Internet services.

Soon after the deployment of browsers and the "Web," the first non-commercial and commercial services started to be delivered by startups. The IMDb (Internet Movie Database) was launched 1990, in 1995 Amazon started its online bookstore

---

[4]Steve Jobs at the Worldwide Developers Conference 1997: http://www.forbes.com/sites/joemckendrick/2013/03/24/10-quotes-on-cloud-computing-that-really-say-it-all/#49a2f53a2102, retrieved 2016-12-16.

[5]Google Presse Center 2006: "Conversation with Eric Schmidt hosted by Danny Sullivan" https://www.google.com/press/podium/ses2006.html, retrieved 2016-11-30.

[6]Larry Ellison on Cloud Computing 2008: https://www.cnet.com/news/oracles-ellison-nails-cloud-computing/

and eBay went live with its online auctioning service. Hotmail introduced the free web-based email in 1996. Google search, Yahoo! and PayPal started in 1998. It was the decade of building a wide range of services for private consumers and developing successful business models to keep them alive.

All these services already used the World Wide Web on top of the Internet to bring their products to the end consumer. Of course, the companies developing and operating the services had to develop computing architectures that would allow for the scalability from hundreds to thousands of users accessing and using their services without any global limitations. At that time, nobody called it "cloud architecture," but it is no question that the major elements of cloud architectures like multitenancy, virtualization and load distribution were already in use. Around 2005, the number of worldwide Internet users had reached 1 billion. At that time and delayed by more than 10 years, enterprises started to use Internet based services. Following the private consumer-oriented services from 10 years ago, several new start-upss targeted enterprises with their products and goods. In 1999, Salesforce started to offer CRM (Customer Relationship Management) as a service adopting the mission statement "No Software!." SuccessFactor released the first version of HCM (Human Capital Management) as a cloud application in 2001 and ServiceNow launched its product in 2003.

It is obvious that a new trend had started based on the Internet and the Web as the global network, the availability of comparably cheap end devices and a standardized, open source browser technology. A huge market for providers of infrastructure and services addressing private consumers and businesses using these technologies began its growth, but nobody referred to it as a "cloud" at that time.

Creation of a Market

Since that time, the cloud business climbed to a volume of more than $280 billion for cloud infrastructure and cloud services. Today, cloud services are delivered to private consumers as well as to companies and organizations. They cover basic services like storage or computing resources in the cloud, but also a large and growing number of software and applications services, media services (audio and video), a dramatically growing market of different social media networks and an exploding market of online retail and travel platforms. This "secondary market" comes to more than $1500 billion counting the segments of e-commerce, online booking and internet advertising.

Similarities and Differences to Other Service Ecosystems

Comparing other service ecosystems with today's cloud services may lead to the assumption that cloud services are the same kind of service ecosystems, only based on new technologies. Although all the major elements of service ecosystems (network, end-devices, business models) can be identified with cloud services, there are some huge differences.

1. Single or Multi-Purpose: today's cloud services provide a large and rapidly growing number of different solutions and applications, we are dealing with a multi-purpose service ecosphere.
2. Speed of Change: the change and development of new aspects of network technology (e.g. IoT) is accelerating dramatically.
3. Number of Players: the number of players in terms of providers of technology and providers of services are much larger than in other service ecosystems.
4. Global Distribution: cloud services and other types of Internet based services are distributed worldwide; regional borders no longer exist.

Technology and Business

In the world of clouds, we distinguish between "cloud computing" as the set of technologies used (Internet, WWW, virtualization, etc.) and the term "cloud services," which describes the distribution of services and solutions delivered and based on those technologies (see Fig. 3).

Cloud computing and cloud services are based on a layered model of technologies consisting of the Internet itself, data centers hosting servers and storage networks using virtualization technologies and finally the World Wide Web as the standardized layer to access Web pages and services. We focused on these technology layers in the previous chapters. In the following sections, we will concentrate on cloud and cloud services as the new paradigm to deliver information technology to users.

From an architectural point of view, the world of cloud services consists of the consumers of a service, the service itself and the cloud resources used to provide the service.

Cloud Consumers

Consumers of a cloud service may be persons using end devices like PCs, notebooks, tablets or smart phones, which are connected to the Internet or software using standardized interfaces to connect and interact with either a cloud service or other cloud services interacting with a cloud service (see Fig. 4).

Cloud Resources

Resources used by a cloud service consist of all types of virtual or physical devices, delivering storage, computing power and network access. The concept of virtualization is essential to the architecture of cloud computing. It guarantees a flexible and scalable usage of computing or storage power for many consumers (see Fig. 5).

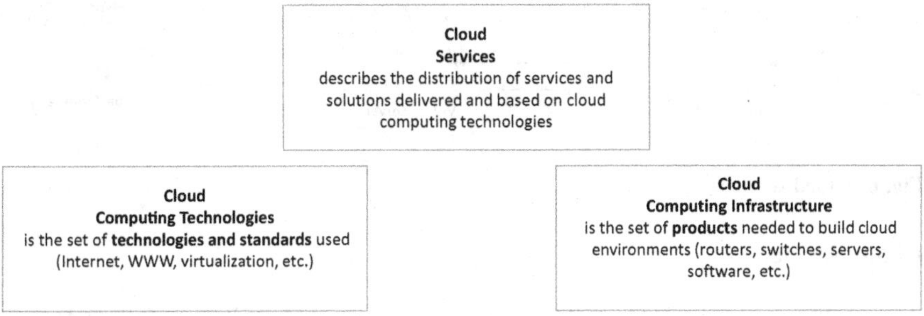

Cloud
Services
describes the distribution of services and solutions delivered and based on cloud computing technologies
Cloud
Computing Technologies
is the set of technologies and standards used (Internet, WWW, virtualization, etc.)
Cloud
Computing Infrastructure
is the set of products needed to build cloud environments (routers, switches, servers, software, etc.)

Fig. 3 Cloud computing themes

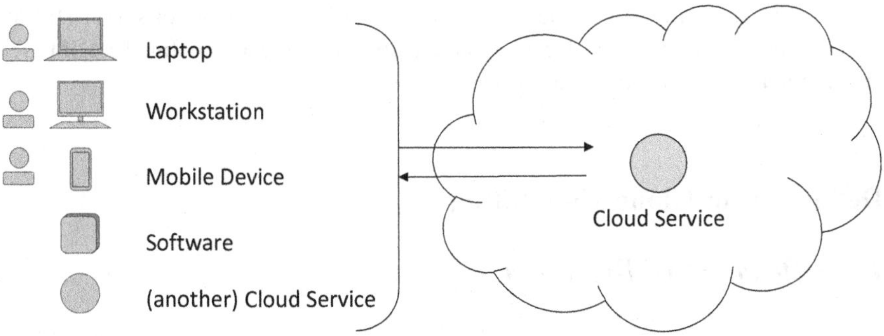

Fig. 4 Cloud consumers

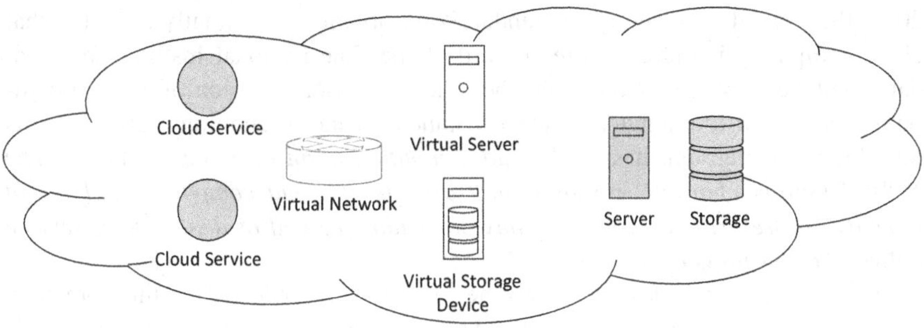

Fig. 5 Cloud resources

The set of virtual or physical resources used to provide a cloud service to consumers is operated by the cloud service provider.

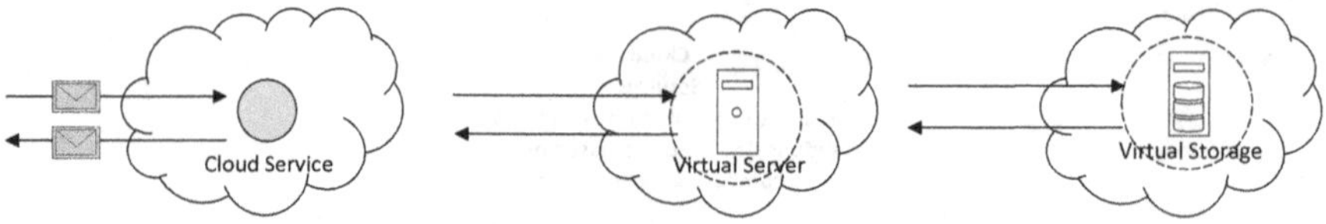

Fig. 6 Cloud services

Cloud Services

A cloud service itself is the service that is provided to the consumer using different types of virtual or physical resources in the background (see Fig. 6).

Depending on the type of cloud service delivered, the service may be an application (software as a service), the access to a virtual storage or server (Infrastructure as a Service) or the access to a platform providing a predefined set of tools and applications (platform as a service).

Definitions of Cloud Computing

The Official NIST Definition

Besides all the discussions about what a cloud service is, a clear definition published by the National Institute of Standards and Technology (NIST) describes five essential characteristics, three service models and four deployment models. In 2011, the official community of standardizing organizations finally accepted that cloud computing is more than temporarily hype. The National Institute for Standards and Technology (NIST) published the first official version of a cloud computing definition (National Institute of Standards and Technology, 2011). In its introduction paragraph, it says: "*cloud computing is an evolving paradigm. The NIST definition characterizes important aspects of cloud computing, [it] is not intended to prescribe or constrain any particular method of deployment, service delivery, or business operation.*"[7]

In response to the question of whether the market would follow this somehow vague statement, we see the constant and steadily growing evolution of new models of cloud services and cloud computing architecture. The basic idea and vision behind this new and accelerating business is, nevertheless, clear and precise. It is simply to use and share IT resources using a huge and well established worldwide network. The not so simple fact is that it is hard to define which type of services could and will be distributed as it can be everything from file and storage services and computing power to email, video streaming, collaboration, shopping and the

[7]*National Institute of Standards and Technology, N. (2011). "The NIST Defintion of Cloud Computing".*

list goes on. The sky is the limit for creative ideas and—sometimes—finding a working business model to create a sort of win-win situation between service provider and consumer.

Using a mail service, a social media network or a cloud storage service follows these principles exactly and it is related to one of the service models. Additionally, it also follows one of the deployment models. As we have seen, this definition is not very old as it was published by NIST in October 2011, but it is worth mentioning that the first cloud services deployed on the Web already had a history of more than one decade.

Essential Characteristics of Cloud Computing & Services

Cloud computing provides an economic and efficient alternative to private data centers, especially for customers, by running web applications and batch processing, simply because of the better scalability of large data centers resulting from massive aggregation and high predictability. Furthermore, data centers can be conveniently located by considering inexpensive power sources and can significantly lower OPEX through the deployment of homogeneous compute, storage and networking solutions. As another bonus, enterprises and end users are largely freed from detailed specifications when using cloud computing.

To understand and enhance our view of what a cloud service is—and what it is not—the National Institute of Standards describes five essential characteristics to define a cloud service in comparison to other types of IT based services:

- On-Demand Self Service
- Broad Network Access
- Resource Pooling
- Rapid Elasticity
- Measured Services

On Demand Self Service

> A consumer can unilaterally provision computing capabilities, such as server time and network storage, as needed automatically without requiring human interaction with each service provider.[8]

As consumer of a cloud service, you should be able to change the set of functions you want to use and decide on-demand on the resources (e.g. storage or computing capacity). The provisioning of new functions or resources can be ordered by the

[8]National Institute of Standards and Technology, N. (2011). *The NIST Defintion of Cloud Computing.*

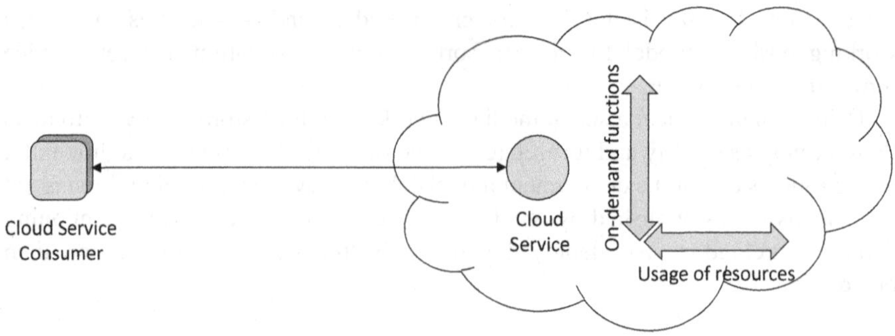

Fig. 7 On-demand self-service

consumer without any human interaction with the provider of the cloud service (see Fig. 7).

This principle follows the idea of self-provisioning and supports the approach of rapid change that follows the actual demand of the consumer. Compared to the traditional organization of IT resources, it makes it possible to consume additional resources or functions within a short time and to reduce the reserved capacity when it is no longer needed.

Broad Network Access

> Capabilities are available over the network and accessed through standard mechanisms that promote use by heterogeneous thin or thick client platforms (e.g., mobile phones, tablets, laptops, and workstations).[9]

The access via a network—and this is meant to be the Internet—is, without a doubt, the basic characteristic of cloud services. The Internet together with standard protocols (TCP/IP, HTML, etc.) and in combination with industry standard devices (PCs, Smartphones and Tablets) is a major condition of building and running cloud computing architectures (see Fig. 8).

Beyond the evolution of the Internet, the World Wide Web and mobile devices were the major drivers of cloud services.

Resource Pooling

> The provider's computing resources are pooled to serve multiple consumers using a multi-tenant model, with different physical and virtual resources dynamically assigned and reassigned according to consumer demand. There is a sense of location independence in that the customer generally has no control or knowledge over the exact location of the

[9]National Institute of Standards and Technology, N. (2011). *The NIST Defintion of Cloud Computing.*

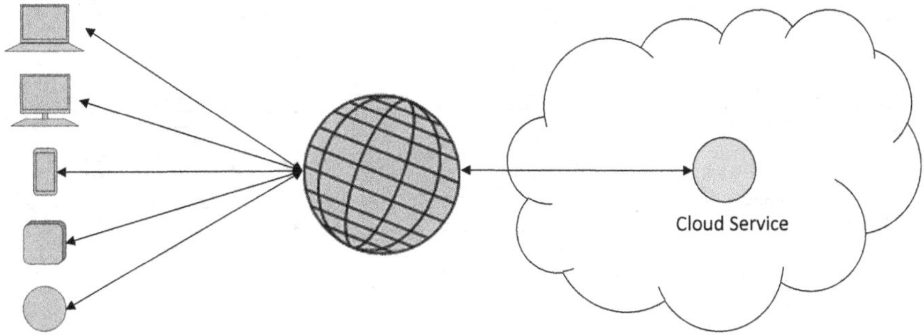

Fig. 8 Broadband network access

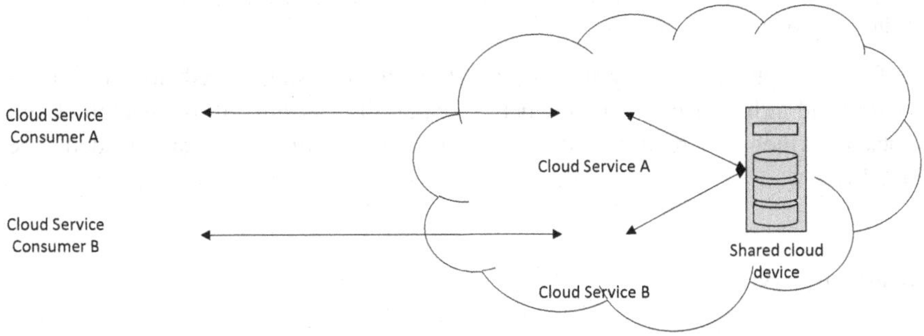

Fig. 9 Resource pooling

> provided resources but may be able to specify location at a higher level of abstraction (e.g., country, state, or datacenter). Examples of resources include storage, processing, memory, and network bandwidth.[10]

Providing services to many consumers demands smart and efficient resource management. The basic technology used to achieve that is virtualization and multitenancy (see Fig. 9).

This principle leads to a better utilization of physical resources but also demands the acceptance of sharing resources with others by the consumer.

Rapid Elasticity

> Capabilities can be elastically provisioned and released, in some cases automatically, to scale rapidly outward and inward commensurate with demand. To the consumer, the

[10]National Institute of Standards and Technology, N. (2011). *The NIST Defintion of Cloud Computing*.

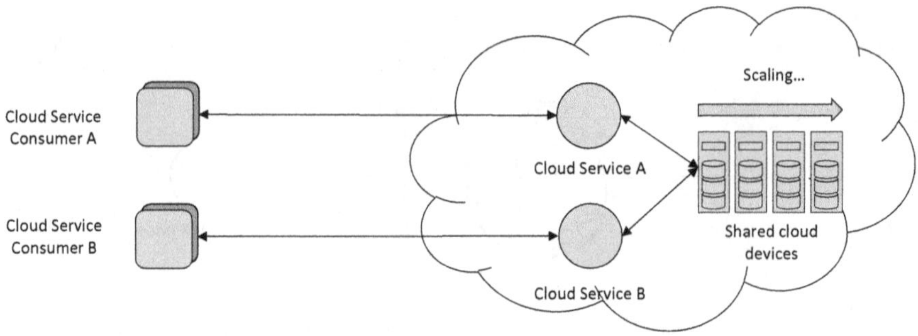

Fig. 10 Elasticity

> capabilities available for provisioning often appear to be unlimited and can be appropriated in any quantity at any time.[11]

The principle of elasticity allows for adding and releasing shared cloud resources on demand. This brings a much better scalability, as only those resources are blocked, which are really needed and only for the time they are needed (see Fig. 10).

Measured Service

> Cloud systems automatically control and optimize resource use by leveraging a metering capability at some level of abstraction appropriate to the type of service (e.g., storage, processing, bandwidth, and active user accounts). Resource usage can be monitored, controlled, and reported, providing transparency for both the provider and consumer of the utilized service.[12]

A cloud service offered by a service provider to a consumer must be measurable to make the extent of usage transparent for both sides. This follows the principle of "pay-what-you-use." Metering the consumption and also provision of a certain service together with the coverage of the agreed service level is basic for remuneration and cost control (see Fig. 11).

Definitions Are Never Complete...

There are some interesting details in this definition. First, it claims that usage of cloud services is based on having access to the "Net" (Broad network access) and the user must accept that the resources or tools he is using are shared with others

[11]National Institute of Standards and Technology, N. (2011). *The NIST Defintion of Cloud Computing*.

[12]National Institute of Standards and Technology, N. (2011). *The NIST Defintion of Cloud Computing*.

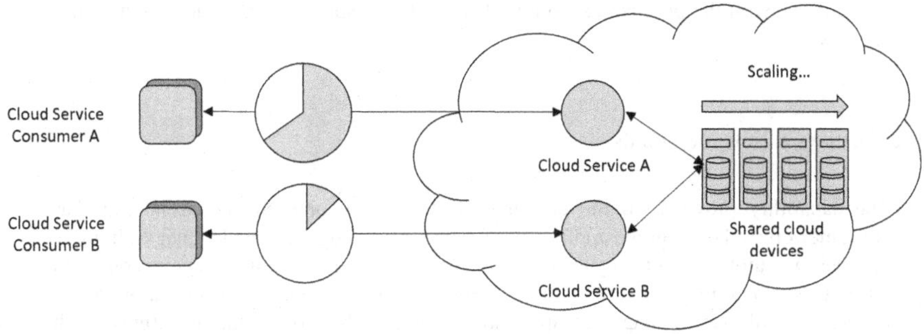

Fig. 11 Measured service

(resource pooling). He must also look after what he is using as a service (On-demand self-service). These characteristics simply move responsibilities and duties to the user of the service. On the other side, cloud services claim to provide a high level of transparency (measured service) and a wide range of consumption options (rapid elasticity). So, this is the deal: if you accept, as consumers, that you must be "in the net," that you are willing to share resources and tools and that you will assume responsibility for what you consume, then and only then—cloud services will give you transparency and elasticity.

It is important to see that the definition says nothing about the price you must pay for any type of service. It is also a common understanding that using the cloud should lead to positive economic effects for consumers. At least this is one of the principles of sharing resources and tools. We will see that the economic models behind cloud services are sometimes more complex and demanding than the technical aspects.

Service Models

The NIST definition also tries to structure the types of cloud services into three categories:[13]

Software as a Service (SaaS).

> The capability provided to the consumer is to use the provider's applications running on a cloud infrastructure. The applications are accessible from various client devices through either a thin client interface, such as a Web browser (e.g., Web-based email), or a program interface. The consumer does not manage or control the underlying cloud infrastructure including network, servers, operating systems, storage, or even individual application

[13]National Institute of Standards and Technology, N. (2011). *The NIST Defintion of Cloud Computing*.

capabilities, with the possible exception of limited user-specific application configuration settings.

Platform as a Service (PaaS)

The capability provided to the consumer is to deploy onto the cloud infrastructure consumer-created or acquired applications created using programming languages, libraries, services, and tools supported by the provider. The consumer does not manage or control the underlying cloud infrastructure including network, servers, operating systems, or storage, but has control over the deployed applications and possibly configuration settings for the application-hosting environment.

Infrastructure as a Service (IaaS)

The capability provided to the consumer is to provision processing, storage, networks, and other fundamental computing resources where the consumer is able to deploy and run arbitrary software, which can include operating systems and applications. The consumer does not manage or control the underlying cloud infrastructure but has control over operating systems, storage, and deployed applications; and possibly limited control of select networking components (e.g., host firewalls).

Even though this definition is only 5 years old, it already seems to be out of date. Many cloud services evolved during the last few years and seem, now, not to fit into one of these categories. Using products like Skype, Google, Wikipedia, Facebook, Amazon Store is a cloud service without doubt, but does not really fit into one of these high-level categories.

Deployment Models

Within the NIST definition, a set of three different deployment models is described.[14]

- Private cloud
- Public cloud and
- Hybrid cloud.

Private Cloud

The cloud infrastructure is provisioned for exclusive use by a single organization comprising multiple consumers (e.g., business units). It may be owned, managed, and operated by

[14]National Institute of Standards and Technology, N. (2011). *The NIST Defintion of Cloud Computing*.

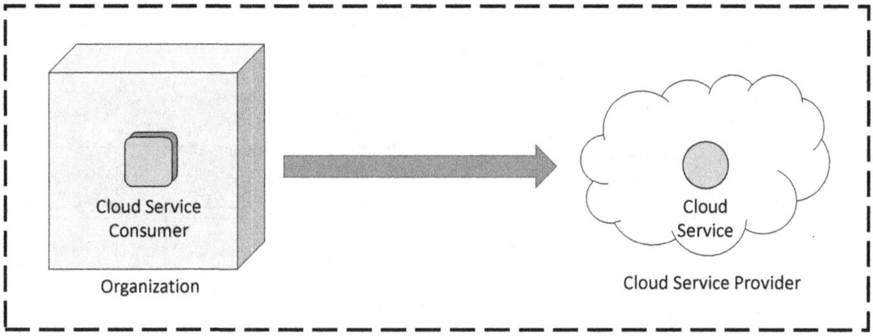

Fig. 12 Private cloud

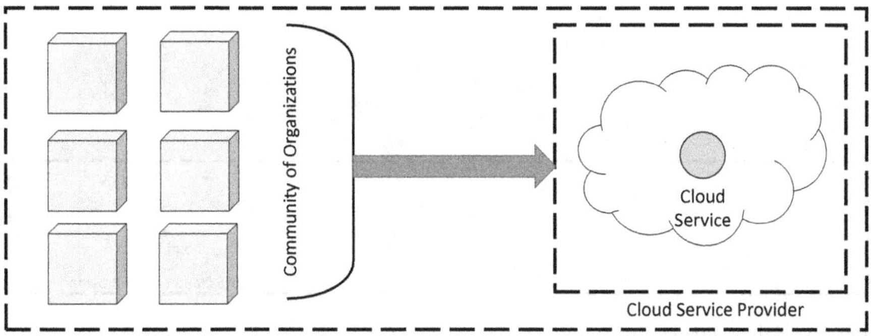

Fig. 13 Community cloud

the organization, a third party, or some combination of them, and it may exist on or off premises." (see Fig. 12).

Community Cloud

The cloud infrastructure is provisioned for exclusive use by a specific community of consumers from organizations that have shared concerns (e.g., mission, security requirements, policy, and compliance considerations). It may be owned, managed, and operated by one or more of the organizations in the community, a third party, or some combination of them, and it may exist on or off premises." (see Fig. 13).

Public Cloud

The cloud infrastructure is provisioned for open use by the general public. It may be owned, managed, and operated by a business, academic, or government organization, or some combination of them. It exists on the premises of the cloud provider. (see Fig. 14).

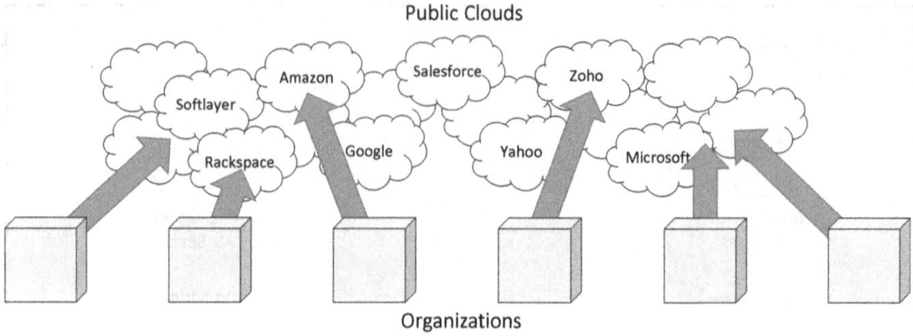

Fig. 14 Public cloud

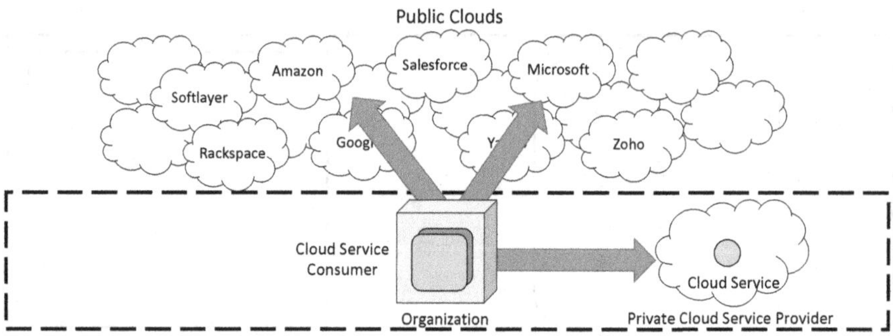

Fig. 15 Hybrid cloud

Hybrid Cloud

> The cloud infrastructure is a composition of two or more distinct cloud infrastructures (private, community, or public) that remain unique entities, but are bound together by standardized or proprietary technology that enables data and application portability (e.g., cloud bursting for load balancing between clouds). (see Fig. 15).

The ITU Cloud Reference Architecture

In 2014, the ITU (International Telecommunication Union) published a more detailed document, "Cloud Computing Reference Architecture". It covers the same service and deployment models as defined in the NIST definition of 2011 but goes into the details of functions and roles for the organization and operation of cloud computing environments. As an extension to the NIST definition, the ITU document is structured into a "User View" and a "Functional view" and describes the relations between users of a cloud computing services and the functions implemented for integration, security, operation, business support and development.

The user view addresses the following cloud computing concepts: cloud computing activities, roles and sub-roles, parties, cloud services, cloud deployment models and cross-cutting aspects.

The functional view describes the distribution of functions necessary for the support of cloud computing activities. The functional architecture also defines the dependencies between functions. The functional view addresses the following cloud computing concepts: functional components, functional layers and multi-layer functions (see Fig. 16).

To build a complete picture of these relations, the cloud architecture is described in four layers: user layer, access layer, service layer and resource layer. For each of these layers the functions are described in detail. The 50-page document is a comprehensive definition of organizational requirements for the planning, integration and operation of cloud computing environments.

The Ownership Model

The concept of ownership was introduced by the authors in a CISCO white paper in 2014 (Oppitz & Tomsu, 2014). We claim that there are eight types of ownership. Combined with different owner types they form a matrix of possible distributions of ownership. This leads to a much more precise definition of different delivery models including (among others) different types of cloud services. We can show that different types of ownership in terms of technical ownership, financial ownership, location ownership, data ownership, process ownership and IP (intellectual property) ownership can be used to describe a clear profile for cloud solution services.

Moving Ownership from Consumer to Provider

Taking over ownership from the customer to the provider opens an additional value for service providers. It moves responsibility from the consumer of a service to the professional provider of a service and frees up the consumer for its core business. This was already a key argument for many customers/enterprises and would gain even more significance over the next few years. The organization of service and business solutions is influenced by the ownership of the IT infrastructure. Speaking about ownership means clarifying which of the partners in the service process (end user, customer, managed service provider, provider or vendor) has control over the IT infrastructure and thus carries the responsibility of managing it. Drilling down into ownership, we see that there are at least eight different types of ownership (see Fig. 17).

It is no question that the major trend is simply moving ownership away from the end user or consumer to centralized providers of technology and services. This has

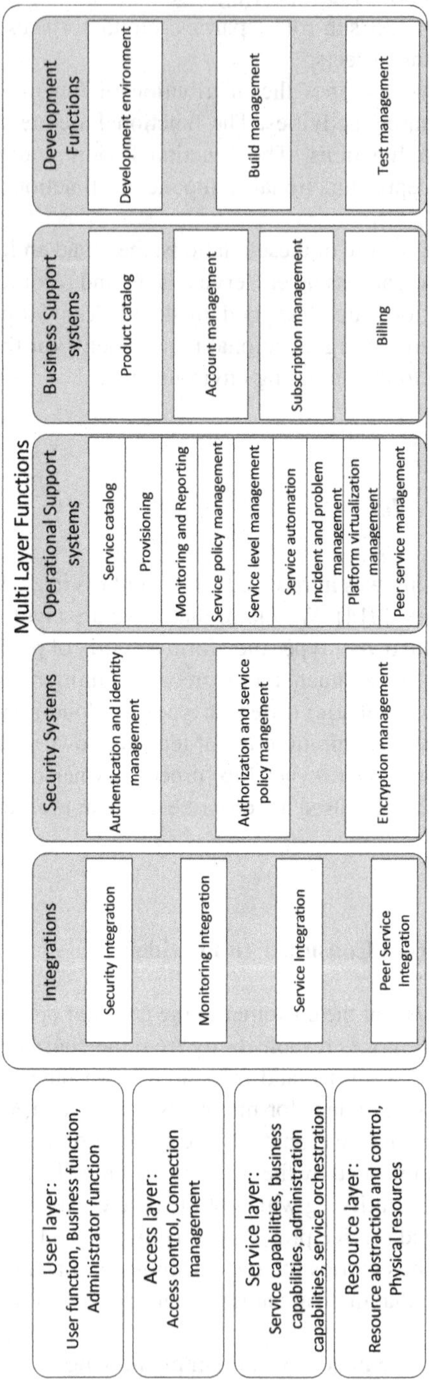

Fig. 16 ITU Cloud Reference Architecture

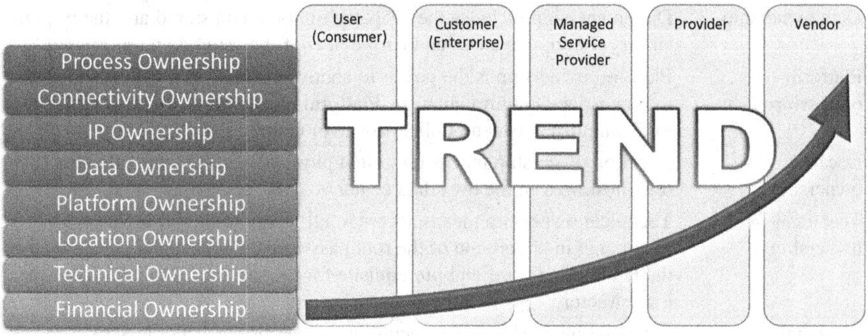

Fig. 17 Ownership model

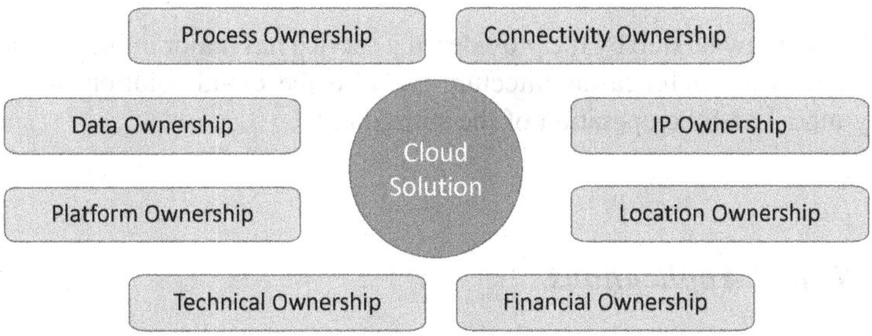

Fig. 18 Ownership types

been—for centuries—a strategy of industrial and economic development following the idea of resource sharing and work distribution.

Ownership Types

We claim that there are eight types of ownership and, combined with owner types, creates a matrix of possible distributions of ownership and leads to a much more precise definition of different delivery models including (among others) the different types of cloud services (see Fig. 18).

Process ownership	Process ownership defines the property right on the business processes and includes the responsibility for completeness and correctness of these processes.
Connectivity ownership	Connectivity ownership defines the property rights for connections between the user or consumer of the IT solution and the provider of the service.
IP Ownership	IP (intellectual property) includes the property rights of the applied technology, patents, knowledge and skills for operating the infrastructure and knowledge and skills regarding the implemented processes.

(continued)

Data ownership	Data ownership includes the property right on data stored and the responsibility for completeness and correctness of data stored and processed.
Platform ownership	Platform ownership is the power to control the platform, having the power to request and confirm changes. Platform ownership is the major attribute to distinguish between public, private or community clouds.
Location ownership	Location ownership means unlimited physical access to the infrastructure (e.g. hosting it in the own data center).
Technical ownership	Technical ownership means access to all layers of the infrastructure (e.g. being in possession of the root password). The technical ownership is the leading indicator and prerequisite for the service responsibility of the infrastructure.
Financial ownership	Financial ownership means being in possession (or being lessee) of the infrastructure. The financial ownership may be with the customer or the provider.

Taking over ownership by the provider of a service has major influences not only on the underlying technical architecture but also the cloud solution offered, the delivery model and the operation of the solution.

Native Cloud Applications

Building applications for the Web and as a cloud services is different from conservative centralized applications or client-server architectures. Traditional software consists of a centralized data base or data repository, and a set of software programs running on a server or split as client-server application between the server and the user's clients. The same architecture can be used for web-based applications operating centralized web-server-side software components and web-based clients consisting of a standard-browser and local code executed by the browser as script (e.g. java-script). This architecture works fine but is not optimized for cloud computing. In a cloud architecture, the computing infrastructure is distributed to an elastic number of virtual machines, which take over processing and data load in a stretchy way depending on the capacity required to fulfill the consumer requests. In this case, an optimized software architecture must be distributable to different virtual machines in different locations and run, in parallel, on an elastic number of so-called nodes containing code as well as distributed data elements.

This type of software architecture is called native cloud application (NCA). For the developer, it is fundamental to break down tasks into separate services that can be executed on several servers. The target of cloud native applications is to separate the infrastructure from the application, thus bypassing the limited reliability of the infrastructure (servers, network, storage) by distributing code and data to redundant "nodes," thereby moving the responsibility for availability and resilience to the software. There are several approaches to implement native cloud applications. They are focusing on frameworks composed of loosely coupled services running in

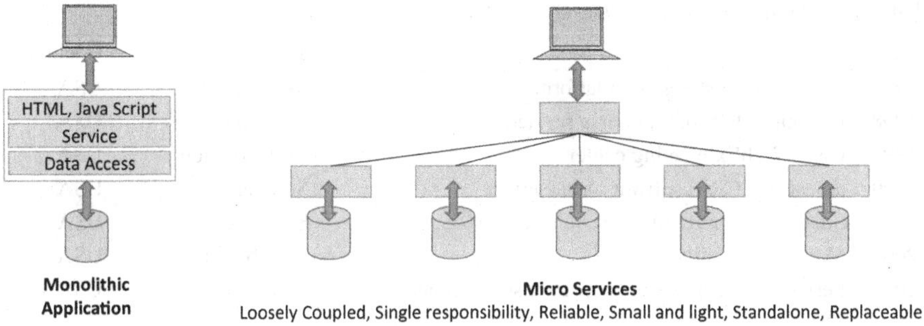

Fig. 19 Monolithic applications versus micro services

containers or as micro-services, enabling a reliable execution of cloud applications on distributed virtual environments (see Fig. 19).

One of these environments—and maybe the most popular now—is "Linux Containers" (LXC), which provides operating system-level virtualization and Docker[15] as a project defining the deployment of applications into containers. It seems that the new cloud native architecture will evolve into another area of software engineering where open source projects lead development.

Most cloud platform providers started to move their environments towards the new software architecture concept. Microsoft offers Docker containers and API management in its Azure PaaS, Amazon provides Docker compatible containers in its AWS Elastic Cloud, and VMWare launched the Linux light weight container ready for Photon distribution. Google as one of the forerunners of distributed cloud native computing, using the container architecture for Gmail, Maps and Google Search. In 2014, Google also announced the optimization of LXC containers with dynamic scheduling of Docker containers along the virtual machines in the Google Compute Engine. IBM offers support for Docker based containers on its IBM Bluemix platform.

Moving Towards the Clouds

The new paradigm and definition of cloud computing has influenced the market of web applications and services. Beginning at the end of the last century, many services were introduced that followed the principles of cloud computing, some of them before the term was coined. The following—not complete—list of services introduced shows the diversity of technologies, applications and disruptive business models covering not only IaaS, SaaS, PaaS but also many new applications that followed the growing demand to use the Internet and the Web as a nexus for new

[15]https://www.docker.com/, retrieved 2016-10-12.

Table 1 Time line of cloud services

Year	Service	Type	Country
1997	Netflix video streaming platform	Media	USA
1998	NetSuite ERP software-as-a service	Software	USA
1999	Cisco WebEx meeting center	Collaboration	USA
1999	Salesforce CRM software-as-a-service	Software	USA
2001	SuccessFactor HCM software as-a-service	Software	USA
2003	Skype video calls	Collaboration	USA
2003	ServiceNow service management software-as-a service	Software	USA
2004	Citrix GoToMeeting collaboration tool	Collaboration	USA
2005	Zoho office applications-as-a-service	Software	India
2005	Softlayer platform-as-a service	Infrastructure	USA
2005	Box file sharing	File Sharing	USA
2005	TechCrunch	Knowledge	USA
2006	Twitter microblogging	Social network	USA
2006	Rackspace infrastructure as-a-service	Infrastructure	USA
2006	WikiLeaks	Knowledge	Iceland
2007	ZenDesk service management software-as-a-service	Software	USA
2007	Spideroak secure file storing	File sharing	USA
2007	SAP cloud solutions	Software	Germany
2007	Google apps for business	Software	USA
2007	Amazon video on demand	Media	USA
2007	Amazon Kindle e-book reader and platform	Media	USA
2007	Statista	Knowledge	Germany
2008	Spotify music streaming	Media	USA
2008	Encyclopedia of Life	Knowledge	USA
2008	Microsoft One Drive file storage	File Sharing	USA
2008	Groupon e-commerce market place	e-commerce	USA
2008	Dropbox file hosting	File Sharing	USA
2008	Apple AppStore software distribution platform	Software	USA
2008	Amazon Web Service infrastructure as-a-service	Infrastructure	USA
2008	AirBnB travel and booking platform	Travel and booking	USA
2009	Quora	Knowledge	USA
2009	Documentcloud	Knowledge	USA
2009	WhatsApp chat	Social network	USA
2009	Wetransfer file transfer service	File Sharing	USA
2009	Uber car booking service	Travel and booking	USA
2009	Kickstarter crowd funding	Crowdfunding	USA
2009	Google mail	Mail	USA
2009	Google docs business apps	Software	USA
2009	Bitcoin cryptocurrency	Cryptocurrency	USA
2010	The World Bank Open Data	Knowledge	USA

(continued)

Table 1 (continued)

Year	Service	Type	Country
2010	SAP HANA data management	Platform	Germany
2010	Pinterest file sharing	File Sharing	USA
2010	Microsoft Azure platform-as-service	Platform	USA
2010	Instagram photo sharing and social network	Social network	USA
2010	Google cloud storage	Infrastructure	USA
2010	Google cloud platform	Platform	USA
2010	Google app engine	Platform	USA
2011	Snapchat video and photo sharing	File Sharing	USA
2011	Microsoft Office365 software-as-a- service	Software	USA
2011	Google+ social network	Social network	USA
2011	Apple iCloud storage service	Infrastructure	USA
2011	Apple Facetime video telephony	Collaboration	USA
2012	WikiData	Knowledge	USA
2012	Oracle Cloud infrastructure-as -a-service	Infrastructure	USA
2012	Google drive file sharing	File sharing	USA
2013	VMWare vCloud Air infrastructure-as-a-service	Infrastructure	USA
2014	Missing Maps	Knowledge	USA
2014	IBM Bluemix platform-as-a-service	Platform	USA
2014	Ethereum cryptocurrency	Cryptocurrency	UK

services. Some of these applications simply optimized existing services or processes, some of them created completely new markets and triggered alternative behaviors in consuming goods and building communities (see Table 1).

Infrastructure as a Service

Providing storage and computing capacity as a cloud service became one of the first applications of cloud service technology for consumers as well as for corporations. The idea was simple but convincing and based on the idea of resource sharing. Buying and operating your own storage or computer is more expensive than using a centralized infrastructure operated by specialized companies. More than that, access to data and computing capacity can be provided location independent using standardized devices like PCs or mobile devices. Starting in the middle of the first decade of the twenty-first century, many offers for infrastructure in the cloud were created by new startups. Amazon, as e-commerce provider "born-on the web," started the first infrastructure-as-a service business but was soon followed by traditional IT companies like IBM, Microsoft and Oracle.

2005 Box.com

Box was developed as a college project for sharing files between users by the University of Southern California student Aaron Levie in 2004. Levie left school in 2005 to manage the company full-time. In 2012, Box received his first lot of funding from General Atlantic in the amount of $125 million. After two more rounds, Box held its initial public offering on the NYSE in 2015. The IPO raised $175 million at a market valuation of $1.6 billion. Today, the company's revenue is about $375 million (2016) with a negative EBITDA of $121 million.[16]

2006 Amazon Web Services

In 2003, Amazon had become one of the most successful retail services for books, CDs, and computer hardware. To operate the large data bases and customer services, Amazon had created an extended data center infrastructure to provide these services to their customers and had started to standardize the hardware, software and data center operations. The idea ended up offering that infrastructure as a service to customers.

A first white paper was created by Rick Dalzell (CIO of Amazon) and Chris Pinkham (VP of IT Infrastructure) and the Web site engineer, Ben Black. CEO Jeff Bezoz was immediately convinced and supported the project. Chris Pinkham was on his way back to his home country, South Africa, and invited Amazon engineer, Christopher Brown, to come with him and work on the new product in South Africa. The team worked together for two years and, in 2006, Amazon released its first version of AWS (Amazon Web Services), covering storage (Amazon S3), computing (Amazon EC2), databases (Amazon SimpleDB), and data flow (Amazon Simple Queue Service). In 2008, the service exited the beta phase and Amazon offered service level agreements and introduced availability zones and redundancy.

By 2010, Amazon had moved all its internal IT operation to the Amazon Elastic Cloud. Amazon Web Service was the first mover into the new infrastructure-as-a-service business and faced several challenges regarding availability and security in the beginning. However, starting early, Amazon could rely on two major benefits: it already had huge experience in operating a large infrastructure for its e-commerce business and—maybe more important—Amazon's knowledge in processing numerous web-customers was tremendously useful for the new business. In 2013, Amazon got a $600 million contract from the CIA beating its competitors, IBM among them.[17] The market began to understand that infrastructure-as-a-service was evolving into a new and rapidly growing business.

---

[16]https://www.nyse.com/quote/XNYS:BOX, retrieved 2017-01-10.

[17]Amazon wins CIA contract, Bloomber, 2013: https://www.bloomberg.com/news/articles/2013-10-07/amazon-wins-ruling-for-600-million-cia-cloud-contract, retrieved 2016-10-12.

2006 Rackspace Cloud

Rackspace was launched by Richard Yoo as a hosting company focusing on customer support and services in 1998. Rackspace has its origins in Cymitar Network Systems, a small internet service provider started by Richard Yoo 2 years earlier out of a small apartment in San Antonio, Texas. The young company was funded by Norwest Venture Partners and Sequoia Capital in 2000 and filed their public offering in 2008 generating $187.5 million at the NYSE. From the beginning, Rackspace followed a multi-layer business model offering a large variety of hosting services from simple web-page hosting to dedicated servers and different types of cloud services.

Though the "Rackspace Cloud" is a competitor to Amazon Web Services and Microsoft Azure, Rackspace has developed a broad band portfolio of cloud offerings and services including management and support for the products of their competitors over the last decade.

In 2010, Rackspace, together with NASA, Dell and Citrix, was the initiator of the OpenStack project, which set the target of creating an open software and enabling the implementation of cloud-computing services running on standard hardware. The OpenStack project was transformed into the OpenStack Foundation in September 2012 to "*produce the ubiquitous Open Source Cloud Computing platform that will meet the needs of public and private clouds regardless of size, by being simple to implement and massively scalable.*"[18] Rackspace became one of the most successful cloud service companies generating a revenue of $ 2 billion (2015). In autumn 2016, Rackspace was acquired for $4.3 billion by the private equity firm, Apollo Global Management, one of the five largest private equity firms.

2007 SpiderOak

Using file sharing services in the cloud automatically led to questions about security and encryption. Several file sharing services focused on that challenge and started to create secure and encrypted data hosting services. One of the first was Spideroak founded by Ethan Oberman and Alan Fairless in Chicago in 2007. The Spideroak services are based on Crypton, a java script based framework for building web applications, where the server doesn't know the contents it's storing on behalf of users. Spideroak services are offered as subscription plans starting at $5 per month for 100GB of storage. The company is extremely focused on security and privacy and became popular when Edward Snowdon recommended Spideroak, citing its better

[18]Open Stack mission statement: http://docs.openstack.org/project-team-guide/introduction.html, retrieved 2016-11-16.

protection against government surveillance.[19] Today, Spideroak is still a privately held company and received two investments: $2 million in seed financing in 2013 and a $3.5 million A-round investment from BW Capital Partners in 2015.[20]

2008 Dropbox

As student at MIT, Drew Houston experienced problems in organizing his files and decided to use cloud technology for his personal use to synchronize files between different desktops. In 2007, he founded Dropbox and, together with Arash Ferdowsi, started to provide a file-hosting service for private users that grew in users to 50 million by 2011. Dropbox was initially financed by Y Combinator and Sequoia, receiving additional funding of $250 million from BlackRock in 2014. Usage of Dropbox services is free for private users. In 2013, Dropbox released Dropbox for Business to compete against the large players in the file sharing field like Amazon. Google and Microsoft. Today, Dropbox has 500 million users and enjoys an estimated value of $10 billion, making it larger than many public companies. Only a very small fraction of Dropbox's active users are paying customers. In June 2016, Drew Houston said: "*... the company has about 150,000 enterprise customers ... But don't expect an IPO anytime soon.*" He hopes that this just means the company has significant room for growth.[21]

2008 Microsoft One Drive

Microsoft OneDrive started in 2008 as SkyDrive available in 38 countries and regions for file storage services and allowed users to sync files and access them from a web browser or mobile device. Files can be shared publicly or with users. OneDrive is included in the suite of Microsoft Office 365 online services with a storage capacity of 1 TB for paying subscribers or 5 GB of free storage.

2009 WeTransfer

Storing data is not always the primary task of cloud technology. In some cases, users or organizations just want to transmit data like large documents or photos to a receiver, but experience limitations with email services. WeTransfer is a service and company based in Amsterdam that offers the transmitting of files that may be much larger than permitted by email systems. The service is free with additional

[19]Edward Snowdon's Privacy Tips, Techcrunch, 2014: upon https://techcrunch.com/2014/10/11/edward-snowden-new-yorker-festival/, retrieved 2016-10-12.

[20]Crunchbase/Spideroak: https://www.crunchbase.com/organization/spideroak, retrieved 2016-10-12.

[21]Interview with Drew Houson 2016: https://techcrunch.com/2016/06/14/dropbox-says-it-is-cash-flow-positive-in-no-rush-to-ipo/

features available in a subscription plan. The company was founded by Bas Beerens and the popular Dutch blogger Nalden (born as Ronald Hans) and the startup reached profitability in early 2015 even as it was sending 1 billion transfers for 80 million users.[22] In the same year, WeTransfer received $25 million in funding by Highland Capital Partners.[23]

2010 Google Storage, App Engine and Cloud Platform

In 2010, Google entered the cloud platform business by introducing its Google storage and, in 2011, its Google App Engine as a platform to create and operate web applications using Google's infrastructure. Compared to other cloud platform services, Google's App Engine supports only a limited scalability and portability within the borders of Google's environment. The storage service and App Engine was used by Google as a starting point into a more extended cloud based platform product. In 2012, Google had grown its business to $50 billion and turned a profit of $10 billion supporting 425 million Gmail users and managing 100 petabytes of web index for its Google search.[24] Larry Page decided to "*. . .take this technology and extended it via Google Cloud Platform so that you can benefit from the same infrastructure that powers Google's applications.*"[25] The Google Cloud was extended by the Google Compute Engine as an infrastructure as a service product in 2012 and by Google Cloud Dataproc, providing Spark and Hadoop as a platform-as-a service in 2016.

2011 Apple's iCloud

At the World Wide Developers Conference in June 2011, Apple announced its new cloud storage service as a follow up to the not so successful MobileMe. iCloud was released later the same year in October. The service is based on storage as a service and provides 5 GB of free storage for owners of either an iOS device or a Mac. Users can purchase additional storage for a total of up to 2TB. On top of this service, many additional applications built for Apple users are provided including backup and restore, Find My Friends, Find My iPhone, iTunes Match, Photo Stream, Photo Library and others. With iCloud Drive, Apple offers a functionality similar to Microsoft's OneDrive or Google Drive.

---

[22]Alex Konrad, Forbes, 2016: http://www.forbes.com/sites/alexkonrad/2016/04/07/dutch-startup-wetransfer-is-taking-on-the-us/#25384a076270, retrieved 2016-10-12.

[23]Crunchbase/Wetransfer: https://www.crunchbase.com/organization/wetransfer, retrieved 2016-10-12.

[24]Larry Page on Google revenue 2012: http://www.dawn.com/news/780915, retrieved 2016-12-06.

[25]Google Cloud Blog 2011: https://cloud.googleblog.com/2012/07/introducing-google-cloud-platform.html, retrieved 2016-12-06.

Revenue from iCloud represents only a small part ($800 million) of Apple's total revenue of $234 billion (2015) but it is an important element of Apple's service strategy. "*...Apple said its customers are now carrying around 1 billion "active" devices, which are Apple gadgets that people interacted with at least once in the past 3 months. "This is an unbelievable asset for us. Because our install base has grown quickly, we have also seen an acceleration in ... what has become one of the largest service businesses in the world." Apple CEO Tim Cook said.*[26] In 2016, the number of iCloud users reached 782 million.[27]

2012 Google Drive

Following Dropbox, Microsoft and Apple, Google management decided to follow the same path and introduced Google Drive in April 2012: "*Just like the Loch Ness Monster, you may have heard the rumors about Google Drive. It turns out, one of the two actually does exist.*"[28] By October 2014, the service had 240 million monthly active users. Like other cloud based file-sharing services, Google Drive is offered as a "freemium" service with 15 GB free and up to 30 TB in additional subscription options. In September 2015, Google reported more than 1 million paying organizations actively using Google Drive.[29]

2012 Oracle Cloud

In June 2012, Oracle's Larry Ellison started announcing its own cloud services based on Oracle operating systems and the Oracle database products. For the following years, Oracle's strategy regarding infrastructure as a service was unclear and announcements concentrated on selling its system and application software suites to corporate customers for implementation on private cloud environments. On July 28th, 2016, Oracle acquired NetSuite, one of the first cloud hosting companies, for $9.3 billion. At the Oracle Open World in September 2016, Larry Ellison announced new Infrastructure-as-a-Service (IaaS) products and a new Oracle database in the cloud that would beat the competition by a factor of 20.[30]

---

[26]David Goldman, CNNTech, 2016: http://money.cnn.com/2016/01/27/technology/apple-services/, retrieved 2016-12-16.

[27]Eddy Cue, February 2016: http://appleinsider.com/articles/16/02/12/apple-music-passes-11m-subscribers-as-icloud-hits-782m-users

[28]Google Drive Announcement 2012: https://googleblog.blogspot.co.at/2012/04/introducing-google-drive-yes-really.html, retrieved 2016-12-16.

[29]Scott Johnston, Director of Product Management, Google Drive, 2015: https://medium.com/@gsuite/making-google-drive-the-safest-place-for-all-your-work-b5248f9b9ddc#.ctqnzk7bq, retrieved 2016-12-16.

[30]Oracle Announcement 2016: https://www.oracle.com/corporate/pressrelease/database-benchmarking-092016.html, retrieved 2017-01-12.

2013 IBM Softlayer

Softlayer was founded in 2005 as a dedicated server, managed hosting and cloud computing provider. By 2011, the company had expanded to hosting more than 81,000 servers for more than 26,000 customers in locations throughout the United States. At the same time, IBM decided to move into the cloud services market using their market power, customer relations and first-class IT infrastructure. The acquisition of Softlayer in June 2013 as the biggest privately held cloud infrastructure provider (IaaS) in the world was a logical step. Together with other already existing cloud services Softlayer was integrated into the new IBM Cloud Service Division to "*...provide a broad range of choices to ... clients, ISVs, channel partners and technology partners*"[31] Today, the IBM SmartCloud brand includes infrastructure as a service, software as a service and platform as a service offered through public, private and hybrid cloud delivery models.

2013 VMWare vCloud Air

VMWare is one of the early movers and shakers in virtualization software and was founded back in 1989. Travelling through several acquisitions and partnerships involving companies like EMC and Cisco, today VMWare is a subsidiary of Dell. Being one of the major players in cloud services from the beginning, VMWare started to plan its own a cloud service product around 2006. Announced in 2008 at the VMWare world conference, it took until 2013 before general availability was announced for vCloud Hybrid Service. Rebranded to vCloud Air in 2014, the service offers three "infrastructure as a service" (IaaS) subscription service types: dedicated cloud, virtual private cloud, and disaster recovery. Like other public cloud providers, vCloud Air supports the concept of regions to increase application performance or in disaster recovery. vCloud Air supports more than 5000 applications and 90+ operating systems.

Software as a Service

Providing email, e-commerce and information services via the web led to a rapidly growing community of mostly private web users. By the end of the 1990s, the Web had arrived in many households but not in companies, corporate usage was mainly through the use of the underlying Internet for building connections between subsidiaries of large companies. Through those years, the idea evolved to offer application services via the Internet and the Web to companies. The idea was not

[31] IBM Press Release 2013: http://www-03.ibm.com/press/us/en/pressrelease/41191.wss, retrieved 2016-12-16.

so fresh and already known as ASP (Application Service Providing) or "Hosted Application" which simply means operating business applications in a data center and providing the usage to corporate users via fixed lines or, later, Internet connections. The arrival of the Web including HTTP, HTML and browsers changed that approach in favor of a new type of business application using the power of centralized services operated on Internet hosts in combination with a browser based HTML client.

1998 NetSuite

NetSuite was founded in 1998 by the former Oracle manager Evan Goldberg as NetLedger and was based on the idea of providing a web-hosted accounting software. The story goes during a discussion with Larry Ellison, Goldberg came up with the idea to develop a web-based sales automation application and Ellison convinced him the best starting point for the new approach was in accounting and ERP instead of CRM.[32] Ellison invested in the new founded company and Oracle later licensed the products NetSuite offered under the name of The Oracle Small Business Suite. NetSuite filed its IPO in December 2007 raising $161 million at a valuation of $1.5 billion. The company was able to increase its revenue between 2009 and 2014 by 149% but still showed annual losses. The 2015, revenue was $741.1 million.[33] In July 2016, Oracle announced it had offered to purchase NetSuite for $9.3 billion. The deal created discussions within the NetSuite investor's community since Oracle executive chairman and co-founder, Larry Ellison, owned nearly 40% of NetSuite.

1999 Salesforce

One of the first attempts to create a Software as a Service business model was made by the former Oracle executive Marc Benioff together with Parker Harris, Dave Moellenhof and Frank Dominguez. Analyzing the market for business applications, they found that CRM (Customer Relationship Management) applications are a type of business software that has a wide field of target customers in different industries and company sizes. In 1999, Salesforce was founded as one of the first companies creating and offering CRM as Software as a Service via the Web.

The team developed the first version of their sales automation software and launched it to the first customers in the fall of 1999. The functions included the typical sales processes like contact and opportunity management, forecasting and

---

[32]Julie Bort, Business Inside, 2012: http://www.businessinsider.com/netsuite-ceo-larry-ellison-invented-cloud-computing-2012-10?IR=T

[33]10-K for NETSUITE INC 2016: https://biz.yahoo.com/e/160224/n10-k.html, retrieved 2016-12-10.

sales reports. Rollout to users and customers was completely different from other CRM applications: Salesforce's logo statement claimed, "No Software," which simple means that the user only needs a browser, as software and data is stored and operated within the premises of Salesforce and access is via the Web. 5 years later, Salesforce was one of the first companies offering something that was called a cloud service or Software as a Service. In 2004, Salesforce was listed on the New York Stock Exchange, raising $110 million at its initial public offering. Today, Salesforce is one of the most successful companies in the SaaS sector generating revenue of $ 6.6 billion (2016) and having a market capitalization of more than $40 billion (2016).

Salesforce also had its origins in Oracle. The NetSuit CEO, Zach Nelson, said the following in an interview in 2012 about a meeting with Larry Ellison and Evan Goldberg: "*There was one other guy in that conversation and that guy was Marc Benioff. Benioff called back two weeks later and said 'I'm going to do that Siebel online thing,' and that became* Salesforce.com. *Larry funded both of those companies.*"[34]

The company was founded in 1999 by Marc Benioff, Parker Harris, Dave Moellenhoff, and Frank Dominguez as a company specializing in software as a service (SaaS) for sales automation. At that time, Siebel was the market leader for CRM (Customer Relationship management) software and Beniof was convinced that CRM was the best place to start with a software-as-a-service product. Harris, Moellnhof and Dominguez developed the first version of the sales automation software, which was launched to customers in the fall of 1999. From the beginning, Salesforce was designed as pure cloud solution: customers could access the applications via browser, the payment model is user-based, and application and data is hosted in the Salesforce cloud. The new service was immediately accepted by the market and salesforce became the first rapidly growing software-as-a-service company. In June 2004, the Salesforce IPO was listed at the NYSE and raised $110 million. Among the first investors was also Orcale's Larry Ellison. Today, Salesforce creates revenues of $6.67 billion (2016) with an EBIT of $64 million.[35] With a market capitalization of around $50 billion, it belongs to the most valuable companies in the Web and Internet business sector.

2001 SAP SuccessFactor

HR (human resources) software was the next field that would be reached by software-as-a service. SuccessFactor was founded as a private company by Lars Dalgaard in 2001, a graduate of Stanford University. Dalgaard was looking for

---

[34]Zach Nelson about NetSuite and Salesforce, 2012: http://www.businessinsider.com/netsuite-ceo-larry-ellison-invented-cloud-computing-2012-10?IR=T

[35]https://www.google.com/finance?q=NYSE%3ACRM&fstype=ii&ei=pZfxVtmRNoSR0ASnha2YBQ, retrieved 2017-01-10.

business to invest in and "*Amid the chaos of the dot-com bust, he acquired the assets of two failed startups and relaunched them as an online business software company called SuccessFactors.*"[36] The company targeted SMBs and large enterprises with its HR software by offering it on a subscription basis. The company filed its public offering in 2007 raising just over $100 million and carrying a market cap above $500 million. In December 2011, SAP AG and SuccessFactors announced that SAP America, Inc. had entered a definitive merger agreement. The acquisition was completed in February 2012 for $3 billion.[37] Today, SAP SuccessFactor is a subsidiary of SAP and creates a revenue of $630 million (2014) and serves more than 1000 customers (2015) out of a network of six globally distributed data centers.

2003 ServiceNow

Service management became the next target for infrastructure-as-service. Around 2002, Peregrine and Remedy were the market leaders in service management software. After a number of acquisitions, both companies are united in BMC Software today. The former Peregrine and Remedy CTO, Fred Luddy, founded ServiceNow in 2003 based on the idea of moving service management into the cloud. In 2007, ServiceNow reported annual revenue of $13 million growing that to $92.6 million and a profit of $9.8 million by 2011, up from a loss of $29.7 million in 2010. ServiceNow filed its IPO in 2012 raising $209.7 million. In 2015, ServiceNow reported annual revenue of $1 billion but also a net loss of $198.4 million.[38] ServiceNow market cap reached $13.56 billion in 2016.

Office as a Service

For the personal user of computers, the set of applications running on their PCs is one of the major elements of daily computing. Word processors, spreadsheet applications or presentation software were, for a long time, the typical local environments of personal computers. Keeping those applications up-to-date and running is vital for the daily private or business user. In the middle of the first decade of this century, the idea evolved to provide these applications out of cloud environments to rid them of complex and costly licensing and update processes.

---

[36]Bob Evans and Lars Dalgaard, Forbes, 2011: http://www.forbes.com/sites/sap/2011/12/08/top-10-reasons-for-sap-acquisition-from-successfactors-ceo-lars-dalgaard, retrieved 2016-12-10.

[37]Eugene Kim, Interview with Lars Dalgaard, 2015: http://www.businessinsider.com/lars-dalgaards-one-interview-question-2015-10?IR=T, retrieved 2016-12-10.

[38]ServiceNow Financial Results 2015: http://www.servicenow.com/company/media/press-room/servicenow-reports-financial-results-for-fourth-quarter-and-fiscal-year-2015.html, retrieved 2016-12-10.

The office application market was influenced by Microsoft as a major player and Apple. Different approaches were tested. New startups began to develop apps for business applications and to offer the usage within subscription plans. Microsoft and others soon understood that the old-fashioned buy-and-install plans for PC software were losing attractiveness and started to change the software they offered to cloud based subscription plans.

2005 Zoho Office Suite

Zoho was founded as AdventNet by Sridhar Vembu in 1996. In 2005, the company launched Zoho Office Suite, which includes a web-based word processor, a spreadsheet program, a mail program, a calendar, contacts list, and other business-oriented programs.[39] The idea behind the Zoho Suite is to provide the typical office functions as a cloud service and thereby compete against Microsoft and Google. Their products have both a free edition and paid editions. The company is based in Chennai, Tamil Nadu, India and has offices in the Philippines, China and Singapore. Since 1996, Zoho went through at least two tough situations during the dot-com bubble in 2000 and the financial crisis in 2008 but managed to survive and is still privately held by its founders. Vembu claims to create revenue of $500 million, serving 100,000 paying customers and 18 million users.[40]

2006 Google Apps

In August 2006, Google launched "Google Apps for Your Domain," a set of apps for organizations including Gmail, Google Talk, Google Calendar, and the Google Page Creato (later replaced with Google Sites). Dave Girouard, then Google's vice president, outlined its benefits for business customers: "*A hosted service like Google Apps for Your Domain eliminates many of the expenses and hassles of maintaining a communications infrastructure, which is welcome relief for many small business owners and IT staffers...*"[41] The Google Apps suite went through several expansions and is offered today as G Suite including Gmail, Google Docs, Calendar, Google Drive and many others.

---

[39]Steve Lohr, NYT, 2009: https://bits.blogs.nytimes.com/2009/08/26/zoho-thriving-amid-the-giants/, retrieved 2016-12-10.

[40]Vishnal Krisnha: Interview with Sridhar Vembu, 2016: https://yourstory.com/2016/04/sridhar-vembu-zoho-corp/, retrieved 2016-12-10.

[41]Google Press: Google Launches Hosted Communications Services, 2006: http://googlepress.blogspot.co.at/2006/08/google-launches-hosted-communications_28.html, retrieved 2016-12-10.

2009 Google Docs and Google Mail

Google Docs has its roots in the Writeley web-based word processor that was launched in 2005 and in Google spreadsheets developed in the Google labs. In 2009, Google combined the two applications, added a presentation program and launched the suite as Goggle Docs, tightly integrated with Google Drive and Chrome. Google Mail (Gmail) was officially released in the same year (2009) after having gone through a beta phase that began in 2004. Today, Gmail is one of the most widely used mail services with 1 billion users (2016).

2011 Microsoft Office 365

Microsoft launched its Office cloud service under the brand name of Office 365 in June 2011 as successor to the Microsoft Business Productivity Online Suite. The new Office 365 brand was also the foundation for the distribution of Office 2013 and the transition from Microsoft's one time license plans to subscription based plans with a cloud sourced distribution and update to Microsoft's business applications: "*For the first-time Microsoft is tempting Office users to rent, not own, software that for decades they've bought as a standalone program.*"[42] In 2016, there were 1.2 billion office users worldwide and 60 million monthly active Office 365 users.[43]

Platform as a Service

Providing storage and computing capacity as infrastructure-as-a service is helpful, but in many cases consumers are more interested in having access to complete environments of tools and software in a cloud environment in order to enjoy the flexibility and the cost advantages of a platform-as-a service. At the end of the first decade, the first cloud products were created by large and traditional IT companies like Microsoft, SAP, IBM and others.

2010 Microsoft Azure

In 2007, Microsoft decided to join the platform-as-a service party and created a new service under the name of Windows Azure, which was later renamed to Microsoft

---

[42]Ian Paul, PCWorld 2013: http://www.pcworld.com/article/2026703/office-365-vs-office-2013-should-you-rent-or-own-.html, retrieved 2016-10-12.

[43]Windows Central, 2016: http://www.windowscentral.com/there-are-now-12-billion-office-users-60-million-office-365-commercial-customers, retrieved 2016-10-12.

Azure. Announced at the Professional Developers Conference in 2008,[44] the service started in 2010 with services for hosting storage, computation and networking, Microsoft SQL services, Microsoft .NET services, Live services that stored, shared and synchronized documents, photos, files and information, Microsoft SharePoint services and the Microsoft Dynamics CRM services cloud. In the following years, the Azure platform was extended by many different types of cloud services including computing, storage, messaging, developer tools, management and machine learning. Microsoft Azure is maintained and operated out of Microsoft's data centers and is based on Windows servers managed by the Microsoft Azure hypervisor.

2010 SAP Hana

Storage and management of large amount of data is one of the key elements of cloud applications. SAP is one of the largest providers of business applications concentrated on the "big data and analytics" layer as it poses one of the most important challenges. Starting around 2008, a team from SAP, the Hasso Plattner Institute and Stanford University had designed and demonstrated a new application architecture named "Hassos New Architecture" or HANA including a high-performance search engine and an in-memory cache engine.

The product was extended in the following years and SAP announced a platform as a service product called the SAP HANA Cloud Platform including analytics, integration, IoT and security services in 2012.[45] The HANA Cloud Platform is offered as a freemium cloud service with a rich set of subscription plans starting at $23 per user per month. Additionally, a managed private cloud service was announced as the HANA Enterprise Cloud service in 2013.[46] In July 2016, SAP reported having 3700 corporate customers for its S4/HANA suite, compared with 900 a year before.

2014 IBM Bluemix

IBM Bluemix was announced on February 2014 and generally made available in June of the same year. Bluemix is a cloud platform-as-a service for software development and operation (aka DevOps).[47] It is based on Cloud Foundry, an

---

[44]Microsoft Azure announcement 2008: https://news.microsoft.com/2008/10/27/, retrieved 2016-12-16.

[45]SAP HANA Cloud platform: https://hcp.sap.com/index.html, retrieved 2016-10-12.

[46]SAP HANA Enterprise Cloud: http://www.sap.com/product/technology-platform/hana-enterprise-cloud.html, retrieved 2016-10-12.

[47]IBM Bluemix Announcement, 2014: http://www-03.ibm.com/press/us/en/pressrelease/43257.wss, retrieved 2016-10-12.

industry standard platform for cloud applications and runs on IBM's Softlayer infrastructure. Beginning in 2007, IBM had focused on a new set of cloud products leading to the IBM Cloud Service Divisions and the acquisition of Softlayer in 2013.

Based on those technology layers, Bluemix was built by IBM teams located in different places within 18 months.[48] Today, Bluemix consists of over 120 services covering computing, storage, network, programing languages and environments, security, data and analytics, internet-of-things and cognitive computing based on IBM's Watson product line.[49] In May 2015, IBM said that 8,000 new Bluemix users are added to the service each week.[50]

Chat, Collaboration and Video

Extending the communication channels from text to voice and video using the Web was a logical step. Broadband access was on its way to households and ADSL connections had spread out and started to replace ISDN. The expansion of network capacity and the availability of better and more mobile end-devices powered the acceptance of collaboration and communication tools. Using the Web for telephone and video conferencing turned out to become a much more usable and inexpensive alternative to conventional telephones. The consumer community welcomed the new opportunities to connect to friends and family via the Web using Skype, Twitter, Facetime and many other services. The business world soon recognized that web-conferencing can reduce travel costs and may boost productivity when working in distributed and international teams.

1999 Webex

Webex has its roots in a software utility for offline web-browsing included in LapLink, a product owned by the company Travelling Software. WebEx was founded in 1996 under the name of ActiveTouch by Subrah Iyar, a software engineer with Intel, Apple and Quarterdeck and Min Zhu, the co-founder of Future Labs and a Stanford University teacher. The name Webex was sold by Traveling Software to the new company in 1999 and ActiveTouch was renamed to Webex.

[48]Ann Fisher: "The best way to develop new ideas at work", Fortune 2014: http://fortune.com/2014/10/29/ibm-innovation/, retrieved 2016-10-12.

[49]IBM Bluemix Catalog: https://console.ng.bluemix.net/catalog/, retrieved 2016-10-12.

[50]Larry Dignan, ZDNet, 2015: http://www.zdnet.com/article/ibm-bluemix-adding-8000-users-a-week/, retrieved 2016-10-12.

The concept of Webex was fueled by an interest in web conferencing and the first product was released in July 1999 named WebEx Meeting Center. [51]

The application is based on the MediaTone platform and supported by the WebEx MediaTone Network designed by Shaun Bryant, WebEx's Chief Network Architect. Largely due to low bandwidth, WebEx struggled to initially attract customers. However, with the shift to broadband technology, things began to move in the right direction and the revenue passed the $1 million mark in 2000. In April 1999, WebEx received its first venture funding of $22 million from a group of 14 investors, followed by a second round in April 2000 raising $22 million from a group of 6 investors including Deutsche Telekom and Oracle. In 2007, WebEx had more than two million subscribers and in March 2007, Cisco Systems purchased WebEx for $3.2 billion.[52] Today Cisco WebEx has a large portion of the web-conferencing market share and more than 3.5 million monthly active users.

2003 Skype

In August 2003, an Estonian software team released a beta version of a video chat software that would become famous under the name of Skype. The software was created by Ahti Heinla, Priit Kasesalu, and Jaan Tallinn, and Skype was founded in 2003 by Niklas Zennström from Sweden and Janus Friis from Denmark. The service was warmly welcomed by a large community of users worldwide. It allowed audio and video conferencing using the desktop and the Internet connection at comparable high quality.

"Let's Skype" became a common term for getting into contact. Skype-to-Skype calls to other users are free of charge, while calls to landline telephones and mobile phones are charged via a user account system called Skype Credit. Two years later, eBay agreed to acquire Skype for $2.5 billion. In 2011, Skype was acquired by Microsoft for $8.5 billion and is a division of Microsoft today. Skype's user base has grown to 300 million and it has reported $2 billion in sales in 2013.

2004 Citrix GoToMeeting

The Californian based IT company, Citrix, began to develop GotToMeeting as a new conferencing application on the Web in 2004. The application was based on already existing Citrix products like GoToMyPC and GotToAssist and expanded to GotToWebinar to support Web based seminars in 2006.

---

[51] ActiveTouch Launches WebEx Meeting Center: https://www.thefreelibrary.com/ActiveTouch+Launches+WebEx+Meeting+Center%3A+The+First+Application...-a055122546, retrieved 2016-10-12.

[52] Matt Marshall, VentureBeat, 2007: http://venturebeat.com/2007/03/15/cisco-acquires-webex-for-29-billion/, retrieved 2016-10-12.

2011 Apple Facetime

Apple Facetime was introduced in February 2011 as a proprietary video calling product for Mac, iPhone and iPad. Facetime specializes in one-to-one video chatting compared to other video-chats like Skype, which allow multi-person video chatting.

New Business Models for Media

Selling books was one of the first successful business models in the e-commerce sector. The largest cost element in this business was logistics. So why not deliver books and other media as digital material.

2007 Amazon Kindle

Selling and distributing books, CDs, and DVDs became a huge business for Amazon since its start in 1995. Amazon had built not only a rich e-commerce portal but also created a huge organization to physically distribute the goods to their customers. It was evident that a transition from physical items to digital content would completely change the game. In 2004, Amazon's founder and CEO Jeff Bezos decided to move in exactly that direction by building the world's best e-book reader before Amazon's competitors Apple or Google could: "*I want you to proceed as if your goal is to put everyone selling physical books out of a job.*"[53] The new product was named Kindle and introduced in 2007.

Compared to other e-book readers at that time, the Kindle focused on screen technology optimized for reading text in different lighting situations and was combined with the Kindle store to purchase e-books for the Kindle and later also other reading devices. The store started with 88,000 titles in 2007 and expanded to 4.6 million titles by 2016. Since 2012, the e-book reader sales have been shrinking from a level of 40 million units worldwide in 2012 to 20 million units in 2015. The Amazon Kindle is still the most popular e-book reader claiming a market share of 75% in 2015.

2007 Amazon Video on Demand

Distributing video on DVDs was one of the major businesses for e-commerce-stores like Amazon.com and others. With the availability of broadband connections

[53]Casey Newton: "The everything book". The Verge, 2014, http://www.theverge.com/2014/12/17/7396525/amazon-kindle-design-lab-audible-hachette, retrieved 2016-12-10.

for private homes and mobile devices, the technology of streaming music and video opened a new field of possibilities. Around 2005, cable modems and Internet connections via cable providers could offer a bandwidth up to 38Mbit/s, ADSL offered by telecommunication providers reached a bandwidth of 10Mbit/s.

Amazon started its first video-download service as Amazon Unbox in the US in September 2007. In September 2008, the service was renamed to Amazon Video-on Demand offering the streaming of 40,000 licensed movies and TV shows instantly on PCs, Sony TV sets or Xbox.[54] In the following years, the service was extended and rebranded to Amazon Instant Video in 2011 and to Amazon Video in 2015. In the same year, the company also started to produce their own shows for its video streaming service, thus opening a new business model for entertainment production and distribution. Today, Amazon Video is available in 200 countries, movies or shows can be rented for 24 hours or bought.

2008 Spotify

Buying and downloading music is nice, but in many cases users just want to hear music and see no need to store files on their devices. With the growing bandwidth of Internet access the option to stream music and video became more important. Spotify was founded Daniel Ek and Martin Lorentzon.

In July 2008, the company announced that it had closed several licensing deals with major music labels: "*We've been looking forward to posting this message for quite some time now. This morning we announced that we have signed ground-breaking licensing deals with a list of companies that include Universal Music Group, Sony BMG, EMI Music, Warner Music Group, Merlin, The Orchard and Bonnier Amigo.*"[55] The Spotify application was launched in October 2008 offering a limited free service and extended services with a monthly subscription fee between $9.99 and $14.99. By end of 2011, Spotify had 2 million paying users, which grew to 6 million in 2013 and 10 million in 2014. Despite the strong growth rate, the company has been unable to turn a profit since its launch and is financed by investments alone. In 2010, Spotify received a small seed funding from Founders Fund, followed by a $100 million in funding based on a $1 billion evaluation in 2011. In 2012, the company got $100 million from Goldman-Sachs, and a third round of funding was closed in June 2015 at a value of $8.53 billion. In 2016, a total of $1.5 billion in additional money was raised through bonds and debt financing and

[54] Amazon News Release, 2008: http://phx.corporate-ir.net/phoenix.zhtml?c=176060&p=NewsArticle&id=1193455, retrieved 2016-12-10.

[55] Spotify News, 2008: https://news.spotify.com/us/2008/10/07/weve-only-just-begun/, retrieved 2016-12-10.

the company reported it had 40 million paying subscribers and more than 100 million active users.[56]

2008 Netflix Video on Demand

Netflix has its roots in a DVD rent and sales business founded by Reed Hastings and Marc Randolph in 1997. Both founders were experienced entrepreneurs and had successfully launched, developed and sold startups. The new DVD business was extremely successful and led to an initial public offering in 2002 raising $82.5 million. In 2003, Netflix had more than 1.1 million subscribers each paying a monthly fee of $19.95 for the unlimited rental of DVDs per mail.

In 2007 and impressed by the video-on demand trend, Netflix decided to enter that market and expanded its business to streaming media services for the US in 2008 and Canada in 2010. The rest of the traditional media market took notice but was not very alarmed. Jeffrey L Bewkes, CEO Time Warner said: "*It's a little bit like, is the Albanian army going to take over the world? I don't think so.*"[57] Three years later, Netflix entered the production business with its first series "House of Cards." Today, Netflix services are available in more than 190 countries and it releases an estimated 126 original series, more than any other media network. Netflix reported serving over 86 million subscribers worldwide in October 2016 generating a revenue of $6.78 billion (2015) and a net income of $122.6 million (2015).

New Business Models for Sharing Resources

At the end of the first decade e-commerce, online travel booking, collaboration and communication had expanded to large, new businesses based on new and disruptive business models. But the race was still on to find new areas of disruption using the capabilities and wide acceptance of the Web. The rapid growth of communities and social media networks proved that platforms and communities have the tremendous power to implement new forms of collaboration and complete new business models based on sharing resources and direct connection between consumers and private or corporate providers of products and services.

[56]Jordan Kahn, 9to5Mac, 2016: https://9to5mac.com/2016/09/14/spotify-40-million-subscribers/, retrieved 2016-10-12.

[57]New York Times, "Time Warner Views Netflix as a Fading Star", 2010: http://www.nytimes.com/2010/12/13/business/media/13bewkes.html, retrieved 2017-02-10.

2008 Groupon

In 2006, Andrew Mason, a web-designer working for a Chicago company owned by Eric Lekovsky was frustrated when trying to cancel a cell phone contract. He started to think about a platform that would allow the leveraging of many people's bargain power and achieve a better pricing of products using social media. He was able to convince his employer, Lekovsky, who provided $1,000,000 in startup funding, a company named ThePoint.com and introduced an online collective action website. The name was later changed into Groupon and the new service launched in November 2008.

Groupon is a platform offering a special deal for a short time. If a certain number of people sign up for the offer, the deal becomes available to all. Within a short time, the revenue grew rapidly and the company was valued at over $1 billion in 2010 entering the group of unicorns (companies with more than 1 billion of market value) in a record time. Groupon raised $125 million from Digital Sky Technologies in April 2010 and reported a monthly revenue of $89 million in January 2011. The initial public offering was filed in 2011 and Groupon Inc raised $700 million valued at almost $13 billion after having rejected an offering by Google at a value of $5.3 billion in 2010. In 2012 Groupon had reached a peak, reporting that they had missed their third quarter plans and causing stocks to fall from $20 in 2011 to $2.93. Today (2016) Groupon is traded at a price below $5, having reported revenues of $3.11 billion (2015) and a net Income of $20.66 million (2015).

2008 Air BnB

Preparing for the IDAS conference in San Francisco in 2007, Brian Chesky and Joe Gebbia came up with the idea to rent out extra space in their loft apartment to conference attendees. To publish their offer, they created a website which became the starting point for AirBnB—Air Bed&Breakfeast. In February 2008, Nathan Blecharczyk joined the team as the technical architect and third co-founder.

The AirBedandBreakfeast website was officially launched in August 2008: "*The site is spare but it does the job. It was pulled together for less than $20,000 in seed capital from friends and family of the founders—San Francisco designers Joe Gebbia and Brian Chesky, and software engineer Nathan Blecharczyk.*"[58]

The name was changed from Airbedandbreakfast.com to Airbnb.com in 2009 and, 2 years later, Airbnb announced its one millionth booking. The total of bookings grew rapidly to five million in January 2012 and ten million in June 2012, 75% coming from outside the US. In July 2012, AirBnB raised $119.8

[58] Erik Schonfeld, TechCrunch 2008: https://techcrunch.com/2008/08/11/airbed-and-breakfast-takes-pad-crashing-to-a-whole-new-level/, retrieved 2016-12-10.

million from a group of venture companies and private investors including Ashton Kutcher. The revenue grew to $250 million in 2013 and to $900 million in 2015.[59] In 2014, a second round of funding closed raising $450 million at a value of $10 billion. In March 2015, AirBnB was valued at $20 billion, growing to $30 billion by September 2016 and raising $850 million from Google Capital.[60]

2008 Waze

Providing location-based services for commuters and car drivers was, for a long time, reduced to mapping and navigation systems integrated in cars. Using the power of the Web and cloud services and the option to integrate status information from different sources including individual drivers was the next logical step.

In 2008, the Israel based startup, Waze, accepted the challenge and developed a cloud service that provides traffic information including travel time and route status automatically submitted by Waze users to other users in real-time. In 2011, the company opened a subsidiary in Palo Alto. Between 2008 and 2011, Waze raised $67 million in three rounds of funding [61] reaching 50 million users in 2013.[62] In June 2013, Waze was acquired by Google for $1.1 billion to integrate social data into its mapping business.

2009 Uber

Uber was founded in 2009 by Garrett Camp and Travis Kalanick, both experienced founders and entrepreneurs. The Uber concept is based on the idea of connecting drivers and consumers using mobile devices and apps thus creating a new global market for personal transportation and bypassing local taxi enterprises and networks. Uber driver use their private cars. Trip requests are submitted by the consumer using the Uber app, which sends the request to the nearest Uber driver. The fare is calculated automatically and the cashless payment is transferred to the driver.

The company received $200,000 of seed financing in 2009 and raised $1.2 million in 2010.[63] Additionally, investments were reported but not disclosed by

---

[59]Kia Kokalicheva, Fortune, 2015: http://fortune.com/2015/06/17/airbnb-valuation-revenue/, retrieved 2016-10-12.

[60]Maureen Farrell, Wall Street Journal, 2016: http://www.wsj.com/articles/airbnb-raises-850-million-at-30-billion-valuation-1474569670, retrieved 2016-10-12.

[61]Crunchbase/Waze: https://www.crunchbase.com/organization/waze#/entity, retrieved 2016-10-12.

[62]Yahoo! Tech News, 2013: https://news.yahoo.com/waze-sale-signals-growth-israeli-high-tech-174533585.html, retrieved 2016-10-12.

[63]Michael Arrington, TechCrunch, 2010: https://techcrunch.com/2010/10/15/ubercab-closes-uber-angel-round/, retrieved 2016-10-12.

Google Ventures in 2013, Chinese search engine Baidu in December 2014, and Toyota in 2016. Since its start, Uber has raised $8.71 billion in total equity funding through 12 rounds from 75 investors.[64] Uber's revenue was $495 million in 2014, $1.5 billion in 2015 and around $2 billion in the first half of 2016. Still, Uber is burning money at a rate of more than $500 million per quarter. Nevertheless, the company's value is estimated at about $62 billion in September 2016.[65]

2009 Kickstarter

Getting people together is one of the major outcomes of social networks and platforms. Bringing investors together with people with creative projects is the basic idea behind crowdfunding. This is the mission statement of Kickstarter. Kickstarter was launched in 2009 by Perry Chen, Yancey Strickler, and Charles Adler as public-benefit corporation. By January 2017, Kickstarter had successfully helped to fund more than 110,000 projects with a total funding volume of $2.8 billion.[66]

2009 Bitcoin, Ethereum and Others

The idea of creating a peer-to-peer transaction network using existing Internet infrastructure has its origins in the open society movement that began to discuss privacy of communication and anonymity on the net as principles and basic rights for users. One of the methods of peer-to-peer transactions are cryptocurrencies. Cryptocurrencies are completely different from normal currencies like USD, Euro or Yen, which are controlled by governments and national or global financial organizations, although there is still an exchange rate between a cryptocurrency and national currencies.

In 2009, the first anonymous cryptocurrency was launched under the name of Bitcoin. The technology behind Bitcoin is blockchain, a peer-to-peer communication protocol based on the concept of a distributed ledger, thus omitting the need to have a centralized institution controlling the transaction flow. Today, Bitcoins traded in the Web have reached a volume of up to $10 billion of the market cap.

Other cryptocurrency projects followed in the next few years. Today, there are more than 600 cryptocurrencies projects released, among them Ethereum, founded in 2014 at a foundation in Switzerland, and provided an open environment to create blockchain applications including smart contracts. We will focus on the concept of

---

[64]TechCrunch, Uber: https://www.crunchbase.com/organization/uber#/entity, retrieved 2016-10-20.

[65]Izabella Kaminska, Financial Times, 2016: https://ftalphaville.ft.com/2016/09/13/2173631/mythbusting-ubers-valuation/

[66]Kickstarter Statistics: https://www.kickstarter.com/help/stats retrieved 2017-01-15.

distributed ledgers, blockchains and cryptocurrencies in the chapter "New Paradigms and Big Disruptive Things".

Social Networking 2

Social networks were one of the major, new applications of the Web in the years between 2000 and 2005. The hype about social communities was constant and got an additional boost with the arrival of smart phones in 2007. Private connection and communication was shifted from the homebased PC to mobile devices and the form of communication changed from text to short messages, symbols (e.g. emoticons), images and video. A new generation of chat, communication and social media tools and services appeared and changed the way people interact. Communication became location and time independent.

2006 Twitter

Since the early 1990s and the availability of GSM mobile phones, text messages using the SMS standard became the most popular standard. In 2006, Jack Dorsey, an undergraduate at New York University, thought of a way to improve that concept by moving towards a messaging or microblogging system that could be used to communicate within a small group. Dorsey met with Odeon, a podcasting publishing company, and presented his idea of something he called "Twttr," a messaging service where the users could send 170-character text messages to their friends (followers.)

A prototype was developed for Odeon by Dorsey and contractor Florian Weber; the full version was launched in July 2006 with Odeon owning the property rights. In October of the same year, a group of Odeon members together with Jack Dorsey acquired Odeon along with its assets including twitter.com. Twitter was sold in April 2007 eventually becoming its own company. The company grew extremely quickly from 400,000 tweeds per quarter in 2007 to 1000 million in 2008. In September 2007, Twitter announced that it had filed papers for an initial public offering. On the first day of trading, in November 2007, Twitter ended with a market capitalization of $24.46 billion. Today, Twitter is one of the most popular messaging services rankling in the top ten of social media platforms.

2009 WhatsApp

In 2007, Brian Acton and Jan Koum, two former employees at Yahoo!, travelled to South America as a break from work. After having bought an iPhone, they soon recognized that Apple's App Store would lead to a completely new platform for mobile applications. They started to discuss creating something that would allow "... *having statuses next to individual names of the people*".

Taking their savings from Yahoo!, they went back to California, hired developers and incorporated WhatsApp in February 2009. The first version of the new chat tool was not very stable and unsuccessful. After Apple had introduced push notifications in 2009, WhatsApp was reworked and released with a messaging component branded, WhatsApp 2.0. The number of users suddenly increased to 250,000. In April 2011, Sequoia Capital invested approximately $8 million in WhatsApp for a share of 15% followed by a second round of $50 million in 2013 at a value of $1.5 billion. The WhatsApp user numbers reached 400 million in December 2013 and 600 million in 2014. In February 2014, Facebook announced the acquisition of WhatsApp for $19 billion, Sequoia received a 50× return on its investment.

2010 Pinterest

Pinterest was developed as a photo sharing platform in 2009 by Ben Silbermann, Even Sharp and Paul Sciarra, and the first version was released in March 2010 reaching 10,000 users by end of the same year. An iPhone app was launched in early March 2011, an iPad app in August of the same year. In January 2012, the site had 11.7 million unique users, making it the fastest site in history to break through the 10 million user mark and the third largest social network in the United States in March 2012, behind Facebook and Twitter.

Rakuten invested $100 million in Pinterest in 2012 based on a value of $1.5 billion. In October 2013, Pinterest received $225 million equity funding that valued the website at $3.8 billion. Today, Pinterest has 150 million monthly users, 85% are female. Pinterest reportedly generated about $100 million in revenue last year and is reportedly expected to generate about $300 million next year.[67]

2010 Instagram

In October 2010, Kevin Systrom and Mike Krieger, both graduates from Stanford University, launched Instagram as a free mobile app for sharing photos. Within 1 year, the site had 100 million users growing to 300 million by end of 2014.

Instagram was supported by a seed financing of $500,000 in March 2010 by Baseline Ventures and Andreessen Horowitz. In February 2011, they received a round A funding of $7 million based on a value of around $25 million. In April 2012, Facebook acquired Instagram for approximately $1 billion in cash and stock options, keeping it as an independently managed company. In December 2014,

---

[67]Kathleen Chaykowski, Forbes, 2016: http://www.forbes.com/sites/kathleenchaykowski/2016/10/13/pinterest-reaches-150-million-monthly-users/#1dbd05151018, retrieved 2017-01-12.

Instagram co-founder Kevin Systrom reported that the site had passed the 300 million monthly users mark. [68] In June 2016, Instagram reached 500 million users.

2011 Snapchat

Two other Stanford students, Evan Spiegel and Bobby Murphy, created Snapchat, an image messaging and multimedia mobile app in 2011. The app was launched in December 2012 focusing on more than photo sharing: "Snapchat isn't about capturing the traditional Kodak moment. It's about communicating with the full range of human emotion—not just what appears to be pretty or perfect."[69] Snapchat is a mixture of private messaging and public content, built around multimedia messages referred to as "snaps" and includes stories and conversations.

In the same year, the company received a first seed financing of $485,000 from Lightspeed Ventures. In 2013, the company raised a total 143.5 million of funding in three rounds. The daily snaps sent grew to 400 million by November 2013 and, in 2014, Snapchat raised $485 million from 23 investors for its Series D round at a value of at least $10 billion. The number of daily video views grew to 10 billion in 2016. In May 2016, the company was valued at $17 billion and raised $1.8 billion equity funding.

2011 Google+

Google+ was launched in June 2011 as a new social network offered within the Google platform. The new service includes photo sharing, chat, messaging, status updates and the creation of different types of relationships to other users in so-called "circles." The number of Google+ users is estimated at approximately 100 million, which is only a very small part of the approximately 2.2 billion Google users.[70]

Sharing Knowledge and Information

Since the successful launch of Wikipedia in 2000 sharing and distributing of knowledge using the Web had gained a lot of momentum. Many open source

---

[68] Jonathan Blake, BBC, 2014: http://www.bbc.co.uk/newsbeat/article/30410973/instagram-now-bigger-than-twitter, retrieved 2016-12-10.

[69] Snapchat news blog, May 212: https://www.snap.com/de-DE/news/page/7/, retrieved 2016-12-10.

[70] Steve Denning, Forbes, 2015: http://www.forbes.com/sites/stevedenning/2015/04/23/has-google-really-died/#7dfb6a4616e9, retrieved 2016-12-10.

initiatives, specialized news and information platforms, and knowledge distribution business models have appeared on the radar screen, some of them extremely successful and highly welcomed by their target communities. Though there is a massive diversity in the approaches used, these platforms share the idea of providing data, information and knowledge in an easy-to-use way. To demonstrate the different approaches and to underline the rich diversity, we have selected a handful of typical projects and businesses. All of them focus on data and knowledge and have found innovative ways either to create a new business or to contribute to changing the world.

2004 Google Books

In 2004, Google launched its Google Print project which was renamed to Google Book Search and would later become Google Books. The basic idea was to start an initiative to digitalize books that were no longer published or available via a high-speed scanning process. After facing two lawsuits, which charged Google of not respecting copyright law, the company could convince a few libraries in the US and Europe to join the initiative. Google estimates the number distinct titles to be about 130 million and stated that they would scan them all. As of October 2015, the number of scanned book titles was over 25 million and scanning speed was raised to 6000 pages in an hour per operator.[71] As an additional service Google introduced the Ngram viewer, a statistical tool, that allows for the analysis of word usage across a book collection.

2005 TechCrunch

TechCrunch was founded as a platform for technology industry news by Michael Arrington and Keith Teare in 2005. The platform became one of the leading information hot spots about technology news based on a subscriber business model and reached revenue of $2.4 million in 2007.[72] Additionally, TechCrunch operates CrunchBase, a startup and technology database with 500,000 data points profiling companies, people, funds, funding and events. In September 2010, AOL acquired TechCrunch for a reported $25 million.[73] The number of subscribers grew to more than 4.56 million in 2011. Today, TechCrunch is one of the most popular data sources for startups, investors and new businesses.

---

[71]Stephen Heyman, New York Times, 2015: https://www.nytimes.com/2015/10/29/arts/international/google-books-a-complex-and-controversial-experiment.html?_r=1, retrieved 2016-10-12.

[72]Michael Arrington, Wired, 2007: https://www.wired.com/2007/06/ff-arrington/, retrieved 2016-10-12.

[73]Mark Sweney, The Guardian, 2010: https://www.theguardian.com/media/2010/sep/29/aol-buys-techcrunch, retrieved 2016-10-12.

2006 WikiLeaks

WikiLeaks became popular through its founder, Julian Assange, when he published a series of leaked documents about the Iraq and Afghanistan wars disclosed to WikiLeaks by the US Army soldier, Chelsea Manning, in 2010. WikiLeaks was founded in 2006 in Iceland by Assange as platform to publish secret information, news leaks, and classified material from various, mostly anonymous sources. Julian Assange said in an interview in 2015: "*WikiLeaks is a giant library of the world's most persecuted documents. We give asylum to these documents, we analyze them, we promote them and we obtain more.*"[74]

After the publication of the Manning material, the US authorities started an investigation into Assange and WikiLeaks with the option to prosecute them under the Espionage Act. In the same year, he also faced sexual assault allegations from Swedish authorities. Assange applied for political asylum in Ecuador, which was granted to him in the Ecuadorian embassy in London. Since then he has been living there. WikiLeaks is still operated and active as a platform for whistleblowers supported and funded by many news media companies.

2007 Statista

Statista was founded by Mattias Protzmann, Friedrich Schwandt and two others in Hamburg, Germany in 2007. The site has been available since 2008 and provides statistics and reports from business organizations, research institutes and government institutions in easy to use, standardized formats for online viewing and downloading. Within a short time, it became one of the most popular sources of statistical data counting over one million statistic data records on over 80,000 topics. With more than five million visitors per month, the service is one of the most frequented data sources in the world.[75] The service is offered in a freemium pricing model, subscription fees for additional services start at 49 € per month. The company forecasts $20 million in revenue for 2016 and is valued at approximately 70 million euros.[76]

---

[74]Julian Assange Interview, Der Spiegel, 2015: http://www.spiegel.de/international/world/spiegel-interview-with-wikileaks-head-julian-assange-a-1044399.html, retrieved 2016-10-12.

[75]Martin Müller, Spiegel Online, 2015: http://www.spiegel.de/wirtschaft/unternehmen/statista-balken-die-die-welt-beschreiben-a-1055683.html, retrieved 2016-10-12.

[76]EU Startups, 2016: http://www.eu-startups.com/2015/12/big-exit-in-germany-stroer-se-acquires-global-data-and-business-intelligence-platform-statista-for-roughly-e-57-million/, retrieved 2017-01-15.

2008 GitHub

Building open source and organizing open source projects needs a common infrastructure for the developer's community. In 2008, Chris Wanstrath, PJ Hyett and Tom Preston-Werner created GitHub as a cloud-based repository for open source projects. Within a short time, the service became one of the most successful platforms for managing open software projects reaching more than 14 million users and hosting more than 10 million repositories in 2016.

In 2012, GitHub received $100 million in funding from Andreesen Horowitz followed by a second round in 2015 receiving $250 million from Sequoia Capital. While the GitHub service is free for private users, the company generates revenue from corporate accounts. Revenue for 2016 is reported to be around $140 million.[77]

2008 Encyclopedia of Life

The Encyclopedia of Life (eol.com) was founded 2007 with the bold idea of providing "*a webpage for every species.*" The project was founded by private investors and supported by several research institutes around the world including the Field Museum, Harvard University, the Marine Biological Laboratory, the Missouri Botanical Garden and the Smithsonian Institution. The site went live in 2008 and started to expand into "*…a global community of collaborators and contributors serving the general public, enthusiastic amateurs, educators, students and professional scientists from around the world.*" [78] By 2011, the project had collected 70,000 entries and it is estimated that completion will take 10 years.

2009 Quora

In 2009, the former Facebook employees Adam D'Angelo and Charlie Cheever developed the idea of creating a knowledge platform based on question-and-answer blog entries. They founded Quora in 2009 and released the site in 2010. Quora users collaborate by editing questions and suggesting edits to other users' answers. "*We thought that Q & A is one of those areas on the Internet where there are a lot of sites, but no one had come along and built something that was really good yet.*"[79]

Between 2010 and 2014, Quora raised $141 million of funding in 3 rounds from 14 investors among them Peter Thiel, Benchmark and Y Combinators. The site

[77]Medium, 2016: https://medium.com/@moritzplassnig/github-is-doing-much-better-than-bloomberg-thinks-here-is-why-a4580b249044#.ojp72dha4

[78]Encyclopedia of Life: http://eol.org/info/the_history_of_eol, retrieved 2017-01-16.

[79]Adam d'Angelo on Quora, BosTinno, 2011: http://bostinno.streetwise.co/2011/02/01/the-mystery-behind-quora-digging-in-with-co-founder-adam-d%E2%80%99angelo/, retrieved 2016-12-10.

became popular growing to 100 million users in 2016.[80] Subscription is free, Quora makes no revenue and finances its operational expenses from investor's money. The future business model may focus on advertising, monetizing the knowledge base or job-and-career matching features.[81]

2009 DocumentCloud

DocumentCloud was launched in 2009 by the journalists Scott Klein, Eric Umansky and Aron Pilhofer to provide an open source platform for sharing primary source documents. It is used primarily by journalists to upload documents they reference in their reporting, making them available as the original source to others. By 2015, DocumentCloud had more than two million documents uploaded.[82] DocumentCloud is supported by many media outlets and has received awards for their contributions to journalism.[83]

2010 The World Bank Open Data

The World Bank is an international financial institution providing loans to developing countries. Its president, Robert Zoellick, said in an interview in 2009: "*Knowledge is power, making our knowledge widely and readily available will empower others to come up with solutions to the world's toughest problems. Our new Open Access policy is the natural evolution for a World Bank that is opening up more and more.*"[84]

Following that strategy, the World Bank introduced its World Bank Open Data project in 2010 and started to collect and publish statistical data under a creative common license. The service is free and offers open access to more than 1000 data items about economic and social topics containing historical and current numbers for countries and regions.

[80]Key Eyung, Venuture Beat, 2016: http://venturebeat.com/2016/03/17/quora-now-has-100-million-monthly-visitors-up-from-80-million-in-january/, retrieved 2017-01-10.

[81]Quora Business Model Options, 2011: https://www.quora.com/What-could-Quoras-long-term-business-plan-be, retrieved 2016-10-12.

[82]DocumentCloud Blog, 2015: https://blog.documentcloud.org/blog/2015/08/a-summer-days-worth-of-updates/, retrieved 2016-10-12.

[83]Zachary Seward, Niemanlab, 2009: http://www.niemanlab.org/2009/09/documentcloud-adds-impressive-list-of-investigative-journalism-outfits/, retrieved 2016-10-12.

[84]The World Bank, SparcOpen 2012: http://sparcopen.org/our-work/innovator/world-bank/, retrieved 2016-10-12.

2012 Coursera

Coursera was formed by Andrew Ng and Daphne Kollerin in 2012 as a platform to provide massive open online courses produced by universities and colleges. In 2013, the platform started to work with Stanford University, Princeton University, the University of Michigan, and the University of Pennsylvania. The courses offered are free, but, as an option, the so-called “Signature Track” can be added by paying a fee and receiving certificates for courses taken. The company is venture backed and raised $146.1 million between 2012 and 2014 in 6 rounds from 12 investors. In 2015, Coursera claimed to have 15 million users, more than both of its major competitors EdEx and Udacity.[85]

2012 Wikidata

In 2012, the Wikimedia Foundation, also hosting Wikipedia, decided to join the open data community and started WikiData as a new project. WikiData’s intention is to provide a common source of structured data under the public domain license (PDL). The content is collaboratively edited and consists of documents organized in items pointing to different types of data sources including Wikipedia articles and describes the type and content of the item. The creation of the project was funded with 1.3 million euros by the Allen Institute for Artificial Intelligence, the Gordon and Betty Moore Foundation and Google.[86]

2014 Missing Maps

In November 2014, a group of humanitarian rescue organizations including the American Red Cross, British Red Cross, Humanitarian OpenStreetMap Team, and Doctors Without Borders recognized that the absence of detailed maps for crisis zones is a major handicap for efficient support. The Missing Maps project is a volunteer crowd sourcing initiative bringing people together to add details to OpenStreet Maps in a so-called “Mapathon”.

---

[85]EdSurge, 2015: https://www.edsurge.com/news/2015-09-08-udacity-coursera-and-edx-now-claim-over-24-million-students, retrieved 2016-10-12.

[86]Sarah Perez, TechCrunch, 2012: https://techcrunch.com/2012/03/30/wikipedias-next-big-thing-wikidata-a-machine-readable-user-editable-database-funded-by-google-paul-allen-and-others/, retrieved 2016-10-12.

References

DeMarco, T. (1995). *Why does software cost so much?* New York: Dorste House Publishing.
MIT Technology Review. (2011, October 11). Retrieved from Who coined 'cloud computing?': http://www.technologyreview.com/news/425970/who-coined-cloud-computing/
National Institute of Standards and Technology. (2011). *The NIST defintion of cloud computing*. Gaithersburg, MD: NIST.
Oppitz, M., & Tomsu, P. (2014). *Cloud service business model transition*. Retrieved from http://www.cisco.com/c/dam/en/us/td/docs/services/Cloud-Services-Business-Model-Transition.pdf
Weinberg, G. M. (1992). *Software quality management*. New York: Dorset House Publishing.

Building Cloud Businesses and Ecosystems

Free enterprise cannot be justified as being good for business. It can be justified only as being good for society.
Peter Drucker (1954) The Practice of Management

Successful business models and a well-developed market are some of the major elements of each service ecosystem. Today, the cloud market, consisting of cloud services and cloud infrastructure, has completely changed the information technology market. From a global market perspective, we speak of a pre-Internet that was dominated by one company: IBM. Starting with the availability of the Internet and further with the World Wide Web, completely new types of enterprises appeared beginning in 1992. It was called New Economy and was based on new types of business models using two different approaches: the mass distribution of Internet based services and the sharing of capacity and resources between a large group of customers. Since then these new ecosystems have gone through an epic development of different pricing models, marketing approaches and partner strategies.

Information technology and its share in global business has grown since the start of computing business in the late 1950s. IT has reached a share of around 10% of the average GDP worldwide. With the introduction of the Internet, the Web and cloud computing technologies, this share began to expand at a higher speed. The term coined to describe that effect is "digitalization," which simply means that information technology has a larger and growing part in each segment of economy and the personal life of consumers. The share of the Internet business in the GDP is growing and has reached a level of between 4% and 6% in 2016 (see Fig. 1).[1,2]

Calculating the world GDP at $73 trillion,[3] the Internet economy share is approximately $3 trillion with the share of the IT industry around 7.3 trillion (see Fig. 2).

[1]OECD, 2011: https://stats.oecd.org/Index.aspx?DataSetCode=REV retrieved 2016-10-12.
[2]Boston Consulting Group, 2012: https://www.bcg.com/documents/file100409.pdf
[3]GDP 2015, Worldbank: http://data.worldbank.org/indicator/NY.GDP.MKTP.CD retrieved 2017-01-15.

M. Oppitz, P. Tomsu, *Inventing the Cloud Century*,
DOI 10.1007/978-3-319-61161-7_12

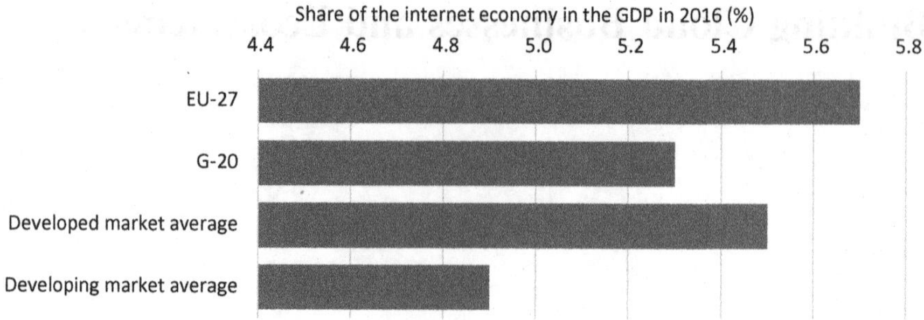

Fig. 1 Share of the internet economy in the GDP 2016 (%)

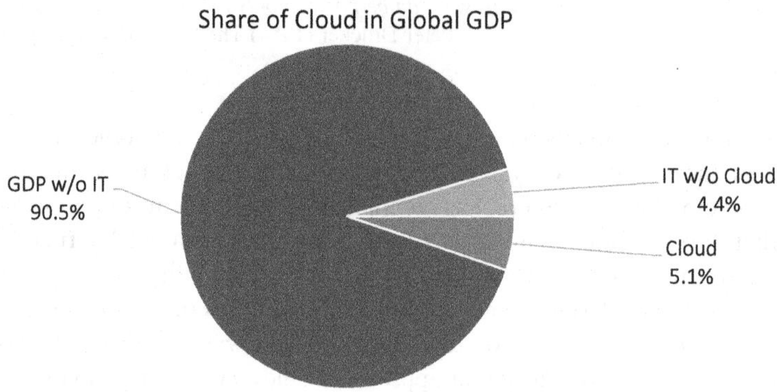

Fig. 2 Internet economy share in global GDP

The cloud market itself consists of several industry segments covering the cloud infrastructure and cloud services and the secondary markets of a large and growing number of "on-top" business models like advertising, e-commerce, travel, music and video (see Fig. 3).

The different segments are covered by numerous providers, some of them focusing on very specific market segments and other delivering a wide range of Internet, Web and cloud based services.

Business Models

Basic Assumptions

Successful business models are one of the five major components of service ecosystems (network, end-devices, provider technology standards, business models) and, thus, also for cloud services. Business models have the major target of creating a win-win situation between providers of a service and the consumers.

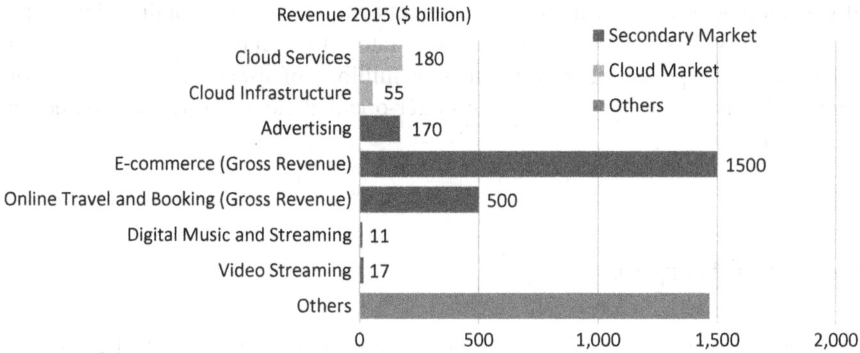

Fig. 3 Internet, web and cloud economy per segment

Provider and Buyer Perspective

The development of such business models always focus on the two parties involved in the new ecosystem. The provider of a cloud service must create a business model, which enables him to develop, produce, operate and offer a service thus creating a competitive advantage for his organization, revenue and profit. The buyer or user of cloud services must pass through a transition changing its value chain from internal operation to outsourcing the service to cloud providers. This transition must have the end goal of creating business value and growing competitive advantages for the buyer's organization.

Consumer and Business

It is evident that provider-centric service offers are already more accepted by consumers than by business customers. Today, consumers have widely adopted ideas like using infrastructure as a service, app stores, consuming music or video via streaming platforms instead of buying CDs or DVDs or using car sharing services. Looking at the list of potential obstacles, it seems that consumer behavior is much more outcome-driven and disregards security threats or independence from their service providers.

Data as New Currency

With the rise of cloud-based services and the creation of large platforms like Facebook, Google or Amazon, it became evident that one of the values of the new businesses is data. Consumers using search engines, e-commerce platforms, and social networks create a huge volume of data, which describes their consumer behavior, preferences, social relations and much more. This data is used by the platforms to create target-oriented and effective advertising messages. Selling this as a service to advertising customers created the huge revenue streams for platforms like Facebook, Google, Amazon and others.

Platforms Are Stronger

The creation of services like social networks soon led to the finding that platforms create additional value for the users if the community is growing rapidly. This effect could be observed with the first Internet communities and continued with the social networks in the Web 2.0 area. Growth rates of social media platforms turned out to follow exponential curves, creating millions of users within a couple of months. Many business models of consumer-oriented cloud services are based on this effect.

Drivers of Acceptance

The basic drivers behind this development are similar and follow the idea of focusing on the outcome and not on the production processes of systems, technology and services:

Growing Complexity
Technical systems and services—like IT—are growing dramatically in complexity. Building and operating complete IT infrastructures and solutions is a highly complex business today. Collecting and maintaining skills, know-how and intellectual property are major components of success.

Distributed Usage
Compared to the IT usage 10 or 20 years ago, the geographical distribution and the number of consumers and end users has grown. This requires sophisticated support organizations and a much higher level of system availability around the clock.

Rapid Change
IT technology is changing rapidly and, at the same time, even accelerating. Following this change and staying on track is one of the major challenges in the industry.

Higher Specialization
To meet those challenges, the industry moves into more specialization, which again leads to a higher distribution of tasks and services to a growing number of service providers and vendors.

Cost Control
Enterprises facing higher cost pressure tend to move to usage or consumption based models (OPEX) instead of long term investments in technology (CAPEX).

Obstacles

It is also evident that several obstacles may hinder the rapid development to a complete provider-centric ownership model.

Access Speed and Availability
Access speed and availability are still limiting factors in many parts of the world, especially in emerging countries, plus there are different requirements depending on the cloud service type, like communication services that need guaranteed high-speed Internet access, low latency and high uptime. Today, wireless is one of the most feasible methods for connectivity, especially if it offers certain bandwidth and availability. The latest deployments of Wi-Fi and/or LTE increasingly help to provide these requirements for cloud services while future enhancements in mobility will play a significant role.

Security
Security is a major threat when outsourcing any kind of IT related processes. Security is also one of the most rapidly changing segments in IT. It is no question that global communication based on internet technology has opened a wide range of security related issues, many of them appearing surprisingly hard to predict and not easy to handle despite precautions.

Rules and Regulations
Related to the security field is the growing number of rules and regulations covering data security and privacy. Many of those regulations are regional (e.g.: European or US) and tend to protect a certain geographical ecosystem. This leads to a global inconsistency and hinders the development of globally applicable services.

Losing Control
Controlling the complete business architecture internally is a major driver against any outsourcing model. In those cases, internal organizations claim to be more efficient and effective than external providers.

Dependence
Being dependent from others is a basic fear of organizations. They tend to rely on internal resources and tools instead of trusting an external provider.

Enablers

Network Access
Many companies are changing their overall IT strategies to embrace cloud computing and to open business opportunities. NIST (National Institute of Standards and Technology) defines cloud computing as a model for enabling convenient, on-demand network access to a shared pool of configurable computing resources like networks, servers, storage (commonly called infrastructure) and applications and services that can be rapidly provisioned and released with minimal management effort or service provider interaction.

As such, cloud computing represents a convergence of two major trends—IT efficiency and business agility—in information technology.

IT efficiency refers to using computing resources more efficiently through highly scalable hardware and software resources.

Business agility is the ability of a business to use computational tools rapidly, to adapt quickly and efficiency in response to changes in the business environment.

The promise of cloud computing is to remove traditional boundaries between businesses, make the whole organization more agile and responsive, help enterprises to scale their services, enhance industrial competitiveness, reduce the operational costs and the total cost of computing, and decrease energy consumption. Thus, cloud computing can provide new opportunities for innovation by allowing companies to focus on business rather than be limited and blocked by changes in technology.

Resource Sharing via Virtualization

Resource sharing and management is a core function required of any man-made system. It affects the three basic criteria for system evaluation: performance, functionality and cost. Inefficient resource management has a direct negative effect on performance and cost. It can also indirectly affect system functionality. Some functions the system provides might become too expensive or ineffective due to poor performance.

The strategies for cloud resource management associated with the three cloud delivery models, IaaS, PaaS and SaaS, differ from one another, but in all cases the cloud service providers are faced with large, fluctuating loads that challenge the claim of cloud elasticity. In some cases, when they can predict a spike, they can provision resources in advance. For example, seasonal Web services may be subject to spikes.

Cost Reduction Through Mass Distribution

The cost of cloud services will decrease significantly because of growing markets and offers. As investments and competition among Internet cloud providers heat up, tech giants are finding new ways to distinguish their products. Amazon, Google and Microsoft recently engaged in a pricing war that drove down the fee for their public cloud services. HP is trying a different track—emphasizing its open-source approach and encouraging businesses, its main customers, to use its product to create their own cloud services instead of relying on others.

While cost reduction of cloud services seems to be an imperative, it might be a good idea to build businesses based on value and not get into price competitions with AWS, MSFT, Google, HP, RackSpace, IBM Softlayer, and others.

Accepted Industry Standards

Standards-based cloud computing ensures that cloud services can readily interoperate, based on open standard interfaces. Standards allow workloads to be readily moved from one cloud provider to a different cloud provider. Services created for one cloud computing environment can be employed in another cloud computing environment, eliminating the need to rewrite or duplicate code. Some of the proposed standards are based on open-source initiatives. This has the advantage of making all the code transparent, available for inspection, and more readily suited for an interoperable environment.

Growing Quality
Experiences by cloud service providers on how to design, build and run their cloud services is growing. Service providers like AWS have been active in the market for many years and have gone through multiple failures while improving constantly.

Primary Cloud Infrastructure and Cloud Service Business

There are two different cloud market segments—cloud services and cloud infrastructure.

Cloud Infrastructure

Major cloud infrastructure components are:

- Computing systems
- Networking
- Storage systems
- Management
- Security

The largest component of cloud infrastructure are computing systems, followed by networking and storage systems and then by management and security products and services. The total revenue from cloud infrastructure reached $55 billion in 2015 and is projected to grow to more than $80 billion in 2018.[4,5] Analyzing the data, we see that the major share (>60%) of cloud infrastructure revenue is dedicated to private cloud infrastructure (corporate clouds or enterprise cloud computing infrastructure) (see Fig. 4).

Cloud Services

Cloud services cover all types of "as-a-service" offers for enterprises as well as for private users. The enterprise cloud services market is growing rapidly, having

[4]Synergy Research Group—SRG: "Cloud Infrastructure Equipment—Totals Worldwide Market Share Report", Data File, 4Q 2013.

[5]Technology Business Research—TBR "Cloud Business Quarterly, Public Cloud Landscape", Fourth Calendar Quarter 2013.

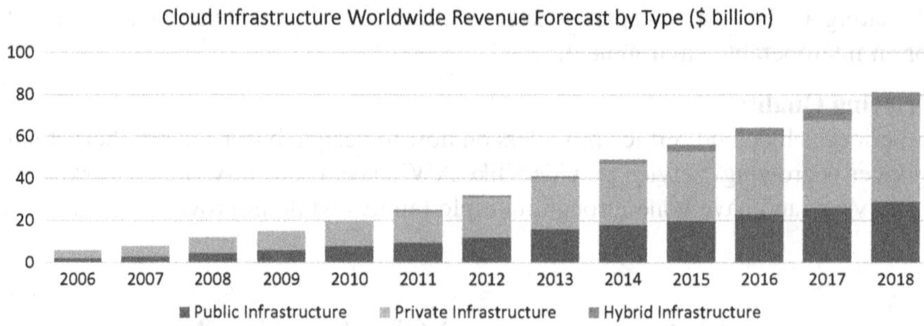

Fig. 4 Cloud infrastructure worldwide revenue forecast

reached $80 billion in 2015 and is forecasted to grow to $140 billion in 2018.[6] In parallel, the public clouds have reached a revenue volume of more than $90 billion in 2015 and will also grow rapidly reaching an estimated revenue of the same size in 2018. Following the projections and growth rates published today, the total cloud services business is estimated to grow to more than $350 billion in 2020 (see Fig. 5).

The public cloud service market will grow to more than $250 billion in 2020 following the estimates and projections of TBR and others.[7] The market segment is also divided into the three service types of cloud services (SaaS, IaaS, PaaS) (see Fig. 6).

IaaS—Infrastructure as a Service: customers pay to gain access to infrastructure (hardware, compute and storage, and/or networking) on a utility basis to run their licensed software. **SaaS—Software as a Service:** providers offer hosted applications to customers on an on-demand or per-use basis; therefore, customers can simply use the application or applications they need when they need them and can avoid the cost of installing and maintaining supporting infrastructures. Typical software includes: CRM, analytics, asset management, knowledge management, document, ITSM, security and content management, and similar software. **PaaS—Platform as a Service:** provide customers the supporting infrastructure to either conduct software application development or run existing software. This includes the operating environment, which manages workflow and collaboration, and the underlying hardware, such as servers and networking.

[6]IDate, DigiWorld Yearbook 2016: www.idate.org/en/News/DigiWorld-Yearbook-2016_949.html retrieved 2017-01-25.

[7]Louis Columbus, Forbes, 2016: http://www.forbes.com/sites/louiscolumbus/2016/03/13/roundup-of-cloud-computing-forecasts-and-market-estimates-2016/#a02ff3b74b07 retrieved 2017-01-20.

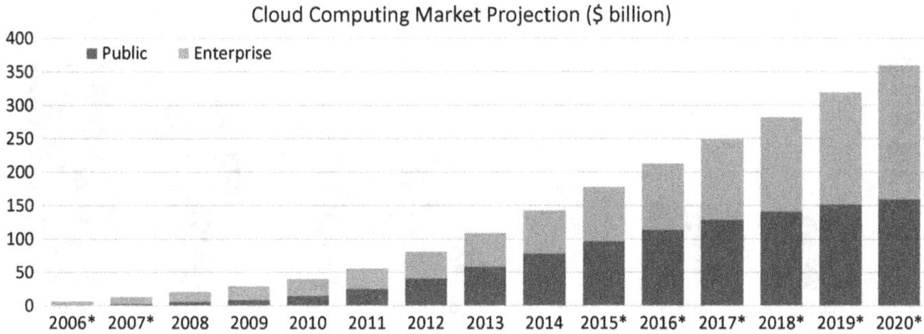

Fig. 5 Cloud computing market projection

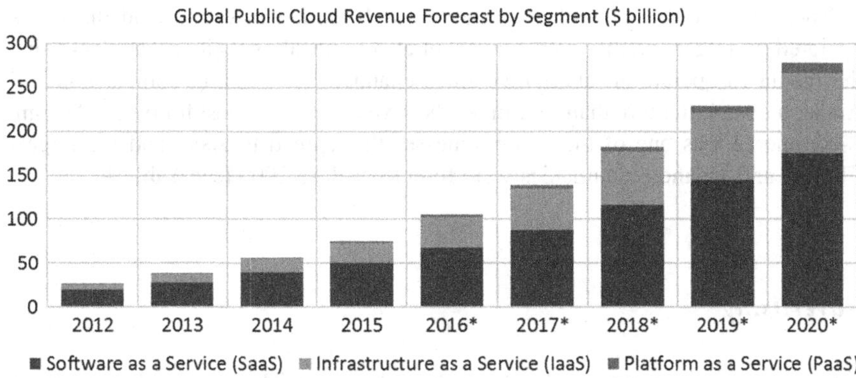

Fig. 6 Public cloud market projection per segment

Secondary Markets

Besides the literal cloud businesses exists a diverse and growing landscape of secondary markets propelling the Internet, Web and cloud business with new business models and offers to corporations and private users. The size of this market is much bigger than the cloud infrastructure and service market and it is growing at a high speed disrupting existing businesses like retail, private transport or travel.

E-Commerce

The e-commerce market is the major contributor to the Internet and Web market. The total volume of e-commerce sales is estimated at nearly $2 trillion in 2016. Projections show that the market is growing and will reach $4 trillion in 2020 (see Fig. 7).

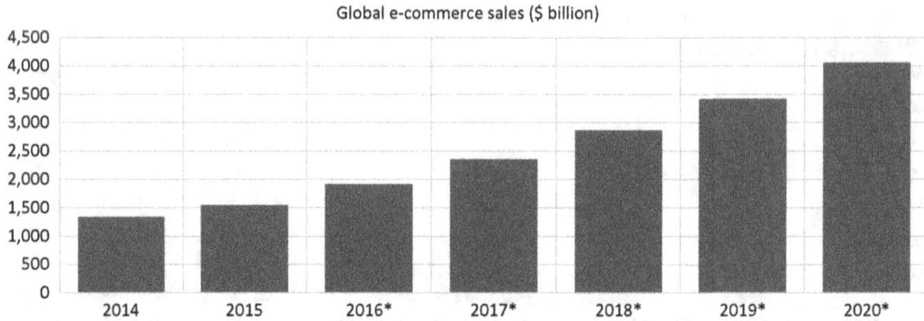

Fig. 7 Global e-commerce sales

There are literally tens of thousands of web market places around the world operated by larger Internet companies, but also by local or regional retailers using the media to operate an alternative sales channel. The leading company is still Amazon.com with more than four times the revenue of its largest follower, JD.com. E-commerce was one of the first businesses that spread to Asia, and the largest Chinese and Japanese companies were founded before 2000 (see Table 1).

Advertising

Advertising is the second market for web applications. The worldwide market for advertising is growing and has a total size of $550 billion, growing to more than $700 billion by 2020. The share of digital advertising within that market is also growing rapidly with a volume of approximately $170 billion in 2015 and reaching a market share of more than 40% or $380 billion in 2020 (see Fig. 8).

Digital advertising is a major revenue source for Internet companies like Google, Yahoo! and Facebook (see Fig. 9).

The share of advertising revenue of these companies is 80% or more (see Table 2).

Travel and Online Booking

The online travel and booking market reached a volume of $500 billion in 2015 and is projected to grow to $800 billion in 2020 (see Fig. 10).

In the US and Europe, the share of online booking for the total travel market is approaching 50% for accommodation bookings and more for transportation bookings. Asia and Latin America are slightly below those numbers but catching up. The major providers of online accommodation bookings are a handful of travel booking

Table 1 Largest e-commerce platforms

Company	Founded	Country	Revenue ($ million) 2015
Amazon.com	1994	USA	107,000
JD.com	1998	China	26,270
Alibaba	1999	China	12,292
Priceline Group	1997	USA	9220
eBay	1995	USA	8590
Rakuten	1997	Japan	6240
Groupon	2008	USA	3119
Zalando	2008	Germany	2330

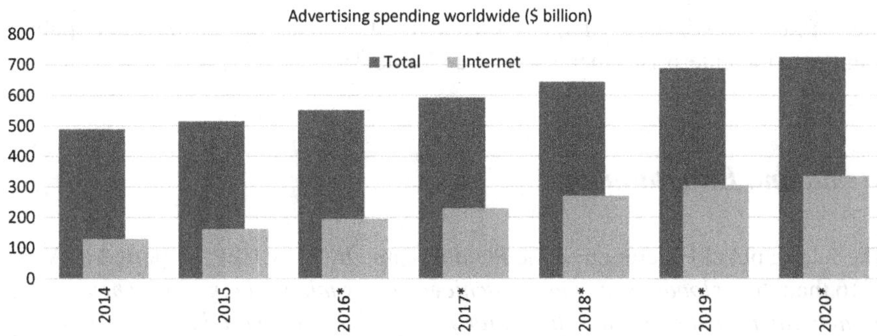

Fig. 8 Worldwide advertising spending

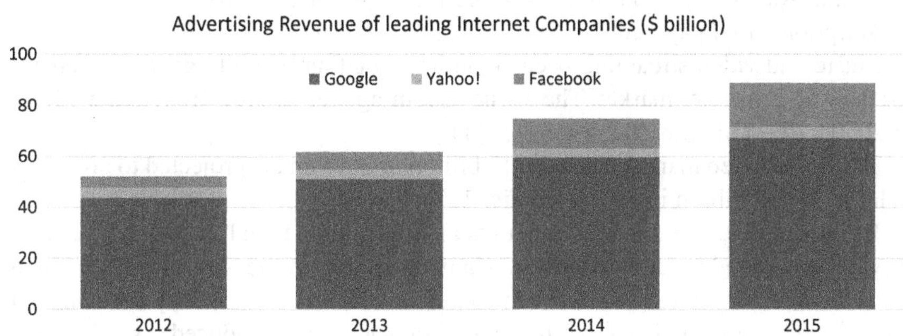

Fig. 9 Advertising revenue of largest Internet companies

Table 2 Advertising revenue of Google, Alphabet and Yahoo ($ million)

	Total revenue 2015	Advertising revenue 2015
Google (Alphabet)	74,990	67,390
Facebook	17,928	17,080
Yahoo	4968	4016

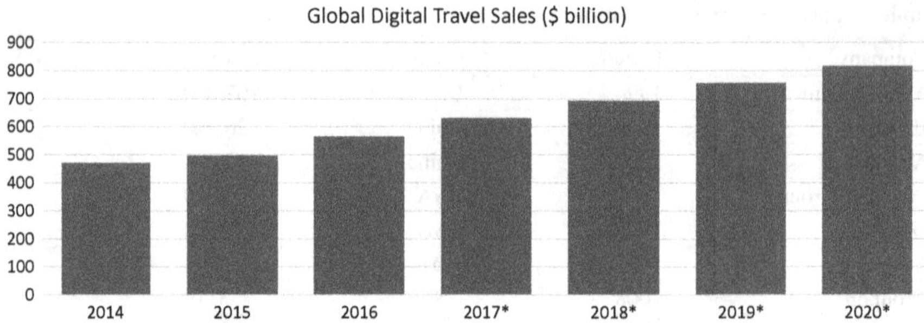

Fig. 10 Global digital travel sales

sites (Priceline, Expedia, Tripadvisor) and, of course, the airlines and public transportation enterprises themselves (see Table 3).

Media and Entertainment

The International Federation of the Phonographic Industry (IFPI) reported in April 2016 that: "*the global music market achieved a key milestone in 2015 when digital became the primary revenue stream for recorded music, overtaking sales of physical formats for the first time.*"[8] The total digital music revenue in 2015 was around $11 billion and is forecasted to grow by $20 billion in 2020. The share of music streaming was $2.9 billion in 2015, more than 80% of that revenue coming from subscription based payments.

Music and video streaming became an important and rapidly growing element for the web consumer market. The music streaming market grew from half a billion in 2010 to $3 billion in 2015 (see Fig. 11).

The home video market reached $17 billion in 2015 and is projected to grow to a volume of $35 billion in 2020 (see Fig. 12).

In both cases, the services are either subscription based or offered as free services supported by advertisement. Many streaming services for music or video are available. The most popular companies are Netflix and Spotify focusing on music and video on demand. In 2007, Amazon also introduced its Video on Demand and Apple its Apple TV (see Table 4).

Streaming services are also offered for free via ad-supported services by numerous Internet-radio platforms or video-platforms. Local or regional TV networks have started to offer their channels as Internet subscriptions to their customers.

[8]IFPI Global Music Report 2016: http://www.ifpi.org/news/IFPI-GLOBAL-MUSIC-REPORT-2016 retrieved 2017-01-20.

Table 3 Largest travel and online-booking platforms

Company	Founded	Country	Revenue ($ million) 2015
Priceline Group	1997	USA	9220.0
Expedia	1995	USA	6670.0
Tripadvisor	2000	USA	1490.0
Trivago	2005	Germany	500.0
Odigeo	2011	Spain	495.6

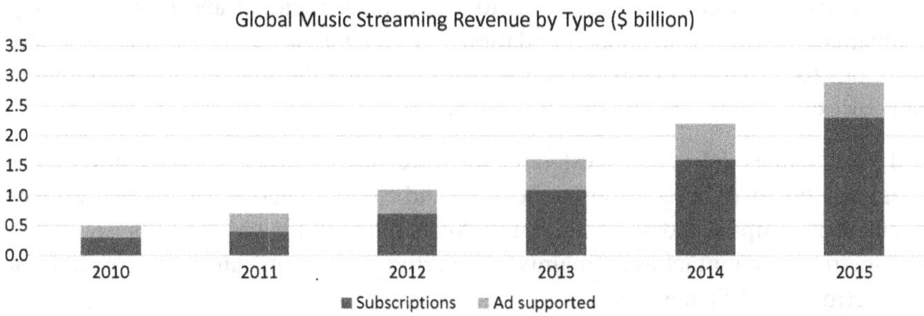

Fig. 11 Global music streaming revenue

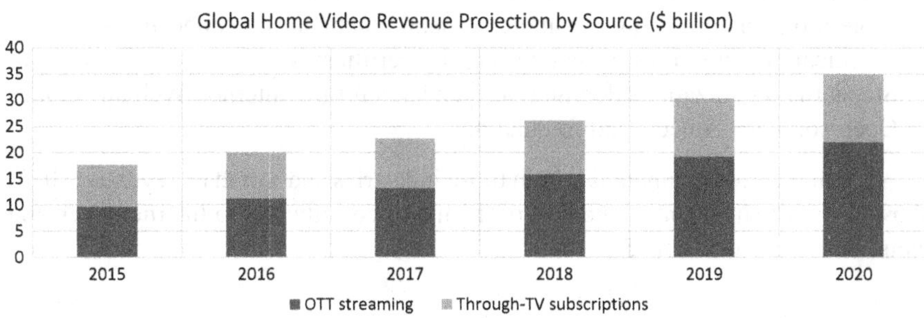

Fig. 12 Global home video revenue projection by source

Table 4 Large media platforms

Company	Introduced	Country	Revenue ($ million) 2015
Netflix	1997	USA	6779.0
Spotify	2006	Sweden	2080.0
Amazon Video	2007	USA	
Apple TV	2007	USA	

Cloud Market Players

With the start of the World Wide Web in 1992 a completely new type of enterprise—born on the Web—started to build large businesses. Today, the three largest IT companies: Google, Apple and Microsoft are either purely Internet based (Google) or deliver hardware or software for the Web and cloud services (Apple, Microsoft). These new companies also rely at least partly on new business models creating indirect business using data as the new currency. The Internet, Web and cloud market is a complex nexus of different local, regional and globally acting companies, organizations and foundations. A categorization of the market is not easy, but for a better understanding we will describe the market in the following categories:

- IT companies delivering hardware, software and services for information technology but also including Internet, Web and cloud related products and services
- Internet companies delivering mainly Web and cloud based services
- Electronic manufacturers mainly delivering processors and components for electronic and IT devices
- Internet service providers delivering the Internet connections to companies and private households
- Backbone service providers operating the Tier-1 backbones of the Internet and cable providers operating the intercontinental (submarine) connections
- Certification authorities providing security certificates
- Standards-organizations defining the standards for the Internet, Web and clouds
- Open software projects and foundations

A list of selected companies within those industries and sorted by revenue (2015) shows diversity in segment and size of the major contributors to Internet, Web and cloud business (see Table 5).

IT Companies

The large IT companies by revenue show a similar picture as in 1980s: one company is far ahead; the rest are far behind. The list also shows the diversity of the leaders in the IT market. Two of them are long standing companies: IBM (1911) and HP (1939) the rest were born between 1968 and 1984. Except for SAP (Germany) all of them are in the US and five are based in the Silicon Valley (Apple, HP, Intel, Cisco and Oracle) (see Table 6).

The largest IT companies providing products and services for Internet and clouds are Apple, Microsoft, IBM, Cisco, Oracle and SAP (see Fig. 13).

Apple

Apples leadership in revenue and profit is the result of the tremendous success of its iPhone and iPad business. Both products started a revolution in the usage of IT and

Table 5 Large cloud market players

Company	Founded	Revenue ($ million) 2015	Type
Apple	1976	233,700.0	IT
Samsung	1938	169,703.0	Electronic
Foxconn	1974	149,000.0	Electronic
AT&T	1893	146,800.0	Telecommunication
Verizon Communications	2000	131,800.0	Telecommunication
Amazon.com	1994	107,000.0	Internet
HP Inc.	1939	51,463.0	IT
Nippon Telegraph & Telephone	1985	94,200.0	Telecommunication
Microsoft	1975	93,580.0	IT
Intl. Business Machines	1911	81,740.0	IT
Deutsche Telekom	1995	76,800.0	Telecommunication
Google (Alphabet)	1998	74,990.0	Internet
Softbank	1981	74,700.0	Telecommunication
Sony	1946	68,508.0	Electronic
Baidu	2000	66,382.0	Internet
Vodafone	1984	51,150.3	Telecommunication
Panasonic	1918	61,350.0	Electronic
Huawei	1987	60,830.0	Electronic
Dell Technologies	1984	59,000.0	IT
LG Electronics	1958	57,350.0	Electronic
América Móvil	2000	56,300.0	Telecommunication
Intel	1968	55,400.0	IT
China Mobile	1997	53,667.8	Telecommunication
China Telecom	2000	52,700.0	Telecommunication
Telefónica	1924	52,400.0	Telecommunication
Hewlett Packard Enterprise	1939	52,107.0	IT
Cisco Systems	1984	49,100.0	IT
Orange	1988	44,600.0	Telecommunication
Oracle	1977	38,230.0	IT
JD.com	1998	26,270.0	Internet
SAP	1972	22,588.0	IT
Facebook	2004	17,928.0	Internet
Tencent	1998	14,905.7	Internet
Alibaba	1999	12,292.0	Internet
Priceline Group	1997	9220.0	Internet
Level 3 Communications. Inc.	1985	8229.0	Tier-1
AWS	2006	7880.0	Internet
Netflix	1997	6779.0	Internet
Expedia	1995	6670.0	Internet
Rakuten	1997	6240.0	Internet
Salesforce	1999	5373.6	Internet
Yahoo	1995	4968.3	Internet

(continued)

Table 5 (continued)

Company	Founded	Revenue ($ million) 2015	Type
China Unicom	1994	4300.0	Telecommunication
Netease	1997	3130.0	Internet
Groupon	2008	3119.5	Internet
Linkedin	2003	3000.0	Internet
Zalando	2008	2330.0	Internet
Twitter	2006	2220.0	Internet
Tripadvisor	2000	1490.0	Internet
Yandex	1997	818.7	Internet

Table 6 Largest IT companies in 2015

Company	Founded	Country	Revenue ($ million) 2015
Apple	1976	USA	233,700.0
Microsoft	1975	USA	93,580.0
Intl. Business Machines	1911	USA	81,740.0
Dell Technologies	1984	USA	59,000.0
Hewlett Packard Enterprise	1939	USA	52,107.0
HP Inc.	1939	USA	51,463.0
Cisco Systems	1984	USA	49,100.0
Oracle	1977	USA	38,230.0
EMC	1979	USA	25,700.0
SAP	1972	Germany	22,588.0
Alcatel-Lucent	2006	France	15,297.0
Nokia	1865	Finland	12,499.0
VMWare	1998	USA	6730.0
HCL Technologies	1976	India	6100.0
Fiserv	1984	USA	4950.0
Juniper Networks	1996	USA	4850.0
CA Technologies	1976	USA	4262.0
Intuit	1983	USA	4200.0
Amadeus IT Group	1987	Spain	3900.0
Symantec	1982	USA	3600.0
Brocade Communication Systems	1995	USA	2263.0
Checkpoint	1993	Israel	1629.0

the Web. Together with Apples strategy of building a user centric platform with the App Store, iTunes and the iCloud services, the company has built a rapidly growing community of customers since 2007 and created a successful business model based on app distribution and media services. Even though the iPhone and the iPad are high priced products, Apple could sell 1 billion iPhones between 2007 and 2016 and 330 million iPads between 2010 and 2016, which delivered up to 60% of Apple's

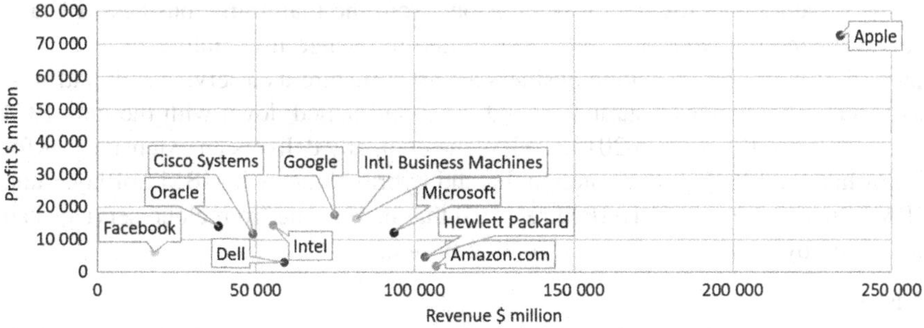

Fig. 13 Top IT companies 2015

revenue. The revenue from the App Store was $28 billion in 2016.[9] The question being discussed in the market over the last few years is if Apple's innovation power is still alive. Apple's PC product, iMac, is criticized and sales are dropping. In the smart device market, Apple's competitors catch up by building their devices on Android and Microsoft OS. Compared to Microsoft, Google, IBM and Amazon, Apple's innovation speed seems to be slowing down now.

Microsoft

Microsoft entered the Web business as a follower but was able to catch up by starting a huge transition from its license based software business to cloud services concentrating around Microsoft Azure and Microsoft Office 365. Target groups for the Microsoft cloud services are corporations as well as private users. Building and maintaining a community of private user helps to expand from free cloud services to paying subscribers. In mid-2016, Microsoft reported a rapidly growing revenue of $12.1 billion[10] in cloud business, which is in the same size as its major competitors, Amazon and Salesforce, but still only around 15% of Microsoft's total revenue.

IBM

Compared to Apple, Google, Amazon and Microsoft. IBM was always focusing on the corporate market. The company went through several major transitions in its more than 100 years of history. Being one of the first to enter the commercial computing market in the late 1950s, it was the first developing and shipping a highly-standardized system of main frame architecture with its IBM 360. Together with a concentrated product management for mid-size computers and servers, IBM could maintain its market leadership in the IT market until the beginning of the new century. The growing competition and dropping profits in the hardware business led

[9]Apple AppStore revenue 2016, ZDNet: http://www.zdnet.com/article/apples-app-store-2016-revenue-tops-28-billion-mark-developers-net-20-billion/ retrieved 2017-01-15.

[10]Microsofts cloud revenue, Business Insider 2016: http://www.businessinsider.de/microsoft-takes-lead-in-the-race-to-10-billion-in-cloud-revenue-2016-8?r=US&IR=T retrieved 2016-10-12.

to the separation of the hardware divisions from the mainframe business, PC and server businesses were sold and IBM started to change itself into a service company. The entry into the cloud business as infrastructure-as-a service and platform-as-service provider was started around 2005 and nailed down with the new IBM Cloud Service Division in 2013. Today, the company delivers an extensive portfolio of IaaS and PaaS for private, public and hybrid clouds with IBM Softlayer and IBM Bluemix. In mid-2016 IBM reported that its revenue from cloud services had increased by 12% year-over-year to $8.2 billion.

Cisco

Cisco is the major provider of cloud infrastructure with its routers, switches, security and network management software. Since the beginnings of the Internet, the company developed and delivered the components for any type of IT network infrastructure and expanded its business along the growth of the Web and cloud computing by focusing on corporate customers in the telecommunication and large enterprise segment. An entry into the market as cloud service provider was never completed thus several attempts were made including offering Cisco Intercloud as a service. Nevertheless, the company is the major provider for Internet and cloud infrastructures and partners with major providers of cloud service solutions like IBM, Amazon, Salesforce and others.

Oracle

Oracle has been the major player in data management business since its beginning in 1977. Thus, the company focuses on data-base products for corporations. Oracle made some major contributions to the Internet. Web and cloud business. Through the acquisition of SUN in 2010, Oracle became the home of the Java open source platform and took over SUN's server business. Oracle's founder and long term CEO, Larry Ellison, played a major role in the foundation of Netsuite and in the foundation of Salesforce in 1998 as the first software-as-service company. Through the acquisition of Netsuite by Oracle in 2016 the company entered the software-as-a service business. Oracle's data management products are offered and optimized for enterprise clouds and Oracle has started to create new public cloud services according to its announcements in autumn 2016.

SAP

SAP is one of the major providers of business software for large and medium sized companies. It entered the cloud business in 2010 with the release of SAP HANA as the foundation for large scale data management. With the acquisition of SuccessFactor in 2012, SAP also entered the software-as-a service market.

Internet Companies

Internet companies are corporations making all or a substantial part of their business on the Internet or cloud segment. They are typically "born on the Web" which simply means their initial business strategy was already based on cloud

Table 7 Largest internet companies

Company	Founded	Country	Revenue ($ million) 2015
Amazon.com	1994	USA	107,000.0
Google (Alphabet)	1998	USA	74,990.0
Baidu	2000	China	66,382.0
JD.com	1998	China	26,270.0
Facebook	2004	USA	17,928.0
Tencent	1998	China	14,905.7
Alibaba	1999	China	12,292.0
Priceline Group	1997	USA	9220.0
AWS	2006	USA	7880.0
Netflix	1997	USA	6779.0
Expedia	1995	USA	6670.0
Rakuten	1997	Japan	6240.0
Salesforce	1999	USA	5373.6
Yahoo	1995	USA	4968.3
Netease	1997	China	3130.0
Groupon	2008	USA	3119.5
Linkedin	2003	USA	3000.0
Zalando	2008	Germany	2330.0
Twitter	2006	USA	2220.0
Tripadvisor	2000	USA	1490.0
Yandex	1997	Russia	818.7

services, even if the term was not yet coined when they were founded. There are many different business models in this list covering a colorful spectrum of approaches. The list—sorted by revenue—also shows a strong focus on the Asian market with 5 companies out of the top 20 coming from Japan or China (see Table 7).

Amazon

Amazon was the first to move to two major market segments. Starting its Amazo.com e-commerce platform in 1995 and expanding it to business segments like e-books and video on demand. It is the largest enterprise in web-based consumer services. With the release of Amazon Web Services (AWS) as a new business unit in 2006, Amazon was also the first to move to the new infrastructure-as-a service segment and has expanded that service to a billion-dollar business during the last decade.

Alphabet (Google)

The success of Google and its top position in the Internet and web-business was based on Google's revolutionary search engine technology released in 1998 and its move to a disruptive business model based on advertisements and the collection of user data to optimize not only search but also the accuracy of user focused ads. In parallel, Google created a rapidly growing community of private and corporate users with its additional services (Gmail, Google Drive, Goggle+, Google App engine), which made the Google platform the largest advertising revenue engine on the market. Google also follows a focused innovation strategy and runs many research projects to build the Web of tomorrow. Google's founders, Larry Page and Sergey Brin, decided to restructure Google in 2015 and founded Alphabet Inc. as the parent company for Google and a dozen other companies concentrating on different segments and research activities including traffic management, augmented reality, autonomous driving and bio technology.

Facebook

Facebook was not the first social network but became the largest in the western world. With one billion daily users, Facebook is the largest Internet community and has a major impact on social media usage and acceptance. Facebook's revenue model is completely based on advertising revenue.

Semiconductor and Electronics Manufacturers

Electronics manufacturers and semiconductor companies deliver the basic elements for computers, networks, PCs, mobile devices, consumer electronics and industrial electronics. The semiconductor and electronics industry has its roots in the US and the major players in the semiconductor industry are still companies like Intel, Texas Instruments or Qualcomm. Driven by the need for mass production the assembling of electronics was outsourced to Asian companies for cost reasons, creating a huge market of electronics manufacturers in Japan, Taiwan, South Korea and China. The largest electronics manufactures by revenue are Asian companies among them Samsung, Foxconn, Sony. Panasonic and LG Electronics (see Table 8).

Infrastructure

Internet Service Providers

Internet service providers (ISPs) are telecommunication companies delivering "the last mile" to the consumer of the Internet and Web, either as landline or mobile services. Many of the literally hundreds of ISPs have their roots in telephone companies, some of them founded more than 100 years ago like AT&T. Another group of ISPs have been founded as Internet companies within the last three decades. Providing Internet services to consumers and corporations became a huge and rapidly growing business also driven by the wave of connection technologies, bandwidth and the growth of the mobile market.

Table 8 Large semiconductor and electronics manufacturers

Company	Founded	Country	Revenue ($ million) 2015
Samsung	1938	South Korea	169,703.0
Foxconn	1974	Taiwan	149,000.0
Sony	1946	Japan	68,508.0
Panasonic	1918	Japan	61,350.0
Huawei	1987	China	60,830.0
LG Electronics	1958	USA	57,350.0
Intel	1968	USA	55,400.0
Qualcomm	1985	USA	25,281.0
SK Hynix	1983	South Korea	15,670.0
Texas Instruments	1930	USA	13,000.0
STMicroelectronics	1987	Switzerland	6866.0
Broadcomm	2005	USA	6824.0
SanDisk	1988	USA	6627.0
MediaTek	1997	Taiwan	6476.6
Infineon	1999	Germany	6199.0
NXP	2006	Netherlands	6101.0
Renesas Electronics	2010	Japan	6047.0
Micron Technology	1978	USA	4573.0
Avago	2005	Singapore	4185.0
AMD	1969	USA	3991.0

The market size is estimated at more than $560 billion in 2015 with an annual growth rate of 9%.[11] The largest companies acting as Internet service providers are AT&T, Verizon, NTT, Deutsche Telekom and Softbank. The revenue numbers used for the sorting in the list below are total revenue numbers of the companies, including not only ISP business (see Table 9).
Internet service providers connect their networks to Tier-1 or Internet backbone providers. Backbone providers operate the international Internet connections, selling their bandwidth to the ISPs.

Backbone Service Providers

The list of Tier-1 or backbone-providers is much shorter that the list of ISPs. There are approximately 20 Tier-1 providers of significant size. The Center for Applied Internet Data Analysis lists in its rankings the Tier-a providers sorted by number of customer networks connected or "Customer Autonomous Systems" (Customer AS).[12] An autonomous system (AS) is a collection of connected Internet Protocol routing prefixes under the control of one or more network operators and used as an

[11]IBIS: Global Internet Service Providers: Market Research Report, 2016: https://www.ibisworld.com/industry/global/global-internet-service-providers.html retrieved 2016-10-12.

[12]CAIDA—Center for Applied Internet Data Analysis, AS Ranking: http://as-rank.caida.org/?mode0=as-ranking&data-selected=42 retrieved 2017-01-22.

Table 9 Large ISPs by revenue

Company	Founded	Country	Revenue ($ million) 2015
AT&T	1893	USA	146,800.0
Verizon Communications	2000	USA	131,800.0
Nippon Telegraph & Telephone	1985	Japan	94,200.0
Deutsche Telekom	1995	Germany	76,800.0
Softbank	1981	Japan	74,700.0
Vodafone	1984	U.K.	51,150.3
América Móvil	2000	Mexico	56,300.0
China Mobile	1997	China	53,667.8
China Telecom	2000	China	52,700.0
Telefónica	1924	Spain	52,400.0
Orange	1988	France	44,600.0
China Unicom	1994	China	43,000.0

Table 10 Largest Tier-1 internet providers by number of customers

Company	Number of AS	Customer AS
Level 3 Communications. Inc.	32	28,948
Cogent Communications	16	22,194
TeliaSonera AB	13	20,812
NTT America. Inc.	29	17,207
Tinet Spa	3	16,967
TELECOM ITALIA SPARKLE S.p.A.	1	13,982
TATA COMMUNICATIONS (AMERICA) INC	2	11,390
Hurricane Electric. Inc.	2	7339
Beyond The Network America. Inc.	3	6506
XO Communications	25	5787
Cable and Wireless Worldwide plc	1	5144
Abovenet Communications. Inc.	2	4074
nLayer Communications. Inc.	2	3985
MCI Communications Services. Inc. d/b/a Verizon Business	30	3827
RETN Limited	2	3690
Qwest Communications Company. LLC	13	3437
Deutsche Telekom AG	25	3164
Closed Joint Stock Company TransTeleCom	29	3141
Sprint	32	2897
AT&T Services. Inc.	12	2794

indicator for the size of Internet service providers, giving a clear definition of the number of customer networks served by the provider (see Table 10).

Some of these companies are ISPs (NTT, MCI/Verizon, Deutsche Telekom, AT&T) some others are focused on backbone services only. The backbone providers operate the Tier-1 Internet connections but do not construct or operate the physical cables. The physical cable network is either provided by local telecommunication providers or—in case of intercontinental connections—by submarine cable operators.

Cable Operators

The land based cable network is owned by the local telecommunication providers or by tier-1-providers. Additionally, there are more than 250 submarine cables operated between the five continents. Submarine cables form the physical backbone of the Internet; carrying more than 99% of the Internet data traffic of around 1 Zettabyte/year. A complete documentation of operators and owners of the global submarine cable network is provided by TeleGeography (see Fig. 14).[13]

The submarine cables are either owned and operated by private tier-1 providers or by consortiums of telecommunication or backbone providers. The consortium's members share the cost of construction and operation of the cable and are allotted specific bandwidth rights. They, in turn, make money by selling either access to this cable system or services/accessibility to their own clients. The number and bandwidth of intercontinental submarine cable connections has increased dramatically in the last decades. Between 2000 and 2010 more than $30 billion was invested in extending the submarine cable network. Cable systems like SeaMe-We-3 (Europe-Australia-East Asia). Southern Cross (Australia-USA) or China-US bridge more than 30,000 km. Bandwidth is also growing: in 2016, a new cable was laid between USA and Japan, providing a new record capacity of 60 Terabits per second by using six pairs of fiber each of them being able to transport 100 Gb/s in 100 different wavelength-bands (60 Terabits/s = 100 Gb/s × 100 wavelengths × 6 fiber-pairs). The cable is owned by a consortium formed by Google, NEC, China Mobile, China Telecom, Global Transit and KDDI and guarantees each of its members a bandwidth of 10 Terabits per second.

Internet Exchange Points

To connect the tier-1 networks with each other and to provide connections points between tier-1 networks and Internet service providers. IXPs (Internet Exchange Point) are operated as network facilities enabling the exchange of Internet traffic. The task of IXPs is to connect "Autonomous Systems" (AS) using the Border Gateway Protocol as defined by IETF.[14] There are more than 700 Internet Exchange Points around the world. A list of IXPs is provided by TeleGeography (see Fig. 15).[15]

Internet Exchange points are owned and operated by various commercial organizations, some of them organized as consortiums of the ISPs and backbone providers who connect to that exchange.

---

[13]TeleGeography: http://www.submarinecablemap.com/#/ retrieved 2017-01-22.

[14]IETF, RFC 4271, 2006: https://tools.ietf.org/html/rfc4271 retrieved 2016-10-12.

[15]TeleGeography: http://internetexchangemap.com/#/ retrieved 2017-01-22.

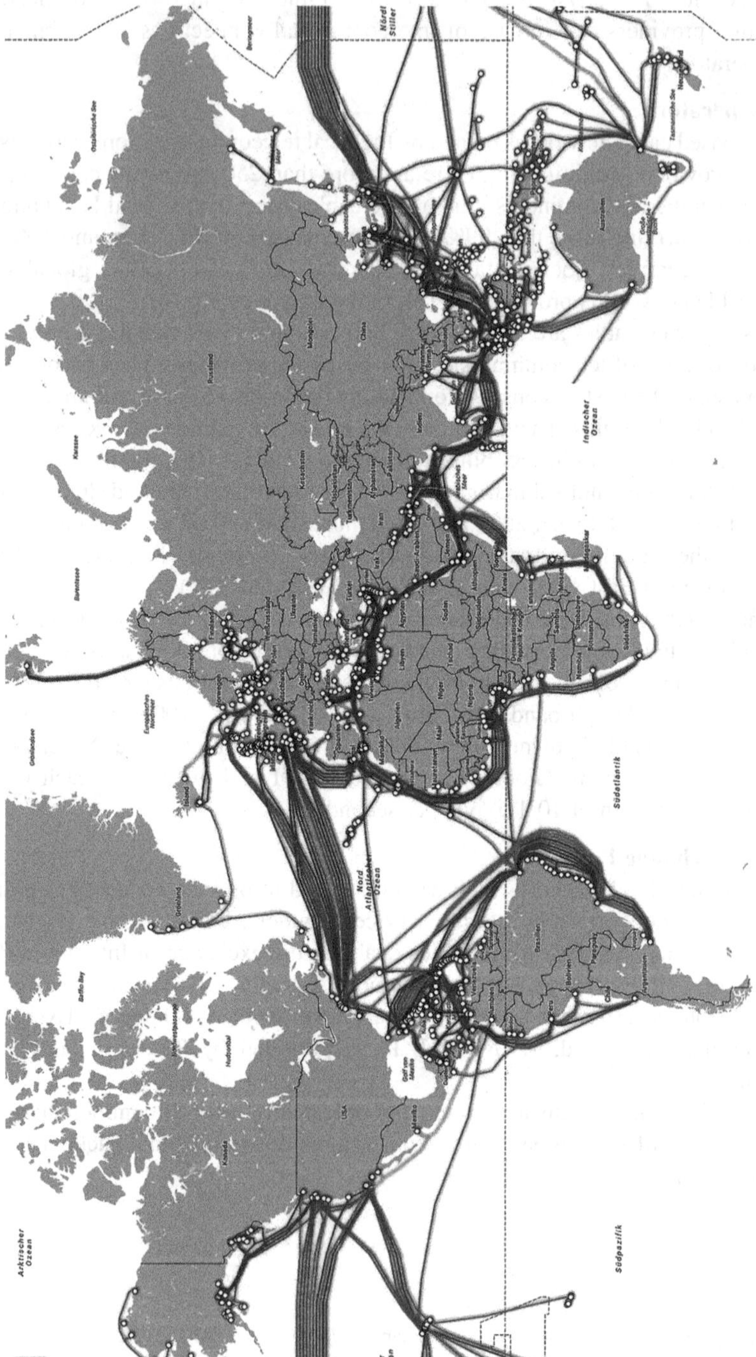

Fig. 14 Global submarine cable connections map (by courtesy of TeleGeography)

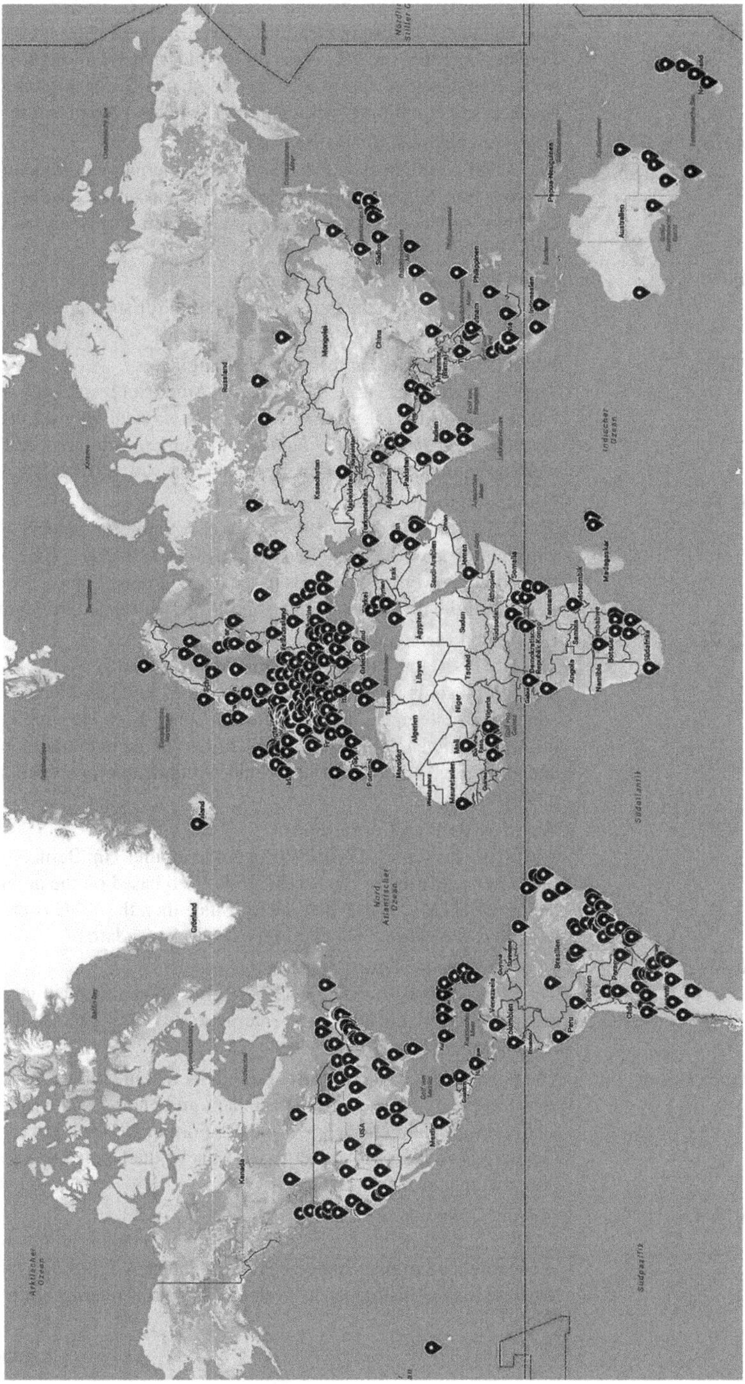

Fig. 15 Internet exchange points map (by courtesy of TeleGeopgraphy)

Table 11 Standard organizations

1865	ITU	International Telecommunication Union Founded in 1865, the ITU is one of the oldest standard organizations. It defines and approves numerous standards for telecommunication and computing including the ITU Cloud Architecture
1918	ANSI	American National Standards Institute The ANSI coordinates US standards with international standards and overseas the development of voluntary consensus standards for products, services, processes, and systems including many standards for computing
1947	ISO	International Standard Organization The ISO operates as the parent organization for standard organization of its more than 160 member countries
1963	IEEE	Institute of Electrical and Electronics Engineers The US based IEEE is a consortium of American companies and engineers working together to "*foster technological innovation and excellence for the benefit of humanity*." Numerous Internet and web-based standards are supported and approved as IEEE-standard
1986	IETF	Internet Engineering Task Force The IETF was founded in 1986 as a consortium of engineers working together to create the standards for the Internet. It became the driving power for standards based on the Internet protocol suite and is still the root organization for all Internet related standards
1992	DMTF	Distributed Management Task Force The DMTF is an industry standards organization led by 14 major technology companies. Its focus is to create the international adoption of interoperable management standards and to support its implementations. Standards created by DMTF include traditional and emerging technologies like cloud, virtualization, network and infrastructure
1994	W3C	World Wide Web Consortium The W3C was founded in 1994 by a group around Tim Berners-Lee to create standards for the World Wide Web based on the architecture of HTML, HTTP and URLs. Since then the W3C is the driving organization behind all types of web standards
1994	NANOG	North American Network Operators Group The NANOG was founded in 1994 to coordinate technical information related to backbone and enterprise networking technologies and operational practices
1996	Open Group	The Open Group is a consortium of UNIX manufacturers. It is certifying body for the UNIX trademark and publishes the Single UNIX Specification technical standard. Former X/Open Portability Guide (XPG). It was founded as a merger of the Open Software Foundation and X/Open
1998	ICANN	Internet Governance Forum ICANN is responsible for coordinating the maintenance and procedures related to the namespaces of the Internet. It performs the actual technical maintenance work of the central Internet address pools

(continued)

Table 11 (continued)

2011	Open Compute	The Open Compute Project (OCP) is reimagining hardware. Making it more efficient, flexible and scalable. It shares designs of data center products among companies. Including Facebook, Intel, Nokia, Google, Apple, Microsoft, Seagate Technology, Dell, Rackspace, Ericsson, Cisco, Juniper Networks, Goldman Sachs, Fidelity, Lenovo and Bank of America
2011	Open Network Foundation	The Open Networking Foundation (ONF) is a non-profit organization dedicated to accelerating the adoption of Software Defined Networks (SDN). ONF works with leading network operators to transform networks into agile platforms for service delivery

Standards and Standard Organizations

Standards used by the Internet, the Web and numerous hardware and software architectures deliver a major contribution to cloud business and cloud services based on Internet and web-architectures. Some of these standard organizations have their roots in the industrial revolutions of the nineteenth century and the early twentieth century. Others have been founded with the rise of computers and the software technology to create common platforms for discussion and the definition of standards (see Table 11).

Open Source Software

Open source software delivers a major contribution to Internet, Web and cloud services. Countless software tools and applications are organized as open source projects following the idea of open software developed by large communities of software engineers and distributed under open license agreements. The open source business consists of open source projects assembling a large crowd of software engineers and creating a new tool or product. There are thousands of open source projects active, the largest open source platforms being Linux and Apache. The most prominent and influential are listed below (see Table 12).

Some of the open source platforms receive support not only from individual contributors but also from large technology companies and governmental groups like Google, Apple, AMD, Microsoft, Fujitsu, the US Government, WordPress, CGHQ and Oracle (Java, MySQL, Open Office).[16] These companies and governments welcome open source initiatives as an innovative contribution to their business and as a future business option. Thus, open source development platforms act as non-profit organizations, and revenue can mainly be generated through

[16]Swapnil Bhartiya, CIO, 2015: http://www.cio.com/article/3017996/open-source-tools/9-biggest-open-source-stories-of-2015.html#slide2

Table 12 Open source projects

1994	MySQL	MySQL is an open-source relational database management system. The name is a combination of "My", the name of co-founder Michael Widenius' daughter and "SQL" (Structured Query Language). The MySQL development makes its source code available under the term of the GNU General Public License. MySQL was created by the Swedish company. MySQL AB, founded by David Axmark, Allan Larsson and Michael "Monty" Widenius. In 2008 MySQL AB was acquired by Sun Microsystems and is today owned by Oracle after the acquisition of Sun by Oracle in 2010. Several paid editions are offered with additional functionality available
1998	Linux Foundation (ex FSG. 1998 and ODSL. 1999)	The Linux Foundation appeared from the merger of the Open Source Development Labs (OSDL) and the Free Standards Group (FSG) in 2007. It sponsors the work on the Linux kernel is funded by leading Linux and open source companies. Including major technology corporations such as Cisco, Fujitsu, HP, IBM, Intel, Microsoft, NEC, Oracle, Qualcomm and Samsung and developers from around the world
1999	Apache Software Foundation (ex Apache Group. 1993)	The Apache Software Foundation is a decentralized open source community of developers to support Apache software projects. Including the Apache HTTP Server. It was formed from the Apache Group
2003	Mozilla Foundation	Mozilla was originally founded by Netscape as Mozilla Organizations in 1998. During restructuring of AOL and Netscape, the Mozilla Foundation was founded in 2003 to ensure the ongoing development and support of Mozilla and the Firefox browser. Today the foundation controls the Mozilla source code repository and coordinates the development of the Mozilla and Firefox code
2003	Wordpress	WordPress is an open source CMS (Content Management System) and first appeared in 2003 as a joint effort between Matt Mullenweg and Mike Little. WordPress is developed by the WordPress core developers and a large community. Today, WordPress enjoys one of the greatest brand strength of any open-source content management system
2004	Ubunto	Ubunto is an open software project developing a Debian-based Linux distribution for PCs. The software itself is open source. Ubunto is published by Canonical Ltd. UK that is offering additional services for Ubuntu
2004	OpenStreeMap	OpenStreetMap is a collaborative project to create a free editable map of the world. OpenStreetMap is motivated by restrictions on use or availability of map information across much of the world and the advent of inexpensive portable satellite navigation devices. It is considered a prominent example of volunteered geographic information

(continued)

Table 12 (continued)

2010	OpenStack	OpenStack is a free and open-source software platform for cloud computing. The software is released under the terms of the Apache License. OpenStack was started as joint venture between Rackspace and NASA and is today managed by the Open Stack Foundation with more than 500 member companies
2013	Open Daylight	The OpenDaylight Project is a collaborative open source project hosted by The Linux Foundation. The goal of the project is to promote software-defined networking (SDN) and network functions virtualization (NFV). The software is written in the Java programming language

services. This is the business of open source distributers like Red Hat for Linux, but also IT companies like IBM, HP, Cisco and Google. Some of those companies act not only as service providers for open source, but also contribute engineers and code to open source projects like IBM by supporting Eclipse, Oracle supporting Java, MySQL and Open Office and Google.[17]

Open Source License Agreements

The foundation for open source was laid by Richard Stallman in 1983 by creating the GPL (General Public License) as first open source license agreement. Stallman also founded Free Software Foundation as a platform for the definition of Public Licenses like GPL and LGPL. Today, there are several slightly different open source license agreements that are used by the producers of open source. All of them have in common that the software, source or also content is free under a few restrictions (see Table 13).

Open Source Business

Depending on the type of the open source license agreement the software can also be adapted and distributed to others. Nevertheless, the open source business has been growing dramatically in the last decade estimated to reach volumes of more than $50 billion in 2016 (see Fig. 16).[18]

[17]Matt Asay, CNET, 2016: https://www.cnet.com/news/worlds-biggest-open-source-company-google/ retrieved 2017-01-20.

[18]PAC: Open Source State of the Art, 2008 https://www.pac-online.com/download/7282/121263 retrieved 2016-10-12.

Table 13 Open source license agreements

1982	Shareware and Freeware	Shareware is provided free of charge to users. Copies can be made and shared without restriction. Freeware is provided to the user without cost but also without the source code being available. The first software called "Shareware" was developed in the beginnings of the 1980s for the IBM PC
1989	GNU GPL	General Public License GPL is a so-called copyleft license. That means that right to freely distribute copies and modified versions of a work with the stipulation that the same rights be preserved in derivative works down the line. Software under the GPL may be run for all purposes. Including commercial purposes and even as a tool for creating proprietary software, for example when using GPL-licensed compilers
1991	LGPL	Lesser General Public License LGPL allows for the integration of software in its own code or products without being required by the terms of copyleft. It is usually used for software libraries so that the usage of open source in a software product does not restrict the license options of the developer
2000	Apache License	The Apache License was created by the Apache Software foundation and is a free- and open permissive license. Compared to a copyleft license, changes (derivative work) can be distributed without using the same license. Unmodified parts of the Apache code must preserve the original copyright, patent or trademark
2005	OSL	Open Software License (2005) OSL is comparable to LGPL, but includes a termination clause intended to dissuade users from filing patent infringement lawsuits
2009	CC	Creative Commons Licence CC is a group of license agreements that are used to enable the free distribution of an author's work. They can be applied to all works falling under copyright law like books, music, movies, pictures or also websites

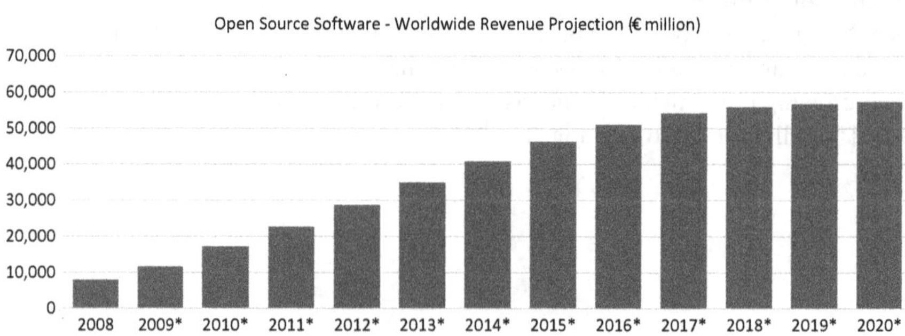

Fig. 16 Open source revenue projection

The business model behind open source is mainly focused on services and not on license revenue. Open source software is basically free for downloading and usage. Distributers of open source products generate revenue, mainly from services. In some cases, additional functionality is offered as a paid license.

Creating Innovation

I have learned fifty thousand ways it cannot be done and therefore I am fifty thousand times nearer the final successful experiment.

Thomas A. Edison

Creative Destruction and Disruption

Innovation is and always was the major driver of economic, technical and social change. The invention of agriculture and the domestication of plants and animals between 12,000 BC and 9000 BC led to the foundation of settlements and, later, to the foundation of cities and the division of labor in ancient societies. This was also one of the drivers of new forms of dominance and control. The development of transportation methods using horses and carts or ships led to the first international trading networks. In the eighteenth and nineteenth centuries the innovation of steam engines triggered the development of industrial production methods, later followed by the second and third industrial revolutions. Growth in production and, simultaneously, growth in population were always triggered by technical innovations followed, in many cases, by social and economic changes. It seems that innovation is always the forerunner of major transitions in social life and new economic structures.

One of the first to describe innovation and its mechanism and effect on society and economy was Peter Schumpeter in his book "Theorie der wirtschaftlichen Entwicklung" published in 1912 while he was a university professor at the Karl-Franzens-University in Graz, Austria. In his book, Schumpeter claims that the major drivers of economic growth are innovative entrepreneurs or pioneers searching for new combinations of different factors in production. Schumpeter also defined five categories of innovation:

- Production of a new good or a new quality of a good
- Introduction of a new production method
- Acquisition of new markets

M. Oppitz, P. Tomsu, *Inventing the Cloud Century*,
DOI 10.1007/978-3-319-61161-7_13

- Acquisition of new sources for raw material or semi-finished products
- Creation of a new market position or creation or break up of a monopoly

Schumpeter worked on this approach for more than 10 years publishing a second edition in 1926. In the meantime, he was also a professor at Columbia University in New York, Minister of Finance for the young Austrian Republic between March 1919 and October 1919, a professor at the University Bonn in Germany and a professor at Harvard by 1932. Even today, Schumpeter's idea of innovation as the root cause for economic growth and social change is one of the widest accepted theories in economics. The technical and economic history of the decade after Schumpeter' work seems to confirm that. Technical innovations like cars, telephones and radio broadcasting had a major impact on the economy and society in the first three decades of the twentieth century. New technologies like electronics, digital computing and nuclear power formed the economy and society of the decades after World War II. Today, the new technologies and business models used in cloud computing and adjacent technologies created many new business models, formed new economic structures and had a huge impact on political and social life.

The effect of innovation in all these developments is easy to see and certain. Innovation, as an effectual driver of change, seems not only to influence but sometimes also to dictate the transition into new economic and social structures. More than that, innovation became a kind of industry that tries to develop processes that identify, create or accelerate innovation to empower economic growth and change. Following that idea, it seems important to drill down into the mechanisms and patterns of innovation a little bit more.

Essentially, innovation means the discovery and introduction of "something new." Schumpeter's definition was more restrictive. He said innovation is a creative destruction of existing elements or factors of production. In 1995, Joseph L. Bower and Clayton M. Christensen[1] coined the term "Disruptive Innovation" as a stronger and more sustainable form of innovation. Disruptive Innovation is not only based on existing elements of production, it replaces existing business models, technologies or markets through innovating new technologies, production methods or through creating new markets.

Analyzing the innovative events in technology, economy or society in the last 200 years, it turns out that most innovations exactly follow that pattern. There are only very few innovations that could claim to be basic and disrupting in terms of creating new fields of technology and new economies. In the last 200 years, the innovation of steam engines, electricity, combustion engines, telegraphs, wireless communication, nuclear power and digital computers are the most important examples. In the future, quantum computing and "real intelligent" machines could be candidates for basic and disruptive new technologies. Most of the other innovative—and sometimes also disruptive—developments are in fact following

[1]Harvard Business Review, Joseph L. BowerClayton M. Christensen, 1995: https://hbr.org/1995/01/disruptive-technologies-catching-the-wave retrieved 2016-1-12.

the pattern of innovative combination. The Internet was a smart combination of computing and network technologies and the definition of a standard that binds these elements together.

The World Wide Web was—as also claimed by its inventor Berners-Lee—the combination of already existing technologies and concepts. Cloud computing is based on the Internet and the stepwise development of faster and smarter organized hardware/software combinations leading to the deployment of virtual servers, storage and networks. Disruptive products, like Apple's iPhone, were a smart combination of the existing iPod technology with cell phone hardware and the usage of wireless Internet access. They also created a new market in a disruptive way. Social networks like Facebook used existing technology combined with a new application and—later—disruptive, new business model.

Technology, Paradigms and Ecosystems

In search for a more structured model for innovation patterns, we find that new technology is a basic element of innovations but without any disruptive effect if it is not combined with a new view on economic, social or physical structures. New basic technologies like digital computing, networks, Internet, wireless communication, and virtualization lead to new technologies like cloud computing, web services, smart environments or augmented reality. All these great technical inventions or developments go hand in hand with paradigm changes.

The Web and cloud computing changed the way we experience the world and helped to introduce the idea of a global village. Smart environments, augmented reality and cognitive computing lead to the new paradigm of an explorable cyberspace within computers and networks. Cloud computing together with virtualization helped to create the new paradigm that sharing is better that owning. Following new technologies and accepting new paradigms, innovations play a role in combining new technologies to develop new business models for consumers and industries.

True disruptive innovation is always a combination of new technologies, paradigm changes and the creation of new business models and ecosystems. Since the introduction of the first commercial digital computers in the late 1950s, we have experienced several of those important changes influencing our way to the cloud century.

The Importance of Paradigm Changes

Innovations need, as a basic starting point, a completely new view on physical, social or economic structures. This is called a paradigm or a paradigm change. Nuclear power is a typical example. The technical development of the nuclear bomb and—later—the use of nuclear power for energy production was based on the change from classical physics to modern physics or quantum physics. Albert Einstein's "Theory of Relativity" is one of the most popular examples of a paradigm change, which not only influenced but also triggered the development of nuclear power technologies. Looking back on the technical and economic innovations in the last 50 years, we can identify several paradigm changes that influenced and triggered technical innovations that led to cloud computing and the move to cloud based services.

- Paradigm: Digital computers are general usable machines
- Paradigm: Software is reusable
- Paradigm: An open society needs open systems
- Paradigm: The network is a common resource
- Paradigm: We live in a global village
- Paradigm: Access to information is free
- Paradigm: Data is value and currency
- Paradigm: Production costs are no longer the driver
- Paradigm: Sharing is better than owning
- Paradigm: The default status of users is mobile
- Paradigm: Services are delivered from a virtual space to a global space
- Paradigm: Peer-to-peer can replace centralized services
- Paradigm: There is a cyberspace in the machine
- Paradigm: Machines (can be) intelligent

We will also see that those paradigm changes are not restricted to technology, but may also have their origin in social behavior, economic structures or are influenced by mainstream politics.

Acceleration of Paradigm Changes

The change to new paradigms triggered industrial revolutions and influenced social, economic and political evolutions in a dramatic and accelerating way. Steam, railway and electricity helped to form the first and second industrial revolutions in the nineteenth century. The telegraph, the telephone, wireless communication and broadcasting powered the third industrial revolution. Digital computing, the Internet, Web and cloud brought us cyber-physical systems and cognitive computing. By plotting the major paradigm changes on a timeline and calculating the time between them on the vertical axis, we see the dramatic concentration of disruptive developments in the last several decades (see Fig. 1).

The speed of paradigm changes is accelerating; the time between major paradigm changes is becoming shorter, creating new opportunities but also challenges for technology, business and society. New paradigms are the first step to new technologies, businesses and social changes.

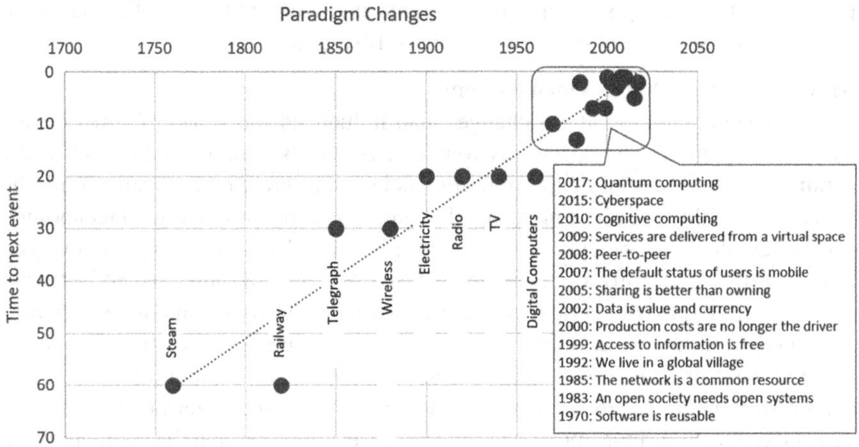

Fig. 1 Paradigm changes

14 Major Paradigm Changes Since 1950

Digital Computers Are General Usable Machines

Maybe the first important paradigm change related to today's economy of cloud computing was the idea of the computing machine. Alan Turing's theoretical approach "On Computable Numbers" created a completely new view of machines. Before Turing, a machine was a mechanical or electrical construction that could manufacture a very specific product. Turing proved that machines can be constructed, which can calculate every function if it is defined. This changed the paradigm of machines from specifically to generally usable and led to the first digital computers constructed in the UK and America in the 1940s. An important part of Turing's approach was the clear definition of what we today call data and program as a set of instructions. Based on this approach, the first programmable and generally usable computers were developed. It was John von Neumann who described the principle technical architecture of modern digital computers.

Software Is Reusable

In the following decade, the technology was improved in terms of efficiency and performance. The next major paradigm change appeared when it became evident that the instruction set or program was more important and had more value than the underlying technical hardware. Creating correct and efficient programs turned out to be critical and the writing, testing and executing of instruction code on different types of hardware became a major bottleneck for the new and emerging technology. Starting in the early 1960s, the idea evolved to strictly separating the architecture of the hardware from the structure of the software to accomplish independence of the software from the hardware. The same software should be executable on different types of hardware, leading to one of the most important paradigm changes for software development: the reusability of software. The consequences of this major paradigm change were—among others—the development of high level programming languages, compilers, programming structured as a method and—on the economic side—the "invention" of the software industry.

An Open Society Needs Open Systems

The next significant paradigm changes also influenced the world of software but had their origins in the social environment. At the beginning of the 1970s, the community of technicians and programmers began to grow dramatically. With the appearance of the first home or personal computers at low prices, digital technology became available for everybody. Programming and software development evolved into a profession, software companies hired technicians and created software development teams and computer science was introduced as a major subject with dedicated departments at universities. At the same time, parts of the US society adopted ideas oriented around open society principles, which required more democratic structures, less regulation and a higher level of self-determination. Movements like the civil rights movement, the anti-war initiatives and the counterculture movements that started in 1968 are well known today.

It is also evident that this social and political paradigm change had a major influence on how technology was seen. Frustrated by the monopolies of computer companies and their licensing policies on software, computer engineers like Richard Stallman started to think about a world of "Open Systems" and software that is not created by a closed community within a computer company, but by an open community making their innovations available to everybody. The "Open System" and "Open Software" movement was created and had a tremendous impact on how software is developed and published. Today, a huge number of system software like LINUX and application software packages are maintained and distributed as open software or open source.

The Network Is a Common Resource

A major paradigm change was still missing. Internet technology was developed by a group of US research organizations and funded by the Department of Defense within the ARPANET-project. Usage was restricted to the members of that community. In the late 1980s, it became evident that the Internet and its underlying technology could be a major driver for economic growth. In 1990, the ARPANET was decommissioned and within the next 5 years the Internet was opened for commercial traffic thus providing a common resource for the industry. Since then, the Internet has become a worldwide resource for communication, which is operated by private tier-1-providers, organized by international organizations and funded by the users paying fees to the national ISP (Internet Service Providers). It can be said that the Internet is still the only globally available and globally organized resource for commercial and private communication services today.

We Live in a Global Village

With the development of global communication and globally available media, the importance of national borders and regional communities changed. This effect was first described as the "Global Village" by the Canadian Marshall McLuhan in his book "The Gutenberg Galaxy" in 1962. In his book, McLuhan references globally available information using electronic communication, but also seems to foresee today's world of personal and public information exchange: *"The next medium, whatever it is it may be the extension of consciousness—will include television as its content...*". Living in the global village reduces the implication of location and distance. In a global village, everybody can connect to and communicate with everybody else. Information is distributed in a wide variety of media formats and available independent of location and time. The paradigm of a global village is the foundation of today's social and media networks, but also shares in the idea of the human right to access free information.

Access to Information Is Free

Within the last decades, information has become a major resource for private and business life. In parallel, the discussion about the ownership and monopoly of information sources and distribution has influenced business models based on information provision and usage. When Jimmy Wales introduced Wikipedia in 2001, one of his ideas was to provide a platform, which is based on the principle of the freedom of knowledge. Providing information to users for free has become a major driver for the World Wide Web. Today, millions of platforms offer access to information and knowledge including search engines like Google, retail platforms like Amazon, streaming platforms like YouTube and nearly every news media provider including television and newspapers. The fact that everybody connected to the Web has free access to terra bytes of content is one of the most dramatic transitions stimulated by the Internet and cloud services being the technical foundation of all this content. Side by side with this development, a dramatic discussion is still progressing about the classification and ownership of information. This discussion focuses on three elements: what exactly is the definition of "public data," what is the proper definition of "private" and where is the border between free information and billable information.

Data Is Value and Currency

Well, in fact, data always was currency. Gathering knowledge about the market, the ecosystem or the threads you always had to face the prerequisite of success. With the introduction of global search engines, retail platforms' and social networks' data became much easier to collect, process and apply. Every time a user is connected to a free service, huge amounts of data are collected and used to trigger individually tailored offers and recommendations. This data is the real value of search engines, social networks and retail platforms. The value and revenue of social media platforms like Facebook depends on user data. Nearly 100% of Facebook's revenue comes from advertising (2015: $17 billion). Google's advertising revenue reached $67 billion in 2015, which was a major part of its total revenue of $74 billion. User data is the currency, which users "pay" to access those types of services.

Production Costs Are No Longer the Driver

Compared to traditional industry manufacturing methods, the production costs of information or software services have been reduced dramatically over the last three decades. Distribution of software or services via cloud technology is negligible in terms of costs and much lower than in the twentieth century. Compared to the 1990s, today's software is distributed via downloading processes instead of distribution using media like tapes, disks, CD or DVDs. Access to cloud services is directed via portals and user-based provisioning, and payment is organized by credit card or web-based payment services. This paradigm led to a new pattern of order and delivery processes like Microsoft's Office 365 cloud services or Apple's AppStore. Offer, order and provisioning processes are completely "web-based" thus reducing operational and production costs on the provider's site and improving flexibility on the consumer's side.

Sharing Is Better Than Owning
Sharing tools and resources has a long history that even includes public transport, telephone services and power distribution on a grid. Finally, that paradigm arrived in the world of data processing triggered by the cloud technology based on the Internet and virtualization. With the introduction of cloud computing technologies, it became evident that tool and resource sharing can be applied to information technology as well and became the leading paradigm change for the application of cloud services.

The Default Status of Users Is Mobile
For a very long time, the delivery of a service depended on location. IT services like computer programs could only be used in a specific workplace or environment. The introduction of technologies like wireless networks and GPS triggered the development of the first mobile devices in the first decade of the twenty-first century. With the introduction of the first smart phone—Apple's iPhone in 2007—it became evident that a huge paradigm change was to be announced. Private as well as corporate users of information services were no longer bound to a specific location or device: their default status became mobile. Access to any type of information, service and knowledge could be independent of location or device: the default status of users was mobile, being stationary became the exception.

Services Are Delivered from a Virtual Space to a Global Space
Until the twentieth century the offer and delivery of a service had two major parameters in the physical world: the location of production and the location of delivery. This is no longer the case. Cloud services are delivered from an (sometimes) unknown location somewhere in the virtual space of the cloud and can be delivered to any location in the real world. Using a cloud service like Amazon Cloud, Google or an app from the AppStore, you never know exactly where the source of the service is located. At the same time, you can use that service wherever you are around the globe. This huge paradigm change led to a dramatic increase in flexibility, but also to many open questions in terms of statuary requirements, tax liability and, finally, data security. Compared to the economic system of the nineteenth and twentieth centuries, nation specific regulations and national borders have lost importance and have not yet been replaced by other—more globally applicable—agreements.

Peer-to-Peer Can Replace Centralized Services
In the first phase of cloud computing, the architecture was based on centralized services provided to distributed users. Back in 2008, a new approach appeared that aimed at the idea of introducing a digital currency named bitcoin. The technical

background described in a paper by Satoshi Nakamoto (a pseudonym) defined a new method of organizing any type of transaction without a central service. The method was called "block chain" and is based on the idea of distributed and synchronized log files (aka ledgers) stored within the premises of each user. Whenever a new transaction is created, the data record will be distributed to each user without involving a central database or service. The main motivation behind that idea was to exclude third parties like banks from currency transactions thereby making them more efficient and quicker. The approach can be applied not only to currency transactions but also to any other type of transaction between people or organizations eventually leading to a network of peer-to-peer activities without any type of centralized third party service involved. In that case, the network or cloud is used as the binding element between users, while central services or service organizations disappear. It is no question that this paradigm change could lead to many new applications without the involvement of third party transaction organizations like banks, insurance companies or public administrators of centralized registers like land registers, contract registers and others.

There Is a Cyberspace in the Machine

Since the beginning of digital computing, the machine has been a somewhat mysterious place. Starting with the wide availability of personal computers in the 1970s, computing evolved into a popular discipline for students, engineers and artists. Using the computer not only for business and science but also for gaming and art became quite normal. At the same time, the first experiments involving new types of user interactions like graphic user interfaces, 3D displays or data gloves empowered the creative drive to design and enter virtual worlds generated by the machine. The term "Cyberspace"—created by the science fiction author, William Gibson, in 1982 was used for those kinds of experiences and experiments. Due to insufficient performance at that time, those experiments had no sustained effect apart from applications for fighter jets powered by the huge funding potential of the defense industry. With the improvement of hardware performance and new interface technologies appearing in the upcoming decades, the theme of virtual and later augmented realities attained a more feasible status. Today, virtual reality and augmented reality applications have a major impact on the gaming industry, but also on other businesses like automotive, aerospace and industry 4.0. Today, "Cyberspace" has become a reality, and there is a virtual world behind the old-fashioned screen that can be used for real world applications.

Machines (Can Be) Intelligent

Intelligence is a difficult word to define. Since the construction of the first computers, there has always been a kind of artificial human brain. Today, we know that human intelligence is something we are not able to describe or re-engineer in a machine. Nevertheless, huge progress has been made in the last decades in creating cognitive computers for certain applications, developing software for speech recognition and text analysis and building a cyber physical system for specific uses in cars, planes and industry. The "holy grail" of artificial intelligence is still somewhere out there. It is the intelligent machine or, as Nick Bostrum calls it, the "Super

Intelligent Machine" which simply means a machine that has general intelligence unrestricted to any dedicated applications. The paradigm is postulated but not proven. It is perhaps one of the big next things.

Innovation as Business

Today, the major players in cloud services are companies that have their roots in the startup culture. Eight of the ten most valuable brands of 2016 are tech companies shaping the cloud business and six are enterprises that started as startups no more than 30 years ago (Google, Apple, Microsoft, Facebook, Amazon, Verizon) (see Fig. 2).

The term "startup" was coined during the last decade of the twentieth century and was used to describe young companies building their business on the new Internet and Web technology. In his book "Zero to One" (Thiel, 2014), Peter Thiel (co-founder and investor in PayPal, Palantir, Facebook and others) describes a startup as a new enterprise that creates a disruptive new business from "zero" without using existing business models or improving them. Thiel distinguishes the zero-to-one approach from a "one-to-n" approach where existing business models or product patterns are improved or adapted. He claims that a true startup follows a completely new way, consequently building not only new products and services but also inventing a new market.

Looking at a list of successful startups that influenced the making of the Internet and Web ecosystems, we can identify several companies following that pattern. All of them delivered a major contribution to the cloud ecosystem: Amazon invented the e-commerce market starting with books and CDs, PayPal invented the digital payment market, Google invented the search engine market by creating a unique technology and later using it to generate wealth from digital advertisement. Uber invented the car sharing market, and Airbnb created the apartment sharing market. It is but also true that startups existed long before either the Internet or the Web was born. Companies like IBM (1911), Hewlett-Packard (1939), Intel (1968), Microsoft (1975) and Cisco (1984) were all founded as startups. Apple (1976) was a true

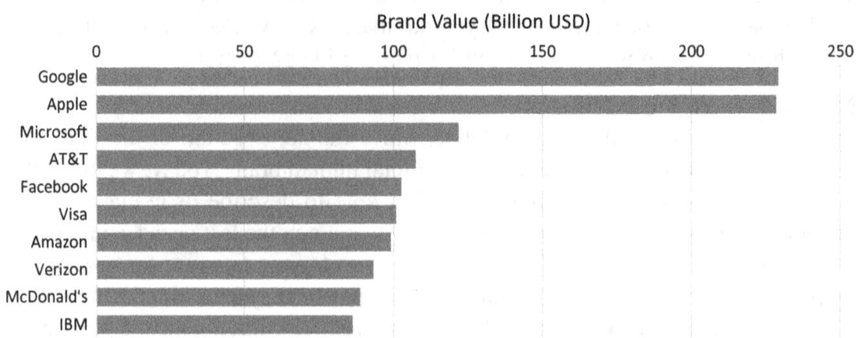

Fig. 2 The most valuable brands 2016 (Forbes list)

startup that created a market for home computers before it disrupted the economy a second time with the iPhone. Going back to the nineteenth century, enterprises like George Stephenson's first railway line, Graham Bell's telephone company or Edison's electric light company were early examples of successful combinations of disruptive technology and financial investments.

Startup Culture

The startup culture has its roots in the last decade of the twentieth century. With the rise of the Web, startups became a rapidly growing phenomenon and created a new culture also called the "New Economy." The investment in new economy companies reached its first peak in 1999 before the burst of the dot-com bubble. At the same time, the "old economy" IT companies had to accept major losses in market shares and the stagnation or shrinking of their revenue. The expectation from investors was clear: generate new businesses based on new technologies and disrupt existing markets to create a rapidly growing company value.

Since then, the startup movement has developed into a well-structured ecosystem with the mission of transforming innovation into an industry and consisting of established partners that work together in an innovation nexus. The major partners in that ecosystems are founders and investors. They share the target of growing company value within a short time.

Startup Life Cycle

Building and investing in startups became a huge business during the last few decades. With the beginning of the Internet and Web business, a comprehensive nexus was created around startups and developed into a structured ecosystem. Today, the basic approach of startup development is a sequence of business development steps defined by investment rounds or series and triggered by a—hopefully—successful execution of the startup's roadmap.

The initial step is performed by the founder's team in developing their business idea. In this phase, support may come from so-called business angels helping to build the business plan, providing the first, initial funding or also delivering their expertise in exchange for "sweat" equity. Once the first product and business idea is on the table, a seed-financing round can be started to develop an initial technical proof of the concept by building a prototype or also by securing test customers or users. If successful, a sequence of investment rounds may follow to expand the business and move towards a sustainable revenue and profit generation situation (see Fig. 3).

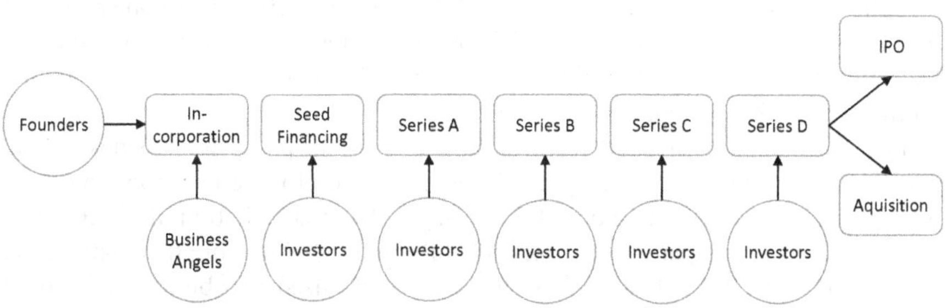

Fig. 3 Startup life cycle

Table 1 Facebook investment rounds

Date	Series	Valuation/market cap ($ million)	Investment round ($ million)	Investor
2004, Sep	Angel	5.0	0.5	Peter Thiel
2005, May	Series A	100.0	12.7	Accel Partners
2006, Apr	Series B	1000.0	27.5	Meritech Capital Partners
2007, Nov	Series C	15,000.0	60.0	Horizons Ventures
2007, Oct	Series C	15,000.0	240.0	Microsoft
2008, Jan	Series C	15,000.0	15.0	European Founders Fund
2008, Mar	Series C	15,000.0	60.0	Horizons Ventures
2008, May	Debt Financing	15,000.0	100.0	TriplePoint Capital
2009, May	Series D	10,000.0	200.0	DST Global
2010, Jun	Secondary Market	41,000.0	210.0	Elevation Partners
2011, Jan	Private Equity	50,000.0	1500.0	DST Global, Goldman Sachs
2012, May	IPO	106,000.0	16,000.0	IPO

The final target of startups is to go public by preparing an initial public offer or getting acquired for a fair price by another company.

Facebook as Role Model

One of the legendary examples of a new business startup is Facebook and even today it serves as a role model for a successful launch of a new and disruptive business model. Since the initial creation of the business idea by Mark Zuckerberg in 2004, Facebook went through 11 investments steps between 2004 and 2012 and finally recorded its public offering in May 2012 reaching a record IPO valuation of $106 billion (see Table 1).

Following the track record of Facebook beginning in 2004, it becomes evident that revenue and profit are not the major motivations for investors in startups. Facebook first created significant revenues no earlier than 2007 and drew its first profits after 2009 (see Fig. 4).

In 2006, Facebook started to offer so-called group pages for businesses and gained 100,000 business users by 2007.The motivation for the Facebook investors was the rapidly growing number of active users that reached 500 million in 2010, thus growing the company value from $10 billion in 2009 to more than $40 billion in 2010. In May 2012, Facebook went public and raised $16 billion based on a

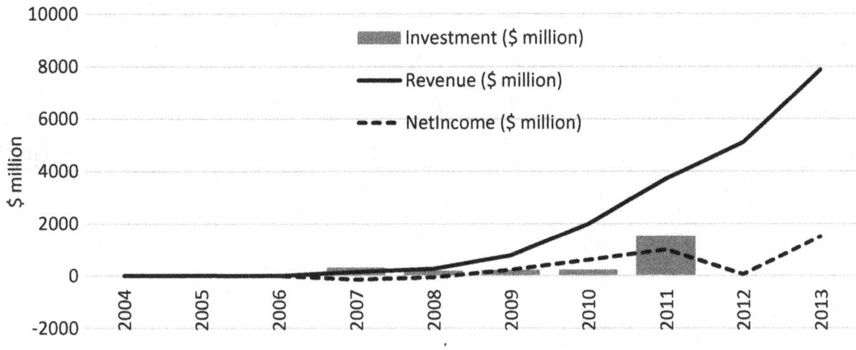

Fig. 4 Facebook investment history

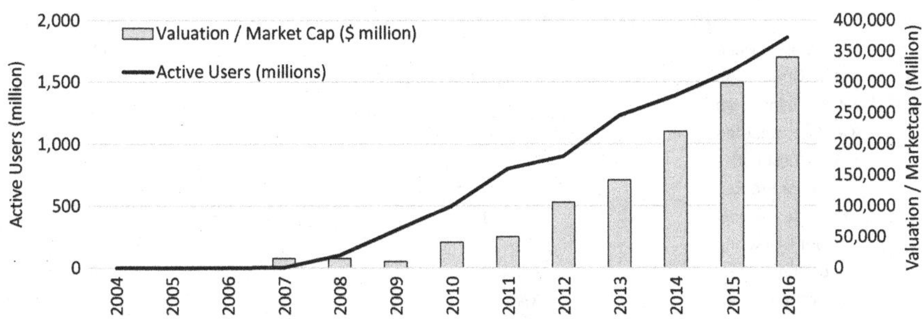

Fig. 5 Facebook user and company valuation

company value of $106 billion.[2] Today, Facebook's revenue is more than $27.6 billion (2016) coming mostly from advertising. The company generates a net income of more than $10 billion (2016), and its market cap is more than $340 billion (see Fig. 5).

Risks and Obstacles

Nevertheless, building or investing in startups is still accompanied by risk and obstacles. Today, the worldwide number of new enterprises claiming to be startups is huge. At the same time, it is also true that failing is a typical element of the startup culture. Betting on a new business idea is combined with a high level of risk. According to a CBInsights report[3]: out of 100 startups reaching the status of

[2]Facebook IPO: http://dealbook.nytimes.com/2012/05/17/facebook-raises-16-billion-in-i-p-o/?hp&_r=0

[3]CBInsights, 2016: https://www.cbinsights.com/research-corporate-innovation-trends retrieved 2016-10-10.

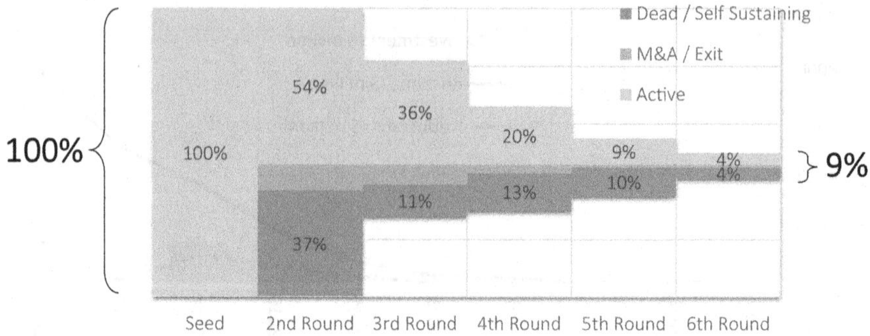

Fig. 6 End state for startups

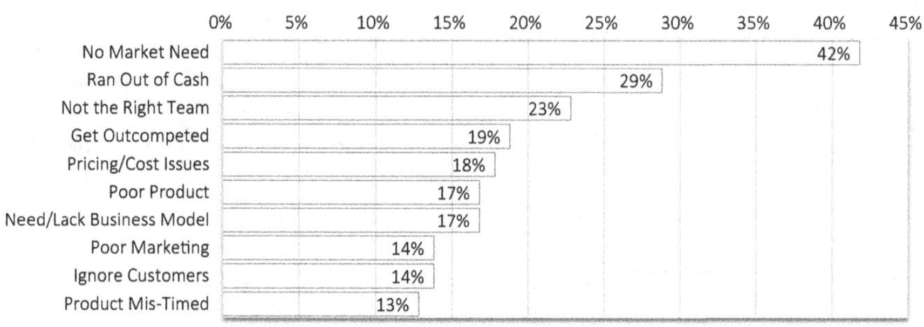

Fig. 7 Top ten reasons why startups fail

receiving a first seed investment, only 9% succeed either in getting acquired or in self-sustaining (see Fig. 6).

Especially in the US, the culture of failure is an important assumption for success. Having learned from failure is recognized as advantageous for newcomers starting a business. The saying: "fail fast, fail early" is known as the principle for surviving a startup's life cycle. According to CBInsight analysis, the most frequent reasons for failure are not only a lack of funding but also failures related to the product idea (no market need, competition, pricing, poor product), the product team and also the missing concentration on essential business tasks (business model, marketing, customers) (see Fig. 7).

Simultaneously, the technical and economic environment for creating new businesses has never been as inviting as today.

Internet and Web as Drivers

Building a new product based on the Internet, Web or cloud architectures has never been as inexpensive as today. The infrastructure costs for development and production dropped dramatically in the first decade of the twenty-first century. In 2000, the cost to launch an IT startup was estimated at $5 million or more. With the wide availability of open source software tools and low price structure as a service offers, the initial cost for tools and infrastructure dropped by the factor of 100 by 2010 and today is estimated to be at or below $10,000. Building a new product still needs creativity but, more than that, passion and time. The barrier for entry in terms of initial investments has never been as low as it is today (see Fig. 8).

The Right Moment to Accelerate

Creating a new and possibly disruptive business idea is not only a question of creativity and technical talent. The track records of successful startups show clearly that timing plays an essential role. Facebook was not the first social network; there had been Friendster, LinkedIn, Xing and others before it. Google was not the first search engine and Amazon had predecessors in e-commerce. Choosing the right moment and the right go-to-market approach were crucial elements in the success of Facebook, Google, Amazon and others. Successful startups, that address the consumer markets, have been successful in choosing the perfect window of opportunity. The typical measure of success is the number of users or customers reached within the first years of the new service or offer. These numbers have dramatically changed since the first network based service was launched (see Table 2).

Today, successful network based services reach significant user numbers within no more than 1 or 2 years. The window of opportunity is small and closes rapidly.

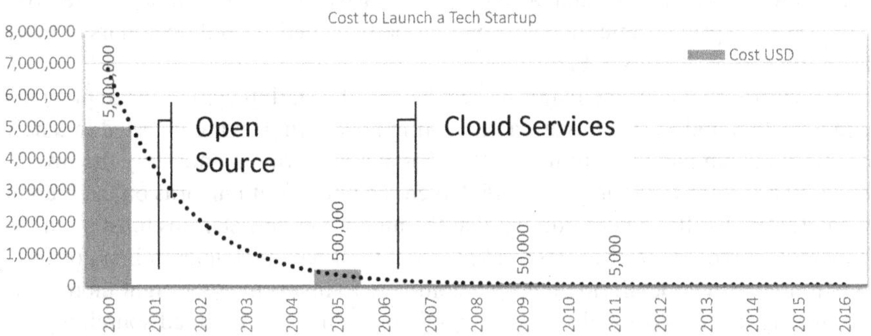

Fig. 8 Barrier to entry for startups

Table 2 Years to reach 100 million users

Product or service	Start	Years to reach 100 million users
Telephone	1870	90
Cars	1885	56
Credit cards	1950	25
Mobile phones	1985	15
Web users	1991	6
Online banking	1994	10
Facebook	2004	4
Smart phones	2007	4
Whatsapp	2010	2.7

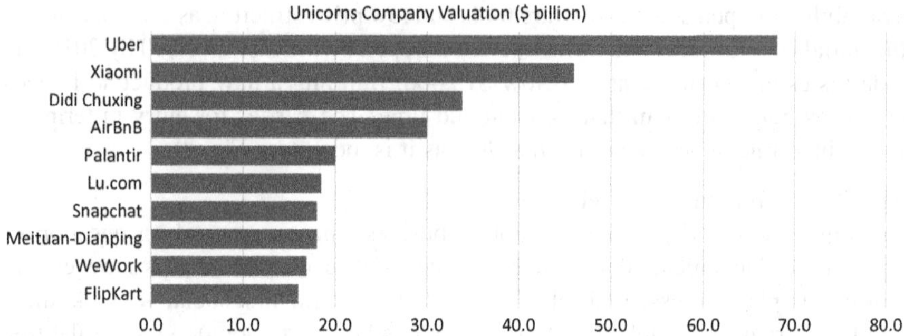

Fig. 9 Company valuation of private startups 2016

For each successful and disruptive business, there are hundreds or thousands that came out too early or missed the train and were too late.

Hunting Unicorns

Since the restarts of the new enterprise business in the first decade of the new century the number of new businesses and companies entering the technology market is vast, many of them struggling for success, but some of them reaching an impressive level of value or market acceptance within a very short timeframe. Those are called unicorns and, as the name suggests, they are rare but impressive. There are 174 private unicorn companies globally valued at a billion dollars or more as of the end of 2016 (see Fig. 9).[4]

Unicorns are one of the major drivers behind the startup culture. Entrepreneurs and investors find their motivation in creating a unicorn. Having invested in a new economy, a company that turns out to become a unicorn may multiply the investment by a factor of 100 or even more. Given the number of unicorns on the market today (around 170) and considering that the number of new startups must be in the hundreds of thousands, the statistical probability of having a unicorn in a portfolio is extremely low. Identifying the right company to invest in at the right moment is the artistry of investors and venture capital companies and is called sourcing.

Successful sourcing is the starting point of investing and a complex and time-consuming process for investors. Since the beginning of startup culture and the creation of the startup ecosystem, a variety of organizational approaches have been developed to support the building of new startups. These partners in the startup ecosystem play the role of incubators or accelerators of startups and find their mission in providing support and consultancy for startups in exchange for equity, money or know-how.

[4]CBInsight, 2016: https://www.cbinsights.com/blog/startup-unicorns-top-countries/?utm_source=CB+Insights+Newsletter&utm_campaign=f3871505f6-Top_Research_Briefs_11_19_2016&utm_medium=email&utm_term=0_9dc0513989-f3871505f6-87759901 retrieved 2017-01-20.

Incubators and Accelerators

In parallel to venture capital firms, a nexus of supporting organizations has been created during the last decade to provide platforms or breeding areas for startups. Those organizations find their interest in not only supporting startups in their initial phase but also in identifying potential investment opportunities and sharing or exchanging knowledge to create new business opportunities. Incubators are organizations funded by universities, research organizations or corporations to provide infrastructure like office space and access to tools and knowledge. Incubators address startups in their initial phase of creating new businesses. They focus on validating and developing an idea until it becomes a business proposition in exchange for equity.

Accelerators focus on slightly later stage startups and target those with a prototype or a beta version of their product or service. Typical accelerators are organized as a program that guides the startups through a structured process helping to develop proof-of-concept projects and connecting the startup to potential business partners or early customers. Accelerator programs are organized as corporate accelerators by corporations to maintain a closer relationship to new technologies and business ideas, or by investment firms to create a network and closer relationship to future investment opportunities. In the best case, the aims of a successful accelerator program are a first investment, first customer projects or partnerships. The number of accelerator programs offered is vast; startups usually must apply and go through a selection process.

Surviving the First Years

Depending on the road map and the product and service created by a startup, a sequence of additional investments is necessary to grow the business. In this phase, venture capital companies, corporate venture capital organizations or private equity all come into play. Analyzing the track record of successful startups like Facebook, PayPal, Google, Amazon or others, a startup may need up to four investment rounds to reach the final goal of having created a profitable, sustainable and rapidly growing business. In this early phase, the success in the market has to reflect the increasing value of the new company and is the essential parameter for the volume of fresh funds received from investors in relation to the percentage of equity that has to be spent for that. The motivation for investors is to create a high return on their investment within a short time span of typically 3–5 years. From the investor's perspective multiples of 5 years or, even better, 10 years are typical investment goals.

At the end of this business development phase, an initial public offering or an acquisition by another company should be the end goal. In some cases, the startup remains privately held like Uber, Palantir or Qualcomm.

Initial Public Offering

Company valuation is a tricky thing. At a certain point in time the company may go public by offering shares at the stock exchange. This is when buyers of shares have the chance to set the price via buying the shares or not. Today, the e-commerce platform Alibaba leads the list followed by Facebook, Google and Twitter (see Fig. 10).

Mergers and Acquisitions

Getting acquired by another company is also an option for startups. Many acquisitions are executed each year by large Internet and IT companies. Google, HP, IBM, Oracle and others develop their businesses, markets and products not only through internal development but also through the acquisition of adjacent companies (see Tables 3 and 4).

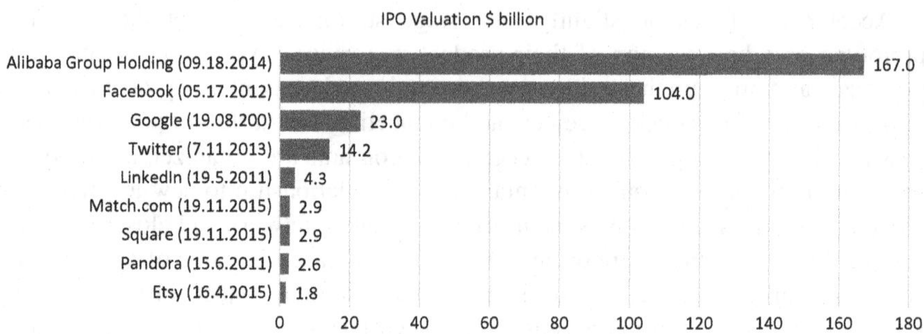

Fig. 10 IPO valuation of the largest IPOs

Table 3 Large IT acquisitions from 2000 to 2016

Year	Company	Acquired by	Value ($ billion)
2016	LinkedIn	Microsoft	26.2
2016	ARM	SoftBank	32.2
2016	Yahoo	Verizon's	4.8
2016	NetSuite	Oracle	9.3
2015	Altera	Intel	16.7
2015	EMC	Dell	67.0
2014	WhatsApp	Facebook	19.0
2013	Softlayer	IBM	2.0
2011	Skype	Microsoft	8.5
2011	Motorola	Google (Alphabet)	12.5
2011	Autonomy	HP	10.3
2010	Sun Microsystems	Oracle	5.6
2008	EDS	HP	15.4
2005	Veritas	Symantec	13.5
2005	PeopleSoft	Oracle	10.3
2002	PayPal	eBay	1.5
2001	Compaq	HP	33.6
2000	Time Warner	AOL	106.0

Table 4 Number of acquisitions by large IT and internet companies

Company	Acquisitions
Amazon	68
Apple	83
Cisco	184
Facebook	61
Google (Alphabet)	200
IBM	94
Microsoft	200
Oracle	120
Twitter	54
Yahoo	114

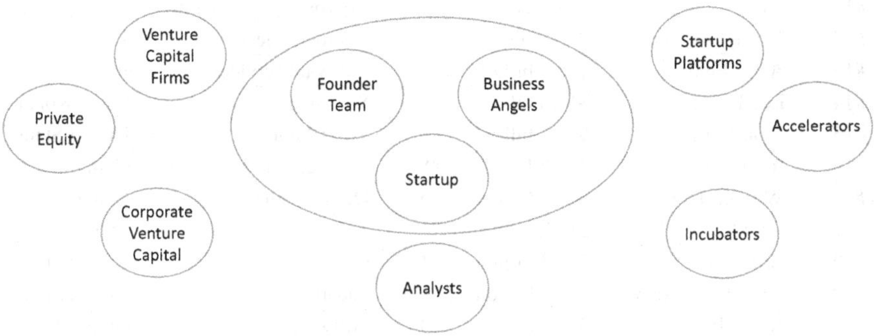

Fig. 11 Innovation ecosystem

The Innovation Ecosystem

Innovation became an ecosystem of its own during the last decades. The innovation ecosystem is grouped around startups and their founders and consists of the nexus of investors (venture capital firms, corporate venture capital, private equity), a supportive structure of accelerators and incubators and various international or global business analysts and business research organizations (see Fig. 11).

Founders

The initial causes of the creation of a disruptive business idea are the founders of a new business. The startup culture developed a clear understanding about the role and mission of founders. They must combine a disruptive business idea with the motivation to "change the world" to create a new market.

It seems to be evident that, in many cases, the founders of a new business are on a team of two or three people combining innovative thinking with the ability to execute and deliver. Typical examples are Apple's founders Steve Jobs and Steve

Table 5 The top 30 tech billionaires worldwide (2016)

Rank	Name	Net worth	Age	Origin of wealth	Country
#1	Bill Gates	$78.0 billion	61	Microsoft	United States
#2	Jeff Bezos	$66.2 billion	53	Amazon.com	United States
#3	Mark Zuckerberg	$54 billion	32	Facebook	United States
#4	Larry Ellison	$51.7 billion	72	Oracle	United States
#5	Larry Page	$39 billion	43	Google	United States
#6	Sergey Brin	$38.2 billion	43	Google	United States
#7	Steve Ballmer	$27.7 billion	60	Microsoft	United States
#8	Jack Ma	$25.8 billion	52	Alibaba	China
#9	Ma Huateng	$22 billion	45	Internet services	China
#10	Michael Dell	$20 billion	52	Dell computers	United States
#11	Paul Allen	$18.5 billion	64	Microsoft, investments	United States
#12	Masayoshi Son	$17 billion	59	Internet, telecom	Japan
#13	Azim Premji	$16.1 billion	71	Software services	India
#14	Lee Kun-Hee	$13.6 billion	75	Samsung	South Korea
#15	Elon Musk	$12.7 billion	45	Tesla Motors	United States
#16	Robin Li	$12 billion	48	Internet search	China
#17	William Ding	$11.7 billion	45	Online games	China
#17	Shiv Nadar	$11.7 billion	71	Software services	India
#19	Eric Schmidt	$11.3 billion	61	Google	United States
#20	Dustin Moskovitz	$10.8 billion	32	Facebook	United States
#20	Hasso Plattner	$10.8 billion	73	Software	Germany
#22	Lei Jun	$9.8 billion	47	Smartphones	China
#23	Jan Koum	$8.8 billion	41	WhatsApp	United States
#24	James Goodnight	$8.7 billion	74	Software	United States
#24	Dietmar Hopp	$8.7 billion	76	Software	Germany
#26	Zhang Zhidong	$8.1 billion	45	Internet media	China
#27	Pierre Omidyar	$7.9 billion	49	eBay	United States
#28	Eduardo Saverin	$7.4 billion	34	Facebook	Brazil
#29	Hiroshi Mikitani	$7.3 billion	51	Online retail	Japan
#30	Gordon Moore	$7.1 billion	88	Intel	United States

Wozniak, Microsoft's founders Bill Gates and Paul Allen, Jeff Bezos from Amazon, Facebook founder Mark Zuckerberg, Oracle's Larry Ellison or Google's founders Larry Page and Sergey Brin. Many of the successful founders of IT and Internet businesses are "tech billionaires" and have a major influence on the Web and Internet business (see Table 5).

Investors

Finding investors to fill the role of business angels, to provide seed financing and for the successive funding rounds of a new business is vital for keeping the innovation ecosystem alive. Since 2008, the low interest rates have eased funding, money is

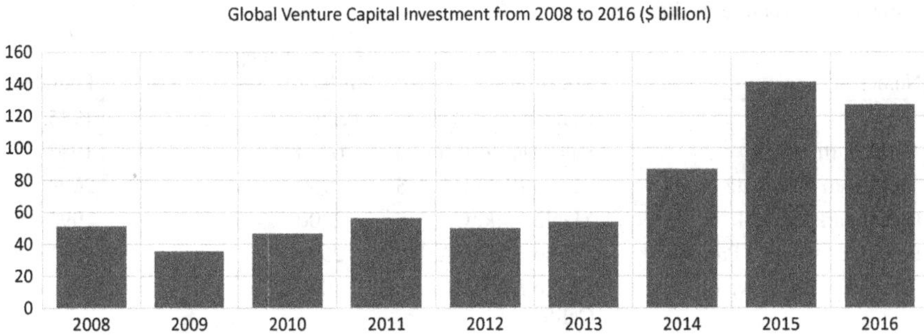

Fig. 12 Global venture capital investment

cheap now and it seems that that global trend will last for the next couple of years. Investment statistics show that the level of global venture capital investing has grown since 2008 and is still on a high level, having reached an all-time high of $140 billion in 2015 (see Fig. 12).

Nevertheless, willingness to invest in high-risk enterprises may change and drop, being temporarily influenced by the global economy. In 2016, criticism about the performance of VC funds was heard and the volume of new investments dropped in 2016 compared to 2015.[5]

Venture Capital Firms

Venture capital firms (VCs) manage funds and invest in new businesses. Many of these VCs have existed for decades and have already funded many startups. There are thousands of VCs worldwide operating in different markets and regions. The following list contains some of the most prominent names and their current estimated size of funds managed (see Table 6).

Corporate Venture Capital

Corporate venture capital is provided by companies interested in investing in new technology. The leading Internet companies have their own investment firms and manage an intensive investment strategy, among them Google, Intel, Comcast, Salesforce, Cisco and Qualcomm[6] (see Table 7).

Private Equity

Private equity firms manage the investment portfolios of private people or families. The top 20 firms are in the US[7] (see Table 8).

---

[5]The World Financial Review: Is Venture Capital in Crisis?: http://www.worldfinancialreview.com/?p=1563 retrieved 2016-10-12.

[6]CBInsight: https://www.cbinsights.com/blog/corporate-venture-capital-active-2014/ retrieved 2016-10-12.

[7]Top 20 Private Equity Firms: https://en.wikipedia.org/wiki/List_of_private_equity_firms#Largest_private_equity_firms_by_PE_capital_raised retrieved 2016-10-12.

Table 6 Top venture capital firms

Name	Location	Funds managed ($ million)	Founded
3i	London	12,700	1945
Accel Partners	Palo Alto, California	6000	1983
Andreesen Horowitz	Menlo Park, CA, US	950	2009
Benchmark Capital	Menlo Park, CA, US	2300	1995
Bessemer Venture Partners	Menlo Park, CA, US	2500	1911
Draper Fisher Curvaton	Menlo Park, California	5000	1985
Greylock Partners	Cambridge, MA, US	2000	1965
IDG Ventures	San Francisco	6800	1995
Insight Venture Partners	New York City	7600	1995
Institutional Venture partners	Menlo Park, CA, US	2900	1980
Intellectual Ventures	Bellevue, Washington	5000	2000
Kleiner Perkins Caufiled & Byers	Menlo Park, CA, US	1500	1972
New Enterprise Associates	Menlo Park, California	17,000	1978
Norwest Venture Partners	Palo Alto, California	6000	1961
Oak Investment Partners	Westport, Connecticut	8400	1978
Sequoia Partners	Menlo Park, CA, US	4000	1972

Table 7 Ranking of the top corporate venture capital organizations

Rank	Investor
1	Google Ventures
2	Intel Capital
3	Comcast Ventures
4	Salesforce Ventures
5	Cisco Investments
6	GE Ventures
7	Qualcomm Ventures
7	Pfizer Venture Investments
9	Bloomberg Beta
9	CyberAgent Ventures

Regions and Hotspots

Although the Internet, Web and cloud business is a global phenomenon, the startups and innovation ecosystem is concentrated around a few hotspots most of them on the east and west coasts of the US. Between 2010 and 2016, the volume of venture capital investments has grown from $40 billion to more than $120 billion in 2016. The geographic distribution in 2010 shows a clear focus on North America with

Table 8 Largest private equity firms by capital raised

Firm	Headquarters	Capital raised ($ million)
The Carlyle Group	Washington D.C.	30,650.33
Kohlberg Kravis Roberts	New York City	27,182.33
The Blackstone Group	New York City	24,639.84
Apollo Global Management	New York City	22,298.02
TPG	San Francisco	18,782.59
CVC Capital Partners	Luxembourg	18,082.35
General Atlantic	New York City	16,600.00
Ares Management	Los Angeles	14,113.58
Clayton Dubilier & Rice	New York City	13,505.00
Advent International	Boston	13,228.09
EnCap Investments	Houston	12,400.20
Goldman Sachs Principal Investment Area	New York City	12,343.32
Warburg Pincus	New York City	11,213.00
Silver Lake	Menlo Park	10,986.40
Riverstone Holdings	New York City	10,384.26
Oaktree Capital Management	Los Angeles	10,147.28
Onex	Toronto	10,097.21
Ardian (formerly AXA Private Equity)	Paris	9805.25
Lone Star Funds	Dallas	9731.81

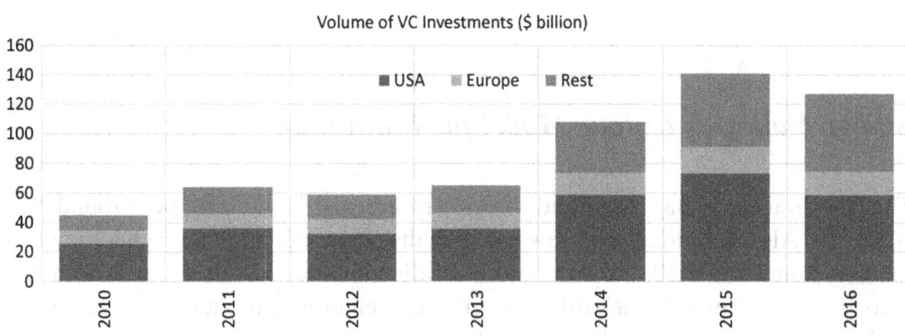

Fig. 13 Development of worldwide venture capital investment

more than 50% of the worldwide investment volume on the east and west coasts of the US. The balance has changed over the following years but the USA still leads, with Europe far behind and Asia catching up (see Fig. 13).

The KPMG Venture Pulse Report from 2016 estimates the total venture capital investment volume in Europe at $16 billion, Asia at $52 billion and America at $59 billion.[8] Within America, the hotspots are the New England region around the

[8]KPMG Venture Pulse Report 2016 Q4: https://home.kpmg.com/at/de/home/insights/2017/01/venture-pulse-q4-report.html retrieved 2017-01-12.

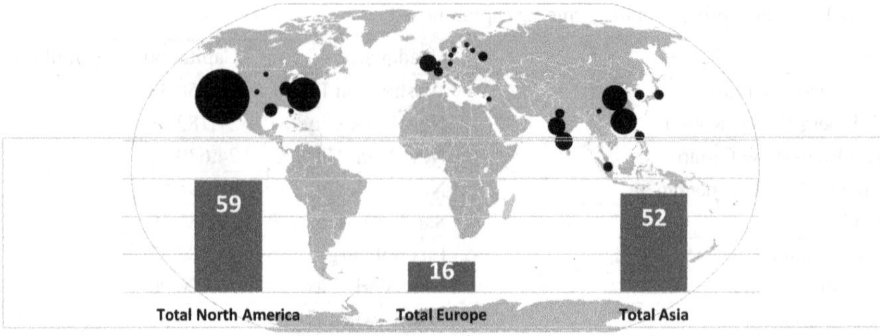

Fig. 14 Distribution of venture capital investments in 2016

traditional technology business areas between Boston, New York and Washington, DC and the west coast with Silicon Valley and the San Francisco and Seattle areas (see Fig. 14).

East Asia is catching up based on the concentration of electronic manufacturing companies in China and South Korea and through the rapid development of local Internet and Web providers and services.

Unfortunately, Europe is definitely not in the lead when it comes to innovative new ventures and this is something that could change dramatically in the near future.

Silicon Valley as Unique Model for Innovation

The San Francisco Bay Area and the electronics and IT melting pot around San Jose, Palo Alto and Mountain View has been the center of computer technology and industry since the 1950s. The tradition of Silicon Valley has its roots in Stanford University and Hewlett-Packard as the first large enterprise founded on the grounds of the university in 1939. The name Silicon Valley was coined in the 1960s when companies like Intel started to develop and to produce the first semiconductor based integrated circuits. Today, 10 of the top 15 IT and Internet companies have their roots and headquarters in California, most of them in Silicon Valley (see Table 9).

The ecosystem of the "Valley" has a long tradition and consists of endless lists of computer companies that have been founded there since the 1960s. This innovative nexus is based on several important universities and research organizations like Stanford University, but also the legendary Palo Alto Research Center PARC and Berkley University. An impressive number of founders were students at Stanford among them the founders of HP (William Hewlett and David Packard), Cisco (Leonard Bosack), Sun Microsystems (Andy Bechtoldsheim), Yahoo! (Jerry Yang and David Filo), Netflix (Reed Hastings), Paypal (Peter Thiel and Ken Howery), Google (Sergey Brin and Larry Page) and many others. Stanford is the only US university besides MIT with the largest number of entrepreneurs.

Table 9 Location of the top 15 IT and internet companies

Company	Founded	Country	Region	Revenue ($ million) 2015
Apple	1976	USA	Silicon Valley	233,700.0
Microsoft	1975	USA	Seattle	93,580.0
Intl. Business Machines	1911	USA	New York	81,740.0
Google (Alphabet)	1998	USA	Silicon Valley	74,990.0
Baidu	2000	China	Beijing	66,382.0
Dell Technologies	1984	USA	Texas	59,000.0
Intel	1968	USA	Silicon Valley	55,400.0
Hewlett Packard Enterprise	1939	USA	Silicon Valley	52,107.0
HP Inc.	1939	USA	Silicon Valley	51,463.0
Cisco Systems	1984	USA	Silicon Valley	49,100.0
Oracle	1977	USA	Silicon Valley	38,230.0
EMC	1979	USA	Massachusetts	25,700.0
Qualcomm	1985	USA	California	25,281.0
SAP	1972	Germany	Waldorf	22,588.0
Facebook	2004	USA	Silicon Valley	17,928.0

At the same time as the growth of the computer and semiconductor industry in Silicon Valley, many venture capital firms decided to follow the trend and settled next to their investments. In 1972, the investment firm Kleiner Perkins Caufield & Byers moved to Sand Hill Road next to Stanford University. Several other venture capital companies followed and Sand Hill Road became the top address for entrepreneurs seeking funds. Today, 10 out of the top 15 venture capital companies have their headquarters around Stanford University and provide services to the growing number of founders blossoming on the fertile grounds of the Valley.

The environment of Stanford University around Palo Alto was also the playground for the first generation of computer wizards. The Homebrew Computer Club in Menlo Park was, along with MIT's Tech Model Railroad Club, one of the hot spots for the pioneers of personal computing in the 1970s.

Today, Silicon Valley is still the focal point for Internet and Web companies. The unique culture of innovation, creativity and disruptive thinking evolved through decades of successful cooperation between universities, research facilities, corporations and startups. What we call "startup culture" today was invented in the Valley, and all attempts to copy that pattern in other regions of the world ended up being a weaker version of the original.

Innovation's Effects

Innovation and the innovation ecosystem is not merely a value-creating machine, it has tremendous impacts on the global economy. As we have discussed at the beginning of this chapter, Peter Schumpeter claimed that innovation is the driving force behind economic growth.

New Jobs

As result of the new economy that built the Internet, Web and cloud as network based service ecosystems, we see the "old economy" business models losing ground and traction. Web and cloud based services trigger change and transition in nearly all industrial segments. Commerce is moving from retail shops to e-commerce, news and media is distributed via the net, advertising is going digital, industrial production methods are automated using the Internet-of-Things and cognitive processes, resources like cars or apartments are shared via web platforms, and personal infrastructure is evolving into smart homes and autonomous cars. This transition, together with the economic crisis of the last decade, led to the loss of traditional jobs. At the same time, new companies and business have started to create new jobs. A report from the Kaufmann Foundation claims that the number of new jobs created by startups is higher than the loss of jobs in the old economy.[9]

Quick Success or Fail

The principle of "fail fast, fail early" applied to the startup culture, leads to the quick creation of new jobs and businesses typically in the first year of a new company. If unsuccessful, the new business is stopped, immediately liberating capital and human resources to be put to more productive uses. Compared to old companies, this has led to a completely new working environment where short-term engagements are offered instead of long-standing employment at a certain company.

Innovation Accelerates Productivity

Productivity and growth of GDP can be improved by increasing inputs of labor and capital. In an expanded model, the "Total Factor Productivity" is added as a third parameter to calculate the influence of higher productivity through better technical processes and methods. Technological growth and efficiency are regarded as two of the biggest sub-sections of Total Factor Productivity. Innovations as the major driver for technical progress directly influence economic productivity. The economic growth of countries like China by more than 10% between 2004 and 2008 is seen as an effect of a much higher total factor productivity through the introduction of efficient production methods.

[9]By Jason Wiens and Chris Jackson, Kaufmann Foundation, 2015: http://www.kauffman.org/what-we-do/resources/entrepreneurship-policy-digest/the-importance-of-young-firms-for-economic-growth retrieved 2017-01-12.

Delayed Effect on Economy

At the same time, it is also true that innovation does not have a short-term effect on economy. The introduction of new and sometimes disruptive production methods requires the sometimes-revolutionary change of existing production and distribution processes. Replacing an existing retail shop infrastructure by e-commerce is a game changer for production and logistic processes. Moving from self-owned and self-driven cars to shared autonomous vehicles needs a major transition on the part of the whole automotive industry. The application of smart agents for customer support could improve customer satisfaction, but requires a completely new type of customer relationship.

Reference

Thiel, P. (2014). *Zero to one*. London: Penguin Random House.

Security and Privacy Challenges

If all that Americans want is security, they can go to prison. They'll have enough to eat, a bed and a roof over their heads. But if an American wants to preserve his dignity and his equality as a human being, he must not bow his neck to any dictatorial government.

Dwight D. Eisenhower.

All human beings have three lives: public, private, and secret.

Gabriel García Márquez: A Life.

Security is a major threat when using any kind of IT related processes. Security is also one of the most rapidly changing fields of IT. It is no question that global communication based on Internet technology has opened a wide range of security related issues, many of them appearing suddenly, and are hard to predict and not easy to handle even with precautions.

During the past decades, major efforts were undertaken to reduce the risk of damage and to build security technologies for the Internet, Web and cloud services. The fight between attackers of private data and the owners or users of this data have gone on for centuries. Security and privacy of information have been concerns since people started to use script and write letters. The growth of the Internet and the Web lead to many new threats and allow new types of criminal organizations to evolve using the global Web to develop destructive criminal businesses.

Good, Bad and Ugly

A cloud service's data security and privacy is based on a set of fundamental characteristics that distinguish "good" from "bad" data traffic and data usage. A "good" message has to be confident, integer, authentic and non-reputable. Data stored should be accessible and used only by clearly defined people or organizations. These principles are not new but became more important with the use of electronic data transport and storage.

M. Oppitz, P. Tomsu, *Inventing the Cloud Century*,
DOI 10.1007/978-3-319-61161-7_14

Fig. 1 Message confidentiality

Fig. 2 Message integrity

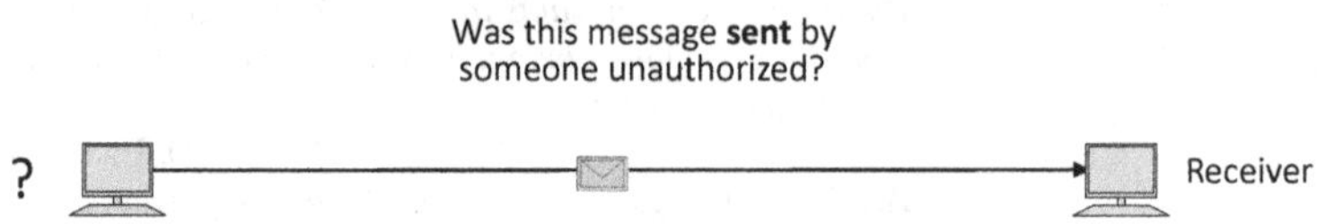

Fig. 3 Message authenticity

Message Confidentiality
Messages are confidential when access is only possible by authorized persons, and these messages cannot be seen by unauthorized parties. Both sides, the sender and the receiver of a message can be sure that the message is a secret between them (see Fig. 1).

Message Integrity
Integrity of data and messages guarantees that the data was not altered somewhere between the sender and the receiver. Again, both sides can be sure that the information sent is the same as the information received (see Fig. 2).

Message Authenticity
Authenticity of messages guarantees for the receiver that an authorized party sent the data. Spam mails or faked emails are typical examples of non-authentic messages (see Fig. 3).

Message Non-Repudiation
Non-repudiation means that a message received is clearly referenced to the sender and the sender cannot repudiate, that he created and sent the message (see Fig. 4).

Data and Service Privacy
Privacy of messages and data is based on the clear definition of a trust zone confined by a trust border. This trust border can guarantee that information exchanged and shared will not leave the trust zone (see Fig. 5).

Trust borders are not easy to define and may be organization borders but could also be geographic borders. Data privacy is the key term for the definition of legal rules and regulations and has become a huge field of discussion with the rise of

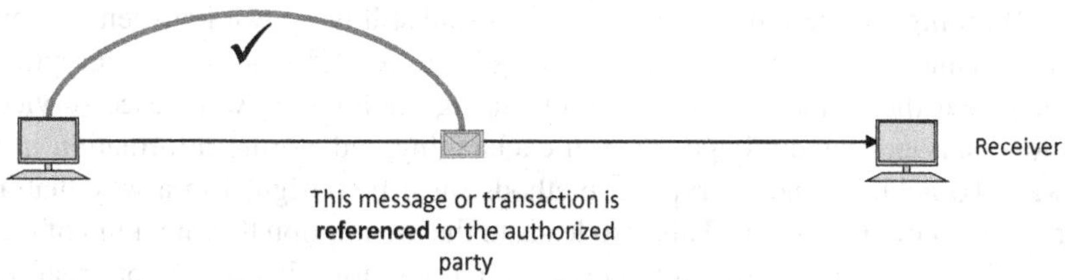

Fig. 4 Non-repudiation

Fig. 5 Data and service privacy

social media platforms, cloud services for company data and the handling of personal data by corporations and federal administrations.

The principles described above have been the principles of secure communication and privacy since people started to exchange written messages. They can be applied to letters, telegrams, telephone calls as well as to data exchange using the Internet and the Web. Sharing secrets in private communication or within a group of individuals has been a part of social, political and economic activity for centuries. Trying to get access to private or secret data has always been a major challenge for groups outside the intended communicators. The motivations behind keeping something a secret or trying to uncover secrets have not changed much over the last few centuries.

Before we analyze the role and status of security and data privacy in today's environment of Web and cloud services, we will follow the history of secret messages and attempts to keep data exchange private through the last centuries.

A Short History of Private Communication: Secret Messages

Safe encryption of information is one of the major obstacles when using information networks. Since the beginning of information transport, the successful encryption of data exchanged between the sender and the receiver has been one of the most contested areas. The question is simple: how can a message been transported between two persons so that nobody else can read it. Today, encryption methods like RSA (Rivest, Shamir und Adleman) and PGP (Pretty Good Privacy) are used. These methods are based on so-called public key encryption and have their roots in number theory and algorithms that are based on very large prime numbers.

Building encryption tools was always—and still is—a war between new encryption technologies and codebreakers trying to successfully break the encryption. It is also clear that persons or organizations using the Internet, web based services and cloud services have a huge stake in exchanging and storing information in a safe way. To achieve that encryption methods must be designed in a way that a third party cannot break them. This also leads to the assumption that any kind of message or information can be kept totally secret, not only for other people or organizations but also for government or intelligence organizations. Taking into consideration that today's web and cloud services are also widely used by criminals or terrorist groups leads to an enormous conflict of interest between those who simply want to protect their company or personal data and those public organizations who feel responsible for the security of countries and societies. The simple fact that strong encryption methods are available to everybody today—individuals, companies, public organizations, but also criminals and terrorists, makes it impossible to draw a border between "good" and "bad" encryption usage. The basic question behind the state of the war between encryption and codebreakers is and always was the same: who is leading now? We will see that this question cannot be answered in today's world.

50 BC: Weak Encryption

The first attempts at the successful encryption of text go back to Julius Cesar. He used a simple method, which is based on the idea of shifting the letters in a text. He shifted them by 2 and wrote C instead of A, D instead of B and so on. In this case, the receiver only has to reverse the shifting to read the text. This simple encryption method is still called "Ceasar-encryption." It is not a very strong encryption method, and breaking that code is a matter of minutes. Later, the idea came about to use not only a simple shift but also to exchange each character with another one based on a different order of the letters in the alphabet called the "key-alphabet." In order to decrypt such a text, you must know the key-alphabet. That seemed to make more sense but there is an easy method to break that code. Each language has a certain frequency of letters. In the English language, the most frequent letters are e, t, o, i, n. Knowing that, one has only to count the letters in the encrypted text and the can easily start to decode the message stepwise by finding letter combinations and words that make sense.

Until the eighteenth century, these so-called one-alphabet encryption methods were cutting edge in encryption technology and during this time the codebreakers were always in favor. In 1586, the usage of a mono-alphabetic encryption led to a dramatic historical event. The Scottish Queen Mary Stuart kept under arrest by the English Queen Elizabeth I was approached by the English nobleman Anthony Babington who pledged to Mary Stuart not only to assist in her escape but also to help her to take the thr one of England. Babington was writing letters to Mary using a mono-alphabetic encryption. At the same time, Sir Francis Walsingham began to organize the first English secret service organization. He was aware of the importance of intelligence and encryption and hired Thomas Phelippes, one of the best crypto analysts in Europe, as his private secretary. Phelippes was able to decipher the complete correspondence between Mary Stuart and the conspirators led by Anthony Babington. The conspiracy was discovered, the conspirators were executed and Mary Stuart was sentenced to death by Elizabeth I. Mary Stuart was beheaded on February 8th, 1587. It is a fact that if Babington had used a stronger encryption method English history would have been different. It is also a fact that at that point in time a stronger encryption method was already made available by Phelippes and Walsingham.

1550 Success for the Encrypters
For many centuries, the simple mono-alphabetic encryption was adequate for secret communication. With the establishment of intelligence organizations—the so-called "black chambers" by the governments of many European countries, the code war seemed to be lost for the encrypters. It was the Frenchman, Blaise de Vigenere, who, around 1550, found a better solution by using a poly-alphabetic encryption. He simply extended the current method by using not one alphabet, but 26 different alphabets for each letter of the alphabet. To encode a text, a table consisting of 26 columns and 26 lines containing 26 different coding alphabets is used. For decoding and encoding text, one must use the so-called Vigenere Square or "tabula recta" together with a keyword. The keyword defines which of the 26 alphabets was used for the letters in the message. Introducing the concept of poly-alphabetic encryption and keywords was a major step in encryption.

For the next few centuries, decryption was impossible if the Vigenere-method was used. I took 300 years until Charles Babbage developed a method to decipher poly-alphabetic encrypted messages. His approach was to look for repeated sequences of letters in the encrypted text. The probability that these parts of the text had been encrypted by the same part of the keyword is very high, which also is a good estimation of the length of the keyword. Knowing the length of the keyword is then used to split the task into single tasks of decryption, which are much easier to break using a frequency analyses. This also leads one to the assumption that the best shield against decryption is a long—or very long—keyword. This can easily be achieved by using another long-text as the keyword for the poly-morphic encryption. Expanding that idea, led to the approach of using a new text or keyword for each message. This method is called one-time-pad encryption and provides a strong encryption. Without having that "one key per message" at hand, it is impossible to decipher a message. There is only one huge disadvantage. If an organization wants to use one-time-pad encryption, you must organize the distribution of the one-time-pads between all senders and prospective receivers. This is easy if only a small group of people exchange messages. In large organizations, however, like companies or armies it would mean distributing tons of paper along with the process of when and how to use this method. Not very efficient.

1900 Encryption to Be Kept a Secret
After Charles Babbage found a method to decipher Vigenere encrypted messages, the score was again on the side of the codebreakers. At the beginning of the twentieth century, communication technologies like the telegraph were already widely used by the industry and, by World War I, the armies. Safe communication

was essential. Several attempts were made to improve existing methods, most of them based on a combination of polymorphic translation and none of them really successful. One of the most famous encryption methods was the German ADFGVX-System introduced in March 1918. The talented French crypto analyst, Georges Painvin, cracked it within 3 months. The Germans did not know that and used ADFGVX for a long time thus creating huge advantages for the French without the Germans being aware of it. This was the typical effect at that time. Breaking a code was one thing, keeping that a secret much more important.

Machines for Encryption and Decryption

1925 Enigma

By the beginning of the twentieth century it was clear that requirements for cryptology had changed a lot: the telegraph and later wireless communication created a mass of messages transported through networks instead of single point-to-point communication. Finding an efficient but also safe way to encrypt information became a question of survival for armies and companies. It was the German, Arthur Scheribus, who started to build mechanical encryption and decryption devices. His "Enigma" was a machine that looked like a typewriter and could encrypt messages through a combination of mechanical and electrical functions.

It consisted of a keyboard for typing the text clearly, a set of 3 (later 4) coding wheels, and an array of light bulbs indicating the encrypted letters by pushing a key. Each wheel had 26 different positions, connected to the next wheel mechanically so that after every 26th movement, the next wheel was moved one position ahead. The wheels were also connected via electrical contacts. One the far-left side, a reflector led the current back through the wheels. By pushing a button on the keyboard, the current would flow through the wheels, was reflected by the reflector and found its way back through the position of the wheels to the array of light bulbs. Additionally, a switch panel on the front of the Enigma was used to patch letters. All you needed to encrypt or decrypt was the machine itself, the starting position of the 3 wheels and the match table for the front patch panel. Counting all possible combinations of the positions of the wheels and the patch panel comes to 10^{25} possible combinations. Enigma production started in 1925. Until the end of World War II, the German military bought over 30,000 Enigmas. The German army, navy and air force used the Enigma for most of their military communication network, believing in Enigmas unbreakable encryption method. This was wrong.

1940 Bletchley Park

At the beginning of World War II, the British intelligence branch was comparably small. One part of it was a group of crypto analysts called Room 40, and most of them were linguists. But time has come to introduce mathematics and logic to crypto analysis as well as mechanical computing. Seeing as the sheer volume of information communicated over German military wireless networks was a huge asset for tactical and strategic warfare, the British government founded the Government Code and Cypher School (GC&CS) as a part of MI6 (Military Intelligence, Department 6). The GC&CS was located at Bletchley Park and was one the most secret organizations in Britain for more than 30 years. By the end of World War II, more than 6000 people worked in Bletchley Park on the encryption methods used in the German military's communications.

Traffic Analysis and First Fake News

Intelligence became a major part of warfare. The intelligence organizations at that time used multiple approaches to collect information about the enemy's plans and activities. Because wireless became one of the major communication methods it was comparably easy to locate the geographic position of senders through radio direction finding and trigonometric measurement. Secondly, the volume of messages sent was analyzed. Combining the data led to knowledge about the enemy's movements, intensity of activities and could be used for predicting the enemy's plans and next steps. Today, volume analysis is also a widespread method of data analysis based on web-data and used by marketing and intelligence organizations. During WWII, the opposing side was aware of that approach and tried to create misleading messages and traffic volume from other geographic positions to mask their real intentions. We find the same idea behind fraud messages in today's social networks creating a false picture of popularity and usage of terms or persons.

Code Cracking Using Computing

Beyond traffic analysis, the major achievement of Bletchley Park was cracking the Enigma code to eavesdrop on the complete German military communication network. It was the English mathematician and computer pioneer, Alan Turing, who one was one of the drivers behind that success. The basics for that accomplishment had their roots in Poland. A group of Polish crypto analysts led by Marian Rejewski found a method to analyze and decipher Enigma encrypted messages in the late 1930s.

The approach of the Polish crypto analysts was based on the fact that no character is translated into the same character by the Enigma machine and on the fact that the prescribed procedure for Enigma encryption was the so call "Spruchschlüssel," which is a 3-letter word sent at the beginning of each message. These 3 letters must be used to change the position of the wheel a second time for the specific message. Rejiwski found a characteristic of Enigma encoded messages. Letters in the encrypted message can be ordered in chains of different lengths starting and ending with the same letter. These chains and their lengths are typical for specific combinations of the 3 wheels. Rejiwski worked for 1 year to build a catalogue of all the possible 105,456 wheel-starting positions and the matching chain lengths. Having that catalogue made it possible to identify the wheel position when the chain length in the encrypted message could be found. After the Germans started to alter the Enigma's basing settings, Rejiwski built an electro-mechanical machine called "Bomba" that made it possible to search for wheel positions automatically. This was the first time a type of computing device was used for crypto analysis. The code breakers had the advantage again. During Herman

Göring's visit in Warsaw in 1934, he had no idea that his complete wireless communication was captured and decrypted by the Polish team.

After 1939, the Polish group got in contact with the British. Both sides managed to deliver most of the Polish material and technical know-how to Bletchley Park, thus creating a major foundation for the later success of the British group. Bletchley Park faced a big challenge: encrypting all the different military networks within the German forces. Bletchley Park developed into a unique combination of part industrial organization built for crypto analysis and part brain trust for applied mathematics and logic. One of the darkest hours came when the Germans decided to add a forth wheel to the navy's Enigmas in order to improve encryption for the U-boat network in the Atlantic. It was a huge success when Bletchley Park was finally able to break the new code. Today, it is no question that the success of the codebreakers at Bletchley Park coupled with the fact that the Germans had no idea their messages were open for the enemy to read led to a reduction of the war by 1–2 years. Keeping the successful decryption of a code system a secret is a major part of the overall success of code breakers. Bletchley Park and the existence of the GC&CS were kept a secret by British authorities until the late 1970s, leading to the first publications about computer usage in WWII. David Kahn published "The Codebreakers," which was the first history of Bletchley Park and the role of computing in breaking the coded German communication network (Kahn, 1967).

Going Industrial: Standard Technology

With the application of computers to economic environments, the problem of safe data transfer and storage moved from military applications to industrial ones. This simply required a clear set of standardized methods without losing security and creating legal frameworks for establishing security regulations within ecosystems. In parallel, governmental intelligence organizations tried hard not to lose focus on encryption and decryption technologies. As we will see, these organizations still play a major role in the global information ecosystem. Starting in the early 1970s, several inventions enabled the development of security products and pushed the creation of legal frameworks in Europe and the US.

1975 Encryption Standards: DES and AES

With the growth of international electronic communication in the 1960s and 1970s, it became evident that encryption of data is major requirement for industry as well as for governmental organizations. Establishing this must be based on a common understanding of encryption methods and standards that are publicly available and accepted. The first attempt was DES (Data Encryption Standard) published on March 17th, 1975, in the US Federal Register. Its origins go back to the early 1970s. In 1972, the US National Bureau of Standards (now NIST National Institute of Standards) identified the need to establish a common encrypting standard. Consulting with the US NSA (National Security Agency) was part of the decision-making process. The decision, made in 1974, was based on the discussion with the NSA and includes encryption algorithms and a key length of 56 bits. IBM developed DES under the surveillance of the NSA evoking discussions about the true intention of the NSA in pushing an industry standard for encryption. For critics, it seemed evident that an intelligence organization would have a major interest in an encryption standard that can only be cracked by them. The DES encryption method is based on an algorithm splitting the data into 64-bit blocks and encrypting each of the blocks in a sequence of 16 transformations using a 56-bit key. In late 1992, it became evident that DES was too weak and would no longer withstand brute-force decryption methods using high performance computers.

In January 1997, the US Department of Defense started the search for a new and better symmetric encryption algorithm also expanding the required key-length from 56 bit to 128 and higher. After 5 years of analysis, the decision was made to use AES (Advanced Ecryption Standard) with the so-called Rijndal cipher introduced by Joan Deaemen and Vincent Rijmen, two Belgian cryptographers. It is a symmetric encryption method; the same keys are used for encryption and decryption and it supports key-lengths of 128, 192 and 256 bits. The encryption and decryption algorithm consists of 4 steps and is based on a sequence of transformations and can encrypt data blocks up to 14 times. The National Institute introduced the AES as a specification for standards (NIST) in 2001. AES is used today in numerous encryption applications like IEEE 8201.11 (Wireless LAN), WPA2 (WiFi), SSH (Secure Socket Shell), IPSec and in various encryption methods used for file encryption. The encryption and decryption process is support by high performance AES-hardware modules in many computer architectures.

With its key length of up to 256 bits, it is more secure against brute force attacks than DES based methods that are only 56 bits in length. Besides the encryption algorithm, which must be a standard and also sophisticated enough to withstand attacks, the length of the key is crucial. Brute-force attacks are based on trial-and-error attempts. They simply try out all combinations of the digits in a key to find the right key-combination. Doing that is relatively simple if you use a key length of 4-bit (16 possible combination) or 16-bit (65,536 possible combinations). A key with a length of 56 bit has 7.2×10^{16} combinations, a 256-bit key has 1.1×10^{77} combinations. For a 10 PFlops super computer (PFlops $= 10^{15}$ Floating Point Operations per second), it would take 339 s to crack a DES encryption using the 56-bit key, but 3.31×10^{56} years to crack a AES encryption using a 256-bit key.[1] With the fastest supercomputers today reaching nearly 100 PFlops ($=10^{17}$ Flops), there is still a way to go (see Fig. 6).

Assuming Moore's Law is applicable to supercomputers, we will reach up to 10^{28} Flops in 2060. It would still take 3.31×10^{44} years to crack a 256-bit DES encryption. The value proposition of DES was clear: if a message is encrypted using a strong key,

[1]Mohit Aurora: "Is AES Encryption Safe?https://kryptall.com/index.php/2015-09-24-06-28-54/how-safe-is-safe-is-aes-encryption-safe retrieved 2016–10–12

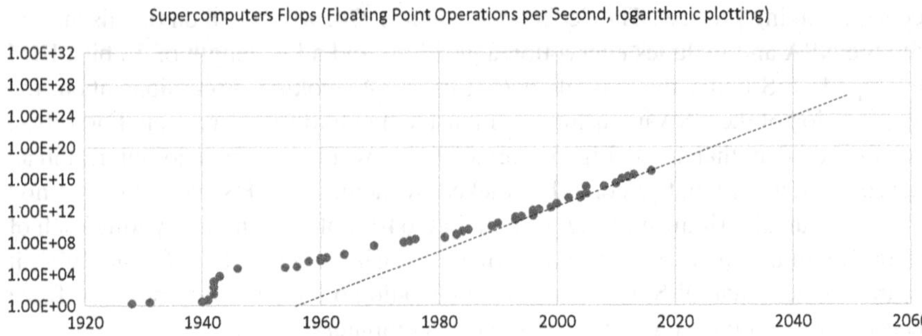

Fig. 6 Supercomputer Flops increase since 1940

it cannot be decrypted or at least it would be too expensive or time consuming to do so. Unfortunately, this was not the complete solution for security and privacy.

1976 The Key Problem: RSA (Rivest, Shamir und Adleman)
Until 1970, the strongest encryption methods were still a one-time pad encryption. Both sides used the same key; the length of the key and the frequency of changing the key provided safety. As we have seen with the one-time pad encryption method, the distribution of the key used to decipher a message by the receiver is one of the major obstacles. Although DES and AES improved the encryption algorithm and used key-lengths that created a highly secure encryption method, the problem was still the key. Using very long keys and using different keys for each message is a solution but not very practical one in widely distributed networks of senders and receivers. This type of encryption is called symmetric key encryptions: sender and receiver use the same key for encryption and decryption of the message. The keys used must be distributed between senders and receivers. In the 1970s, the idea arrived to use 2 different keys: one for encrypting the message, the other one for the decryption. In 1974, Whitfield Diffie a crypto analyst and freelancer in data security visited the IBM Thomas J. Watson research laboratory to talk about different strategies of asymmetric encryption. One of the participants was Alan Konheim. He later remembered that he had talked to another guest about the same topic a while ago. The name of the guest was Martin Hellman, professor at Stanford University. Diffie entered his car and took the 5000-kilometer trip to California to meet Hellmann. This was the beginning of asymmetric public-key encryption.

The problem sounds simple: is there a way for two people to exchange messages using different keys? Diffie and Hellmann knew that there must be a solution, because they could imagine an experiment: Alice and Bob (names always used for crypto analytic experiments) want to exchange secret messages. Alive wants to send a message to Bob. She puts the message in a small box and locks it with a padlock. She sends the box to Bob and keeps the key. After having received the box, Bob simply adjusts a second padlock on the box, keeps the key and sends the box back to Alice. Alice removes her padlock, sends the box back to Bob and Bob unlocks the box with his key. It sounds great and it's also good news: there is a way of exchanging secret messages without exchanging keys! In a real cryptographic world, this method would not work—there is nothing like safe boxes—but it shows that the principle is not impossible.

To build an encryption method for data, Diffie and Hellman decided to look at mathematical functions. What they were really looking for were functions which are "one-way" functions, that means one can calculate the result from the input parameters, but it is impossible to recalculate the input parameters from the result. This is exactly the missing element you need to lock your message. Diffie, Hellmann and Merkle discovered that modulo-functions could be used to exchange keys by exchanging only the result of a function. The approach was innovative but still a little complex, it takes two steps to exchange the parameters and calculate the key. Diffie started to think about a more efficient way. His new idea was to use a combination of public and private keys, now called an asymmetric encryption.

The idea was again simple. Alice publishes her public key to everybody. With this key, everybody can encrypt a message and send it to Alice. The decryption is only possible with a private key and Alice only knows this private key. In the world of boxes and padlocks this would mean that Alive is sending open padlocks to everybody. If somebody wants to send a secret message to Alice, he must lock it with the padlock and only Alice can open it. In digital encryption, this would demand an encryption function, which is a one-way function. Diffie published his idea in 1976.[2]

Ron Rivest read the paper from MIT. Ron talked with his colleagues Leonhard Adleman and Adi Shamir and they decided to start the hunt for a matching one-way-function. What they found is a solution based on number theory and prime numbers. One of the major challenges in number theory is prime factorization, which is simply the task of finding, for a given value, all prime numbers that are the factors of the number, e.g. 1, 2, 3, 5 are the prime factors of 30. This is comparable easy for small numbers, but the effort grows exponentially for large and very large numbers. Today, numbers with more than 300 digits (10^{30}) are used as keys. The solution is perfect: the prime number is used as the public key; the factors are the private keys. The RSA method, named after Rivest, Shamir and Adleman was published in 1978.[3] It is still one of the fundamentals of encryption today.

Using asymmetric encryption is based on the concept of having two keys for each message. The public key of a receiver is known to everybody and can only be used to encrypt a message that is sent to the receiver. The private key is only known to the receiver and can only be used to decipher the message. To organize that process, each receiver of messages like a person, a computer system, a web service or a cloud application must have a public key, which can easily be published to everybody. A public key is a very large number—up to 300 digits—and is

[2]Whitfield Diffie, New directions in cryptography", IEEE, 1976: http://ieeexplore.ieee.org/document/1055638/ retrieved 2016–10–12

[3]Ron Rivest, Adi Shamir, and Leonard Adleman: "A method for obtaining digital signatures and public-key cryptosystems", ACM, 1978: http://dl.acm.org/citation.cfm?id=359342 retrieved 2016–10–12

calculated by multiplying two prime numbers. These two prime numbers are the private key. The assumption is that recalculating the private key numbers from a public key is impossible or at least so time consuming that it would take years on a super computer. Finding the two prime numbers is called factorizing. Factorizing comparable small numbers is relatively easy: take the number 181 which has the prime factors 11 and 17. It takes only a few seconds to calculate that by testing it on a calculator or computer.

The effort grows for large and very large numbers. On March 18th, 1991, the RSA Laboratories put forward a price for factorizing large numbers. They published a list of semi prime numbers: numbers, which have two prime numbers as factors with a length between 100 decimal digits (330 binary digits) and 617 decimal digits (1048 binary digits). All numbers with less than 220 decimal digits could be factorized by 2012, the smaller ones within the first year of the challenge. Since the beginning of the challenge, no number with more than 220 decimal digits could be factorized. This could lead to the conclusion that using semi prime numbers as public keys is really a very strong method for asymmetric encryption. Also, knowing that the processing power of computers is growing rapidly and that technologies like quantum computing will accelerate that evolution empowers the discussion about the how long this stronghold will hold. For now, it seems that encrypters using asymmetric encryption are on the safe side of this equation and asymmetric keys have become the foundation of secret message transport.

Having a secure encryption method available is a great thing, but making it applicable for everybody soon led to huge discussions. The question about who should be allowed to keep data secret and private became vital for federal intelligence organizations seeing safe encryption used by citizens as a threat. The demand for secure data exchange and secure connections was growing and soon led to the first easy-to-use applications as we will see in the next sections.

Building Secure Connections for the Internet

1980 Malware and the First Virus

In the 1980s, computer viruses started their career as unwanted attachments to the use of information technology. It is disputed what the first computer virus was, but "Elk Kloner" is a hot candidate. The virus was programmed by the 15 year old, Richard Skrenta, and infected computer games on Apple II computers, displaying a black screen with a poem on it. In the following years, the number of viruses increased, most of them being harmless jokes. Soon the first dangerous programs appeared deleting data or crashing the system and starting the war between providers of anti-virus programs like McAfee and Kaspersky. Those early anti-virus programs could recognize virus software by its code and blocked the execution. The virus developers reacted with the design of polymorphic viruses that changed their code with every new infection, a strategy that was copied from biological viruses.

After the release of Microsoft's Windows 95 and the Office software, the virus developers started to embed virus code in Office documents which, again, changed the counter strategy of anti-virus programs and led to the extension of virus checks from executable programs to text-files. The distribution of viruses at that time was limited to files, either program-files or text-document.

1991 Anti-Virus Software and Firewalls

By 1991, countless computers were invaded by hundreds of virus programs and corporate PC infection was becoming a serious problem. Consequently, many projects were started to find solutions for the rapidly growing problem. The anti-virus industry started to work together with organizations like the Computer Anti-Virus Research Organization (CARO) founded in 1991. Symantec released its first version of Norton Anti-Virus in the same year. Eugene and Natalia Kaspersky in Russia founded Kaspersky Lab in 1994. Around 2000, the new type of protection was widely available and effective if installed and activated. The anti-virus applications followed a simple but elaborate strategy: the anti-virus software contained a database with all known code-patterns of malware, enabling it to immediately detect an attack by matching the code. The number of malware and, thus, the number of patterns grew rapidly. The IT-Security Institute AVTEST initially reported the number of malware code patterns at 333,000, which grew to more than 5 million in 2007 and more than 500 million in 2016.[4]

Another type of protection, called "Firewall," was implemented around 1988 filtering IP packages and blocking data from unwanted senders. A second and third generation of firewalls was introduced in 1990 and 1994. By 2000, firewalls were used as gateway computers between the Internet and an internal net and could filter and block data on the network layer and the application layer if put in place, was customized properly and running.

1991 Safe Mail Encryption: Phil Zimmermann and PGP

With the commercialization of the Internet in the mid-1980s and the introduction of the Web in 1991, data traffic expanded rapidly and users of email or websites started to demand private and secure communication.

In 1991, the American software engineer, Phil Zimmermann, developed the idea of using asymmetric encryption for secure mail transfer. He wrote a program that could be installed on PCs and allowed the sender of an email to encrypt the mail using his private key and the public key of the receiver. The mail receiver used his private key to decrypt the message.

The program used the RSA algorithm and Zimmermann published it together with the source code under the name PGP (Pretty Good Privacy) for free download on a FTP-server: *It was on this day in 1991 that I sent the first release of PGP to a couple of my friends for uploading to the Internet. First, I sent it to Allan Hoeltje, who posted it to Peacenet, an ISP that specialized in grassroots political*

[4]AVTEST, Malware Statistics: https://www.av-test.org/en/statistics/malware/ retrieved 2017-01-20

organizations, mainly in the peace movement. Peacenet was accessible to political activists all over the world. Then, I uploaded it to Kelly Goen, who proceeded to upload it to a Usenet newsgroup that specialized in distributing source code. At my request, he marked the Usenet posting as "US only". Kelly also uploaded it to many BBS systems around the country. I don't recall if the postings to the Internet began on June 5th or 6th."[5] Soon, the US authorities started a criminal investigation of Zimmermann for violating the Arms Export Control Act. At that time, RSA was categorized as strong encryption and the US government regarded cryptographic software using keys with more than 40 bits as a munition thereby subject to the arms trafficking export controls. PGP used minimum key-lengths of 128 bit and was classified as illegal by the US authorities. At the same time, PGP quickly became very popular and many people used it to for secure email transfer. For several years, it seemed that Phil Zimmermann was in danger of incrimination for violating US export law but the investigation was closed after three years without filing criminal charges. Zimmermann founded PGP Inc. in 1996 and proposed OpenPGP as a standard in 1997 to IETF. Network Associates acquired PGP Inc. in 1997. In 2002, a new PGP Corporation was founded acquiring the PGP assets from Network Associates, and PGP Corporation was later acquired by Symantec in 2010 for $300 million.[6] Phil Zimmermann left PGP in 2001. He was honored many times as a major influencer of the Internet and as the "hero for freedom."[7] Today, Symantec offers PGP and an open version is available from OpenPGP.[8]

1994 Safe Transport: SSL and TLS

With the rise of the Web and browsers, it became evident that secure communication is a valid element of browsing through the Web. In 1994, only 9 months after the release of the first widely available web-browser, Mosaic, Netscape released the first version of SSL (Secure Socket Layer) for secure and encrypted transport of information between the browser and the web-host. Microsoft followed in 1995 and since then, SSL encryption is a standard feature within each browser. SSL is based on asymmetric encryption between the browser and the web-server hosting the web pages and uses DES for the encryption of the data blocks. It was approved as a standard by the IETF in 1999 and renamed to TLS (Transport Layer Security). Today, TLS is used not only in HTTP as HTTPS but also in many other products and protocols including:

- POP3S for POP3 (Post Office Protocol)
- SMTPS for SMTP (Simple Mail Transfer Protocol)

---

[5]Phil Zimmermann, 2001: http://www.philzimmermann.com/EN/news/PGP_10thAnniversary.html retrieved 2016–10–12

[6]Computerworld, 2010: http://www.computerworld.com/article/2517739/security0/symantec-buys-encryption-specialist-pgp-for--300m.html retrieved 2016–10–12

[7]Heroes of Freedom, Reason, 2003: http://reason.com/archives/2003/12/01/35-heroes-of-freedom retrieved 2016–10–12

[8]Open PGP: http://openpgp.org/ retrieved 2017–03–02

- NNTPS for NNTP (Network News Transfer Protocol)
- SIPS for SIP (Session Initiation Protocol)
- IMAPS for IMAP (Internet Message Access Protocol)
- XMPPS for XMPP (Extensible Messaging and Presence Protocol)
- IRCS for IRC (Internet Relay Chat)
- LDAPS for LDAP (Lightweight Directory Access Protocol)
- FTPS for FTP (File Transfer Protocol)
- OpenVPN (Virtual Private Network)

1995 Security Certificates and Certification Authorities

Organizing the worldwide usage of asymmetric key encryption requires several organizations working together. This is the certification ecosystem. It consists of public organizations creating the standards and quite a few companies acting as certification authorities, creating public keys and certificates and then selling them.

The level of security provided by encryption using public and private keys depends on the reliability of the certification authority. If a certification authority is undermined, the security of the entire system can be lost. One approach is to claim a certificate for another person. Using this certificate for sending messages makes the receiver believe that the sender is a trusted person or organization. Thus, the proper identification of the organization or person that claims a certificate is one of the critical steps in the process. Hackers use this method to obtain certificates from CAs using a false identity.

A certificate issued together with the public and private keys by a certification authority must prove the existence of the person or company and the approved right of ownership of the domain for which the certificate is issued. Industry organizations act to organize and inform public key infrastructure. Most of these organizations were founded by a group of CAs or browser vendors to facilitate and create a common understanding about the creation, distribution and implementation of asymmetric key encryption.

As one of the first certification authorities, Thawte was founded in 1995 by Mark Shuttleworth. VeriSign acquired it for $575 million in December 1999 and, in 2010, it was acquired by Symantec. Today, Symantec is ranked in the top 5 in certification authorities creating a revenue of $6.676 billion (2014). Melih Abdulhayoğlu founded Comodo in the U.K. in 1998. The company is privately held. Identrust is a New York based bank consortium acting as a public key certificate authority whose members include over sixty of the largest banks in the world (see Table 1).

1995 Safe Internet Protocol: IPSEC

In 1993, the Internet and the Web had started to spread out into business and private use. The question of secure Internet communication was raised and led to several research activities at Columbia University and AT&T Bell Labs. The first version of the IPSec protocol was developed in 1994 for BSD, HP-OX and Sun OS at Trusted Information System, a company focusing on security products. Another project was started at the Naval Research Laboratory. To concentrate all these

Table 1 Market share of certification authorities (W3Techs survey, January 2017)

Company	Usage (%)	Market share (%)
Comodo	13.8	44.0
IdenTrust	6.4	20.3
Symantec Group	5.2	16.5
GoDaddy Group	2.5	8.1
GlobalSign	1.6	5.2
DigiCert	0.6	2.1

activities, the IETF introduced a project group in 1995 to create an open standard for secure and encrypted Internet communication. IPSec is based on additions in the Internet Protocol (IP) covering the authentication between the agents operating the communication, the negotiation of cryptographic keys and the encryption of IP packets. It is used to create a secure IP connection between a pair of hosts, between a pair of gateways or between a gateway and a host. Compared to TSL, the IPSec protocol is used to secure the IP connection while TLS is used to secure and encrypt the data transported by the applications using TCP/IP.

1995 Safe Networks: VPN

Using IPSec and TLS provides secure communication between a user and a web service (TLS) on the application level, or between a host or network gateways (IPSec) on the IP level. These security technologies can also be used to build virtual and secure private networks consisting of a defined number of hosts or gateways and users. In this case, the Internet is used as transport media, but the connections between the participants of the VPN are "tunneled," thus protecting the data from any external attack. VPNs are used in corporations or organizations to allow employees to securely access the internal intranet from their home office or while travelling or to connect separated locations through a secure virtual private network.

1997 Email, FTP and IRC as Carrier

With the spread of Internet users and the rapid growth of email communication, the distributers of malware started to use email or FTP as the channel to populate viruses. By 1997, the number of web-users crossed the 100,000 mark and would increase to 1 million in the following 10 years. Most of these web-users operated their own mail account. In February 1997, the virus "ShareFun" started a new chapter and infected Microsoft's e-mail program, MSMail, to send itself as an attachment to all entries in the contact list of the infected PC. Other malware attacks followed, like "Happy99" and "Melissa" in 1999 that infected millions of PCs worldwide within a short time and led to the shut-down of email servers in large corporations like Microsoft, Intel and Lockheed. A 32-year-old programmer, David Smith, was identified as the publisher of "Melissa" and sentenced to 20 months in prison and had to pay a $5000 fine being the first person to be prosecuted for writing a virus. The penalty was not too high compared to the damage caused by "Melissa," which was estimated at $400 million.[9] The new distribution strategy was extremely

[9]Nikolai Bezroukov: Melissa Worm, 1994: Melissa: http://www.softpanorama.org/Malware/Malware_defense_history/Ch05_macro_viruses/Zoo/melissa.shtml retrieved 2017-01-12

successful. In 2000, the virus "Loveletter" was distributed by a schoolboy from Manila and caused damages estimated at $10 billion.[10,11] Around 2001, 90% of all infections of computers were caused by e-mail. The new type of malicious email was called spam-mail and soon conquered a large part of the global email traffic. In early 2002, the portion of spam was around 25% of the total email traffic but increased to 80% and more within 5 years. The peak was reached in 2008 when more than 90% of the global email traffic was spam.[12] With the introduction of effective spam-filters, the percentage decreased again and is at a level of around 50% today (2016).

Beginning in 2001, the distributers of malware started to use the browser itself for infection. Security leaks in Microsoft's Internet Explorer were used to infect a computer by transporting the harmful code via web pages viewed by the browser's user. This started the era of infected web pages and led to additional anti-virus measures in the code of browsers. In January 2003, a new type of virus known as "Slammer-Worm" infected Microsoft's SQL-servers taking advantage of software vulnerability in the server's code. Microsoft knew the problem and a patch had been delivered 6 months before, but not installed by many users. "Slammer" caused a worldwide slowdown of the Internet instigated by overloaded routers operating under the burden of traffic from infected SQL-servers.

The clear majority of security violations in the first decade of the Web between 1992 and 2000 had no organized criminal background, the developers and distributers of viruses were "black hat" hackers finding their motivation in inducing a worldwide effect on technology in order to garner publicity. This would change in the mid-2000s.

2002 TOR Network

In 1990, the US Naval Research Laboratory started to think about a secure method of communicating with agents of the US intelligence organizations while those agents were in the field. Paul Syverson, Michael Reed and David Goldschlag called the idea onion routing and it was developed by DARPA in 1997. An onion serves as a rudimentary example of the principles of onion routing. Layers, like the skins of an onion, encapsulate a message passed through an onion network. Whenever the message passes a network node a single layer is peeled away uncovering the next destination of the message. After the final layer is decrypted, the message is passed to its destination. This method keeps the sender of the message anonymous because each node including the final destination only knows the location of the immediately preceding and following nodes. Passing messages through an onion-network guarantees the anonymity of the sender. The vulnerability of an onion transmission is reduced to the last or exit node. This node must decrypt the final layer and deliver

[10] The Guardian, 2000: https://www.theguardian.com/world/2000/may/05/jamesmeek

[11] Jonathan Strickland: 2017: http://computer.howstuffworks.com/worst-computer-viruses2.htm retrieved 2017–01–27

[12] Mail Statistics, Symantec: http://www.symantec.com/de/de/security_response/landing/spam/ retrieved 2017–01–25

the message. A compromised exit node is thus able to acquire the raw data that is transmitted, potentially including passwords, private messages, bank account numbers, and other forms of personal information.

A first version of TOR (The Onion Routing) was launched in 2002 by Syverson, and the computer scientists Roger Dingledine and Nick Methewson. The Naval Research Laboratory released the code under a free license in 2004. The Electronic Frontier Foundation (EFF) provided funding for the project. The Tor project was founded by EFF in the beginning; other sponsors followed including the US International Broadcasting Bureau, Human Rights Watch, University of Cambridge and Google. Later funding was provided by the US government, which let Roger Dingledine conclude that: *"the United States government can't simply run an anonymity system for everybody and then use it themselves only. Because then every time a connection came from it people would say, "Oh, it's another CIA agent." If those are the only people using the network."*[13] Today, the Tor-project offers free download of the Tor-browser and claims to provide: "*Anonymity Online: Protect your privacy. Defend yourself against network surveillance and traffic analysis.*"[14] The TOR-browser is used for browsing the "Dark Web," which consists of web-pages that cannot be found using a normal or as it is called "clean-net" browser and connection. Discussions have been around about the security of Tor regarding attacks or surveillance by intelligence organizations. The fact that the software was developed using government funding led to the reasonable suspicion that anonymity is provided, except from governmental intelligence organizations.

2005 Crime Attacks Security

The era of cybercrime started to grow around 2005. Beginning in 2004, an increasing number of malware spread out and attacked private data like bank accounts. Cybercrime began to expand into an industry of its own, although not covered by official statistics.

Around 2010 other types of cyberattacks on infrastructure or industry from mostly unknown organizations gained publicity and seemed to increase as well. Today, this is known as Cyberwarfare and it has evolved into a battlefield between states and political organizations. One of the first cyberwar-attacks became known to the public as "Stuxnet" in 2010 and seemed to be an attack by a military organization on the Iranian nuclear power industry infecting Iranian computer networks using virus technologies.

Since 2005, cybercrime and cyberwarfare have become prominent notions whenever Internet, Web or cloud services were discussed. The new technology was criticized because of its vulnerability and marked as the cause of illegal activities and the trigger for the raise in a new type of criminal industry.

---

[13]Roger Dindledine, 2004: https://pando.com/2014/07/16/tor-spooks/ retrieved 2016–10–12

[14]The TOR project. https://www.torproject.org/ retrieved 2016–10–12

Creating Standards and Best Practices

The ongoing discussion about Internet and web-security led to the formation of several consortiums and associations that provide analysis and propose best practices in solving security issues for the industry.

2005 CA/Browser Forum
The CA/Browser Forum was founded in 2005 as a consortium of certification authorities and web browser vendors to define and promote standards for Internet security. The Comodo Group, one of the leading certification authorities, organized the first meeting.

2008 Cloud Security Alliance
The CSA was founded in 2008 in the US as a non-profit association of individuals "*...dedicated to defining and raising awareness of best practices to help ensure a secure cloud computing environment.*"[15] CSA organizes many project teams that concentrate on different aspects of security and operate the cloud security provider certification program and a cloud security user certification.

2009 Common Computing Security Standards Forum
To promote industry standards for public key certification, the founder of Comodo, Melih Abdulhayoğlu, founded the Common Computing Security Standards Forum (CCSF) in 2009. CCSF is an association of security software vendors, operating system providers, and Internet browser software creators. Its goals are to "*Mitigating the risk of malicious intent and software, creating standards for the industry for detecting and identifying malware.*"[16]

2013 Certificate Authority Security Council
Founded in 2013, the largest certification authorities, the Certificate Authority Security Council (CASC), is an industry organization to ensure the compliance of certification authorities according to best practices: "*Membership in the CASC is available to publicly trusted SSL certificate authorities that meet the CASC's reputation, operation, and security requirements.*"[17] The CASC requires its members to undergo an extensive annual audit to check for compliance with the basic requirements of security standards.

---

[15] Cloud Security Alliance: https://cloudsecurityalliance.org/about/ retrieved 2016–10–12

[16] Common Computing Security Standards Forum: http://www.ccssforum.org/objectives.php retrieved 2016–12–10

[17] CASC: https://casecurity.org/casc/ retrieved 2016–10–12

Security Today

Going through the list of encryption and security technologies developed since the 1970s and looking at the activities and efforts made by security associations today, it seems that Internet and web-security is no longer anissue. With AES and public key encryption, a strong encryption methodology is available that is used in safe transport (TSL) and safe connections (IPsec and VPNs). Associations and councils (CSA, CCSSF), CASC) are currently working on the development of better standards and provide certification procedures to guarantee the compliance of products and services with industry security standards. Firewall and anti-malware applications and spam-filters are available and widely used.

Nevertheless, Internet and web security is a constantly growing threat for corporations and private users. The 2016 CSA report[18] lists the top 12 threats to cloud security:

- Data breaches
- Weak identity, credential and access management
- Insecure APIs
- System and application vulnerabilities
- Account hijacking
- Malicious insiders
- Advanced persistent threats (APTs)
- Data loss
- Insufficient due diligence
- Abuse and nefarious use of cloud services
- Denial of service
- Shared technology issues

The costs caused by security violations are estimated at between $300 billion and $500 billion per year and is rapidly growing. The costs caused by cybersecurity attacks can be sorted into the following categories:

- Damage and destruction of data
- Stolen money
- Lost productivity
- Theft of intellectual property
- Theft of personal and financial data
- Blackmail
- Embezzlement
- Fraud

[18]CSA, Treacherous-12, 2016: https://downloads.cloudsecurityalliance.org/assets/research/top-threats/Treacherous-12_Cloud-Computing_Top-Threats.pdf. Retrieved 2017–01–12

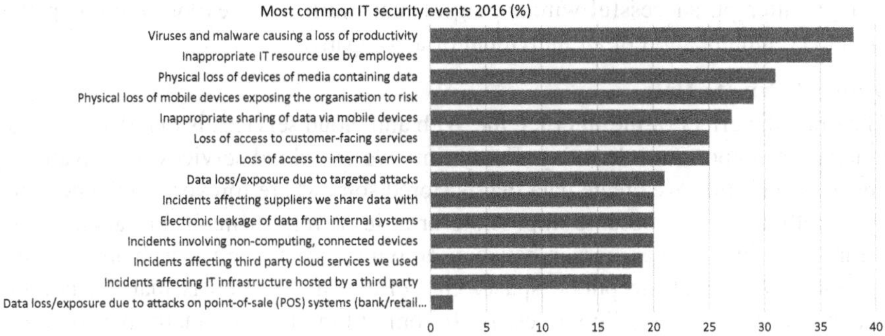

Fig. 7 Most common IT security events in 2016 (Kaspersky Lab)

A report from Crowd Research Partners shows that the general security concerns grew from 45% in 2015 to 53% in 2016.[19] As specific security concerns the respondents mentioned data loss and leakage risk (40%), unauthorized access (50%) and hijacking of accounts (44%). Kaspersky Lab published, in its security report from 2016, the most common IT security events showing malware and viruses on top of the list (see Fig. 7).[20]

The constantly growing menace caused by security threats and incidents has various root causes that are linked together: the dramatic growth of data volume, the value of private data, lack of awareness for security measurements and the unfair advantage for attackers.

Explosion of Data Volume and Data Distribution

Since the introduction of the Web, the volume of personal and corporate data transported and stored has exploded. The volume of data created per time unit has been growing from 100 Gigabytes (10^9) per day in 1992 to 2.5 Exabytes (10^{18}) per day in 2015.[21] To illustrate how large these numbers are, consider that an Exabyte is the same as one Quintillion and used to refer to the mass of earth in tons or the number of molecules in the human brain.

Attacking that mass of data to cause harm, to collect valuable information or to start a criminal business has gained attractiveness to hackers, federal intelligence organizations and criminal organizations. Today, a successful attack on a company's data collection or on message data is much more effective than 10 years earlier. The number of data records stolen by a single attacker may number in the hundreds of millions, phishing emails target hundreds of thousands of users in

[19]Crowd Research Partner, Security Report 2016: http://www.crowdresearchpartners.com/portfolio_item/cloud-security/ retrieved 2017–01–25

[20]Kaspersky Lab: Security Report 2016: https://securelist.com/analysis/kaspersky-security-bulletin/76858/kaspersky-security-bulletin-2016-executive-summary/ retrieved 2017–01–25

[21]vcoudnews, Walker Ben, Every Day Big Data Statistics, http://www.vcloudnews.com/every-day-big-data-statistics-2-5-quintillion-bytes-of-data-created-daily/, 2015, retrieved 2017–01–13

a single attempt, successful wiretapping of mail traffic may expose the complete private communication of an individual or a company.

Value of Private Data
The dramatic effect of the Internet, the Web and cloud services is that data plays a much more important role today. Every time we use cloud services for private or business activities, we create data that is stored somewhere forever. To define data as something private became impossible or at least very difficult. Privacy simply means drawing a clear border between activities we perceive as being in our "private space" and the public space. One effect of the Web is that the border between private and public spaces has become blurred. It is exactly that type of private data that has gained enormous value not only for platforms like Google and Amazon to empower their businesses, but also for intelligence organizations or criminals. The successes of search engines like Google, e-commerce platforms like Amazon and social networks like Facebook have their foundations in analyzing user data and converting it into money via smart advertising.

Lack of Awareness and Measurements: Secure Encryption Is Not Enough
The secure and strong encryption methods developed in the 1980s and 1990s deliver an important element to keep data private when they are in place and properly used. At the same time, the distribution and complexity of communication and data storage architectures provide a huge number of options to attack message flows or data stored. Having a strong lock at your front door has no effect when the intruder finds his way through a weak backdoor or uses "social engineering" to gain access to private territory. One of the most successful strategies to gain access to restricted areas was already described by Homer when the noted Trojan horse was used by the Greeks to conquer the city of Troy. Instead of wooden horses today's attackers use phishing mails, contaminated web pages or fake mails. Measurements to safeguard against these types of security threats must not only be based on encryption but cover a wide range of security measures including comprehensive technical measures, social behavior and legal policies and frameworks.

Unfair Advantage for Attackers
Implementing security measurements and promoting individual awareness to hold off attacks against privacy need long-term strategies. Technical infrastructure must be implemented, operated and updated continuously. The organizational and individual behavior must be questioned, adapted and improved constantly. This refers to private users as well as organizations and corporations. Equally, the legal framework of states should be aligned with a clear policy to protect citizens and industries against unwanted effects caused by security and privacy violations. At the same time, attackers have the advantage of operating in an unregulated environment and can apply new technologies or develop new attack strategies in a very short time. Criminals as well as intelligence organizations use software and knowledge about weaknesses in operating systems, browsers or network management software. The information about those "weak spots" is traded for high prices and either federal intelligence organizations or talented hackers develop the software

used in attacks. Momentarily, attackers seem to have a head start over defense measures, as we will see in the next sections of this chapter.

Cloud Security Threats and Typical Patterns

From a technical point of view, various tactics are used to get access to communication, data or applications. Potential threats try to use weaknesses in a system or of the individuals operating the systems to breach privacy or cause harm. During the last 20 years, a wide variety of different attack patterns have been developed by criminal organizations or intelligence organizations to get access to restricted areas or private content. Some of the most frequently used attack patterns are described below.

Traffic Eavesdropping

Traffic eavesdropping is used gain access to the content of messages, like email, in order to read and copy the content without changing it and, more importantly, without leaving a record that the messages were seen by an unauthorized person. Eavesdropping is an effective method to collect private data for future use, e.g. bank account or credit card data (see Fig. 8).

Typical attack points are email-servers or any other type of message routing service including wireless networks with or without weak encryption. Wiretapping, as a traditional method to access private communication, is also still used by federal intelligence organizations through the physical tapping of submarine cables. Eavesdropping has a very limited effect when messages are encrypted using a strong key.

Malicious Intermediary

If the message is not only read but also changed, we call this a malicious intermediary (see Fig. 9).

This is the case when messages are corrupted, e.g. by changing the amount or the target account number in a bank transaction or by infecting the message with harmful code to infect the IT system with malware.

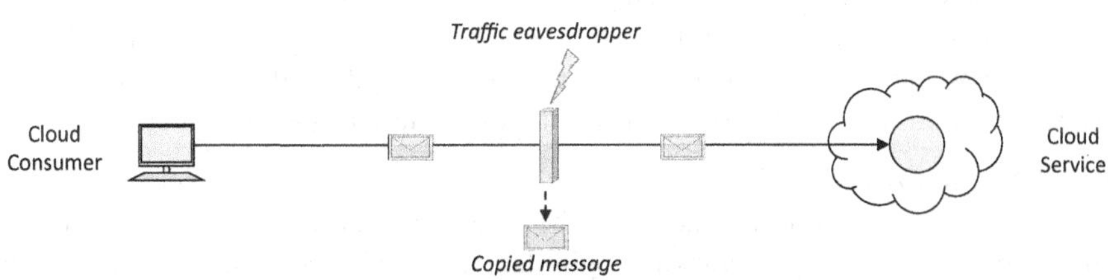

Fig. 8 Traffic eavesdropping

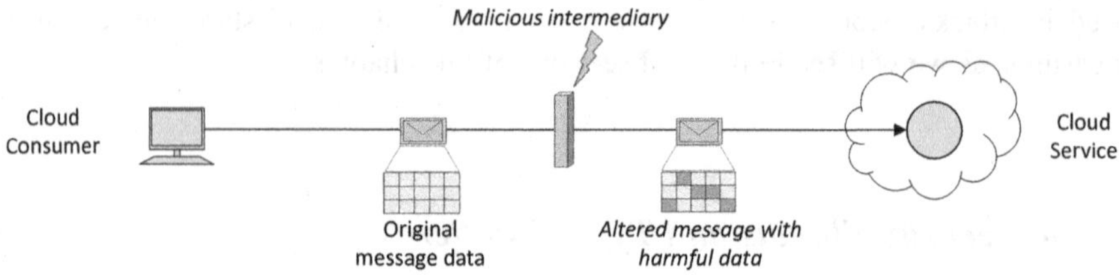

Fig. 9 Malicious intermediary

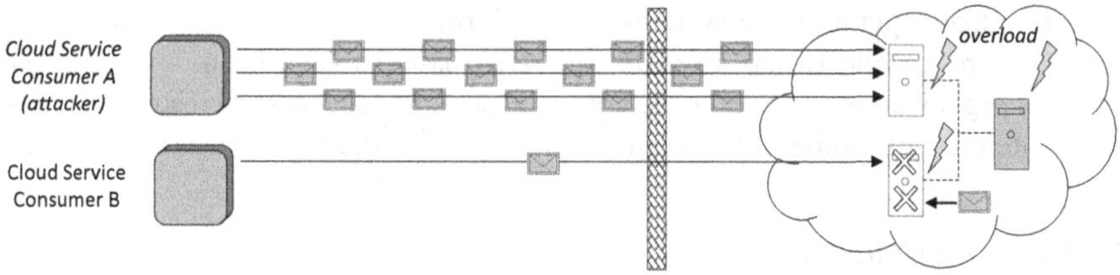

Fig. 10 Denial of Service (DoS)

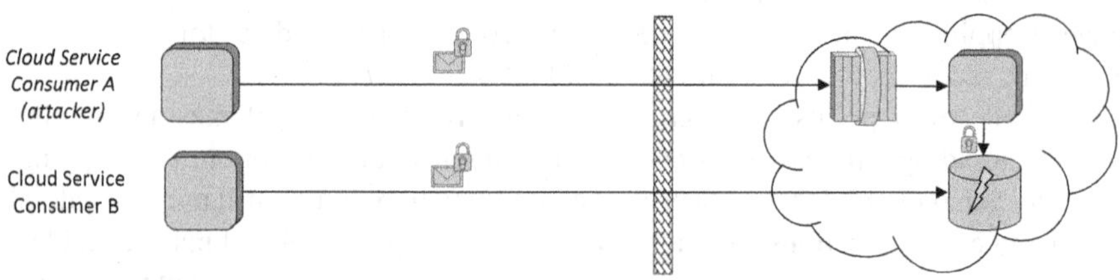

Fig. 11 Insufficient authorization

Denial of Service (DoS)

Denial of service attacks are attempts to overload IT resources like web pages with a huge number of messages or service requests. The desired effect for the attacker is that the service will break down (see Fig. 10).

Used by criminal organizations, DoS had been a common method used for blackmailing companies.

Insufficient Authorization

The goal of authorization attacks is to get access to resources like data bases, which is normally not granted. Successful authorization attacks can mostly be traced back to system weaknesses like patchy settings of user credentials or bad code that allows the infiltration of the software via the user interface (see Fig. 11).

Virtualization Attack

Virtualization attacks use weaknesses in the access procedures like weak passwords and are based on stealing or cracking login-data from authorized users. The attacker

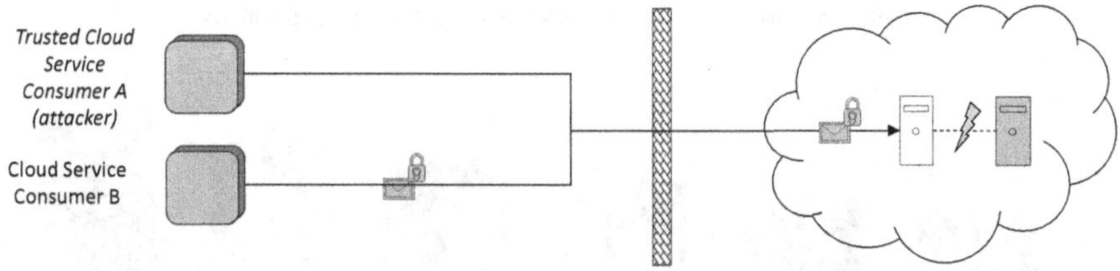

Fig. 12 Virtualization attack

Table 2 Attack types

Type of attack	Percentage (%)	Methods used
Web application attacks	24	Insufficient authorization Virtualization attack
Malware viruses	19	Insufficient authorization Malicious intermediary
Application specific attacks	19	Traffics eavesdropping
DoS/DDoS attacks	9	Denial of service
Reconnaissance	9	Traffic eavesdropping Insufficient authorization Virtualization attack

then poses as the authorized user to gain access to resources like data or programs (see Fig. 12).

A report from NTT Security estimates the occurrences of different types of attacks and ranks web application attacks, malware viruses and DoS-attacks as the top 3[22,23] (see Table 2).

Web application attacks are executed via the user interface of the web-application using insufficient authorization to insert SQL-code (SQL Injection) or malicious scrip code (Cross Site Scripting—XSS). The aim of those attacks is to get access to databases and steal or corrupt data.

Malware viruses became very popular with the increasing number of ransomware appearing over the last 2 years and are the technical foundation of a new business model for criminal organizations (see also the section entitled "Cybercrime").

Application specific attacks try to collect information about a computer system and use traffic eavesdropping as the method. Sifting through data packages transported through a network can lead to viable information about system settings that can be used for future attacks.

Denial of service attacks became popular because they are, in most cases, quickly advertised to the public. Web pages of companies or organizations that

---

[22]Calyptix Security 2016: http://www.calyptix.com/top-threats/top-5-cyber-attack-types-in-2016-so-far/ retrieved 2017–01–20

[23]NTTSecurity, Quarterly Threat Reports: https://www.solutionary.com/ retrieved 2017–01–20

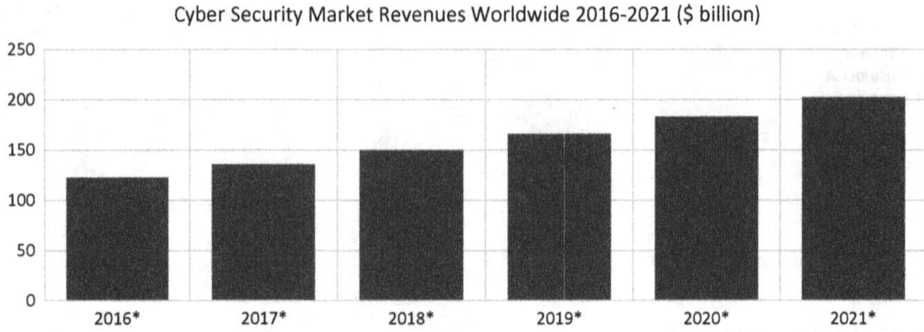

Fig. 13 Cyber security market revenues projection

are under a DoS attack gain immediate popularity. The goal of criminally motivated DoS attacks is the blackmailing of companies.

Reconnaissance attacks are a tactical move aimed at collecting classified information (passwords, system parameters) about the victim in preparation for an attack. These types of attacks are frequent but, in many cases, not easy to recognize. Reconnaissance can be accomplished via eavesdropping, insufficient authorization or even without targeting any IT aspect of the victim but instead via social communication methods.

Security Market

In response to the increasing potential threats and the growing number of actual events, the market for security software and services is predicted to grow at an annual rate (CAGR) of more than 10% reaching more than $200 billion in 2021 (see Fig. 13).[24]

The cyberwar not only creates damage for corporations and individuals but it also creates new jobs. The field of cybersecurity is one of those rare sectors where the unemployment rate is 0% with the industry offering millions of jobs in the upcoming years. A report from Cisco mentions about one million cybersecurity job openings worldwide and an increased demand of 6 million by 2019.[25]

[24]Markets and Markets, 2016: http://www.marketsandmarkets.com/PressReleases/cyber-security.asp retrieved 2017–01–20

[25]Cisco: Cybersecurity Talent, 2016: http://www.cisco.com/c/dam/en/us/products/collateral/security/cybersecurity-talent.pdf retrieved 2017–01–25

Threat Agents and Their Motivations

Cybersecurity and cybercrime have evolved into a huge range of activities. The nexus of criminal organizations, government organizations, individuals and corporations is widely distributed and complex. As we have seen in the section above, the economic impact is enormous and seems to grow at an increasing rate. Going to the roots of this conflict raises questions about roles and motivations. From a structured point of view, there are five types of suspects behind a cyberattack all with different motivations and different targets (see Fig. 14).

Hackers

Starting with the traditional "hacker," we find his motivation for doing something, which is beyond the borders of legality and satisfies his or her desire to cross the frontier between regulation and freedom. Typical hackers are not interested in causing damage; they want to go to (virtual) places that are restricted, just to demonstrate that it can be done. Leaving out monetary or political motivations, hackers want to be the "nice guys" who illustrate what is possible. Targets for hackers are large administrations or corporations that offer maximum satisfaction when the "hack" is successful. Hacking as a culture has a long tradition going back to computer enthusiasts in the 1970s associated with MIT's Tech Model Railroad Club, the Palo Alto Hombrew Computer Club or The German Chaos Computer Club. Some of the celebrities of today's IT economy started their career as hackers like Bill Gates, Steve Jobs and Steve Wozniak.

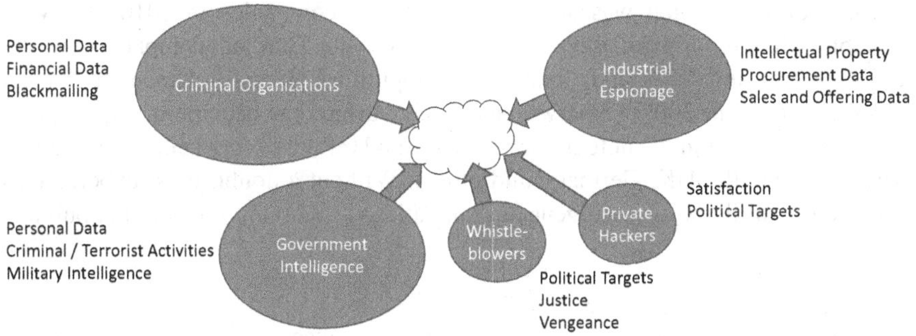

Fig. 14 Threats agents

Whistleblowers

A special category of hackers follow a political or personal target in making internal or classified information public in order to compromise individuals, organizations or political systems. They find their motivating force in acting as whistle-blowers that deliver justice. Popular examples are Chelsea Manning (born as Bradley Manning), an American soldier who handed over internal documents about the Iraq war to Julian Assange to have them later published on the Wikileaks platform, or the former CIA agent, Edward Snowdon. Threat agents in this category are also people that find their motivation in publishing internal information to satisfy their own desire for revenge due to unjust treatment by colleagues or by a company. Threat agents in this category primarily act from within as a trusted attacker or malicious insider.

Cyberwarfare

The third category of attackers is intelligence organizations that see themselves as protectors of a specific legal structure and ideology. Intelligence organizations like the NSA, the UK's MI6 or the German BND see their mission in monitoring and analyzing the activities of other states, terrorists or criminal organizations in order to predict threats and collect as much information as possible. To do so, these intelligence organizations are equipped with special rights and use them. Cyber-attacks by intelligence organization have a long tradition and now have begun to evolve into cyberwar scenarios. One of the first cyberwar attacks was documented as "Moonlight Maze" in September 1999. It was an attack on US government networks where: "*. . .hackers apparently working from Russia have systematically broken into Defense Department computers for more than a year and plundered vast amounts of sensitive information.*"[26] Since then, a vast number of cyberattack events having their origin in government intelligence operations were reported including "Stuxnet," which attacked the Iranian nuclear power industry; "Titan Rain," which was likely a Chinese attack on US industry defense contractors in 2003; "Aurora," which was an attack on US IT companies in 2010 also with possible Chinese origins; the infiltration of private German companies by the UK's GCHQ and NSA and, finally, the tapping of German chancellor Angela Merkel's phone in 2009. Cyberwar and cyberwarfare has become a major operational mission within the defense departments and military organizations of the US, Russia, China, the UK, Germany and others. Without a doubt, most cyberattacks executed by military or intelligence organizations remain unknown to the public.

---

[26]SFGate: Russians Seem To Be Hacking Into Pentagon, 1999: http://www.sfgate.com/news/article/Russians-Seem-To-Be-Hacking-Into-Pentagon-2903309.php retrieved 2017–01–20

Industrial Espionage

Gaining access to the data and information of competitors has always been an option for improving business chances. Industrial espionage is one of the important elements of cybercrime. Companies interested in getting access to the internal information of their competitors like their offers, financial information, and technical specifications has led to private intelligence agencies that are specialized in cyberattacks and hacking. There seems to be a large market for that type of service. Service providers offering cyberattacks against competitors can be easily found on the nefarious markets of the dark web.

Cybercrime

Making money from criminal activities is not restricted to bank robbery or car theft. Using the Internet and the Web for those types of business models has become a rapidly growing ecosystem. Much of the damage caused by cybercrime was produced by criminal data theft, fraud or blackmail. A variety of business models were developed by criminal organization in the last couple of years targeting corporations as well as individuals.

Data Theft

The theft of large collections of personal data is one of the most popular methods. The number of reported data breeches in companies in 2016 was more than 700.[27] Criminals steal data for different reasons. Besides industrial espionage, data also serves as currency. Personal data containing names, addresses, phone numbers and bank accounts or credit card numbers can be easily sold on the black market. There are standard price lists for personal data records offered on the dark web. This data is then used to break into bank or credit card accounts or as basic data for phishing mails. In 2015, the number of records exposed through data breeches increased by 97% as reported by the Identity Theft Resource Center.[28] Insurance companies have started to offer insurance packages to cover the damages caused by data theft. A Juniper Research Study from May 2015 estimates that the worldwide cost of data

[27]DMR: Cyber Security Statistics, 2016: http://expandedramblings.com/index.php/cybersecurity-statistics/ retrieved 2017–01–20

[28]Business Insider, 2016: http://www.businessinsider.de/cyber-attacks-are-costing-companies-mil lions-of-dollars-heres-how-they-can-mitigate-those-costs-2016-2 retrieved 2017–01–20

Table 3 Largest data thefts between 2014 and 2016

Company/Organization	Year	Records Stolen
Friend Finder Network	2016	412,000,000
Voter Database	2014	191,000,000
MySpace	2015	164,000,000
VK	2015	100,544,934
Dailymotion	2016	85,200,000
Anthem	2015	80,000,000
Securus Technologies	2014	70,000,000
Philippines' Commission on Elections	2015	55,000,000
Turkish citizenship database	2015	49,611,709
Weebly	2016	43,000,000
AshleyMadison.com	2014	37,000,000
Mail.ru	2015	25,000,000
US Office of Personnel Management (2nd Breach)	2014	21,500,000

breaches will reach $2.1 trillion in 2019, having increased from $500 billion in 2015 with the average cost of a data breach in 2020 estimated at $150 million.[29]

In the last three years (2014–2016), the number of cases where data was stolen has increased dramatically. The largest incidents contain millions of stolen data records from companies around the world (see Table 3).

Targeted organizations are social network companies, administrations, and industry and medical service organizations. In most cases, the data includes names, email addresses and phone numbers. In some cases, more critical data like bank account numbers, passwords, financial information or information about sexual preferences were included. The business model behind this is very simple: data has a certain value on the black market. Depending on the data items stolen and the origins of this data, there are fixed prices for such personal data. In 2011, a research study of the pricelists for data sold on the black-market was released[30] (see Table 4).

The buyer will gain access to credit card or bank accounts or use the data as input for fake mails or ransomware targets.

Ransomware

The second type of rapidly growing criminal money generation is blackmailing. Corporate IT infrastructure is hacked and data is either deleted or encrypted. The company is then confronted with blackmail with the criminals asking for a ransom

[29]Juniper Research https://www.juniperresearch.com/press/press-releases/cybercrime-cost-businesses-over-2trillion

[30]Sebastian Kexel: "Management in der Malware Industrie", Grin-Verlag, 2011.

Table 4 Typical pricelist for data on the black market

Data		Price per record
Visa Classic	USA	$18.00
Visa Classic	Canada	$29.00
Visa Classic	EU and Asia	$85.00
Visa Classic	Russia	$2.50
Paypal Account		$5.00
Paypal Creditcard		$10.00
Bank Account HSBC		$250.00
Bank Account Barclays		$290.00
Bank Account Abbey Bank		$500.00

to get the data back or decrypted. Over the last four years, this method has been industrialized using ransomware software. In June 2016, the FBI reported that more than $18 million has been paid by US companies as ransom for blackmailing after a data encrypting cyberattack.[31] The ransom is to be paid in cryptocurrency to keep the receiver anonymous and companies have started to open Bitcoin accounts to prepare for such a scenario.[32]

Fake Emails

Fake emails are a widely-distributed method of collecting money by convincing the receiver that he could gain access to a huge fund somewhere in Africa or Asia but must first transfer a comparably small amount to cover operational expenses before the large sum is paid. The method seems to be effective when using large numbers of personal data records to send the fake emails automatically. If one out of ten thousand addressees can be fooled, the business was a success.

Phishing Mails

Phishing mails are used to get access to the private data of the addressee. The mail is a faked message and should trick the victim to either click on a link to a corrupted site or enter his or her personal data like account passwords. Sending phishing mails

[31]ARS Technica, 2015: http://arstechnica.com/security/2015/06/fbi-says-crypto-ransomware-has-raked-in-18-million-for-cybercriminals/ retrieved 2016–10–12

[32]Brave New Coin: U.K. cod bitcoins to pay ransoms, 2016 http://bravenewcoin.com/news/large-uk-businesses-holding-bitcoin-to-pay-ransoms/ retrieved 2016–10–12

can be a successful business. With a success rate of below 0.1% and calculating the costs of the input data (see above) and the costs of the infrastructure like botnets, the profit is estimated between 30% and 70%.[33]

CEO Fraud

A method that is based on a more personal type of communication has evolved in the last two years and attacks any company that uses email. The target person is an employee who manages money transfer operations, and the schema is based on fake emails that seem to have been sent by the CEO requiring a money transfer to an external account. The so-called CEO fraud has caused huge damages over the last few years. The FBI also posted a warning reporting that between 2013 and 2016 complaints globally from 17,642 victims had been received with losses of more than $2.3 billion.[34]

Dark Web and Deep Web

The dark web uses the same Internet as the Web, but consists of websites and services that are not visible on the "bright" web. Websites on the dark-web can only be accessed using specific software, configurations or authorization, like Tor-browsers. The dark web consists of websites that partly offer suspicious content or goods and services beyond the border of legality like weapons or drugs. In comparison to the dark web, the deep web is the term for all websites that cannot be indexed by search engines for different reasons (dynamic pages, password protected pages, non-HTML pages, archives, etc.). The dark net has existed since the beginning of the Internet, there have always been hosts and services that were hidden and only accessible by special software or protocols, thus forming the "dark web." The dark web as it is known today gained popularity with the first TOR browser, which was released in 2002 and made the dark web easier to access using it as a "TOR-Network." When the cybercrime community looked for a place to do business, the dark web was already there.

[33]Sebastian Kexel: "Management in der Malware Industrie", Grin-Verlag, 2011

[34]FBI: CEO Fraud, 2016 https://www.fbi.gov/contact-us/field-offices/phoenix/news/press-releases/fbi-warns-of-dramatic-increase-in-business-e-mail-scams retrieved 2017–01–12

The Growth of Cybercrime

The estimated cost caused by cybercrime in the US was calculated at $300 billion by the Wall Street Journal in 2013.[35] In November 2015, IBM CEO, Gini Rometty, said: "*Cyber crime is the greatest threat to every company in the world*".[36] In October 2016, the Menlo Park based firm, Cybersecurity Ventures, published its predictions for a 5-year period from 2017 to 2021 estimating the worldwide costs in 2015 to be $3 trillion and growing to $6 trillion by 2021.[37]

Since the mid-2000s, cybercrime has evolved into an industry of its own. The cybercrime ecosystems consist of a worldwide distributed but only loosely coupled network of operation groups, malware developers, data dealers, software dealers and coordinators that bring together partners to execute an attack. There are several reasons behind the rapid development of cybercrime over the last 10 years.

First, there is a constantly growing target market. The increasing number of web-users and social media networks created a rapidly growing pool of data stored within the premises of corporations or cloud providers. From a criminal point of view, it is like the vault of a bank that is constantly refilled with gold. The number of possible attack targets grew from 1 billion web-users in 2015 to more than 3 billion in 2016.

Secondly, technology for executing attacks has developed at high speeds since 2000. Today's software packages for creating a botnet infrastructure to run a phishing mail attack are comparable inexpensive and can be rented from criminal providers of cloud infrastructures. The dark web as the market place for software, data and resources is an additional foundation for the cybercrime ecosystem. With the rise of cryptocurrencies like Bitcoin, the criminals also received the perfect solution for anonymous money transfer.

Finally, the level of security provisions for the target groups has become much better in the last few years but has not improved with the same speed as the infrastructure, organizational maturity and skills of the attackers.

Defense and Security Policies on Different Levels

Cybercrime is a global phenomenon, based on the global spread of the Internet and the Web. It seems that the problem is not the lack of defense technologies like

---

[35]Wall Street Journal, 2013: http://www.wsj.com/articles/SB10001424127887324328904578621880966242990 retrieved 2016–10–12

[36]Forbes, 2015: http://www.forbes.com/sites/stevemorgan/2015/11/24/ibms-ceo-on-hackers-cyber-crime-is-the-greatest-threat-to-every-company-in-the-world/#5fff995e3548 retrieved 2016–10–12

[37]EIN News, 2017: http://www.einnews.com/pr_news/350071526/cybersecurity-economic-predictions-2017-to-2021 retrieved 2017–01–25

firewalls, anti-virus software or spam-filters but the swiftness with which large corporations and organizations can implement and constantly improve their defense infrastructures. The race is on between criminals and victims and it seems that in the last two years, criminals have the advantage. Due to the global distribution of criminal networks that organize themselves using the Web and the dark web and act without geographical or political borders, it is difficult to get them under control without cross-border or international cooperation and action. Analysis performed during the last couple of years shows a concentration of cybercrime activities in Russia, China and South America spreading out from there to the rest of the world. Legal measures and executive actions will stay ineffective if not agreed upon on a global basis.

The major target of security policies is to reduce the risk of loss or harm to data and privacy. This could be done by implementing security controls as countermeasures using security mechanisms like firewalls, security software and other components but also by creating nationwide and international measures and policies to protect individuals, industries and communities. The latter part turned out to be more challenging than the technical tools and infrastructures. Intelligence organizations as well as criminal organizations act in an unregulated and borderless space. Furthermore, different states or communities have different political approaches regarding the importance and handling of data privacy and security. We will drill down into the political and social challenges in the chapter "Social and Politics."

Reference

Kahn, D. (1967). *The codebreakers*. New York: Scribner.

Changes in Society and Politics

> *We must plan for freedom, and not only for security, if for no other reason than only freedom can make security more secure.*
> Sir Karl Popper, The Open Society and Its Enemies (1945)

The creation of new technologies for production and communication have not only influenced economic structures but also go hand in hand with social impacts. In comparison to other, older service ecosystems based on networks, cloud computing and the World Wide Web have a much more dramatic influence on social behavior. Cloud computing and cloud services form the first real, global service ecosystem, and create new services and applications every week, day and hour. Cloud computing has reached literally each corner of the globe and each social class, creating countless socially relevant effects. For the first time, a global community has been created, being borderless and "digital." With this paradigm shift, a new type of space on top of private space and public space was created, it's a hybrid area where new rules are developed and change quickly.

The transition to a digital cyberspace has huge impacts on the relation between enterprises and individuals on one side and states and legal organization on the other side. Legislative entities around the world try to develop strategies to achieve the desired level of freedom on the Web by implementing security measures to protect enterprises and individuals from misuse and damage. At the same time, governmental organizations aim to use this data to empower their defense structures and fulfill their mission of protecting the society.

The borderless information flow, communication and interaction options break information monopolies held by the legislative or the—sometimes controlled—mass media. This leads to unexpected effects, where information, intelligence and knowledge bypasses borders and traditional instances. The self-organization of political movements, criminal or terrorist organizations or streams of refugees have all evolved in a dramatic way during the last decade.

M. Oppitz, P. Tomsu, *Inventing the Cloud Century*,
DOI 10.1007/978-3-319-61161-7_15

The Purpose of the Web and the Clouds

The goal of the Web is to serve humanity. We build it now so that those who come to it later will be able to create things that we cannot ourselves imagine.
Tim-Berners Lee, 2010 [1]

The Internet, the Web and cloud services created a completely new nexus compared to traditional forms of personal and business relations. This results in a network structure that spans the globe dissolving existing borders and forming a cyberspace that is not endless but borderless. It moved the consumer from a peripheral position to a virtual epicenter in the Web providing him or her with its own individual reference framework. Creating a digital existence as an individual or company and interacting with others is based on links, relations and digital profiles, which are formed individually. Those relations are relative and not bound to absolute reference systems. Geographical locations of people or organizations lost their importance, roles of individuals and organizations now depend more on their activities, and even time has lost influence, as the net is simultaneously everywhere. It seems that the Web has led to a transition from a world of an absolutely determined reference system to a relativity of digital existences.

In 2014, the founder of the World Wide Web, Sir Timothy Berners-Lee, said:

The web evolved into a powerful, ubiquitous tool because it was built on egalitarian principles and because thousands of individuals, universities and companies have worked, both independently and together as part of the World Wide Web to expand its capabilities based on those principles.[2]

Three years later, it has become evident that the brave new world is not perfect. Increasing problems regarding privacy, security and fair use of the new media and infrastructures has reached a peak of frustration and discontent. On the 28th birthday of the Web in March 2017, Berners-Lee added:[3]

- We've lost control of our personal data
- It's too easy for misinformation to spread on the web
- Political advertising online needs transparency and understanding

It seems that we are right in the middle of a thunderstorm "consisting" of the digitalization of industries and personal life, cybercrime, cyberwarfare and the creation of new social structures. In this stormy weather, geographical borders seem to fade away, globalization's effects influence economic systems and social life. Traditional industries face challenges when transitioning to digital business models, hunted by startups born on the Web. Privacy and the border between

---

[1]Tim Berners-Lee, 2010: https://www.cnet.com/news/tim-berners-lee-the-web-is-threatened/ retrieved 2017-03-10.

[2]The Guardian, 2014: https://www.theguardian.com/technology/2014/aug/24/internet-lost-its-way-tim-berners-lee-world-wide-web retrieved 2017-02-10.

[3]The Guardian, 2017: https://www.theguardian.com/technology/2017/mar/11/tim-berners-lee-web-inventor-save-internet retrieved 2017-03-28.

private and public space is constantly redefined and changes with every new type of service offered on the Web. The decision to refuse being a part of the global village seems to be no longer a valid option for many individuals and businesses. Services and decisions are outsourced to cognitive machines, creating uncertainty on the consumer side.

The legal frameworks build by states and governments to protect their citizens seem to be no longer effective and efficient. As a result, citizens and consumers have lost trust in centralized organizations like governments or banks, having experienced the lack of protection and accountability seen in the financial crisis of 2008.

The Web as an Amplifier

Steam and electricity amplified the strength of humans via railroads, telegraphs and telephones, which changed the perception of geography and distance. Both led to the industrial revolution in the eighteenth and nineteenth centuries and also had a major impact on not only the economy but also on politics and social life. Industrialization was a major trigger in the creation of large enterprises and empowered the establishment of democratic systems. With the invention of digital computers in the 1940s and the shift to generally usable computer architectures in the 1950s, a new type of amplification was introduced. Digital computers were perceived as amplifiers of "intelligence" through their ability to calculate, process and store data with more capacity than humans. The Austrian computer pioneer Heinz Zemanek claimed it is important to see that the converse is also true: computers are also amplifiers of nonsense and stupidity.[4]

Analyzing the Web from that perspective, it is a multiplier of personal connections and thus can amplify emotions. The Web not only amplifies communication, data processing and information distribution, it is also a rapidly growing amplifier of emotions. There are 3.5 billion web users worldwide today, representing nearly 50% of the world's population with access to more than 1.1 billion websites. An estimated number of 2.5 billion users use social media networks to communicate and exchange information, opinions and emotions. For many years, the Web and cloud-based services have expanded from North America and Europe to the rest of the world. China and India are the largest online markets today, each of them with more Internet users than the USA followed by Brazil and Japan. Internet penetration and access speed rankings are also led by countries like South Korea, China or Japan. The Internet, the Web and cloud ecosystems are a global phenomenon. The growth rates are impressive: in 2017, there will be more Internet traffic than all the prior years combined.

[4]Heinz Zemanek, Lecture on "The Rise and Fall of Cybernetics", 1978, notes by the author.

With a network consisting of cable connections and a rapidly growing capacity for wireless connection based on more than 1 billion Internet hosts, the Internet and the Web are the largest technical creations ever built.

Digitalization

Digitalization or the "Digital Revolution" is the term coined for huge changes in business, administration and private life triggered by the massive growth of computer usage beginning in the twenty first century. In the last decade of the twentieth century and the first decade of the twenty first century, the worldwide capacity to store digital data has expanded from 1.6 E+13 MB in 1993 to 3.0 E+14 MB in 2007.[5] Thus, storage capacity has grown by a factor of 1000 over a stretch of 7 years. At the same time, the computing capacity of general purpose computers including supercomputers, servers, PCs and smart phones has expanded from 4.5 E+9 MIPS (Mega Instructions per Second) in 1993 to 6.4 E+12 MIPS in 2007 (see Fig. 1).

Powered by networking infrastructure and services, data distribution was moved from analog to digital technologies. From 1993 to 2007 the share of digital broadcast and telecommunication increased from below 1% in 1993 to more than 20% in 2007 and is expected to reach more than 99% by 2020 (see Fig. 2).

This development had a major influence on business, politics and society. Within the very short period of 20 years, digital technologies empowered by the Internet, cloud computing's new services and business models changed the lives of people, the strategies of companies, influenced international politics and created new types of social communities.

Speed of Change

The number of Internet users reached more than 3.4 billion in 2016 and is projected to reach 50% of the world's population within the next year. The number of social media users has grown since 2000 to more than 2.5 billion. The number of websites exceeded 1 billion in 2015. These trends seem to continue and the growth rates are outperforming the worldwide population growth rates (see Fig. 3).

With the rise of the commercial Internet in the 1980s, the introduction of the World Wide Web in 1992 and this rapidly growing new economy, the digital revolution has gained speed. The steadily growing network and computer capacities

---

[5]Martin Hilbert, Priscila López: The World's Technological Capacity to Store, Communicate, and Compute Information, 2011: http://science.sciencemag.org/content/332/6025/60 retrieved 2017-01-20.

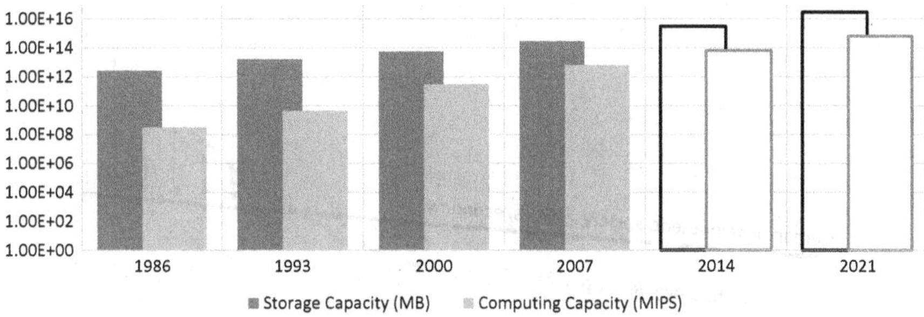

Fig. 1 Growth of worldwide storage and computing capacity (logarithmic plotting)

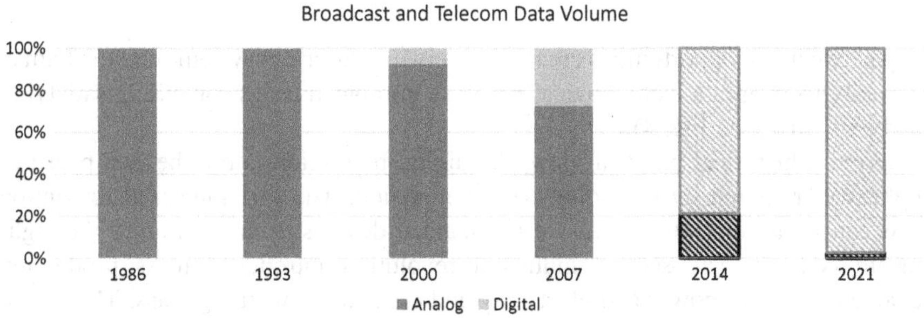

Fig. 2 Broadcast and telecom data volume, share between analog and digital

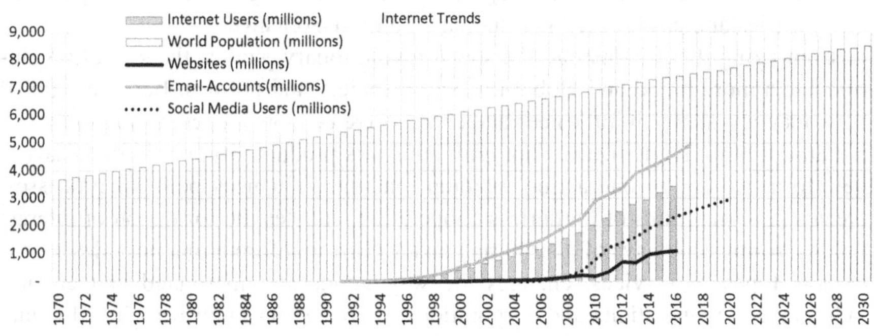

Fig. 3 Internet and web trends and projections

have provided a fertile foundation for new technologies like the Internet of Things, big data and analytics and cognitive computing. These technologies and their resulting new products and services are the building blocks of the digital revolution. One of the major challenges is the ongoing acceleration of technical developments compared to our ability to assimilate the consequences, opportunities and threats from these changes. Compared with the exponential growth of digital capacities and the subsequent new services now offered, the development of social structures, political strategies and legal regulations has been slow. This has led to a growing

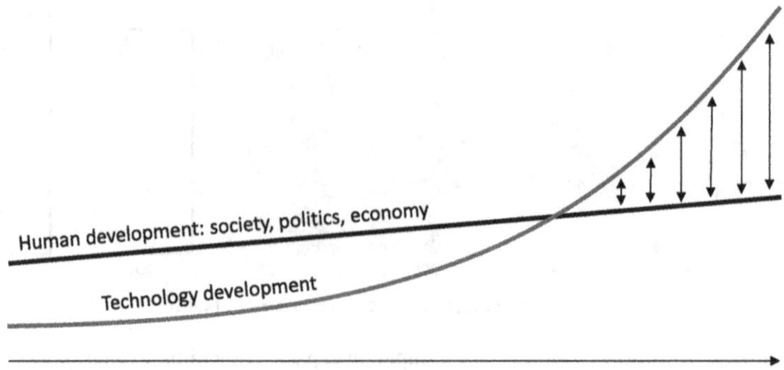

Fig. 4 Human and technology development

gap between the opportunities created by network based ecosystems and the human capacity to integrate technological progress into our lives as something satisfying and enriching (see Fig. 4).

From a historical point of view, the digital revolution is not the first phase of upheaval triggered by technology and innovation. The first industrial revolution was caused by the introduction of mechanical devices to production in the eighteenth century. The second industrial revolution introduced mass production enabled by steam power and electricity and created the working class. This led to an explosion in industrial productivity but also to the formation of new political systems based on social democratic ideas. The third industrial revolution, together with mass communication technologies, was one of the triggers in the globalization of industries and empowered international cultural exchange.

Comparing these industrial and social revolutionary phases, the digital revolution has a more broad and global impact on society and politics. Based on global network ecosystems, it influences and changes the economy, society and politics in a dramatic way. Digitalization not only improves productivity, but it also leads to new forms of social interaction, opinion making, influences political processes, creates new structures of borderless cooperation, enables the formation of interest groups, forms new ways of distributing and accessing information, provides access to new types of services, changes the way goods are distributed and enables individuals, organizations and corporations to create and publish their individual identities.

Quality of Life

The overall effect is two sided. Internet, Web and cloud services provide a huge number of services that improve the quality of life at low or zero cost to the consumer. Communication with remote friends or beloved family members is easy, exchanging or getting access to important or vital information is simple, selecting and buying the best product at the lowest price is likely, and staying informed and referencing numerous sources of information is unpretentious.

On the other side, the use of the Internet and Web has its risks. Being part of the digital community requires giving personal data to the provider of a service and, for example, requires users to make individual profiles where their activities are visible in a digital public space. The Web has formed its own rules about public and private spaces, and they are completely different from the notion of privacy and public space that we had before the Internet. Finally, personal identity took on a new meaning with the introduction of the Web and cloud based services. The Internet and the Web has created a new nexus of digital identities which may have complex relations to the real world. Trust has a new meaning and borders between trusted and untrusted communities, groups or service providers are no longer related to the geographical borders between countries and their legal systems.

We will go through those different aspects within the following chapters.

Digital Social Networks

One of the most visible effects of the Internet and the Web is the dramatic success of social networks. Since the introduction of the first social network services, the number of active users has grown rapidly reaching more than 2.5 billion in 2017 (see Fig. 5).[6]

Social networks like Facebook, WhatsApp, Instagram, Twitter and others became some of the killer applications on the Web. The initial purpose of social networks was to provide platforms for borderless communication to individuals, organizations and corporations, based on a business model that created revenue from advertising. Within a couple of years, it turned out that social networks had created some painful side effects.

Digital Interest Groups

In the last 5 years, social networks have developed to become a widely-used media for interest groups of all kinds. Creating communities and using the Web as a communication platform was recognized as a low-cost and highly effective method of spreading news, sending political messages and building national and international networks of supporters and activists. The spectrum includes hobbyists as well as social movements, political parties, revolutionary movements and terrorist organizations and criminals. In 2009, Twitter became one of the important communication channels for Iranian protesters after the government shut down other communication channels.[7] The political rebellions in Egypt in 2011, known as the

---

[6]Statista 2017: https://www.statista.com/statistics/272014/global-social-networks-ranked-by-number-of-users/ retrieved 2017-03-10.

[7]The Atlantic: The Revolution Will Be Twittered, 2009: https://www.theatlantic.com/daily-dish/archive/2009/06/the-revolution-will-be-twittered/200478/ retrieved 2017-01-20.

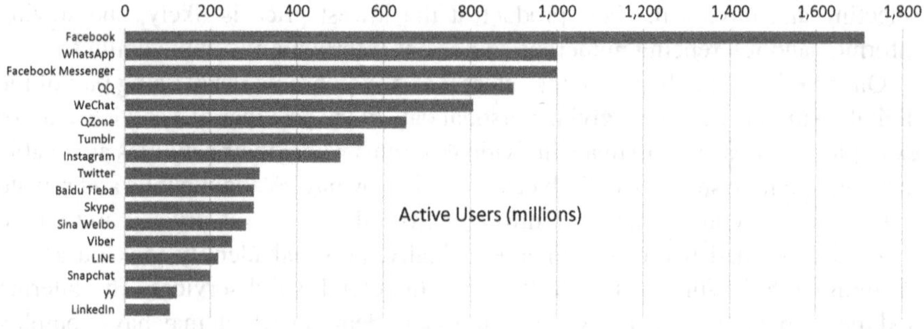

Fig. 5 Social networks, active users January 2017

Arab Spring, were one of the first movements that used social media and messaging services for their organization and communication. These political rebellions were later coined the "Facebook revolution." One of the protesters said: "*We use Facebook to schedule the protests, Twitter to coordinate, and YouTube to tell the world.*"[8] In Turkey during the Gezi Protests in 2013, Facebook and Twitter were used to "broadcast" what was happening on the streets when mainstream media, mostly controlled by the government, was silent on police brutality.

The rapidly increasing refugee streams from the Middle East and Africa to Europe reached a climax in 2015 and were supported using mobile phones and social media. For refugees coming from war-torn Syria or Iraq, the smartphone is as essential as food and shelter, providing vital information about routes, obstacles and enabling them to stay in contact with others.

In the last 5 years, terrorist organizations have started to use social media as a communication and promotion channel. Islamic terrorist groups like ISIS have developed competences and skills using and applying social media networks to a much broader extent than their forerunners. The success of ISIS in promoting their messages and recruiting supporters in different countries is based on the calculated use of social networks and media like Facebook, Twitter and YouTube videos.

Digital Propaganda: Fake News and Trolls

Compared to analog forms of social communication, digital social networks turned out to be a great platform not only for distributing and sharing views and thoughts but also creating opinions. Creating opinions on social media platforms is enforced by the effect of so-called echo rooms. Participants in social media networks tend to connect to others which share the same social, cultural, ethnic or political views. New posts referencing new information sources are shared within that community very quickly, and comments and reflections usually follow the same stream of sentiments or opinions thereby creating an effect of positive reinforcement that echoes the participant's own views. This effect is also used by organizations that follow this strategy of reinforcing a political statement using echo rooms.

[8]Fawaz Rashed, 2011: https://www.theguardian.com/world/2016/jan/25/egypt-5-years-on-was-it-ever-a-social-media-revolution retrieved 2017-03-15.

In the last few years, social media networks gained popularity among interest groups and political parties which used them to distribute their unfiltered messages and bypass traditional media like newspapers or TV-channels. Politicians started to use messaging services like Twitter to communicate with their voters. From the politician's position, using social media is a good way to get attention without exposing oneself to criticism or reflection from third parties or journalists. Furthermore, direct messaging produces a feeling of exclusivity for the target group receiving the original message from the celebrity.

Consequently, the effect of echo rooms and targeted message distribution was recognized as a powerful tool for reinforcing or changing opinions. Once a community is trained to listen to their preferred social media network, it becomes possible to influence and pollute the community with goal-oriented or fake news. The producers of this fake-news are called "Internet-trolls." The method is used today by political parties as well as interest groups to draw attention to their targets and move public opinion towards their preferred direction in order to win voters at the next election.

Since 2015 it seems that the production of fake news has become a business and a powerful tool in forming opinions. In 2016, the Russian administration was suspected of hiring groups of Internet-trolls located in St. Petersburg to influence elections in the USA and Germany.[9] The Harvard political scientist Gary King stated in a 2016 study that the Chinese government creates up to 440 million pro-government posts per year to influence political opinions in China.[10] In a 2016 study for the NATO Strategic Communications Centre of Excellence it was reported that the Russian military intervention in Ukraine "*demonstrated how fake identities and accounts were used to disseminate narratives through social media, blogs, and web commentaries in order to manipulate, harass, or deceive opponents.*"[11]

Digital Violence: Hate Speech and Content Moderators

Social networks are characterized by users who are not personally confronted with their communication partners. Just sitting in front of the screen makes it much easier to transgress emotional and ethical boundaries. In the worst case, the effects of hate posts that are targeted towards specific users are still felt by a larger community of connected users. Hate speech or abusive pictures have become a painful social effect of social media platforms and messaging services.

Though there is no international data available, it seems that the problem is growing. Various countries have introduced laws against hate speech and posts on

[9]Russian Internet Trolls: http://www.businessinsider.de/russia-internet-trolls-and-donald-trump-2016-7?r=US&IR=T retrieved 2017-01-20.

[10]Gary King, et al.: How the Chinese Government Fabricates Social Media Posts for Strategic Distraction, not Engaged Argument, 2017: http://gking.harvard.edu/files/gking/files/50c.pdf retrieved 2017-03-20.

[11]NATO StratCom COE, Internet Trolling as a hybrid warfare tool: the case of Latvia, 2016: http://www.stratcomcoe.org/internet-trolling-hybrid-warfare-tool-case-latvia-0

social media platforms. Those regulations target the sender of the message as well as the provider of the social media network and obligate them to delete the message immediately. It is also evident that the interpretation of what counts as hate speech or, at least, crude and abusive language differs from culture to culture and from country to country. In some cases, they must match with the criteria of specific human rights declarations on free speech. Differences are sometimes cultural (e.g. blasphemy may have a very specific meaning in some Arabic countries), and sometimes political (e.g. Nazi propaganda in Germany or Austria). Also, the position on racial discrimination or sexual orientation is different in different cultures. An international definition still does not exist, and various organizations like the "International Covenant on Civil and Political Rights" have tried to create an international framework matching the different views on freedom of speech and its relation to discriminating content on the Web.

In Europe, the European Union took on the challenge in 2016 and introduced a code of conduct. Three of the largest social media providers (Facebook, Google, Microsoft and Twitter) jointly agreed to the code obligating them to "*review the majority of valid notifications for removal of illegal hate speech*" posted on their services within 24 hours.[12] Some providers of social media networks like Google or Facebook operate so-called content moderator centers to remove crude or abusive texts or pictures. Those centers employ up to 100,000 moderators, most of them located in the Philippines and they do one of the dirtiest jobs in the Web business.[13]

Distinguish Between Information, Advertisement, and Fake News

Our existence in the digital social community depends on our ability to discern between facts and falsehoods or nonsense and relevance. With the dramatic growth of information and messages, which are distributed by social networks, emails and knowledge platforms, this talent became a question of survival on the Web. A large portion of information and content on the Web is either incomplete, misleading or simple false. Filtering relevant content from the rest is hindered by smarter distributers of fake news or smart algorithms applied by the social network platforms to distribute ads or offers within search results or blogs.

In a 2016 study the Stanford Graduate School of Education reported "*a dismaying inability by students to reason about information they see on the Internet. Students, for example, had a hard time distinguishing advertisements from news articles or identifying where information came from*". Around 82% of middle-schoolers couldn't distinguish between an ad labeled "sponsored content" and a real news story on a website.[14]

---

[12]The Guardian 2016: https://www.theguardian.com/technology/2016/may/31/facebook-youtube-twitter-microsoft-eu-hate-speech-code retrieved 2017-02-20.

[13]Wired, The Laborers Who Keep Dick Pics and Beheadings Out of Your Facebook Feed, 2014. https://www.wired.com/2014/10/content-moderation/ retrieved 2016-10-20.

[14]Stanford Graduate School of Education, Stanford researchers find students have trouble judging the credibility of information online, 2016: https://ed.stanford.edu/news/stanford-researchers-find-students-have-trouble-judging-credibility-information-online retrieved 2017-01-10.

Developing skills and behaviors needed to critically analyze information on the Web is a major challenge for today's users and future generations.

Privacy, Identity and Security

Within the last decade, the Web created a completely new perception of roles and identities in a virtual universe. The individual user as well as organizations or enterprises are confronted with the need to build and manage digital identities and profiles in the borderless space of the Internet. In the traditional analog world of personal communication, printed media and physical libraries, the single individuum was the consumer of centralized collections or distributions of information. Libraries or newspapers understood themselves as the epicenter of information, the user acting as the consumer from a peripheral position. Interaction and communication with others took place in physical locations, the participants enjoying a face-to-face discourse within clearly defined borders of publicity.

Users and Their Virtual Private Space
The Web completely changes that model of interaction, it moves each user into his or her own center of a virtual universe, providing the user with an identity, address and profile and promising a virtual, individual space to interact with the rest of a seemingly infinite cyberspace. From a web user's perspective, the Web seems to be endless and arranged around the user's web identity as the center of this universe. Borders between private and public space are redefined, and being "public" and visible on the Web has created a new form of public space, which is completely different from what we experienced as citizens of an analog world. Furthermore, the Web runs a completely different model of data sustainability. On the one hand, information put on the Web is short-term in that its timeliness is restricted and up to 70% of information published on the Web is out-of-date within 4 months. On the flipside, the Web does not forget, and each piece of data put on the Web or cloud service is stored somewhere, not easy to delete and can surprisingly pop up in a completely different context.

Living as Virtual Identity
Living a life as a virtual identity in this environment requires different strategies compared to the clear borders between private and public space in the physical world. Conflicts arise when consumers of cloud services like social networks, search engines or e-commerce platforms claim that they act within a virtual private space reserved only for them and clearly separated from the rest of cyberspace. In most cases, using the Web and cloud services as consumers automatically generates a public character defined by personal profile data and an increasing volume of other data as well. It is a fact that true privacy does not exist on the Web.

Information put on the Web by users is called private by the user but in reality, this information is taken as value and virtual currency by the providers of the services. Many of the business models applied by cloud service providers are based on data collection and the monetizing of user data. As we have seen in the chapter called "Security," these large data collections are also targets for criminal organizations trying to gain access to private data.

Misuse of the virtual public space also takes place when users underestimate their level of privacy and use social networks for personal statements including hate speech or offenses. This effect is facilitated by the lower barrier of aggression seen when someone hides behind his or her digital identity.

Protecting Privacy

The discussion on privacy and identity is closely connected to security and sometimes unclear with respect to the different views on privacy of data. From a web consumer's point of view, the responsibility for privacy and security of data and the protection of identities is on the provider of the cloud service. Like in other areas of social or business life, users expect protection from their government's legal code, rules and regulations. The past couple of years have shown that these expectations were not met by the authorities.

Momentarily there are different strategies under discussion to solve the data privacy and security conflicts:

- Guaranteed anonymity
- Intensified data privacy laws
- Strong identity management

Guaranteed Anonymity

Guaranteed anonymity is based on the idea that a user of a service has the right to stay anonymous, which simply means that no personal data like an Internet-address, name, address, bank account data or other data elements must be provided by the user to enjoy the service. This strategy is used by most of the services in the darknet where connection protocols like TOR are used to hide identities (see also chapter "Security"). Anonymity is also one of the key features of cryptocurrencies like Bitcoin (see chapter "Future Paradigms and Big Disruptive Things"). In both cases, anonymity is implemented based on the idea of anonymity as a human right. Simultaneously it is also true that in both cases anonymity is used by criminals to hide illegal activities. Moving to a world of anonymous users on the Web would protect the individual user but also eliminate each form of legal control and lead to a free and anarchic net.

More Rules and Regulations

The strategy applied momentarily is characterized by the ongoing intensification of national and international rules and regulations regarding data privacy and security. This strategy assumes that national and international authorities have to take over the responsibility of protecting consumers on the Web from damage and harm. Though governments had started to create compliant legal frameworks in the 1980s, the effect was limited. We will drill down into this fight for rules and regulations a little bit later.

Strong Identity
Strong identity management is a solution that restricts the use of certain services to consumers who have provided a well-defined and authorized identity. The model is already used by security certifications offered by web-sites and authorized by certification authorities. Extending this concept to all types of participants in the cloud ecosystem would lead to a Web that is much more secure but also controllable by authorities. What is more, centralized authorities would take over the governance of credentials and the identity issues of individuals and organizations.

Trust and Borders

Building New Trust

Trust is a major pillar for living together in communities and ecosystems. Having confidence in fellow citizens and providers of services or products empowers the sense of a high quality of life. There is a widely-spread consensus in social science that "*Confidence is one of the most important synthetic forces within society*."[15] Trust contributes to economic growth and efficiency in market economics. It also contributes to the provision of public goods, social integration, co-operation and harmony, personal life satisfaction, democratic stability and development, and even to good health and longevity. Building and nursing trust and confidence is empowered by personal relations between people, personal experiences with the deliverers of goods and services and by affirmative relations to authorities. Well known geographical borders and boundaries of laws and regulations help individuals and businesses to form a clear picture of trust borders.

The cyberspace created by the Internet, the Web and global cloud services had changed that picture within two decades. Fostering trust and confidence in a virtual world without well-defined borders and hardly controlled by reliable rules and regulations evolved to be a major challenge. Research reports show that the level of trust in a society correlates with elements like sharing the same culture, language or religion, keeping personal contacts and confidence in the legal system.[16,17] A higher level of trust in fellow citizens also correlates to the economic success and welfare of a community.

---

[15]Georg Simmel, Kurt H. Wolff "The Sociology of Georg Simmel", 1950: https://books.google.at/books?id=Ha2aBqS415YC&lpg=PA345&dq=Simmel%20secret%20and%20secret%20societies&hl=de&pg=PA318#v=snippet&q=trust%20is%20one%20of%20the%20most%20important&f=false retrieved 2017-03-10.

[16]Esteban Ortiz-Ospina and Max Roser "Trust", 2016: https://ourworldindata.org/trust retrieved 2017-01-20.

[17]Glaeser, E. L., Laibson, D. I., Scheinkman, J. A., & Soutter, C. L "Measuring trust", 2000: https://dash.harvard.edu/bitstream/handle/1/4481497/Laibson_MeasuringTrust.pdf?sequence retrieved 2017-02-15.

All these supporting elements for creating a satisfactory level of trust do not exist or are weakly pronounced in the digital world of cyberspace. Relations in social networks are not based on personal acquaintances but on digital profiles, business relations between consumers and cloud providers must rely on the quality of a website and exchange data instead of handshakes. The terms and conditions applied are sometimes unclear and their compliance with local rules are questionable.

Consumers of cloud services and businesses moving into the new world of digital relations to their customers or business partners are challenged to develop new skills for building confidence and trust. Building new trust is a basic assumption of embracing the vast possibilities of the Internet, the Web and cloud services. Consumers, societies and businesses accepting the digital network-based ecosystem provided by the Web as an opportunity to improve their quality of life and increase their business success must develop an awareness of new forms of trust and mistrust to survive the transition into the cloud century.

Trust in Information

Being in a center of a virtual network leads to an overwhelming supply of information offered by search engines, social networks and messaging services. Most of this information is created automatically based on the consumer's activity record or personal profile and serves as a major revenue source for the providers of those services. Cloud service providers like Google, Amazon or Facebook create a large part of their revenue through advertising using smart algorithms to send advertising messages directly to the target group. Evaluating the content and distinguishing between ads and desired content is up to the consumer. A comparable effect takes place when retrieving information actively from digital encyclopedias, data collections or news channels. Not each Wikipedia article is complete and correct, news blogs may contain filtered or fake information and data collections may be incomplete or disputable.

Compared to the analog age of information distribution based on personal conversation, local print media and a small number of local radio and TV channels, the information offered on the Web is like drinking from a firehose: one gets wet and stays thirsty. Getting the most out of this information flush could lead to a better understanding of the world we live in, contribute to social awareness and form a higher level of cooperation and creative thinking, but it also demands highly developed skills in the assessment of Web content, messages received, blog posts and sources of information. The talent of judgment and criticizing what is presented before us will be key competencies in the future.

Trust in Persons

Even though building personal relations in a global village is one of the major achievements of social networks, the saying *"Don't talk to strangers"* seems to be nonexistent in the world of cyberspace. Social networks, news media posts and blogs makes it easy to lower the barrier of entry into conversations and entrap users to give away personal information to unknown persons, groups or organizations. It is sometimes fascinating and disturbing to see what kinds of intimate information or personal opinions are freely offered by users to an unknown community or communication partner. This behavior is caused by an often complete misestimation of the borders of private space on the net and demonstrates that trust between communication partners in cyberspace is much higher than in the real world. Misjudging the importance of trust and thus acting as if the confidential behavior of communication partners is a given assumption on the net is one of the major causes for conflicts leading to financial damages or personal harm.

Trust in Cloud Services and Providers

Living in a digital world as individuals or enterprises means being connected to and exchanging information with other individuals or enterprises. Personal data is communicated or stored in cloud applications, in the data collections of communication providers, Internet service providers, telecommunication providers and all sorts of e-commerce platforms, search engines and social networks platforms.

Compared to the traditional analog world, a trusted relationship between the consumer and provider is no longer influenced by direct personal contact but by the digital appearance and judgements of third parties. Furthermore, providers like cloud infrastructure providers or social media providers cannot easily be found in a trusted legal framework or within certain geographical borders, as they deliver services and content from a virtual space. Building confidence between partners in the cloud ecosystem demands different methods of building trust like the application and execution of strictly defined service levels, terms and conditions. Relying on well-defined legal frameworks that cover requirements and expectations on both sides of the partnership is a helpful foundation. Such frameworks, in their current state of incompleteness, are one of the major obstacles in the rollout of efficient cloud ecosystems.

Trust in Authorities

Building and executing applicable and internationally agreed upon legal frameworks for the cloud and network-based ecosystem is the main way to create confidence and trust. National and international authorities were moved to the role of protectors from damage, misuse and injury for cloud ecosystem partners. It is also a fact that, due to the rapid development of the Web and cloud ecosystems, those authorities were not able to fulfill every expectation. The rapid change in technology and the economy left the comparably slow processes of regulation-building for organizations far behind. Trust in authorities as the custodians and executers of cloud service-related regulations decreased and the responsibility of protecting individuals and corporations is currently moving away from authorities and back to consumers and corporations. The failure of international legislative agreements, like Safe Harbour, led to damaging effects for the cloud ecosystems as seen in the increasing resistance to using cloud services outside certain legal or geographic borders.

As we will analyze later, the fight for international and applicable rules and regulations is still ongoing. It is affected by a completely new definition and relevance of borders associated with the Internet, the Web and cloud-based ecosystems.

Creating New Borders

The fundamental concept behind the Internet and the Web is one of global cooperation and borderless connections. The Internet and Web have dissolved existing political, geographical, legal and cultural frontiers. For centuries, geographic borders determined the daily life of individuals and were the reference line that formed communities, states and nations. Language and the cultural identity of societies had a strong reference to geographic locations, and physical borders were primary used as orientation lines. Political, economic or cultural conflicts usually occurred along borders sometimes marked with fences or walls and perceived as strongholds against unwanted foreign influence or interference. With the rise of global trade, political borders also became the separating lines between different economies and were used, together with custom barriers, to create trade agreements and enable commercial relations consistent with national interests.

In the twentieth century, various social and political movements created visions of a more borderless society driven by the painful experience of military conflicts and wars. National and international peace movements tried to see political and cultural borders as life-enhancing interfaces between nations instead of battlefronts, and political movements like socialism or communism found their vision in building an international community based on social principles. After World War I and World War II, the movement towards international organizations like the United Nations Organization intensified. In Europe, the European Union started to build a political and economic community of nations. The forming of multilateral trading agreements seemed to dissolve trading barriers and the globalization and liberalization of economies significantly changed the distribution and production of goods.

The introduction of the Internet and the Web changed this scene in a dramatic way. Web and cloud services began to bypass traditional borders creating a virtual and basically borderless universe. In contrast to the physical world of geographic borders, national systems of law and regulations, language barriers and cultural identities, within a short time the Web created an alternative world dissolving traditional borders and forming a virtual space for social interaction and economic activities. The Internet, Web and cloud services has begun to realize the dream of the global village and fulfills the requests from the builders of the Internet and the Web to provide a powerful, ubiquitous tool built on egalitarian principles and enabling thousands of individuals, universities and companies to work together independently. At the same time, the loss of geographical, cultural and political borders as reference points are some of the major challenges of a cloud-oriented society. Traditional borders are replaced by new borders, some of them artificial and unwanted by users and some of them highly desirable to protect individuals and enterprises from damage. The struggle for a consensus in defining, building and protecting those virtual borders has become an ongoing trial for governments, social communities and political movements.

Compared to its initial, visionary goal, today Internet and Web use is dictated by a heterogeneous collection of localized and national restrictions and a certain level of uncertainty regarding protection from damage and privacy violation for individuals and corporations.

Culture and Language

In the era before the Internet, the borders between cultures or languages were defined by geographical borders. Building personal or business relations using the Web and cloud services is no longer supported by reference points like a location or home country. Communicating with other individuals in a social network may cross cultural borders, which can be a fascinating and enhancing experience but can also lead to misunderstandings and cultural clashes. Building business relations with customers or providers of services somewhere in the cloud is a virtual experience and not comparable to making a deal with a personal handshake.

The Internet and Web created a unique culture of communication, which is influenced by ideas from network communication ethics, and uses English as the common language thus narrowing the communication channels for some people. In parallel, graphics, pictures and videos started to complement or replace text as the method of communication thus providing a more global language that is easier to use. The sharing of picture and video content on corporate websites and social media networks is increasing dramatically. By 2019 more than 80% of the network data volume will be video having replaced text as the major communication method.

This mashup of different cultures in the global network is also accompanied by clashes between different approaches to ethical principles. Differences may originate from religion, social rules about communal living and different political histories. The picture of a couple sitting hand in hand at the beach may be perceived of as "romantic" in Sweden but offensive in Saudi Arabia. A picture of Adolf Hitler as a profile photo on a blog is illegal in Germany and Austria but seen as a creative marketing joke in Indonesia. Because the concept of time travel does not match up with Chinese philosophy, movies like "Back to the Future" are problematic in China. Supporting Darwin's theory and criticizing the unrestricted right to carry a gun may lead to emotional discussions in some parts of America.

The supporters of a global and free Internet and Web as well as national authorities, international organizations and the providers of Web and cloud services are working hard to find solutions to these dilemmas. The solutions are always different and not completely satisfying. Approaches go either in the direction of fighting for unlimited freedom on the net or implementing the intersection of all cultural and social norms as fundamental law on the net. The first would lead to an extreme form of liberalization and anarchy, the second to an internationally accepted but limited space of interaction reduced to pictures of pets, landscapes and dishes (as long as they do not involve pig, beef, snakes or dogs).

We will go through the different approaches of national governments and supporters of a clearly defined ethics of communication on the net in the following sections.

Censorships and Freedom on the Net

Freedom of speech as a principle can be assumed to be a fundamental human right when using a worldwide communication platform. In reality, this is not the case and governments tend to execute their local interpretation of freedom when restricting use and access to the Web. In a 2016 study, the independent watchdog organization Freedom House reported that: *"Two-thirds of all internet users—67 percent—live in countries where criticism of the government, military, or ruling family are subject to censorship."*[18] In the same study, it was reported that between 2015 and 2016:

- In 46 countries, bloggers had been arrested or imprisoned for political or social content.
- In 32 countries, political, social, or religious content had been blocked.
- In 26 countries, pro-government commentators manipulated online discussions.
- In 25 countries, technical attacks were made against government critics and human rights organizations.
- In 24 countries, social media or communications apps had been blocked.
- In 22 countries, bloggers or users had suffered physical attacks or had been killed.

During the same time period, new laws or directives increasing censorship were passed in 20 countries. Localized or nationwide Internet shut downs took place in 15 countries. Calculating a "Freedom of the Net" (FOTN) score, the list of countries with a high level of censorship is led by China, Iran and Syria[19] (see Table 1).

Surveillance by Authorities

National intelligence organizations use their directive to protect citizens from crime and terrorism to bypass or bend laws and regulations governing privacy. Furthermore, most of the rules and regulations regarding data privacy and the security of democratic states include a certain right to access private data on the net or stored in clouds whenever such information is in a nation's security interests. Several countries have created specialized intelligence organizations and invested in infrastructure to operate surveillance and analysis of Internet data. The USA, UK, China and Russia currently lead in surveillance technologies.

The France-based organization "Reporters without Borders" reported on the status of Internet censorship together with surveillance activities of countries in its 2014 report, "Enemies of the Internet."[20] In the category of surveillance, the USA, Russia and the United Kingdom are referred to as leaders in shadowing the

[18]Freedom House 2016: https://freedomhouse.org/report/freedom-net/freedom-net-2016 retrieved 2017-01-10.

[19]Freedom House, 2016: https://freedomhouse.org/report/freedom-net-methodology retrieved 2017-01-10.

[20]Reporter without Borders: "Enemies of the Internet", 2014: http://12mars.rsf.org/2014-en/enemies-of-the-internet-2014-entities-at-the-heart-of-censorship-and-surveillance/ retrieved 2016-02-12.

Table 1 Top 10 countries in the FOTN score

Country	FOTN score (1...100)
China	88
Iran	87
Syria	87
Ethiopia	83
Cuba	79
Uzbekistan	79
Vietnam	76
Saudi Arabia	72
Bahrain	71
Pakistan	69

Internet and Web activities of users and enterprises. Differing from censorship, surveillance takes place in democratic countries as well as in countries with authoritarian governments. The report claims that: "*Three of the government bodies designated by Reporters Without Borders as Enemies of the Internet are located in democracies that have traditionally claimed to respect fundamental freedoms: the Centre for Development of Telematics in India, the Government Communications Headquarters (GCHQ) in the United Kingdom, and the National Security Agency (NSA) in the United States.*"

In the USA, the National Security Council (NSA) is the driving force behind data collection and analytics of private data on the Internet. The NSA has engaged in massive tracking operations and surveillance of the domestic communication since at least 2001 by intercepting phone calls and Internet communications.[21]

The surveillance activities of the UK's GCHQ (Government Communications Headquarters) became public through the Edward Snowdon papers and included wire taping of transatlantic cables, the collection of private connection data from UK citizens and companies and illegal access to the networks of companies in other countries.[22] Both organizations work together by occasionally sharing information, resources and funds.

The Fight for Rules and Regulations

The discussion about the level of regulations for the Internet and Web is ongoing and is between governments and political communities on one side and interest groups that support the freedom and observance of human rights on the net.

---

[21]Electronic Frontier Foundation, "NSA Spying on Americans"; https://www.eff.org/de/nsa-spying retrieved 2017-01-20.

[22]The Guardian, 2013: https://www.theguardian.com/uk/2013/jun/21/gchq-cables-secret-world-communications-nsa retrieved 2016-12-12.

Non-Profit Organizations

A variety of international organizations support approaches to form a clear understanding of freedom and regulations for the use of the Internet and the Web. These organizations refer to the principles of privacy and freedom as claimed by the Declaration of Human Rights in article 12 on privacy : "No one shall be subjected to arbitrary interference with his privacy, family, home, or correspondence" and article 19 on freedom of expression: *"Everyone has the right to freedom of opinion and expression; yet not everyone receives it. This right includes freedom to hold opinions without interference and to seek, receive and impart information and ideas through any media and regardless of frontiers."*[23]

Those organizations provide reports on Internet censorship or surveillance in different countries and try to deliver a comprehensive picture of the current situation regarding the balance between freedom of use and expression and regulations and restrictions.

1941 Freedom House
Freedom House was founded by Wendell Willkie and Eleanor Roosevelt in 1941 and is an "*. . .an independent watchdog organization dedicated to the expansion of freedom and democracy around the world.*"[24] Starting in 2009, Freedom House has published the "Freedom of the Net"-report providing analysis and numerical ratings regarding the state of Internet freedom in countries worldwide.

1985 Reporters Without Borders
Reporters without borders (RSF) is a France-based organization founded in 1985 by Robert Ménard, Rémy Loury, Jacques Molénat and Émilien Jubineau that promotes and defends freedom of information and freedom of the press: "*At the turn of the 21st century, nearly half of the world population still lacks access to free information. Deprived of knowledge that is essential for managing their lives, denied their very existence, they are prevented from living in pluralist political systems in which factual truth serves as the basis for individual and collective choices.*"[25] RSF publishes reports with its "Enemies of the Internet"-report on Internet surveillance and censorship for up to 50 countries, turning the spotlight on the governments of countries such as China, Burma and North Korea who try to control the Internet.

1990 The Electronic Frontier Foundation
The Electronic Frontier Foundation (EFF) is an international non-profit organization founded in 1990 by Mitch Kapor, John Perry Barlow, and John Gilmore to provide funds for the legal defense of individuals from what it considers abusive legal threats: "*Even in the fledgling days of the Internet, EFF understood that protecting access to developing technology was central to advancing freedom for*

[23]United Nations, Declaration of Human Rights: http://www.un.org/en/universal-declaration-human-rights/ retrieved 2017-01-15.

[24]Freedom House: https://freedomhouse.org/about-us retrieved 2017-02-15.

[25]Reporters without Frontiers: https://rsf.org/en/our-values retrieved 2016-12-12.

all.”[26] The EFF focus on free speech, transparency, privacy and the fair use of the Internet, Web and cloud services. It publishes reports on privacy violations by authorities primarily in the USA but also in other countries.

1990 Privacy International
The London based non-profit organization, Privacy International, was founded by Simon Davies in 1990. Its mission is to *“investigate the secret world of government surveillance and expose the companies enabling it.”*[27] Privacy International publishes numerous reports on the state of privacy and surveillance in different countries and analyzes new technologies focusing on risks in their application.

2004 Open Net Initiative
The Open Net Initiative was a non-profit organization publishing analysis and reports on Internet surveillance and censorship between 2004 and 2014. During this time, it frequently created status reports on the Internet censorship activities of up to 50 countries.

Political Initiatives

With the increasing use of digital data transfer and storage in the 1980s, the question of rules and regulations regarding the fair use and the protection of private data against misuse became more important. The introduction of the Internet and the Web added a new dimension to the problem: the increasing volume of international data transfer and storage of data in different countries required international solutions and agreements. Approaches to create national and international legal frameworks turned out to be different in many countries and the discussion on a unified set of laws and regulations is still ongoing. Most of the countries introduced data privacy laws in the 1980s, and a 2015 report by Graham Greenleaf lists 109 countries with data privacy regulations.[28]

Europe and the European Union
European countries and the European Union were early innovators and started to form a comprehensive legal framework in the 1980s. The Convention for the Protection of Individuals with regard to Automatic Processing of Personal Data was concluded within the Council of Europe in 1981. In 1995, the EU Directive 95/46/EC on the protection of personal data was enacted, and was implemented by the EU member states over the following years. The directive described clear principles for the use of private data:

- Data subjects should be given notice when their data is being collected;

[26]Electronic Frontier Foundation: https://www.eff.org/about retrieved 2017-01-15.

[27]Privacy International: https://privacyinternational.org/ retrieved 2017-01-12.

[28]Graham Greenleaf, Global Data Privacy Laws 2015: https://papers.ssrn.com/sol3/papers.cfm?abstract_id=2603529 retrieved 2017-01-10.

- Data should only be used for the purpose stated and not for any other purposes;
- Data should not be disclosed without the data subject's consent;
- Collected data should be kept secure from any potential abuses;
- Data subjects should be informed as to who is collecting their data;
- Data subjects should be allowed to access their data and make corrections to any inaccurate data; and
- Data subjects should have a method available to them to hold data collectors accountable for not following the above principles.

USA

Compared to Europe, data privacy in the USA is not highly legislated, and rules and regulations about data privacy and use are split into various laws focusing on areas like health data, financial data or private communications. Some states have more stringent laws than others, such as the California Online Privacy Protection Act (CalOPPA), which is the first law in the United States that specifically requires websites to post a privacy policy.

Safe Harbour as the First Attempt at an International Regulation

The mismatch regarding data privacy laws in different countries led to the so-called Safe Harbour initiative in 2000. As part of the EU Data Protection Directive, the EU recommended that *"Companies operating in the European Union are not allowed to send personal data to countries outside the EU unless they guarantee adequate levels of protection."* Such protection can either be at a country level (if the country's laws offer equal protection) or at an organizational level (where a multinational organization produces and documents its internal controls on personal data).

In July 2000, the European Commission decided that US companies complying with the Safe Harbor principles could transfer data from the EU to the US. US companies (or subsidiaries of EU companies in the US) had to opt for a Safe Harbor registration with the US Department of Commerce and go through a certification process. The Safe Harbour principles were based on the 1995 EU directive and defined that:

- Individuals must be informed that their data is being collected and about how it will be used.
- Individuals must have the option to opt out of the collection and forward transfer of the data to third parties.
- Transfers of data to third parties may only occur if the other organizations follow adequate data protection principles.
- Reasonable efforts must be made to prevent loss of collected information.
- Data must be relevant and reliable for the purpose it was collected for.
- Individuals must be able to access information held about them, and correct or delete it if it is inaccurate.
- There must be effective means of enforcing these rules.

By September 2015, approximately 5500 US companies have joined the Safe Harbor register (IBM, Microsoft, General Motors, Amazon.com, Google, Hewlett-Packard, Dropbox and Facebook and others).

The Patriot Act Conflict

In June 2011, Microsoft U.K.'s managing director, Gordon Frazer, admitted that cloud data, regardless of where it is, is not protected from the US Patriot Act. The question put forward was:

Can Microsoft guarantee that EU-stored data, held in EU based datacenters, will not leave the European Economic Area under any circumstances — even under a request by the Patriot Act? Frazer explained that Microsoft must comply with local laws and "*customers would be informed wherever possible,*" but he could not provide a guarantee that they would be informed, providing a gagging order, injunction or U.S. National Security Letter permits it. He said: "*Microsoft cannot provide those guarantees. Neither can any other company.*"[29] The Microsoft statement was also supported by research papers in the following months underlining how the Patriot Act allowed U.S. law enforcement to bypass European privacy laws.[30]

On June 18, 2014, the case "*Europe v Facebook group*" was brought to an Irish high court by the Austrian data privacy activist Maximilian Schrems who then referred the case to the Court of Justice of the European Union. On October 6, 2015, the Court of Justice of the European Union ruled that national supervisory authorities still have the power to examine EU-US data transfers and thus the Safe Harbor framework is invalid for several reasons under EU law. Data-sharing with countries deemed to have lower privacy standards, including the US, is prohibited. The brave attempt to create an international framework covering data privacy and security rights on the Internet seemed to be dead.

General Data Protection Regulation and the EU-US Privacy Shield

Following the October 2015 decision, the European Union started to reform the Data Protection Directive under the name of General Data Protection Regulation also including an improved version of Safe Harbor under the name of EU-US Privacy Shield. The new agreement would guarantee that the data protection rights of European citizens would be observed by US authorities, companies and intelligence agencies.

Negotiations between the European Union and the USA are still ongoing. An internationally agreed on and executable data privacy framework is still far away.

Reference

Popper, S. K. (1945). *The open society and its enemies*. London: Routledge.

[29]CDNet, 2011: http://www.zdnet.com/article/microsoft-admits-patriot-act-can-access-eu-based-cloud-data/ retrieved 2016-09-12.

[30]Joris van Hoboken, Axel Arnbak, N.A.N.M. van Eijk, Cloud Computing in Higher Education and Research Institutions and the USA Patriot Act, 2012: https://papers.ssrn.com/sol3/papers.cfm?abstract_id=2181534 retrieved 2016-10-20.

Internet of Things

The Internet of Things, sometimes referred to as the Internet of Objects, will change everything—including ourselves
Dave Evans, Cisco's Chief Futurist

We consider the enablers of the IoT (Internet of Things) to include factors like falling hardware costs, the deployment of IPv6 addresses, new connectivity options especially wireless networks as well as advanced cellular networks for enhanced machine-to-machine communications.

We see how the Internet Protocol Stack is reused for the IoT and how cloud solutions help to lower costs while increasing flexibility. The combination of big data with cloud computing provides the right platform to exploit the required insights into all these new data, consequently causing one of the next, big disruptive changes in the cloud century.

We will discuss the seven level IoT Reference Model that allows the design of IoT ready architectures, taking into account that data is generated by many kinds of devices, processed in different ways, sent to different locations and finally handled properly by various applications. We will identify the security requirements for a successful IoT implementation based on this reference model.

Finally, we will have a look at some of the most popular IoT solutions each of them creating interesting and promising ecosystems like parking space management, precision agriculture, building and home automation systems, manufacturing and Industry 4.0 as well as media, data capture and big data.

IoT in a Nutshell

IoT Definition

The Internet of Things (IoT) (McEwen & Cassimally, 2014) refers to physical objects (or things) embedded within electronics, software, sensors and connectivity devices that are networked or connected in order to enable the exchange of data with other connected devices, operators, users, etc., thus achieving a greater value

M. Oppitz, P. Tomsu, *Inventing the Cloud Century*,
DOI 10.1007/978-3-319-61161-7_16

and more services for users. These IoT devices form a computer network grid, in which each thing is uniquely identifiable through its embedded computing system with a unique IP address and is able to interact with existing Internet and networking infrastructures (Greengard, 2015).

The IoT offers advanced connectivity of systems and devices as well as services, which go beyond normal machine-to-machine (M2M) communications, by covering a variety of protocols, domains and applications (Kellmereit & Obodovsky, 2013). The interconnectivity of these smart devices (sometimes also referred to as smart objects) has paved the way to automation in many fields and also is the basis for smart grid and smart cities infrastructures.

First Appearance of IoT

The term "Internet of Things" was already mentioned and envisaged as far back as the late 1990s by MIT professors (one of them being Kevin Ashton, cofounder and executive director of the Auto-ID Center at MIT in 1999), who described a future world, where things (they meant devices and sensors) are connected and able to share data. This new data from sensors and devices would suddenly provide new insights into businesses, processes, environments, etc., being previously unavailable or out of reach. The main goal behind this idea was (and still is today) to analyze all the data sent from the connected things, then feed applications with the results and, in turn, provide output decisions to the same or other devices with the aim of providing completely new possibilities for business processes, environments and our daily life.

Interestingly, the first Internet appliance had already appeared in the early 1980s. It was a Coke machine at Carnegie Melon University that could be connected to the Internet by programmers, who could then check the status of the machine and determine whether or not there would be a cold drink waiting for them, should they decide to make the trip down to the machine.

IoT Evolution and Promise

Over the following years, while an IoT as we envisage it today was not available due to technological limitations, a number of visionaries began talking about similar ideas. Like many other technologies, it took almost 10 more years before all the technological requirements had matured and the idea of the IoT finally started to take off. Today experts have divergent estimates about the number of devices (objects) in the IoT that range from 50 billion (Cisco's IBSG[1]) to 200 billion (Intel[2]) by 2020 or, as the Internet of Things Consortium[3] (IOTC) founded in 2012 estimates, the number of the IoT endpoints will grow to 30 billion by 2020.

[1]Cisco, Internet Business Solutions Group (IBSG), The Internet of Things, White Paper, http://www.cisco.com/c/dam/en_us/about/ac79/docs/innov/IoT_IBSG_0411FINAL.pdf, 2011, retrieved 2016-12-30.

[2]Intel, A Guide To The Internet Of Things, http://www.intel.com/content/www/us/en/internet-of-things/infographics/guide-to-iot.html, 2016, retrieved 2016-12-30.

[3]IOTC, Internet of Things Consortium, http://iofthings.org/#home, 2017, retrieved 2017-01-05.

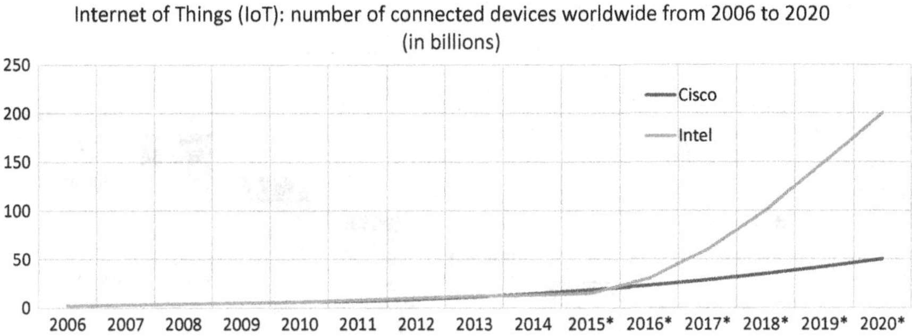

Fig. 1 Number of connected IoT devices 2006–2020 (values extrapolated)

All these estimates ranging from 30 to 200 billion things in the IoT by 2020 tell us basically the same fact: there will be a huge growth in connected devices in the coming years and this will definitely be the foundation of one of the next big disruptions significantly impacting all our lives (see Fig. 1).

Most things will be connected via Wireless Personal Area Networks (WPAN)[4] based on IEE 802.15 or Wireless LANs (WLAN)[5] (see Fig. 2).

Intel estimates that by 2025 the total global worth of IoT technology would reach $6.2 trillion with 40.2% coming from business and manufacturing, 30.3% from healthcare, 8.3% from retail and 7.7% from security (see Table 1).

Things in IoT

The things in the IoT can refer to a multitude of different devices like automobiles with built-in sensors, biochip transponders on farm animals, heart monitoring implants, field operation devices that assist fire-fighters in search and rescue and many more devices. Usually these devices can collect data based on well-known and existing technologies and channel this data autonomously between other devices utilizing wireless technologies such as WiFi, Bluetooth, 4G or NFC. They also generate large amounts of data from diverse locations that need to be aggregated quickly, thus making better and faster storage, processing and intelligent, advanced analysis of these data all mandatory.

IoT Versus IoE Versus InternetofYourThings—Different Names for the Same Concept

There exist many slightly different variations of the term "Internet of Things," especially when interest groups or vendors try to add even more functionality to what is already defined in the IoT. For example, the networking gear vendor, Cisco, started talking about the Internet of Everything (IoE[6]) and defined the IoE (Bradley,

[4]IEEE, WG802.15—Wireless Personal Area Network (WPAN) Working Group, https://standards.ieee.org/develop/wg/WG802.15.html, 2017, retrieved 2017-01-24.

[5]IEEE, IEEE 802.11 Wireless Local Area Networks, http://www.ieee802.org/11/, 2017, retrieved 2017-01-24.

[6]Cisco, Internet of Everything, http://ioeassessment.cisco.com, 2016, retrieved 2016-12-30.

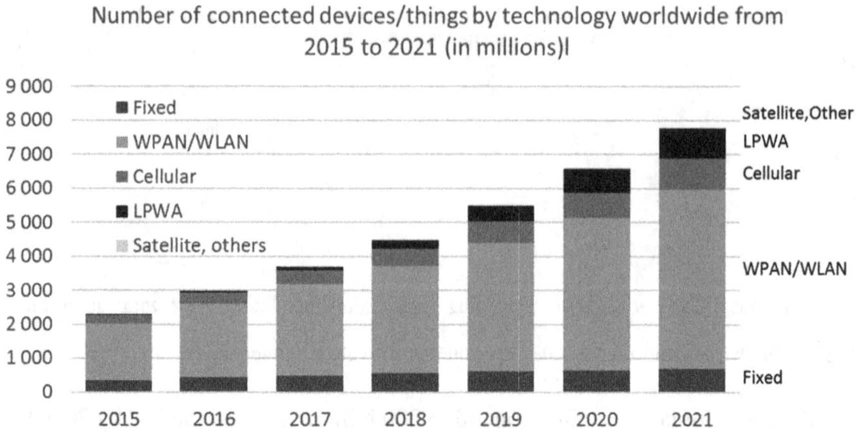

Fig. 2 Number of connected devices by technology 2015–2021

Table 1 IoT technology share in different sectors and applications by 2025 (Intel)

Sector	Application	Percentage
Business and manufacturing	Real-time analytics—supply chains, equipment, robots	40.2
Healthcare	Portable health monitoring, electronic recording, pharmaceutical support and safeguard	30.3
Retail	Smartphone purchasing, inventory tracking, consumer choice analytics	8.3
Security	Facial recognition, biometric data, remote sensors	7.7

Reberger, Dixit, & Gupta, 2013) as the bringing together of people, processes, data, and things to make networked connections more relevant and more valuable, and turn information into actions that create new capabilities, richer experiences, and unprecedented economic opportunity for businesses, individuals, and countries. Without a doubt this defines a superset of the IoT and can easily cause some confusion, while intended to cause better sales of the IoE approach compared to the plain IoT (see Fig. 3).

A similar approach comes from Microsoft when it talks about the InternetofYourThings (Edson, 2015), in order to differentiate it from the plain, native IoT and market their own IoT strategy. Microsoft says that the Internet of Things makes a difference to businesses by bringing the things together that matter most in "your" businesses, thus creating the Internet of Your Things.[7,8] This builds on familiar infrastructures already in place, using familiar devices and services in new ways and incorporates the right technology to ultimately help "you" use data to

[7]Microsoft, Internet of Your Things, http://internet-of-your-things.eu, 2016, retrieved 2016-12-30.

[8]Microsoft, Internet of Your Things, https://channel9.msdn.com/Events/Ignite/2015/BRK2572, 2015, retrieved 2016-12-30.

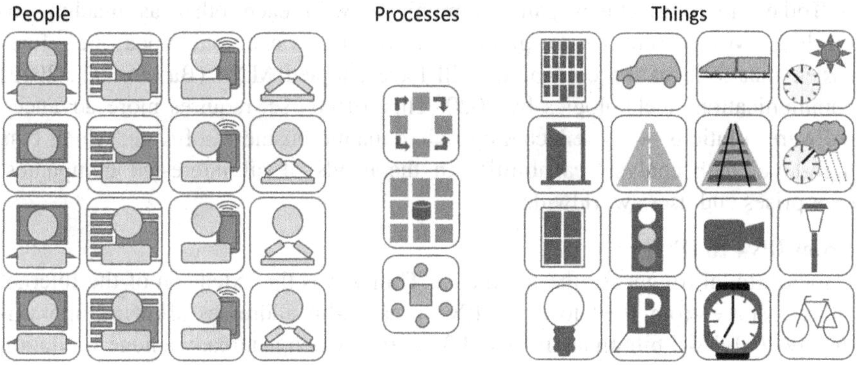

Fig. 3 People, processes and things

create insights and make more informed business decisions. Again, this in principle describes what IoT stands for, this time described by the best use with Microsoft products.

The terms IoE and InternetofYourThings are two examples of marketing definitions. If we really think about the content and what it describes in the best-case scenario, it adds some vendor specific features to what is already defined in general for the IoT. There might also exist some more individual IoT definitions by vendors or user groups in the future. Hence, throughout the rest of this book, we will only use the generic term "IoT," as most of the rest of the industry does.

Enablers of IoT

The basic ideas and concepts behind the IoT are nothing new and, as we already discussed, date back to the 1980s, but why did it take until the mid-2010s before the IoT started to take off? There are multiple factors that have recently converged which allowed for and accelerated the adoption of IoT scenarios in businesses and everyday lives.

Falling Hardware Costs, Advanced Capabilities
The costs of the components of the Internet of Things[9] such as microchips, GPS sensors, accelerometers, surveillance cameras, and most sensors and actuators we use today have dramatically fallen as volumes in production have increased. On top of that, the cost reduction of the microchips running many devices are more realistic today than a decade ago, allowing us to run more advanced applications and software than ever before.

[9]Postscapes, IoT Hardware Guide, http://www.postscapes.com/internet-of-things-hardware/, 2016, retrieved 2016-12-30.

Today, more machines can communicate with each other as machine-to-machine (M2M) solutions are going mainstream. For example, Vodafone[10] forecasted that 50% of the companies will have adopted M2M (Bassi et al., 2013) communications technologies by 2020. This, of course, requires more advanced software solutions being reflected by rich, dynamic business software, which puts high-level data analysis capabilities in the hands of all interested companies, enterprises and users worldwide.

From IPv4 to IPv6

Another initial trigger for the Internet of Things was the extension of the Internet address space from IPv4 to IPv6. IPv4 uses 32-bit addresses allowing approximately "only" 4.3 billion addresses (4.3×10^9), which is already a huge limitation if one thinks about the growing use of personal devices (smart phones, tablets, PCs) by a worldwide population of more than seven billion. All these devices require an IP address for identification, not to mention the sheer number of 50 billion devices (smart objects) due in the IoT near future (Minoli, Building the Internet of Things with IPv6 and MIPv6, 2014).

In contrast, IPv6 addresses are represented as eight groups of four hexadecimal digits separated by colons, for example 2001:0db8: 85a3:0042:1000:8a2e:0370:7334, allowing 2^{128}, or approximately 3.4×10^{38} addresses, or more than 7.9×10^{28} times as many as IPv4. This huge number is, again, not endless and rumors that claim it would allow us to give each atom in the universe a unique ID are simply wrong, because the number of atoms in the universe is approximately 10^{80}. However, the IPv6 address space is large enough to think about giving each thing we use (or will use in the future) its own unique identification as a IPv6 address (see Fig. 4).

Both the Internet of Things and the extended range of IPv6 addresses are complementary as only IPv6 can provide the sheer number of hardware addresses needed for the many devices implemented in sensor networks by the IoT. Further to this additional address space provided by IPv6, there are also several other improved features[11] that support IoT deployment like overcoming the need for NAT (Network Address Translation) limitations, the enabling of IP security, better mobility support, stateless address auto-configuration as well as improved multicast options just to name the most important features.

New Connectivity Options

It took wireless networking[12] to allow for the full bloom of the IoT solutions. As long as these solutions were limited to wired connections, a real mobile approach for the IoT was just out of reach. With the broad deployment of wireless local area

[10]Vodafone, Vodafone M2M Barometer 2015, https://www.vodafone.de/media/downloads/press-releases/150729-vf-m2m-report-2015.pdf, 2015, retrieved 2016-12-30.

[11]Ziegler S., et al., IoT, IEEE, The Case for IPv6 as an Enabler of the Internet of Things, http://iot.ieee.org/newsletter/july-2015/the-case-for-ipv6-as-an-enabler-of-the-internet-of-things.html, 2015, retrieved 2016-12-30.

[12]LinkLabs, The Complete List Of Wireless IoT Network Protocols, https://www.link-labs.com/complete-list-iot-network-protocols/, 2016, retrieved 2016-12-30.

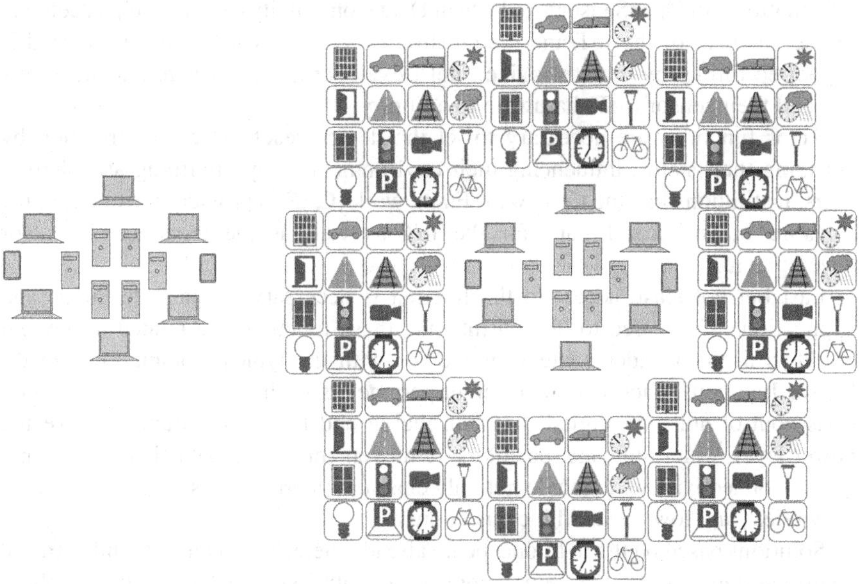

Fig. 4 Internet and the Internet of Things

networks and the advanced cellular networks, an unprecedented increase in additional networking capacity became available to support M2M connections, especially as soon as mobile operators started to embrace the IoT (Yan, Zhang, Yang, & Ning, 2008).

Using the Internet Protocol Stack for IoT

This led to the second basic idea behind the IoT: if we are able to identify things in a unique way, then we will be able to use Internet technology like the Internet Protocol (IP) stack to connect these things and to build new types of solutions based on the smart interactions between things and applications.

In the old Internet, connections were built between components like servers, PCs, routers and switches all being typical computer devices. In the IoT, we think about the integration and connectivity of many items like cars, home infrastructure, production machines or other common items. The fast-growing technology of microminiaturization allows for the design and production of items that easily integrate, making them a potential part of the IoT. The basic functions that make a "thing" a part of the IoT are:

- Sensors
- Actors
- Connectivity

Sensor functions are used to collect the important data and information used by applications like temperature reading, position, movement or any other valuable information. An OpenSensor[13] IoT Open Data Community exists to help accelerate the deployment of the IoT and the creation of smarter workspaces. This trend is supported further by initiatives like Intel Developer Zone,[14] where manufacturers can add their sensor (IoT) hardware to a library.

Actor functions are necessary to let the device react to new information by changing their status, influencing their environment or by informing and altering users. Institutions like the C++ Actor Framework (CAF[15]) provide free open source software in an actor library for the IoT as well as the tools to ease actor development.

Last but not least, access to the Internet is necessary to integrate things and applications. Compared to the old Internet, these functions are related to the real world in a much wider context, as we see with the typical computing devices. Sensor functions react to very different parameters in the real world. Actor functions control not only digital data but analog and technical parameters like the power level, water drainage, position and movement of mechanical structures and so on. Connectivity will include not only classical Internet access but also all types of wireless and low bandwidth connections.

Solutions based on the IoT concept are already there. They can be found in traffic control systems, logistic systems, production control systems, agriculture applications, etc. as discussed later in the section on "IoT solution samples".

Cloud Solutions Offering Lower Costs and Increased Flexibility
The availability of cloud computing and cloud services, cloud storage and processing power became more affordable and available, which expanded the capabilities of analyzing the large amounts of data produced by the IoT. Examples of these new, cloud-based solutions are Microsoft Azure, Amazon Web Services, etc. Modern IoT scenarios that incorporate cloud-based storage, analysis and other tools provide additional benefits for scalability and flexibility and are the basis for businesses when they start or expand their IoT solution.

To modern businesses, data has become critical. This data can have different formats, values, retention requirements and traffic patterns, and comes from different sources like intelligent sensors, devices, services, etc. and is sent via different protocols. The combination of advancements in connectivity, processing power, form factors, operating systems and applications enables us to build cloud solutions that fully unlock the value of the Internet of Things.

[13]OpenSensor, https://www.opensensors.io, 2016, retrieved 2016-12-30.

[14]Intel, Intel Software Developer Zone, https://software.intel.com/en-us/iot/hardware/sensors, 2016, retrieved 2016-12-30.

[15]CAF, C++ Actor Framework, https://www.actor-framework.org, 2016, 2016-12-30.

IoT Protocols and Standards

Today, many connectivity options for the Internet of Things exist, many of these technologies being well known such as Bluetooth, ZigBee, as well as 2G, 3G and 4G cellular options. There are also many new and emerging options for home automation applications and White Space[16] TV technologies. It depends on the application, range, data requirements, security, power demands and the battery life as to which technology or combination of technologies is ultimately the optimal choice for IoT connectivity. We will discuss the major communication technologies, protocols and standards for the IoT as it runs today in this section.

It is interesting to mention that the number of IoT connections will significantly grow more over the next years as opposed to the number of machine-to-machine (M2M) connections (see Fig. 5).

Cellular IoT Standards

Mobile IoT[17] was coined by the Global System Mobile Association (GSMA)[18] and was launched in 2015. It describes a standard and secure option for operator managed IoT networks. The main goal is to offer LPWA (low power wide area) networks specifically designed for IoT applications at low cost with low data rates and offering long battery lives and can work in remote locations. Mobile IoT can be seen as an attempt by mobile operators to extend existing cellular networks to support billions of new IoT devices and offer complete IoT connectivity, thus generating additional income for these operators.

Low Power Wide Area Networks (LPWA)
Mobile IoT aims at accelerating the commercial availability of Low Power Wide Area (LPWA) solutions in a licensed spectrum. Its goal is to develop and standardize new GSM[19] technologies that support devices requiring mobility, long range, low power consumption, low cost and security and are backed by more than 60 of the world's leading mobile operators, OEMs, as well as chipset, module and infrastructure companies.

The GSMA encouraged the 3rd Generation Partnership Project (3GPP)[20] to specify LPWA networks that would be ideally suited for connecting the billions

---

[16]TechRepublic, White Space, http://www.techrepublic.com/article/white-space-the-next-internet-disruption-10-things-to-know/, 2014, retrieved 2017-03-15.

[17]GSMA, Mobile IoT, http://www.gsma.com/connectedliving/mobile-iot-initiative/, 2016, retrieved 2017-01-10.

[18]GSMA, Global System Mobile Association, http://www.gsma.com, 2017, retrieved 2017-01-10.

[19]ETSI, GSM, Global System for Mobile communication, http://www.etsi.org/technologies-clusters/technologies/mobile/gsm, 2017, retrieved 2017-01-10.

[20]3GPP, 3rd generation Partnership Project, The Mobile Broadband Standard, http://www.3gpp.org, 2017, retrieved 2017-01-10.

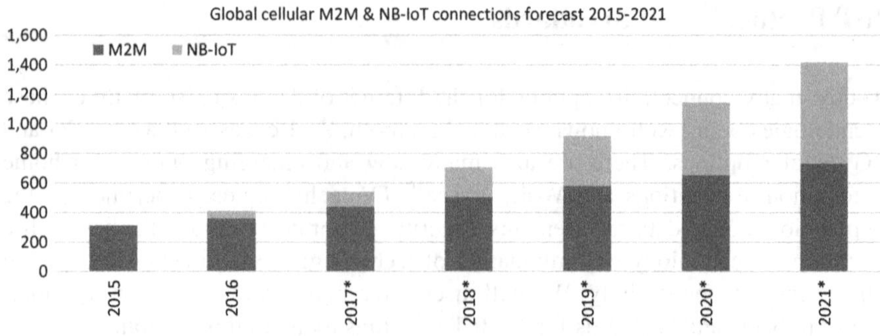

Fig. 5 Global cellular M2M & NB-IoT connections forecast 2015–2021

of devices in the IoT and are designed to support a variety of vertical industries while supporting new types of applications that cannot be supported or at least not optimally supported by currently existing mobile technologies.

The diverse IoT applications have very different requirements resulting from the fact that there are three licensed 3GPP standards:

1. Extended Coverage GSM for IoT (EC-GSM-IoT[21])
2. Long Term Evolution for Machines (LTE-M[22])
3. Narrow-Band Internet of Things (NB-IoT[23])

All solutions are specifically designed to support:

- Very low power consumption with a battery life exceeding 10 years
- Optimized for brief messages typically the length of an SMS
- Allow for an extremely low device unit cost—only a few dollars
- Allow for good indoor and outdoor coverage that enables connectivity in rural as well as underground areas
- Support easy installation on current networks by reusing existing cellular infrastructures as much as possible
- High scalability allowing large numbers of devices over large areas
- Support secure connections appropriate for IoT application requirements
- Allow for integration into a mobile operator's IoT platform

LPWA specifications were completed in June 2016 and they were included in 3GPP Release 13 which already allows full commercial implementation today.

[21]GSMA, EC-GSM-IoT, Extended Coverage—GSM IoT, http://www.gsma.com/connectedliving/extended-coverage-gsm-internet-of-things-ec-gsm-iot/, 2017, retrieved 2017-01-10.

[22]GSMA, LTE-M, Long Term Evolution for Machines, http://www.gsma.com/connectedliving/long-term-evolution-machine-type-communication-lte-mtc-cat-m1/, 2017, retrieved 2017-01-10.

[23]GSMA, NB-IoT, Narrow Band IoT, http://www.gsma.com/connectedliving/narrow-band-internet-of-things-nb-iot/, 2017, retrieved 2017-01-10.

Extended Coverage GSM for IoT (EC-GSM-IoT)
This is a LPWA technology based standard using EGPRS[24] and it is optimized for high capacity, long range, low energy and low complexity cellular systems for IoT communications. Optimizations for traditional GSM networks can be achieved with software upgrades and the first launch is in 2017. EC-GSM-IoT networks can coexist with 2G, 3G as well as 4G networks and use all security and privacy mobile network features.

Current members of the EC-GSM-IoT group are Broadcom, Cisco, Ericsson, Gemalto NV, Intel, KDDI, LG Electronics, MediaTek, Nokia, Oberthur Technologies, Ooredoo, Orange, Samsung, Saudi Telecom, Sierra Wireless, Telit Communications and VimpelCom.

Long Term Evolution for Machines (LTE-M)
LTE-M uses LTE-MTC LPWA technology which was published by 3GPP as Release 13[25] and refers to the substandard LTE CatM1 that is suitable for the IoT. It can reuse the LTE[26] installed base and specifically allows a battery life of up to 10 years or more. It is paired with very low modem costs compared to standard EGPRS. Similar to EC-GSM-IoT, it can coexist with current mobile networks and reuse security as well as privacy features. LTE-M will be commercially launched in 2017.

Current members of LTE-M are AT&T Mobility, Bell Mobility, Bermuda Digital Communications, Broadcom, China Mobile, China Telecom, China Unicom, Etisalat, Gemalto, Huawei Technologies, KDDI Corporation, KT Corporation, Nokia, NTT Docomo, Oberthur Technologies, Orange, Samsung Electronics, Sequans Communications, Sierra Wireless, SingTel Mobile Singapore, SoftBank, Telecom Italia, Telefónica, Telenor Group, TeliaSonera, Telit Communications, Telstra, T-Mobile Austria, u-blox, Verizon Wireless and Vodafone.

Narrow-Band Internet of Things (NB-IoT)
Similar to the other two LPWA technologies, this standard was developed for IoT devices and services again with optimized power consumption in mind, but it also allows for spectrum efficiency and deep coverage. In order to achieve these goals, new physical layer signals and channels are used specifically for meeting the requirements of extended coverage, either in rural or deep indoor settings, while making the technology generally simpler than today's GSM/GPRS. Another big goal was ultra-low device complexity. As with the other LPWA standards, it can coexist with standard mobile networks and reuses the security and privacy features. Commercial launch will happen throughout 2017.

---

[24]Telecom ABC, EGPRS, Enhanced General Packet Radio Service, http://www.telecomabc.com/e/egprs.html, 2005, retrieved 2017-01-11.

[25]3GPP, Evolution of LTE in Release 13, LTE-MTC, http://www.3gpp.org/news-events/3gpp-news/1628-rel13, 2015, retrieved 2017-01-11.

[26]3GPP, LTE, Long Term Evolution, http://www.3gpp.org/technologies/keywords-acronyms/98-lte, initiated 2004, retrieved 2017-01-11.

There is even a GSMA NB-IoT forum[27] that supports the accelerated adoption of 3GPP based Narrow Band IoT technology and ensures uniform operability of these solutions.

The NB-IoT standard easily has the broadest support from all the LPWA standards. The current members of this forum are AIS Thailand, Bell Mobility, Blackberry, China Mobile, China Telecom, China Unicom, Cisco, Deutsche Telekom, Du, Etisalat, Ericsson, Gemalto, Globetouch, Huawei, Intel Corporation, KDDI Corporation, KT Corporation, LG Electronics, LG U+, MediaTek Inc., Megafon, Mobileum, MTS, Nokia, NOS, NTT Docomo, Oberthur Technologies, Oi, Ooredoo, Qualcomm, Quectel, Safaricom, Samsung, Sequans Communications, Sierra Wireless, SingTel, SK Telecom, Starhome Mach, Summit Tech Multimedia Communications, Syniverse Technologies, TDC A/S, Tele2 Group, Tele2 Russia, Telekom Austria Group, Telecom Italia, Telecom Personal S.A., Telefónica, Telia Company, Telit, Telstra, T-Mobile, True Move H Universal Communication, Turk Telecom, Two Degrees Mobile, u-blox, Verizon Wireless, Vodafone and ZTE Corporation.

LPWA Market and Expected Growth

LPWA networks are specifically designed for different markets like agriculture, manufacturing, utilities, transport and clothing. The market opportunity for LPWA growth is huge, especially as soon as the IoT is brought up to scale with around five billion LPWA connections expected by 2022 and having a global connectivity value of $7.5 billion. These numbers will be even larger as soon as additional revenues from value-added services such as big data analytics and security are taken into account.

Industrial IoT (IIoT) and Standards

What Is the Industrial Internet of Things

The Industrial Internet of Things (IIoT) often referred to as "Industrial Internet" is the network of devices connected by different communications technologies resulting in a system that can monitor, collect, analyze and deliver precious new insights not possible before today. The reason for this is to provide faster and smarter business decisions to industrial companies.

IIoT is fundamentally changing the way industries work by either enabling predictive analytics or driving visibility and control over industrial control system environments. This prevents cyber-attacks and delivers real-time production data for uncovering additional capacity in production plants. This requires a combination of machine-to-machine (M2M) communication with technologies like

[27]GSMA, GSMA NB-IoT Forum, http://www.gsma.com/connectedliving/wp-content/uploads/2016/11/Presentation-02.-Global-MIoT-Summit-Tokyo-Nov-2016-Chairs-slides-v0.2.pdf, 2016, retrieved 2017-01-11.

industrial big-data analytics, cyber security and human machine interfaces (HMI), as well as supervisory control and data acquisition (SCADA).

General Electric (GE)[28] coined the term IIoT in late 2012 and it estimates that the market for IIoT could reach $225 billion by 2020. This is underlined by the significant investment GE made and by the fact that it created a new software business called GE Digital, which is dedicated fully to IIoT.

As such, GE was also one of the founders of the Industrial Internet Consortium,[29] with other founding and contributing members being Bosch, EMC, Huawei, Intel, IBM, SAP and Schneider.

IIoT Standards

The commonly considered most important technologies for data and information management in IIoT are Data Distribution Services, introduced by the Object Management Group (OMG), and Open Platform Communications Unified Architecture (OPC-UA) introduced by the Open Platform Communication (OPC) Foundation.

Data Distribution Service (DDS)

In the late 1990s, the US Department of Defense (DoD) introduced the concept of a Global Information Grid (GIG), which would provide the necessary set of capabilities for collecting, processing, storing, managing and disseminating information on demand for combatants. At this time, there were only client server technologies like CORBA (Common Object Request Broker Architecture) or COM+ (Common Object Model)/DCOM (Distributed Common Object Model). These were not suitable for the implementation of the GIG, because they induced tight coupling, were highly sensible to faults and had scalability and performance issues.

DDS was introduced in order to overcome the limitations above and support the data sharing requirements of GIG for real-time awareness, operational intelligence and mission planning. Many consider the GIG to be the precursor of the IoT. Since then, DDS has been used worldwide as the major technology for network-centric systems and functionalities needed by the GIG, including air traffic control, aerospace applications as well as simulation systems.

Open Platform Communications Unified Architecture (OPC-UA)

Long before OPC-UA, Open Platform Communications (OPC) already existed back in the 1990s. At these times, no standard existed for the automation industry to control hardware and field devices. This required client applications to have embedded drivers and protocols for all devices they needed to interact with. This is why OPC was introduced in 1996 with the intention of shielding client applications from the details of automation devices and, for the first time, provided a standard

---

[28]GE Digital, Everything you need to know about the Industrial Internet of Things, https://www.ge.com/digital/blog/everything-you-need-know-about-industrial-internet-things, 2017, retrieved 2017-01-04.

[29]Industrial Internet Consortium, http://www.iiconsortium.org/index.htm, 2017, retrieved 2017-01-04.

interface between control hardware and field devices. There was a rather lengthy evolution of OPC including a new definition of OPC Data Access standard relying on SOAP/XML but it unfortunately had a lot of overhead and was also not compatible with a number of industrial use cases.

Thus, the OPC community finally introduced a new standard for handling data and information management in IIoT, the Open Platform Communication Unified Architecture (OPC-UA).

What Is the Better Solution for IIoT—DDS or OPC-UA?

Both DDS and OPC-UA support information management in distributed systems as well as information modeling, where DDS uses relational data modeling and OPC-UA uses object oriented modeling. These are the only similarities between both solutions. As a conclusion, both address a similar problem in very different ways with DDS being more data centric and OPC-UA more device centric.

Hence, DDS is better suited for large-scale data integration as we find it in GIG-like systems and supports east-west data sharing between edge applications (we will see this when discussing Fog Computing) as well as north-south communication between devices, edge nodes and cloud applications in IoT systems.

OPC-UA is optimized for device interoperability and sensor integration as well as embedded devices and PLCs (programmable logic controllers) into larger systems. As OPC-UA is mostly used in the automation industry, its biggest asset is also its ecosystem. In 2016, Microsoft introduced the new open-source cross-platform OPC-UA support for IIoT.[30]

Other IoT Standards and Communication Technologies

Bluetooth

Bluetooth[31] is a very important technology used for short-range communication and is already widely used today in computing and consumer product markets. There is even a new Bluetooth Low Energy (BLE) also called Bluetooth Smart protocol that is optimally suited for IoT applications.

The Bluetooth standard is the Bluetooth 4.2 specification that works in the 2.4 GHz frequency range, covers distances from 50 to 150 m (Smart/BLE) and delivers data rates of up to 1 Mbps.

[30]Microsoft, Internet of Things, Microsoft introduces new open-source cross-platform OPC-UA support for IIoT, https://blogs.microsoft.com/iot/2016/06/23/microsoft-introduces-new-open-source-cross-platform-opc-ua-support-for-the-industrial-internet-of-things/, 2016, retrieved 2017-01-05.

[31]Bluetooth, https://www.bluetooth.com, 2017, retrieved 2017-03-15.

ZigBee

ZigBee is very similar to Bluetooth, having a large installed base in the industrial area today. ZigBee comes in two flavors:

- ZigBee Pro[32] offering full wireless mesh, low-power networking
- ZigBee RF4CE[33] for low-power, low-latency control for a wide range of products including home automation.

Both are based on the IEEE 802.15.4 standard protocol. A big advantage of ZigBee/RF4CE is the low power consumption, high security, scalability, and high robustness even with high node counts, making it an ideal solution for sensor networks in M2M and IoT applications.

The current ZigBee standard is ZigBee 3.0 based on IEEE 802.15.4, operating in the 2.4 GHz frequency range, covering distances from 10 to 100 m and allowing data rates of 250 kbps.

Z-Wave

Another low-power communication technology that ideally can be used for IoT communication is Z-Wave. It is optimized for low-latency communications of small data packets that are used in home automation and supports full mesh networks without the need for a supervisor or coordinator node, while being scalable up to 232 devices.

The Z-Wave Alliance[34] standardizes Z-Wave, and the corresponding ITU-recommendation is ITU-T G.9959.[35] It operates at the 900 MHz frequency range, has a range of 30 m and can operate with data rates up to 100 kbps.

WiFi

WiFi is a well-established standard for wireless LANs and hence does not require any further explanation. It is standardized by the WiFi Alliance[36] and its current incarnation is 802.11n, which offers data traffic in the range of hundreds of megabits per second.

The current standard is based on IEEE 802.11n, frequency bands are 2.4 GHz and 5 GHz and the range is up to 50 m with data rates ranging from 150 to 600 Mbps.

6LowPAN

6LowPAN (IPv6 Low-power wireless Personal Area Network) is an IP based technology and defines a network protocol for encapsulation and header

[32]ZigBee Alliance, ZigBee Pro, http://www.zigbee.org/zigbee-for-developers/network-specifications/zigbeepro/, 2017, retrieved 2017-03-16.

[33]ZigBee Alliance, ZigBee RF4CE, http://www.zigbee.org/zigbee-for-developers/network-specifications/zigbeerf4ce/, 2017, retrieved 2017-03-16.

[34]Z-Wavealliance, http://z-wavealliance.org, 2017, retrieved 2017-03-15.

[35]ITU-T G9959, https://www.itu.int/rec/dologin_pub.asp?lang=e&id=T-REC-G.9959-201501-I!!PDF-E&type=items, 2015, retrieved 2017-03-15.

[36]WiFi Alliance, WiFi, http://www.wi-fi.org, 2017, retrieved 2017-03-15.

compression mechanisms. It does not define a specific frequency band or a physical layer and is based on the IPv6 protocol stack. 6LowPAN was specifically designed for home and building automation and used to communicate with devices in a cost-effective way via a low-power wireless network.

It is standardized in RFC 4944,[37] can use any low-power wireless technology and does not specify any range or data rates.

Thread

Thread is a new, IP-based IPv6 network protocol that is designed for home automation and is based on 6LowPAN. It is not a specific IoT application protocol, but it is designed as a complement to WiFi especially in home automation setups.

It was launched by the Thread Group,[38] makes use of different standards like IEEE 802.15.4 for the wireless air-interface protocols, IPv6 and 6LeWPAN and guarantees resilient IP-based solutions for the IoT.

NFC

NFC (Near Field Communication)[39] defines a technology that enables safe two-way interactions between devices, extending the capabilities of contactless cards. It enables information sharing between devices up to 10 cm apart.

NFC is based on the standard ISO/IEC 18000-3,[40] operates at the 13.56 MHz frequency and supports data rates up to 420 kbps.

SigFox

SigFox is another wide-range technology that operates in terms of the range between a WiFi network and cellular network. For many M2M applications, which run on a small battery with low data rate requirements, it is ideally suited because cellular is too expensive and WiFi has too limited a range.

The standard is SigFox,[41] it operates at 900 MHz, the range is between 30 and 50 km in rural areas, 3 and 10 km in urban environments and data rates are typically 10–1000 bps.

Neul

Neul[42] uses parts of the TV White Space spectrum and delivers highly scalable coverage while offering low power consumption. It accesses the high-quality UHF spectrum that became available because of the transition from analogue to digital TV transmission.

[37]IETF, RFC 4944, LowPAN, https://tools.ietf.org/html/rfc4944, 2007, retrieved 2017-03-15.

[38]Threadgroup, Thread, http://www.threadgroup.org, 2017, retrieved 2017-03-15.

[39]NFC Forum, http://nfc-forum.org, 2017, retrieved 2017-03-15.

[40]ISO IEC, ISO/IEC 18000 Part 3, http://read.pudn.com/downloads134/ebook/572218/ISO%2018000-3.pdf, 2002, retrieved 2017-03-15.

[41]Sigfox, https://www.sigfox.com, 2017, retrieved 2017-03-15.

[42]Neul, http://www.neul.com/neul/, 2017, retrieved 2017-03-15.

The standard is Neul, frequencies used are 900 MHz for the industrial, scientific and medical (ISM) band, 458 MHz (in the UK) and 470–790 MHz (White Space), covers a range of 10 km and can support data rates from a few bps up to 100 kbps.

LoRaWAN
This is similar to SigFox and Neul and is designed to provide low-power WANs specifically needed for the IoT and M2M communications, as well as smart city and industrial applications.

The standard is LoRaWAN,[43] it uses various frequency bands and the range is from 2 to 5 km in urban areas and up to 15 km in suburban environments with data rates ranging from 0.3 to 50 kbps.

Cost of IoT Connectivity

Today, all these connectivity options have certain price points that make them more or less useable for different IoT applications. These range from a few cents per connectivity module to approximately $50 per module. It can be expected that these numbers will change significantly over the next decade as certain technologies become more mainstream (see Fig. 6).

How the IoT, Cloud and Big Data Play Together

One of the key characteristics of the IoT is that the connected things produce enormous amounts of information, thus those data streams need to be connected to cloud solutions to enable effective processing and analyzing. This way enterprises and organizations can optimize their business processes and increase the number of informed decisions, which allows them to identify new revenue opportunities, while better understanding and predicting customer and partner behaviors. Since all this data from the things in the IoT can either arrive in a predictable and structured way or in an unstructured manner, an automatic framework is necessary for data to ingress and process, which includes filters, rules, triggers, etc.

Big Data and the IoT
The collection of massive data sets is so large and complex that they finally became too difficult to process on applications and database management premises. Effective processing of this amount of data requires flexible and scalable computing models that allow seamless growth as the needs of customers and applications grow over time (Stackowiak, Licht, Mantha, & Nagode, 2015).

[43]LoRa Alliance, https://www.lora-alliance.org, 2017, retrieved 2017-03-15.

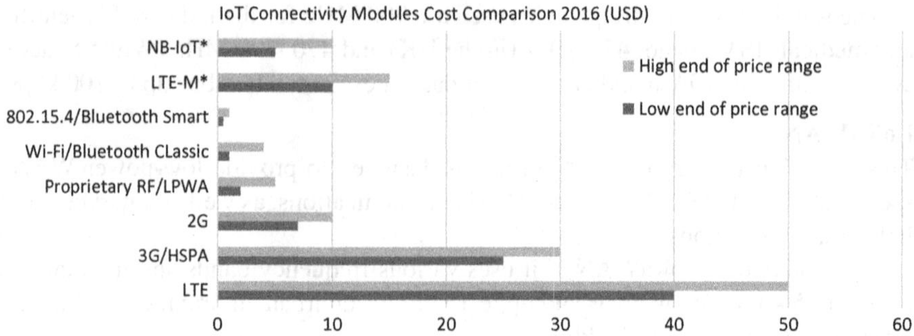

Fig. 6 IoT connectivity modules cost comparison 2016 (USD)

Big data needs to be seen in a contextual way because it needs to be combined with the many assets, sources and data sets and requires processing with the right tools, engines and architectures in order to take real advantage of the customer data before finally applying the right optimizations to the customer's assets. The cloud, in combination with storage, computing, optimized applications and analyzing tools, allows one to draw the required insights from the collected data. The cloud, in combination with the IoT, became the real game changers, finally enabling quite possibly the biggest disruptive change since the introduction of the World Wide Web and the Internet browser.

Cloud Computing and the IoT

The reason to use cloud computing for the IoT is that it offers scalable data collection, processing and analysis capabilities, which promise flexible solutions for different business needs. With cloud solutions, businesses can store and process huge amounts of data either latently or in real time, and apply the necessary rules and structures for computation. It is easy to see that a powerful IoT solution needs to be based on cloud computing technology with a flexible consumption-based pricing structure, along with private or public cloud computing models. These factors will enable the necessary new offers to the market that were nonexistent before the combination of all the described technologies became feasible (Zhou, 2012).

Cloud computing allows us to use more data that comes from different sources and across different business assets and devices, arriving either as structured, unstructured or in variable ways since this is one of the characteristics of the IoT. All this data can also arrive on a regular basis or simply intermittently, but all of it needs to pass through filters, rules and certain kinds of triggers, which calls for the intelligent processing of data. An example of a software package that is able to handle these requirements is Microsoft Azure,[44] which enables the IoT services to improve efficiency, enable innovation and transform the actions of customers.

[44]Microsoft, Microsoft Azure, https://azure.microsoft.com/en-us/?b=16.52a, 2017, retrieved 2017-01-02.

The IoT becomes even more valuable as soon as data from different lines of business assets and devices can be combined with data from other systems in the specific business. Using cloud based solutions like Microsoft Azure,[45] Google Cloud Platform[46] or Amazon Web Services (AWS[47]) for storage and analysis allows for the combination of data from different sources without the worry about capacity constraints or significant costs, which usually result when built into the customer owned infrastructure.

IoT Reference Model

As we have derived so far, the IoT combines several areas like networks, computing, applications and data management. All these branches require the IoT ready architectures and all of them require a specific communication and processing model. It is a fact that even today we still have no standard of how these models could look like nor are there clearly demarcated lines between the IoT devices and systems and non-IoT devices and systems. In the best case, we can assume that if data is generated by machines or devices and sent across networks, we are dealing with an IoT system, but generalizations of the IoT can quickly become inappropriate, as there are many exceptions.

From what we have discussed so far, we understand that due to the sheer complexity of the IoT there will be many different implementations based on complex combinations of hardware and software. It becomes obvious at this stage that it would be helpful to have a kind of reference model for the IoT in order to build a common vocabulary and clarify what functions individual products can contribute through leadership and market credibility.

There are many proposals for an IoT reference architecture. Jim Green presented one of the most complete and powerful proposals by Cisco at the Internet of Things World Forum (IoTWF[48]) in 2014.[49] Meanwhile, an IoT World Forum Architecture

[45]Microsoft, Microsoft Azure, Your app. Your framework., Your platform. All welcome., https://azure.microsoft.com/en-us/overview/what-is-azure/, 2017, 2017-01-02.

[46]Google, Google Cloud Platform, https://cloud.google.com/?utm_source=google&utm_medium=cpc&utm_campaign=2016-q4-cloud-emea-gcp-bkws-freetrial&gclid=CjwKEAiA2abEBRCdx7PqqunM1CYSJABf3qvaktWUu-Ay6LSUlMvJL3WR51RA52Efj_VfiqeUOzINXxoCX7Hw_wcB, 2017, retrieved 2017-01-02.

[47]Amazon, Amazon Web Services (AWS), https://aws.amazon.com/free/?sc_channel=PS&sc_campaign=acquisition_AT&sc_publisher=google&sc_medium=english_cloud_computing_nb&sc_content=cloud_solutions_p&sc_detail=cloud%20solutions&sc_category=cloud_computing&sc_segment=158528574845&sc_matchtype=p&sc_country=AT&s_kwcid=AL!4422!3!158528574845!p!!g!!cloud%20solutions&ef_id=V8av8QAABOvCUCeu:20170102112232:s, 2017, retrieved 2017-01-02.

[48]IoTWF, Internet of Things World Forum, http://www.iotwf.com, 2017, retrieved 2017-01-02.

[49]Cisco, The Internet of Things Reference Model, http://cdn.iotwf.com/resources/71/IoT_Reference_Model_White_Paper_June_4_2014.pdf, 2014, retrieved 2017-01-02.

Table 2 IoT World Forum Architecture Committee

Ayla	BITSTEW	CISCO	EPRI	FANUC
Gartner	General Electric	GM	GRUNDFOS	IBM
IDAHO POWER	IDC	Intel	Itron	KAAZING
ORACLE	PANDUIT	Parc	QUALCOMM	RENESAS
Rockwell Automation	Salesforce	SAMSUNG	SAP	SmartThings
	STARBUCKS COFFEE	THALES	TTTECH	

Committee has been established by a number of vendors driving this architecture (see Table 2). We will have a deeper look on this approach throughout the next sections of this chapter.

The IoTWF also has a Steering Committee[50] with the main mission of establishing premier leadership and providing high-level direction for accelerating the adoption of the IoT.

Multilevel IoT Reference Model

What we need for the IoT is a comprehensive, multilevel model, which takes into account the fact that data is generated by multiple kinds of devices, processed in different ways, sent to different locations and, finally, handled by many applications. There is no restriction to either the scope or locality of the IoT reference model components and this is, of course, true for all levels of the architecture (see Fig. 7).

This results in an IoT model with seven levels, covering necessary IoT functions from the edge to the center, while providing policy, control and data as well security for all these layers as required by the functions.

We will describe the single levels and their functionality in more detail later, but for now we begin with a quick overview. It is important to understand that IoT data flows in both directions: in a control pattern, information flows from level 7 (center) to level 1 (edge) and in a monitoring pattern, the flow of information is in the opposite direction.

IoT Reference Model Bridges OT and IT

While the IoT will certainly include many new things and devices, one of the IoTs' major principle is to connect and incorporate existing infrastructures. Data becomes more valuable as soon as it becomes more portable and can then be accessed by applications that can fully exploit this data. The IoT adds new ways to connect and position more intelligence and autonomy at the network edge and enables sophisticated analysis and business logic for the data collected there. The IoT does not eliminate or replace existing infrastructures; it enables users to make better use of these infrastructures while simultaneously adding new capabilities on top of them.

---

[50]IoTWF, IoTWF Steering Committee, http://www.iotwf.com/steering_committee/info, 2017, retrieved 2017-01-02.

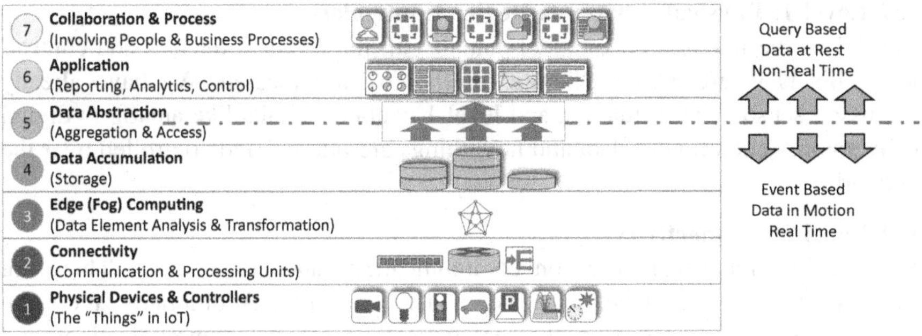

Fig. 7 IoT reference model

The IoT definitely accelerates the pace of change in the OT (Operations Technology) world. Gartner IT Glossary defines OT[51] as: "*Hardware and software that detects or causes a change through the direct monitoring and/or control of physical devices, processes and events in the enterprise.*"

As soon as Ethernet technologies and mobile networks evolved and quickly became adopted, cost points rapidly fell to amazingly low levels, and we have now reached the point where the Internet of Things can finally become a reality. This all means that OT cannot ignore IT any longer, simply because the opportunities are too great. Sensors attached to remote, battery powered nodes using wireless mesh networks will make the sensor data immediately available to analytics and applications on another continent. The remote sensor has become a node in this new Internet of Things. The value of the sensor has increased exponentially, as the data it produces can be immediately processed and analyzed. The IoT simply provides a connectivity stack that carries data via modern networks and modern applications.

As a result, the IoT does not only bridge OT and IT, it also bridges the event-based data we usually deal with in our infrastructures and with query based data, the modern applications now have to deal with event based data as soon as it is transformed and appropriately stored in data centers. We are talking about the conversion of data in motion to data at rest or real time to non-real time data.

IoT Levels

The IoT reference model is based on seven levels. Note that the seven layers of the ISO network reference model should not be confused with the seven levels of the IoT reference model!

[51]Gartner, Gartner IT Glossary, Operations Technology (OT), http://www.gartner.com/it-glossary/operational-technology-ot/, 2017, retrieved 2017-01-02.

IoT Level 1: Physical Devices & Device Controllers

Level 1 of the IoT reference model contains all the physical devices and controllers—this is where all the "things" of the IoT are located. We have already discussed sensors and actuators, and level 1 contains details like analog to digital conversion, generation of data and how things are queried and controlled over the network.

IoT Level 2: Connectivity

Level 2 functionality focuses on communications and connectivity. There are transmissions between level 1 devices and the network. Additionally, there are transmissions across networks that are typically east-west communications and between the network (level 2) and the low-level information processing at level 3. It has to guarantee reliable delivery across the network and is based on existing network infrastructures.

We find traditional switching and routing on level 2 as well as the implementation of various protocols and the necessary translation between protocols. Further security on the network level is implemented at level 1 and through self-learning networking analytics.

IoT Level 3: Edge (Fog) Computing

Level 3 Edge (Fog) Computing[52] functionality focuses on converting network data flows into suitable information for storage and higher level processing of north-south communications including data filtering, cleanup, aggregation, packet content inspection, combination of network and data level analytics and threshold and event generation. As a result of level 3, high volumes of data are analyzed and transformed in order to make information understandable for higher layers (see Fig. 8).

It is important to note that on level 2 and level 3 a lot of conversion takes place from various industrial equipment protocols to industry standards. There are protocol plug-ins for the different devices (things) that use various protocols on the connectivity level (level 2), which are connected via device APIs to level 3 functionality and subsequently via standardized or proprietary APIs to other intermediate nodes, forming the complete IoT network. These can, of course, be physical or virtual nodes in data centers (see Fig. 9).

Converting packets to information is an essential task performed on level 3. Data to information transformation is done by detecting and aggregating events, data is normalized based on rules, sensors are polled dynamically, unstructured data must be put in context in order to allow for its interpretation. Caching of data and information as well as controlled distribution needs to be performed. Special

[52]Bonomi F., et al., ACM DL (ACM Digital Library), Fog computing and its role in the Internet of Things, http://dl.acm.org/citation.cfm?id=2342513&CFID=883969205&CFTOKEN=22808225, 2012, retrieved 2017-01-02.

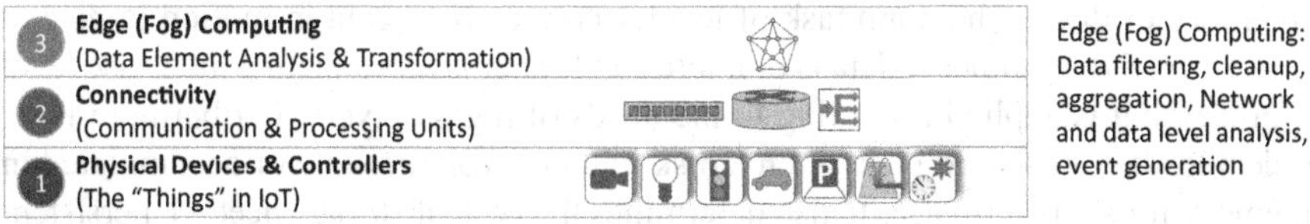

Fig. 8 Level 3 Edge (Fog) Computing

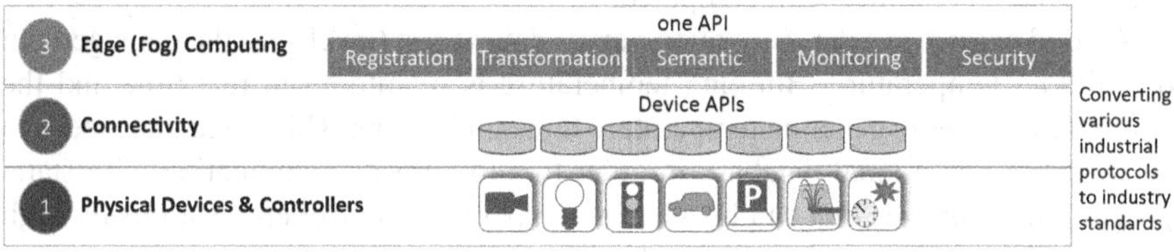

Fig. 9 Level 2 and Level 3 connectivity and data element analysis example

emphasis needs to be put on data in motion,[53,54] use of cases as data reduction and compression (Jorge, 2004) or sensor virtualization[55] and plug & play. Data in motion software[56] runs either on server platforms and blade servers or on hardened edge routers.

IoT Level 4: Data Accumulation

The purpose of Level 4 is to make network data useable by applications. Typically, networks are built to move data, hence data is moved through the network at rates and organizations determined by network devices, which is characteristic of an event driven model.

Level 1 devices do not include computing capabilities themselves, while some computational activities do occur at level 2, like protocol translation or application of network security policy. Any additional computing tasks like packet inspection can be performed at level 3. When we have distributed computational tasks close to the edge of the IoT, using heterogeneous systems distributed across multiple management domains we usually refer to this as fog computing. In combination with fog services, this forms a distinguished characteristic of the IoT.

Typically, most applications do not need to process data at network speed and thus can deal with data at rest (unchanging data) that resides in some sort of storage

---

[53]INAP (Internap), Data in motion vs. data at rest, http://www.internap.com/2013/06/20/data-in-motion-vs-data-at-rest/, 2013, retrieved 2017-01-02.

[54]IBM, IBM developerWorks, 5 Things to Know About Big Data in Motion, https://www.ibm.com/developerworks/community/blogs/5things/entry/5_things_to_know_about_big_data_in_motion?lang=en, 2017, retrieved 2017-01-02.

[55]Monali S., et al., IJCSIT (International Journal of Computer Science and Information Technology, Vol. 5 (1), http://ijcsit.com/docs/Volume%205/vol5issue01/ijcsit20140501151.pdf, 2014, retrieved 2017-01-02.

[56]SAS, SAS Insights, Big Data, Understanding data in motion, http://www.sas.com/en_us/insights/articles/big-data/data-in-motion.html, 2017, retrieved 2017-01-02.

(memory or disk). The main task of level 4 is to convert data in motion to data at rest. Level 4 determines if data is of interest to higher layers and, in which case level 4 processing is applied according to the needs of higher layers. Furthermore level 4 decides if data should be kept on a disk in a non-volatile state or accumulated in memory for short-term use. It also determines the type of storage needed, persistent file systems, big data systems or relational databases. Finally, level 4 ensures the proper organization of data for the required storage system and determines if data needs to be recombined or recomputed.

A good example of data in motion versus data at rest (level 4) is the management of building temperature. Building temperature is monitored in real-time and the temperature sensors generate temperature threshold events. This event data needs to be stored in a relational database in order to allow for its combination with static programmed controls that can be used by temperature control applications through query-based data consumption. On the other side, periodic temperature measurements throughout the building generate streaming data that is fed into big data repositories so it can be used by analytics through batch-oriented data consumption. This describes the two different data consumption models applied in this example: query-based for event data and batch-oriented for streaming data, both handled and converted appropriately on level 4.

IoT Level 5: Data Abstraction

Level 5 of the IoT reference model has the task of abstracting the data interface for applications (Frank & Timothy, 2013). There is a good definition of data abstraction from IT Definitions[57]: "*Data Abstraction: The concept of representing important details and hiding away the implementation details is called data abstraction. This programming technique separates the interface and implementation.*"

Since IoT systems need to be scaled to corporate or global standards, they require multiple storage systems to accommodate IoT device data as well as data from traditional enterprises like ERP, HRMS, CRM and other systems. Dedicated servers provide views of data in the manner that applications need by combining data from multiple sources, which helps to simplify the applications. These servers also facilitate filtering, selecting, projecting and formatting of the data on behalf of the applications by taking into account differences in data shape, format, semantics, access protocol and security.

IoT Level 6: Application

Level 6 is the level where applications run,[58,59] in order to allow reporting, analytics and control (Dominique & Vlad, 2016). Typically, these applications interact with

---

[57] defit, IT Definitions, Data Abstraction, https://www.defit.org/data-abstraction/, 2017, retrieved 2017-01-03.

[58] webofthings, Welcome to the Web of Things, http://webofthings.org, 2017, retrieved 2017-01-03.

[59] webofthings, Web of Things vs Internet of Things, http://webofthings.org/2016/01/23/wot-vs-iot-12/, 2017, retrieved 2017-01-03.

level 5 and data at rest, thus not having to operate in real time. These applications can vary based on vertical markets, the nature of device data and business' needs, hence it is obvious that application complexity will vary widely.

IoT Level 7: Collaboration and Processes

Level 7 is where collaboration and processes involving people and business processes finally take place (Fawzi & Kwok, 2015). This is particularly important to mention as without level 7, which exists to create actions and often requires people and processes, the IoT system would in many cases not be able to generate any value.

The idea is that applications execute business logic and thereby empower people to make decisions. It is not the application making the decision, it is the people working with that application who eventually make the decisions. Frequently the actions taken require multiple people; hence there is also collaboration involved on level 7.

Cyber Physical Systems (CPS) describes the systems of collaborating IoT elements that control different objects, either mechanical or electronic.[60] As we have already discussed, the communications are carried via the Internet, but also other data infrastructures. A CPS communicates with machines as well as human beings and is able to make autonomous decisions and control logistical processes. This is also the reason why CPSs based on the IoT are often seen as the 4th Industrial Revolution or, Industry 4.0. We will discuss this later in this chapter.

IoT Security

IoT Security and the IoT Reference Model

Since security is a very important aspect of the usability and acceptance of the IoT, we will have a brief look at where the different security aspects are handled in the IoT reference architecture.

On level 1, we are dealing with secure content as it comes from silicon that is present and used in the devices (things). On level 2, secure network access is required, implemented by the networking hardware and protocols being used, while on level 3 we need to rely on secure communication by encryption and protocols. On level 4, we must deal with tamper resistance ensured by software, and on level 5, we need to guarantee secure storage offered by the hardware and software we deploy for data abstraction. On level 6, authentication and authorization is controlled via special software for the different applications, while on Level 7, identity management is used to control access to collaboration and processes.

IoT Security and the IoT Security Foundation

Security is such an important prerequisite for the IoT and its general acceptance that there is even a IoT Security Foundation (IoTSF[61]) that was launched on September

[60] Horus, Design and Governance of Collaborative Business Processes in Industry 4.0, http://www.horus.biz/en/design-and-governance-of-collaborative-business-processes-in-industry-4-0/, 2015, retrieved 2017-01-03.

[61] IoTSF, IoT Security Foundation, https://iotsecurityfoundation.org, 2017, retrieved 2017-01-03.

15th, 2015. The IoTSF aims to make the IoT safe to connect in order to allow the benefits of the IoT to be fully realized.

As the IoTSF states, "*IoT is vast and has many security related issues*" and it has established to date, five working groups to work on the different IoT security aspects and areas.

- Working Group 1—Self Certification System—determines requirements for low-cost, accessible and readily actionable systems of self-certification for improved security quality in IoT products
- Working Group 2—Connected Consumer Products—defining security best practice guidelines for different consumer devices
- Working Group 3—Patching Constrained Devices—ensuring the update and maintenance of IoT systems over their lifecycle
- Working Group 4—Framework for Vulnerability Disclosure—defines what has to happen when a security vulnerability is identified in a product or service
- Working Group 5—IoT Security Landscape—end-to-end and system-wide high level mapping of IoT applications identifying security vulnerabilities and determine potential future IoTSF work

Some important outcomes so far are the IoT Best Practice Guidelines[62] and IoT Best Practice User Mark.[63]

IoT Reference Model Status

It makes lot of sense to use a reference model for the IoT in order to establish a common view about all the included technologies, and to speak the same language so that everybody working in the field can efficiently communicate and collaborate, which enables easier standardization of the concepts and terminologies around IoT.

The IoT reference model defines the functionalities required by all the levels we have discussed and, beyond that, highlights concern that should be addressed before the real value and potential of the IoT can be leveraged by the industry. Time will tell how quickly the industry and customers will adopt such a model, but it is very likely that broad acceptance will be found within the next few years, which in turn will help the IoT to accelerate and potentially enable it to become the next big thing since the invention of the Internet and the Internet browser.

[62]IoTSF, IoT Best Practices Guidelines, https://iotsecurityfoundation.org/best-practice-guidelines/, 2016, retrieved 2017-01-03.

[63]IoTSF, Best Practice User Mark, https://iotsecurityfoundation.org/best-practice-user-mark/ 2016, retrieved 2017-01-03.

IoT Solution Samples

Today we can envisage many areas that will benefit significantly from the deployment of IoT solutions. Environmental monitoring is an area where the IoT has already been used successfully, typically using sensors to monitor air or water quality, atmospheric or soil conditions, earthquake and tsunami warning systems or the movement of wildlife between habitats. Another one of these early adoptions is infrastructure management, which uses the IoT to monitor and control operations of urban and rural infrastructures like railway tracks, bridges, wind-farms etc., and monitor events or changes in structural conditions in order to minimize risks. Energy management allows optimizing overall energy consumption through the integration of sensing and actuation systems, where sensors will be integrated into all forms of energy consuming devices like switches, power outlets, bulbs, televisions etc. Another example is the medical and healthcare system, where IoT devices can be used to enable remote health monitoring and emergency notification systems for blood pressure, heart rate or even advanced monitoring with specialized implants like pacemakers or hearing aids.

As of today, there are already many IoT implementations available by some big players like Microsoft,[64] Google with Weave,[65] Android Things[66] or many others, much more than what is listed here not to mention many emerging technologies and methods.

For a better understanding of the options and capabilities of the IoT solutions, we will use the following examples to illustrate the structure and components of such systems.

Parking Space Management

Parking space management is a typical example of the development of smarter cities that optimize the usage of limited parking space, reduce search time for drivers and divert traffic away from the main streets.[67,68] A simple parking management system will help the driver find the nearest or cheapest parking place in a short time. It consists of three elements:

---

[64]Microsoft, Internet of Things, https://blogs.microsoft.com/iot/, 2017, retrieved 2017-01-05.

[65]Google, Weave, https://developers.google.com/weave/, 2017, retrieved 2017-01-05.

[66]Google, Android Things, https://developer.android.com/things/index.html, 2017, retrieved 2017-01-05.

[67]Baratam. M Kumar Gandhi, et al., Indian Journal of Science and Technology, Vol9(17), A Prototype for IoT based Car Parking Management System for Smart Cities, ISSN 0974-5645, 2016, retrieved 2017-01-03.

[68]Jog Yatin, et al., International Journal of u- and e- Service, Science and Technology, Vol.8, No.2 (2015), http://www.sersc.org/journals/IJUNESST/vol8_no2/25.pdf, 2015, retrieved 2017-01-03.

- Information component for the driver
- Device built into each parking space
- Application organizing the best pairing of searching drivers to available parking spaces

The information component is a built-in module in the car's traffic information system or delivered as a smart phone app able to inform the driver about the nearest or cheapest parking and direct him to that position. In both cases, the communication to the parking space management application is based on mobile phone technology.

The device built into each parking space is able to indicate if the space is available, occupied or reserved. It may also have an actor to indicate that this place is reserved (e.g. using a red light). These devices are special purpose modules, communicating via wireless protocol to a proxy and are able to act as sensor and actor.

The parking space management application implements the process of guiding a driver to his preferred place and may also offer reservation and manage parking fees. It has to collect the status of all parking space sensors in near real time, distribute that information to the driver's information component and process requests from drivers for a reservation of a parking space (see Fig. 10).

Precision Agriculture

Agricultural science and technology is rapidly becoming one of the world's fastest growing and exciting markets. It is driven by global changes: a rising population and an increasing need for food, rapid development of emerging economies with aspirations of living a western lifestyle and growing geopolitical instability surrounding shortages of land, water and energy. The Food and Agriculture Organization (FAO) of the United Nations published an article about what will result in 2050: three times as many mouths to feed,[69] which shows the issues we will face concerning food production as well as keeping our worldwide resources intact. In order to make that happen, precision agriculture based on the IoT will be mandatory.

Today, we see a technology revolution in agriculture management[70] and it has gathered momentum through the incorporation of innovative technologies and complex processes (Zhang, 2015). Examples include a Global Navigation Satellite System (GNSS) for farming machine guidance (Hofmann-Wellenhof,

[69]FAO, Food and Agriculture Organization of the United Nations, 2050: A third more mouths to feed, http://www.fao.org/news/story/en/item/35571/icode/, 2016, 2017-01-03.

[70]Beecham Research, Towards Smart Farming—Agriculture Embracing The IoT Vision, https://www.beechamresearch.com/files/BRL%20Smart%20Farming%20Executive%20Summary.pdf, 2014, retrieved 2017-01-03.

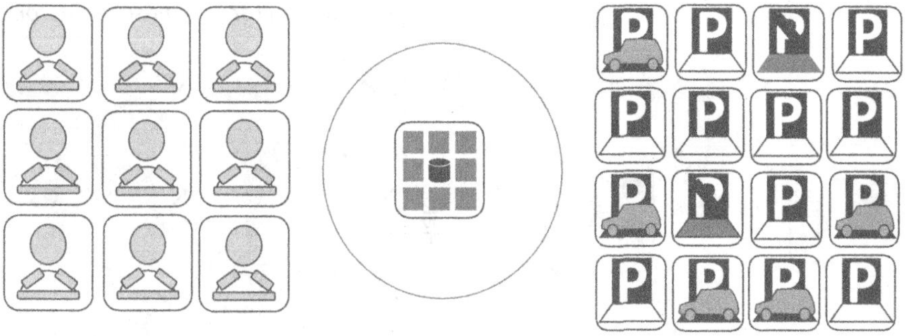

Fig. 10 IoT parking space management

Lichtenegger, & Wasle, 2008; Rao, 2010), a rich portfolio of both field and on-board sensors and actuators, improvements in weather forecasting, analytics to back decision support systems that guide planting and harvesting decisions, and evaluations of the results. A large constellation of companies big and small dots the landscape. Competition is intensifying, and farmers are well aware of the potential gains as well as concerned about their rights.

The agribusiness landscape is dotted by major players in key domains (seeds, fertilizers, pesticides, manufacturers of agricultural machinery, and insurance companies), alongside an ecosystem of minor players like agricultural consultants. Alternative approaches like organic farming or non-GMO (genetically modified organisms) have gained momentum both in popularity among farmers and also in influencing regulations and becoming a growing economic driver in today's farming practices. While there is a trend towards integration and consolidation in the agriculture market, the ability to find and implement solutions remains largely fragmented.

Precision agriculture will offer farmers several benefits, like allowing a unifying view and actual control point of the farm processes and operations, thereby maintaining ownership and sharing data in a controlled way. It guarantees frictionless access to the diverse players in the agricultural landscape—farmers, OEMs, seeds, fertilizers, agro-chemical suppliers, organic farming organizations, non-GMO suppliers, research institutions, regulatory bodies, and ecosystem partners.

Thus, agriculture management encourages the creation of services that add value to the end-to-end food chain. These services help to promote improvements in farming practices and increase the quality of the produce. At the same time, they reduce any negative impact on the environment and help farmers to better manage their investments. There are already many platforms available for this—thingworx Smart Agriculture[71] being one of the most complete in 2017 (see Fig. 11).

[71]thingworx, Smart Agriculture, Solving IoT Challenges for Smart Agriculture, https://www.thingworx.com/ecosystem/markets/smart-connected-systems/smart-agriculture/, 2017, retrieved 2017-01-03.

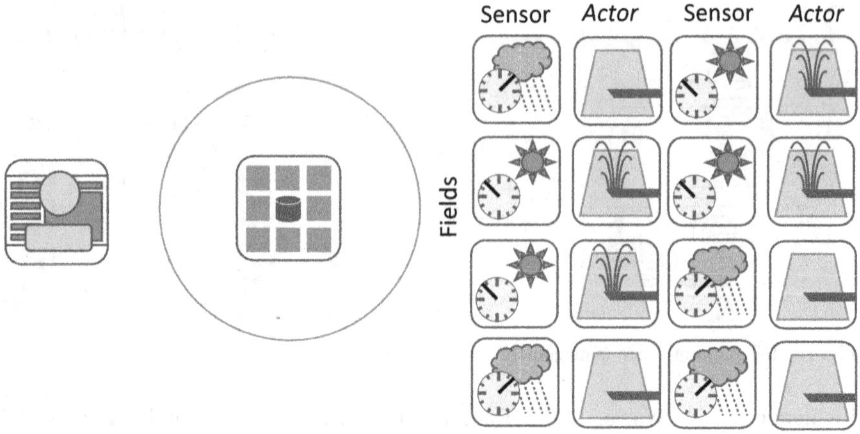

Fig. 11 Precision agriculture

Building and Home Automation Systems

The IoT can be used to monitor and control electrical, electromechanical and electronic systems used in all kinds of buildings ranging from public to private and industrial institutions to residential. Building automation systems are typically used to control lightning, heating, ventilation, air conditioning, appliances, communication systems, entertainment, home security devices and, in so doing, improve energy efficiency, security, comfort, and convenience (Montgomery, 2014). Today many smart home and building automation startups already exist,[72,73,74] and furthermore security[75] systems will benefit greatly from the IoT being applied to them. In the meantime, a broad offering of products and services are available.

A typical scenario is demonstrated by imagining that your water could already be warming up in the early morning hours, just right before you take your morning shower. Of course, the system has been previously informed of your wake-up time from your calendar entries in your personal devices, which are all perfectly synced via cloud services. As soon as you enter the shower, your coffee machines turns on and produces your favorite morning café, all controlled via sensors in your bed and

[72]Postscapes, IoT Home Guide, http://www.postscapes.com/internet-of-things-award/connected-home-products/, 2017, retrieved 2017-01-04.

[73]safewise, What is home automation and how does it work?, http://www.safewise.com/home-security-faq/how-does-home-automation-work, 2017, retrieved 2017-01-04.

[74]CB Insights, Smart Home Market Map: 67 Startups In Home Automation, Smart Appliances, And More, https://www.cbinsights.com/blog/smart-home-market-map-company-list/?utm_source=CB+Insights+Newsletter&utm_campaign=858ef521c5-Top_Research_Briefs_10_15_2016&utm_medium=email&utm_term=0_9dc0513989-858ef521c5-87759901, 2016, retrieved 2017-01-04.

[75]safewise, The Best Home Security Systems in 2016, http://www.safewise.com/best-home-security-system, 2016, retrieved 2017-01-04.

bathroom. Sure, this system would also recognize your wife's body sensor—being either a watch or even some implanted sensor—and prepare the right morning café for her as soon as she leaves the shower a few minutes later. While you sip your café you can skim through the latest news reports which accord with your preferences and listen on your entertainment system in the kitchen. Your wife, meanwhile, can enjoy her favorite morning workout in the living room. As soon as either of you leaves a room, the lights would dim in order to save energy, and the home security system would lock your house entrances and windows automatically once everybody has left. The system would also activate the security alarm that, of course, is connected via multiple paths to either the police or a security company or both (see Fig. 12).

Manufacturing and Industry 4.0

With the IoT we have the possibility of controlling and managing manufacturing equipment, allowing asset and situation management, enabling manufacturing process control, entering smart manufacturing and industrial applications.[76] This allows for the rapid manufacturing of new products, while responding dynamically to product demands and optimizing the manufacture's production through supply chain management in real-time using networking machines, sensors and control systems (Gilchrist, 2016).

Control systems help to automate process controls, operator tools and service information systems in order to optimize plant safety and security and allow for predictive maintenance, statistical evaluation and measurements for maximized optimization of the whole process. This manufacturing revolution is based on the Internet of Things and is often referred to as Industry 4.0. Based on the concepts of the IoT, cyber-physical systems and the Internet of Services, Industry 4.0 facilitates a vision of the smart factory.[77]

The term "Industry 4.0" refers to the fourth industrial revolution and originated from a project in high-tech strategy by the German government in 2011, which promoted the computerization of manufacturing. As shown in Fig. 13: From Industry 1.0 to Industry 4.0," the first industrial revolution was the mechanization of production using water and steam power, the second industrial revolution introduced mass production with the help of electrical power, while the third phase was characterized by the use of electronics and IT for an even higher degree of automated production (see Fig. 13).

[76]McKinsey&Company, The Internet of Things and the future of manufacturing, http://www.mckinsey.com/business-functions/digital-mckinsey/our-insights/the-internet-of-things-and-the-future-of-manufacturing, 2013, retrieved 2017-01-04.

[77]IoT Evolution, How Industry 4.0 and IoT are connected, http://www.iotevolutionworld.com/m2m/articles/401292-how-industry-40-the-internet-things-connected.htm, 2015, retrieved 2017-01-04.

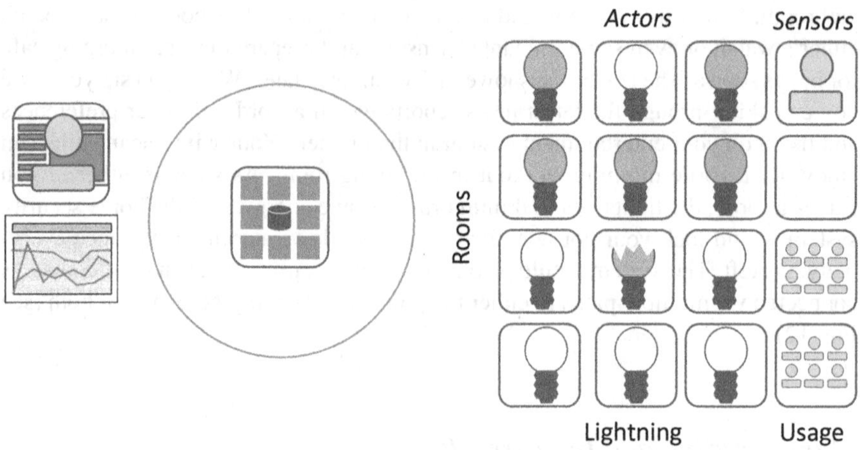

Fig. 12 Home automation

Cyber Physical Systems Versus Embedded Systems

A cyber-physical system (CPS) is a system of collaborating computational elements that control physical entities which are the integrations of computation, networking and physical processes (Alur, 2015; Lee & Seshia, 2015). The concept map of cyber physical systems prepared by CyberPhysicalSystems.org[78] gives us a comprehensive overview of all the different areas of CPSs.

Today, a precursor generation of cyber-physical systems can be found in areas ranging from aerospace, automotive, chemical processes, civil infrastructure and energy to healthcare, manufacturing, transportation, entertainment, and consumer appliances. This generation is often referred to as embedded systems.

In embedded systems, the emphasis tends to be more on the computational elements, and less on an intense link between the computational and physical elements. But, in fact, the cyber-physical systems that are an essential part of Industry 4.0 are objects or things as defined by the IoT.

The concept is seen in the modularly structured smart factories of Industry 4.0 that use such cyber-physical systems to monitor processes, create virtual copies of the physical world and make decentralized decisions. The cyber-physical systems communicate and cooperate with each other as well as with humans in real time, just like the things in the IoT. The Internet of Services[79] offers both internal and cross-organizational services that can be utilized by participants of this value chain.

[78]CyberPhysicalSystems, Cyber-Physical Systems, http://cyberphysicalsystems.org, 2017, retrieved 2017-01-04.

[79]Fraunhofer, Communication and Knowledge—Internet of Services / Internet of Things, https://www.fraunhofer.de/en/research/fields-of-research/communication-knowledge/internet-of-services-internet-of-things.html, 2017, 2017-01-04.

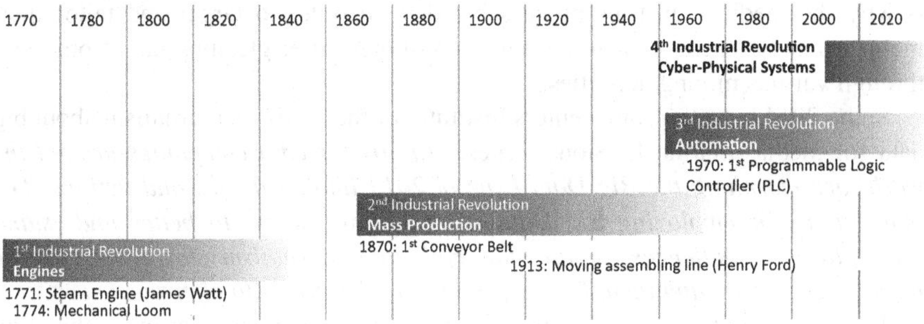

Fig. 13 From Industry 1.0 to Industry 4.0

IoT in Retail Market

There are a lot of possibilities for retail using the IoT. Devices that can record where shoppers focus their attention in stores or scan objects that customers use in order to reorder items at home are just some possibilities. Some retailers already put sensors in shopping carts to track where customers walk through the stores, not to mention the sensors placed on products in order to prevent theft.

Meanwhile, there are even robots that can interact with customers on the shopping floor. Fellow Robots,[80] which was founded at Singularity University,[81] is one good example of this next generation of retail. Another example is Simbe Robotics,[82] which delivers new visibility options for retailers as to the status of their merchandise in stores. Tally performs repetitive tasks like auditing shelves for out-of-stock items or low stock, misplaced items as well as pricing errors.

Media, Data Capture, IoT and Big Data

This describes an example of the IoT that is not quite as obvious as most of the other examples but it is nevertheless a very interesting and promising one and will have a significant influence on all of our lives—whether we like it or not. When looking deeper into the mechanisms used for media processes, we understand how the IoT, media and big data are interconnected (Stackowiak et al., 2015).

As the industry is moving away from the traditional approach of using specific media platforms like newspapers, magazines and television shows they are instead looking to reach consumers with new technologies. These technologies approach their target audiences at optimal times and in optimal locations with the main goal of conveying messages or content that is in line with the consumer's mindset. This

[80]fellow Robots, http://fellowrobots.com, 2017, retrieved 2017-01-05.
[81]Singularity University, https://su.org, 2017, retrieved 2017-01-05.
[82]Simbe Robotics, http://www.simberobotics.com, 2017, retrieved 2017-01-05.

is why the media industries process big data in order to target and tailor their messages and content so as to appeal to consumers after gleaning data from them through various mining activities.

In the 2014 report[83] from Reuters Institute for the Study of Journalism about big data for media, Martha L. Stone writes: "*In 2014, media companies around the world are morphing the Big Data hype of 2013 into strategies and actions. The opportunity for employing Big Data strategies are many: to better understand cross-platform audiences, create powerful data journalism stories, streamline business processes and identify new products and services to offer customers*".

This is a revolutionary step forward, as with the IoT technologies in combination with the right analyzing tools, it opens up the opportunity to measure, collect and analyze consumer behavior and statistics in a way never possible before and, finally, cross-correlate this data in order to significantly improve the marketing of products and services. This is a typical example of the IoT and Big Data working closely together, helping to transform the media industry, companies and governments and allowing advertisement and media to reach a completely new level beyond the conventional mechanisms used by the industry before.

References

Alur, R. (2015). *Principles of cyber-physical systems*. Cambridge, MA: The MIT Press.

Bassi, A., Bauer, M., Fiedler, M., Kramp, T., Kranenburg, R. V., Lange, S., & Meissner, S. (2013). *Enabling things to talk*. Berlin: Springer.

Bradley, J., Reberger, C., Dixit, A., & Gupta, V. (2013). internetofeverything.cisco.com.

Dominique, G., & Vlad, T. (2016). *Building the web of things*. Greenwich: Manninf.

Edson, B. (2015). www.InternetofYourThings.com

Fawzi, B., & Kwok, W. (2015). *Collaborative internet of things (C-IoT): For future smart connected life and business*. Hoboken: Wiley.

Frank, C., & Timothy, H. (2013). *Data abstraction and problem solving with C++*. Boston: Pearson.

Gilchrist, A. (2016). *Industry 4.0 – The industrial internet of things*. New York: Springer.

Greengard, S. (2015). *The internet of things*. Cambridge, MA: MIT Press Essential Knowledge Series.

Hofmann-Wellenhof, B., Lichtenegger, H., & Wasle, E. (2008). *GNSS – Global navigation satellite systems*. Wien: Springer.

Jorge, H. (2004). *Dimension reduction and data compression, chapter structural reliability, Vol. 17 of the series lecture notes in applied and computational mechanics pp. 81–105*. Berlin: Springer.

Kellmereit, D., & Obodovsky, D. (2013). *The silent intelligence – The internet of things*. San Francisco, CA: DND Ventures LLC.

Lee, E. A., & Seshia, S. A. (2015). *Introduction to embedded systems – A cyber-physical systems approach*. Raleigh: lulu.com.

[83]Stone Martha, Reuters Institute for the Study of Journalism (University of Oxford), https://reutersinstitute.politics.ox.ac.uk/sites/default/files/Big%20Data%20For%20Media_0.pdf, 2014, retrieved 2017-01-04.

McEwen, A., & Cassimally, H. (2014). *Designing the internet of things*. Chichester: Wiley.
Minoli, D. (2014). *Building the internet of things with IPv6 and MIPv6*. Hoboken, NJ: Wiley.
Montgomery, L. (2014). *Home automation: A complete guide to buying, owning and enjoying a home automation System*. Cork: BookBaby.
Rao, G. S. (2010). *Global navigation satellite system – With essentials of satellite communications*. New Delhi: Tata McGraw Hill.
Stackowiak, R., Licht, A., Mantha, V., & Nagode, L. (2015). *Big data and the internet of things*. Berkeley, CA: Apress.
Yan, L., Zhang, Y., Yang, L. T., & Ning, H. (2008). *The internet of things: From RFID to the next-generation pervasive networked systems*. New York: Auerbach Publications.
Zhang, Q. (2015). *Precision agriculture technology for crop farming*. Boca Raton: CRC Press.
Zhou, H. (2012). *The internet of things in the cloud – A middleware perspective*. London: CRC Press.

Fog Computing

Imagination is more important than knowledge
Albert Einstein

Fog Computing can be seen as Cloud Computing but close to the ground, helping to create automated responses as an added value. Fog Computing uses a large number of nodes in a widespread and heterogeneous environment mainly connected via wireless access and mobility and is largely optimized for real time applications due to its low latency and location awareness. Fog Computing describes the new layer between the edge devices and the cloud and introduces a new form of distributed computing that is optimized for applications to run as close as possible to the sensors, devices (things), people and processes.

Fog Computing provides plenty of disruptions for the IoT and brings a very significant change by organizing endpoints into systems, which is massively different in scope compared to having a person behind each device. These new IoT systems are mainly self-efficient systems without any direct human interaction and are dynamically organized into macro endpoints. Popular examples are Smart Traffic Lights or Smart Connected Vehicles, Smart Buildings and Smart Grid.

Fog Platform requirements need to cover the edge to cloud (core) interplay and this distributed infrastructure requires a completely new Fog Computing architecture that imposes new networking requirements that would not be possible without the usage of the latest emerging technologies.

Fog Computing in a Nutshell

Fog Computing Origin and Definition

In meteorology, the term "fog" is used for clouds that are close to the ground. Similarly, Fog Computing concentrates on the edge of the network. It was Cisco that actually coined the term "Fog Computing" at the beginning of the 2010s. Flavio Bonomi, who worked for Cisco at the time, can be seen as one of the

M. Oppitz, P. Tomsu, *Inventing the Cloud Century*,
DOI 10.1007/978-3-319-61161-7_17

inventors of Fog Computing. He and his team published the first articles about Fog Computing at Sigcomm in 2012.[1] Bonomi meanwhile became CEO of Nebbio Technologies,[2] which was a young Silicon Valley startup founded in 2015 with the vision of applying the Fog Computing paradigm to industrial automation and other related IoT verticals.

Meanwhile the OpenFog Consortium[3] was established in November 2015 by founding members Dell, Intel, Cisco, Microsoft, ARM and Princeton University to develop open reference architecture as well as drive Fog Computing business value.

Fog Computing Versus IoT Versus Cloud Computing

Fog Computing (or sometimes also called Edge Computing) can be seen as an extension of the IoT and Cloud Computing as it extends to the edge of the network via a distributed computing, network and storage infrastructure, thereby enabling new applications and services. It is defined through a heterogeneous and widespread geographic environment consisting of an extremely large number of nodes, using mainly wireless access and mobility, low latency and location awareness as well as streaming and real time applications.[4]

In the past 10 years, we have seen the migration of computation and data storage to clouds and it is important to understand that cloud services have become mainstream for individual users today. Over the next 20 years, we will see increasing communication opportunities between devices in the fog, which means that the cloud is increasingly becoming the fog or parts of the cloud will be in the fog.

Therefore, fog networks will rejuvenate the Internet of Things by networking appliances, schoolbooks, clothes, medical devices and any other object to other objects and the Internet. There will be future incarnations of what we know today as connected wearables like smart watches, smart glasses, health and fitness monitors etc. that will not only change the consumer experience but also the complete business environment.

We have already discussed the expected number of devices in the IoT to range from 50 to 200 billion by 2020 (see chapter on the IoT), but now the number of communicating devices in the fog will be even larger or at least in the upper range of these estimations. This requires disruption in many areas and the designing of a hierarchical distributed architecture, extending from the edge to the core of fog computing.

---

[1]Bonomi Flavio, et al., Sigcomm, Fog Computing and its Role in the Internet of Things, http://conferences.sigcomm.org/sigcomm/2012/paper/mcc/p13.pdf, 2012, retrieved 2017-01-07.

[2]Nebbio Technologies, http://www.nebbiolo.tech, 2017, retrieved 2017-01-07.

[3]OpenFog Consortium, https://www.openfogconsortium.org, 2017, retrieved 2017-01-07.

[4]Cisco, Fog Computing and the Internet of Things, https://www.cisco.com/c/dam/en_us/solutions/trends/iot/docs/computing-overview.pdf, 2015, retrieved 2017-01-07.

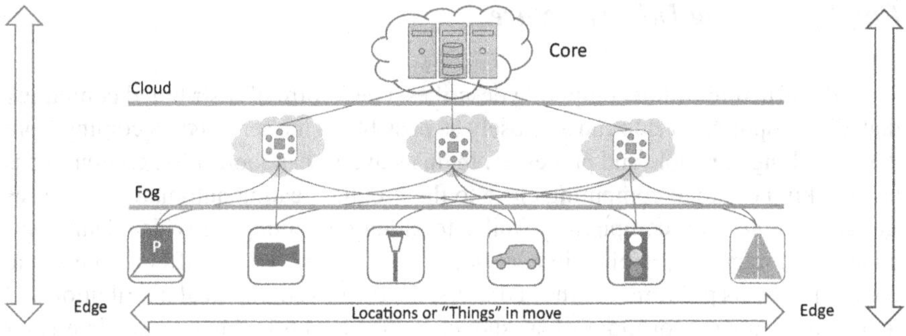

Fig. 1 Fog between edge and cloud

Fog Computing Versus Edge Computing

Both Fog and Edge Computing push intelligence and processing power closer to the source of data where the sensors are located. The key difference between both approaches is where intelligence and computing power is finally placed (see Fig. 1).

Fog computing pushes the intelligence into the local area edge network, being a part of the complete network architecture. Data processing is done in a fog node or IoT gateway. This is also where the term "fog computing," introduced by Cisco, originates—it describes a layer of computing at the edge of the network that allows pre-processed data to be quickly and securely transported to the cloud (the data center). To be more precise, the data from the control system program is sent to the Open Platform Communication (OPC[5]) server, also known as the protocol gateway. It has the task of converting the data into a protocol like MQTT[6] (a machine-to-machine IoT connectivity protocol) or HTTP which is understood by Internet systems. After doing so, the data is sent to another system, which can be either a fog node or the IoT gateway on the edge LAN, collecting data and performing analysis and high level processing. It can also store data for later transmission over the WAN.

Edge computing pushes the intelligence and processing power as well as communication capabilities of an edge gateway or appliance into programmable automation controllers (PACs).

Today both terms are used interchangeably but actually edge computing is the older expression predating the fog computing term. There are advantages of both implementations, but fog computing is positioned to be more scalable while also giving a better view of a more distributed area at the edge as there are usually multiple devices feeding back to the edge network.

---

[5]OPC Foundation, https://opcfoundation.org, 2017, retrieved 2017-01-09.

[6]MQTT.ORG, http://mqtt.org, 2017, retrieved 2017-01-09.

Fog Computing Infrastructure

The infrastructure of fog computing describes a new form of distributed computing that allows applications to run as closely as possible to the sensed data coming from devices (things), people and processes. In this way, we can look at fog computing as being a kind of cloud computing close to the ground creating automated responses as added value. Fog computing, similar to cloud computing, provides data, computation, storage and application services to end users, but it differs from cloud computing by its proximity to the end users, the dense geographical distribution and its inherent support for mobility as shown in Fig. 1. Fog is this intermediate layer between the edge and cloud.

IoT Mandates Transition from Cloud to Fog

Cloud computing offers an economic and efficient alternative to private data centers (DCs), especially for customers running web applications and batch processing. This is due to the better scalability of large DCs resulting from massive aggregation and high predictability. There are also other significant advantages, as mega DCs can be conveniently located by taking into account inexpensive power sources and, at the same time, can significantly lower OPEX through the deployment of homogeneous computing, storage and networking solutions. As another bonus, enterprises and end-users are largely freed from detailed specifications when using cloud computing.

New Applications Requiring Fog Computing

All the advantages that were true for cloud computing are no longer true for latency sensitive applications having stringent delay requirements that can be expected from many if not most devices in the IoT. Especially IoT deployments require geo distribution,[7] mobility, location awareness and low latency, thus needing a new platform called fog computing. This enables new breeds of applications and services, allowing the right combination and interworking between cloud and fog,[8] where fog computing extends cloud computing to the edge of the network.

[7]Uni Stuttgart, IPVS, Geo-distributed cloud & edge computing, https://www.ipvs.uni-stuttgart.de/abteilungen/vs/forschung/topics/Geo-distributed_Cloud_Edge_Computing, 2017, retrieved 2017-01-11.

[8]Open Fog Consortium, Fog Computing and Mobile Edge Cloud Gain Momentum, http://yucianga.info/wp-content/uploads/2015/11/15-11-22-Fog-computing-and-mobile-edge-cloud-

Both fog and cloud can use the same resources such as networking, computing and storage, and share many of the same principles and ideas such as virtualization and multi-tenancy. Contrary to conventional cloud applications and services, these new types of real time sensitive applications and services[9] require Time Sensitive Networking (TSN[10]) and specifically require low latency and predictability, fast mobile applications as found in connected vehicle or connected rail, geo distributed applications for pipeline or environmental monitoring and, finally, large scale distributed control systems such as smart grid,[11] connected rail[12] and smart traffic light systems.[13]

In the chapter on IoT solution samples, we already discussed how the applications behind these samples, i.e. those generating large amounts of data, usually require real-time characteristics. With applications now asking for fog computing, even more data is generated at the edge, hence it becomes clear that effective and efficient analysis of this data should be done at the edge or close to the edge to guarantee scalability. This means that, contrary to typical IoT applications requiring connectivity to the cloud, we are now dealing with special types of applications requiring the novel characteristics of fog computing, which once more underlines that notion that cloud and fog computing are complementary and do not replace each other.

Furthermore, the original IoT applications from the beginning of the 2010s, that typically involved a person behind each device, are shifting towards devices organized as system applications like transportation and connected vehicles, precision agriculture, industrial automation, healthcare, intelligent building, smart grid etc., each requiring a new paradigm such as fog computing. This trend will continue over the next several years until at least the mid-2020s (see Fig. 2).

---

gain-momentum—Open-Fog-Consortium-ETSI-MEC-Cloudlets-v1.pdf, 2015, retrieved 2017-01-11.

[9]Intel, IoT@Intel, Time Sensitive Networking to Fundamentally Change the Way Real Time Industrial Systems Are Designed, https://blogs.intel.com/iot/2016/08/03/industrial/, 2016, retrieved 2017-01-11.

[10]IEEE, IEEE 802.1, Time Sensitive Networking Task Group, http://www.ieee802.org/1/pages/tsn.html, 2017, retrieved 2017-01-11.

[11]SmartGrid, What is the Smart Grid? https://www.smartgrid.gov/the_smart_grid/smart_grid.html, 2017, retrieved 2017-01-11.

[12]IBM, Connected Trains: how IoT is driving the future of rail, https://www.ibm.com/blogs/internet-of-things/connected-trains-rail-travel/, 2016, retrieved 2017-01-11.

[13]IJIRSET, International Journal of Innovative Research in Science, Engineering and Technology, Vol. 5, Special Issue 10, May 2016, Innovative Technology for Smart Roads by Using IoT Devices, https://www.ijirset.com/upload/2016/iccstar/34_O015_CAMERAREADY.pdf, 2016, retrieved 2017-01-11.

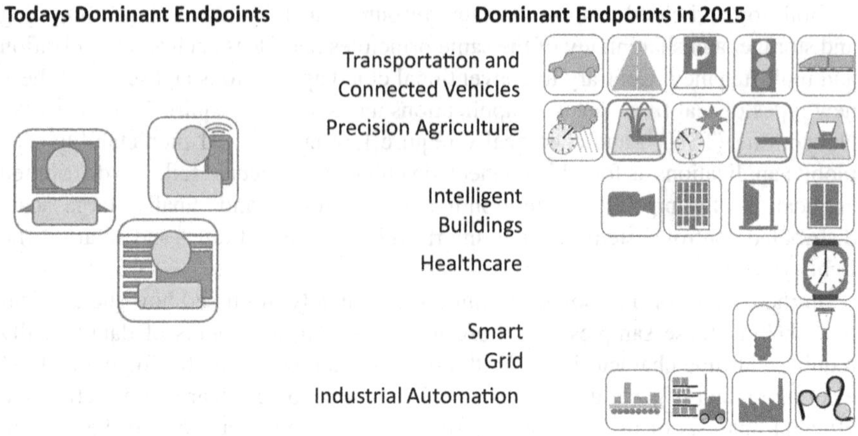

Fig. 2 Transitions in IoT

Fog Computing as Enabler for IoT Success

One of the main promises of IoT is to bring the advantages of cloud computing down to earth for everybody to use in, for example, vehicles, homes as well as any kind of workplace. We have already discussed that this requires smart, connected devices and the guarantee that the gateways enabling this functional reality are reliable with almost undisrupted uptime.

These gateways also handle the majority of processing work before data is passed to the cloud. Hence, Fog Computing needs to meet the requirements of reliable, low latency responses through processing at the edge that also better deals with high traffic volume through the use of smart filtering as well as selective transmission.

Technology matured with the availability of mobile technologies and the processing power of these devices allows for the change from the old hub and spoke architecture designs to mesh, distributed edge-computing architectures (see Table 1).

Fog and Cloud Relationship

Many cases of use require a broad interplay between the edge and core of the IoT environment, as data generated at different levels of the IoT has different requirements and uses at different time scales. Closer to the edge, fog computing is used and closer to the core, we see the use of cloud computing. Further the typical use cases for fog computing described later in this chapter show that we must deal with totally different time scales ranging from milliseconds to months, as shown in the smart grid example below (see Fig. 3).

Table 1 Fog and cloud computing

Requirements	Fog computing	Cloud computing
Location of service	At the edge of the local network	Within the internet
Latency	Low	High
Delay, jitter	Very low	High
Support of mobility	Supported	Limited
Security	Defined	Undefined

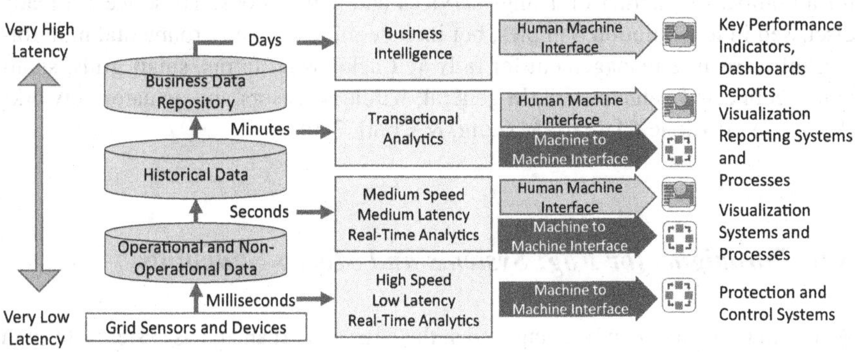

Fig. 3 Different time scales from fog to cloud

There are machine-to-machine communications at the edge happening in the millisecond to sub second levels, because we are dealing with real time processes and hence also need real time analytics. At the next level, we have to deal with second to sub minute processes required for the machine to machine processes as well as the human to machine interactions needed for visualization, and activation/deactivation of appliances. The next level operates in minutes or days used by transactional analytics needed for visualization, reporting systems and processes. Finally, moving from the area of technical operations to the area of enterprise operations, we are dealing with a time frame of days to months as these transactions are needed for business intelligence to show and create key performance indicators, dashboards and reports.

The higher we climb the levels, the larger the geographical areas covered become. In the case of business intelligence, this can be a whole country or even larger whereas, at the sub-second level, we have to deal with the local control and stability of the grid and this is where the real-time requirements arise.

Fog Computing System Level Approach

The disruptive paradigm in fog computing results from the fact that smart things are attached to fog devices and the fog devices are either interconnected and/or linked to the cloud. This adds completely new dimensions to the IoT as it allows incorporating the massive number of smart things at the edge in an optimal way, while at the same time efficiently applying big data analytics to the cloud.

These characteristics essentially enable the Fog to be the appropriate platform for a number of Internet of Things services and applications. These were already discussed in IoT Solution Samples, but include things like environmental monitoring, infrastructure management for railway tracks, wind farms, smart grids, smart cities, connected vehicles and, in general, wireless sensors and actuator networks (WSANs) (Verdone, Dardari, Mazzini, & Conti, 2008).

New Paradigms for Fog: Systems and Macro Endpoints

When smart devices can be connected at the edge and organized as systems through fog computing, this gives the conventional IoT another boost as the number of connected devices we have discussed so far is too low. By 2025 the expected growth of devices organized as systems is estimated to grow to a trillion, which means 10–20 times increase resulting from all these new systems and consisting of things and smart devices at the edge. Looking at the architectures we envisaged so far for the IoT, it becomes obvious that using the same architecture for the expected amount of edge devices as well as the same communication and analytics models, is no longer up to scale and requires completely rethinking the existing paradigms.

The first and most significant change we have already discussed is the organization of endpoints into systems, which massively changes the original scope of having a person behind each device. These new IoT systems will become self-efficient systems without any direct human interaction and will be organized into macro endpoints.

Fog Platform Requirements

It becomes obvious from the use cases we have discussed that most fog computing cases share a number of common characteristics. The most important common attributes of fog platforms are geo-distribution, low and predictable latency, multi-agencies orchestration, consistency, multitenancy and interplay between the fog and cloud.

Geo-distribution

Fog computing adds an additional dimension to the big data from the IoT that generally is described by five dimensions like volume, variety, velocity, value and veracity. As we have seen from the nature of many IoT use cases like smart cities, environmental monitoring, connected vehicles, connected rail, STLs etc., we need to introduce a sixth dimension, geo distribution, for a complete description. For example, wind farms require wide and dense geo-distribution throughout the wind farm, as is the case with precision agriculture, smart grids, industrial automation, transportation and connected vehicles.

Geo-distribution is an obvious attribute that requires the flexibility of fog computing, as many cases of use cover both widely distributed devices across regions and also dense distribution in certain areas. Smart Traffic Light Systems (STSLs) are a typical example of wide distribution in order to allow efficient traffic control, but it also needs density as seen, for example, at intersections and ramp access to highways.

All these geo-distributed devices (sensors) quickly amount to large numbers and this is one of the main reasons to move processing of the data to where it occurs—at the edge. This is what fog computing adds to the equation by offering a distributed intelligent platform at the edge, and performing distributed computing, networking, storage and processing. A successful IoT deployment thus requires a solid architecture from the edge to the core,[14] covering networking, storage and computing as well as new players (partners) that bring successfully ideas about new cases of use to life.

Low and Predictable Latency

Low and predictable latency is another key requirement demanding a fog computing approach and is specifically critical at the intersections of STLs or at the turbine level of wind farms. In many cases, it may also be crucial for smart grids, industrial automation and many examples of transportation and connected vehicles, while precision agriculture might not have these stringent, low latency requirements.

Multi-agent Orchestration

Multi-agent orchestration dictates that all the agencies running a system must be able to coordinate and control policies in real time. This is crucial for STLs, smart grids, industrial automation and transportation and connected vehicles, but less crucial for wind farms (Nik, 2014).

Consistency

Consistency describes the goal of getting a reasonable degree of consistency between many of the data collection points as well as actuators, if they are involved. This is required for smooth running STLs, wind farms and all other cases we have

[14]OpenFog Consortium, OpenFog Architecture Overview, https://www.openfogconsortium.org/wp-content/uploads/OpenFog-Architecture-Overview-WP-2-2016.pdf, 2016, retrieved 2017-01-13.

described so far. Consistency is highly reflected in proper geo-distributed systems and vice versa (Nik, 2014).

Multi-tenancy

For economic and efficient operation, the fog must support multi-tenancy, meaning it needs to support multiple client organizations without mutual interference. Multi-tenancy is needed for both fog and cloud.[15] In addition, the fog needs to allow strict service guarantees for mission critical systems such as STLs, transportation, connected vehicles and smart grids. Wind farms, precision agriculture and industrial automation might not need this multi-tenancy requirement as they might be owned by closed organizations, but different business models could also require multi-tenancy.

Interplay Between Fog and Cloud

Data that is generated by fog platforms has to be collected, analyzed and treated at different time scales. Most importantly, there is real time information (data) generated by the sensors where an action is immediately taken like in STLs. This precludes the general use of the cloud and makes fog computing a must.

Near real time data can be collected by the system and ingested in a data center or cloud for more deep analysis and this data typically extends over time periods like days, months and even years.

The example of STLs perfectly shows these different requirements, as data is collected from sensors and vehicles at intersections and also traffic information at different collection points. While intersection data is mostly collected in real time requiring fog computing, other data can easily be near real time and allow for further and non-time sensitive processing in the cloud. The same is true for the other cases we mentioned, some will require higher amounts of fog computing and less cloud access, while others can have a different distribution pattern.

Further Fog Platform Characteristics

Several attributes do not commonly apply to every case of use like, for example, mobility that is a key attribute for SCV and connected rail, but does not apply to wind farms, smart grids or industrial applications. This shows that fog computing mandates very heterogeneous platform requirements that ideally would support a wide range of verticals instead of specific solutions for each vertical, but it is important to note that different verticals can effectively use different attributes of the platform.

[15] thecloudtutorial, Ho Rickly, Multi-tenancy and cloud computing, http://thecloudtutorial.com/multitenancy.html, 2017, retrieved 2017-01-13.

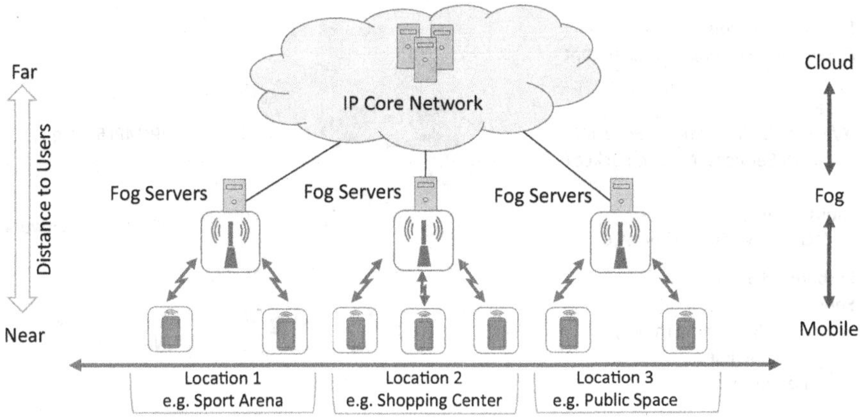

Fig. 4 Fog computing hierarchy

Fog Computing Architecture

Distributed Fog Infrastructure

The fog and cloud solutions are complementary in the way that fog extends cloud-computing residing in the data center to the edge and endpoints. While cloud computing uses mainly homogeneous physical resources that are usually managed in a centralized way. Fog computing uses a distributed infrastructure of heterogeneous resources that, in turn, require a distributed management approach (Bonomi, Milito, Natarajan, & Zhu, 2014). By introducing an intermediate Fog layer between cloud computing, the core network and the mobile devices at the edge, leads to a three-layer Mobile-Fog-Cloud Hierarchy, which can solve this problem (see Fig. 4).

The Fog layer contains the distributed Fog servers that are present on the local premises of the mobile users. Fog servers are virtualized devices running their own built-in data storage, computing and communication parts. This allows Fog Computing to place the necessary computing, storage and computational resources close to the mobile users, thus allowing a fast rate of services available to the mobile users because of short distances and high-speed wireless connections. Fog servers can either be static in fixed locations like WiFi[16] access points or mobile e.g. on moving vehicles.

[16]WiFi Alliance, WiFi, http://www.wi-fi.org, 2016, retrieved 2016-12-27.

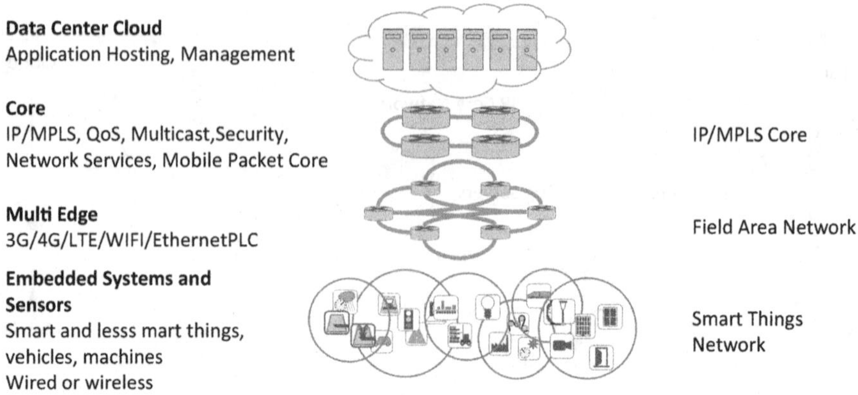

Fig. 5 Fog computing distributed infrastructure for IoT

Fog Architecture Network Infrastructure View

The network infrastructure of this distributed fog infrastructure ideally consists of (see Fig. 5):

- A data center in the cloud providing application hosting and management
- A core network based on well-established state of the art technologies like IP/MPLS with all the required extensions such as QoS, multicast, SDN, security, network services and a mobile packet core
- A multi-service edge or field area network supporting 3G/4G/LTE and WiFi wireless access technologies and Ethernet/PLC for wired connectivity, containing the already discussed Fog Servers that are positioned close to the mobile devices at the edge
- A network of the embedded systems and sensors comprising all smart (and less smart) things, vehicles and machines, built either on wired or wireless technologies

Emerging Technologies Enabling Fog Computing

Fog computing is only becoming a reality today as there are a number of new and emerging technologies available that can be appropriately applied in a layered architecture. Most importantly these are Network Virtualization (SDN), Network Function Virtualization (NFV) and 5G mobile technologies (see Fig. 6).

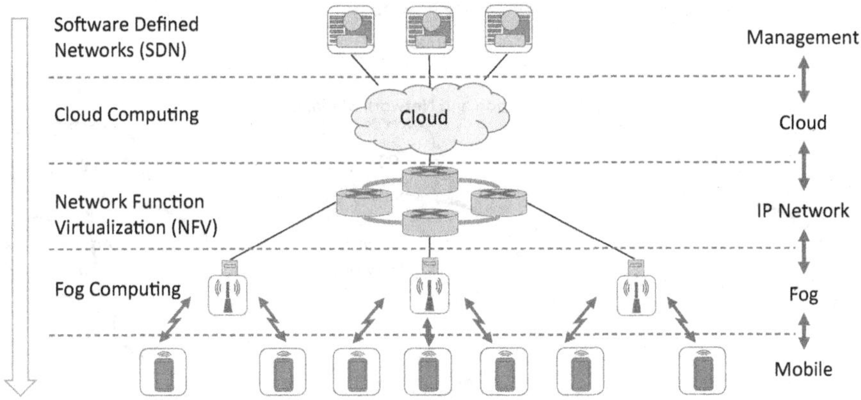

Fig. 6 Fog computing and emerging technologies

- Software Defined Networking (SDN)[17]: Fog computing needs frequent synchronization with the cloud in order to update data and manage the infrastructure. SDN enables a seamless management and programmability of the entire network from the edge to the core.
- Network Function Virtualization (NFV)[18]: NFV enables the network virtualization of location-based applications at the edge devices in order to provide the required services to the localized mobile users. This is in addition to the classic virtualization of network functions like routers and switches.
- 5G[19] Technologies: The main goal of Fog Computing is to offer location-based application services to mobile users. Existing wireless access networks like WiFi or emerging wireless technologies like 5G can be adapted to Fog Computing with virtualized architectures.

Fog Computing Solution Samples

In the following section, we will describe a few of these future, Fog based IoT systems, outlining their functional requirements and show how many sensors and actuators will work together in meaningful ways. These macro endpoints need to

[17]IEEE, Network Softwarization, A Think Tank on Network Softwarization (from SDN, NFV, Cloud-Edge-Fog Computing . . . to 5G), DAN linking Cloud with Fog, http://ieee-sdn.blogspot.co.at/2015/01/sdn-linking-cloud-with-fog.html, 2015, retrieved 2017-01-13.

[18]ETSI, NFV Network Functions Virtualization, http://www.etsi.org/technologies-clusters/technologies/nfv, 2016, retrieved 2016-12-27.

[19]IEEE, SDN, Gupta Lav, et al., Mobile Edge Computing – An Important Ingredient of 5G Networks, http://sdn.ieee.org/newsletter/march-2016/mobile-edge-computing-an-important-ingredient-of-5g-networks, 2016, retrieved 2017-01-13.

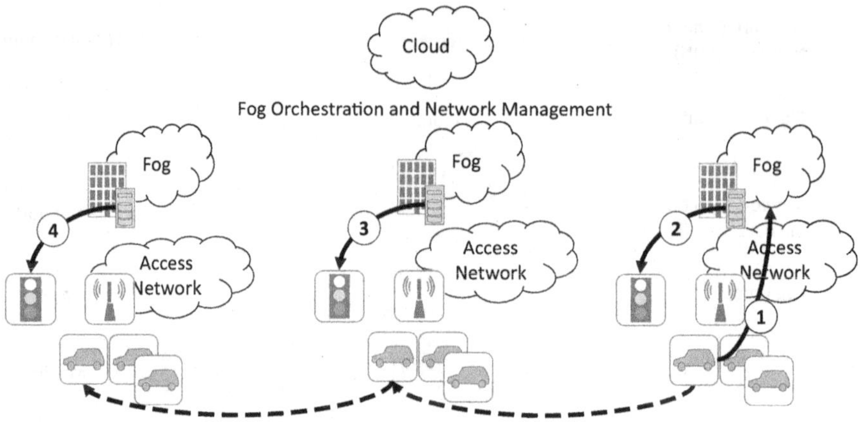

Fig. 7 Fog computing in STLs and SCVs

deliver basic as well as real-time decisions at the lowest tiers while summarizing the rest of the data for further decisions and analytics in higher hierarchies of intelligence. This reinforces our definition of fog computing—pushing intelligence and processing power closer to where data originates and where the sensors are located.

Smart Traffic Lights (STLs) and Smart Connected Vehicles (SCVs)

The STL,[20,21] and SCV[22] cases are shown in Fig. 7. By sensing, for example, ambulance lights, an STL system can automatically detect the approaching ambulance via video cameras and automatically change streetlights to open lanes for the ambulance to pass through the traffic. Also, these smart streetlights can interact with sensors and detect the presence of pedestrians and bikers, measure the distance and speed of approaching vehicles and allow for the smoother control of traffic. Furthermore, intelligent lights can turn traffic lights on and off as sensors detect the movement of any kind of traffic. In addition, neighboring smart lights that serve as fog devices can coordinate green traffic waves (see Fig. 7).

Smart Connected Vehicles (SCVs) use sensors for communicating between each other, while the whole vehicle (the higher-level intelligence) communicates over

[20]Cisco, Smart Traffic Management With Real Time Data Analysis, http://www.cisco.com/c/en_in/about/knowledge-network/smart-traffic.html, 2016, retrieved 2017-01-12.

[21]Create tomorrow, Barcelona: leading the world in a smart city revolution – 4, http://www.createtomorrow.co.uk/en/live-examples/barcelona, 2017, retrieved 2017-01-12.

[22]Industrial Internet Consortium, Connected Vehicle Urban Traffic Management Testbed, http://www.iiconsortium.org/connected-vehicle-urban-traffic-management.htm, 2015, retrieved 2017-01-12.

the Internet with other vehicles or with roadside units (RSUs)[23] using remote control and management systems mainly located in data centers. This same hierarchical principle is valid for smart cities, smart lights, industrial automation, environmental monitoring etc. This is the main reason for using a system view instead of an individual view of sensors and actuators and this in turn will have multiple consequences.

Smart connected vehicles (SCVs) will become much more important by 2025, set to reach 94 million by 2021 due to BI Intelligence,[24] which will require support for fast mobility as well as lowest and predictable latency. All these shifts in applications and application behavior demand fog-computing capabilities that we have outlined and they require special attributes from the fog platforms as well.

Wireless Sensor, Actuator Networks (WSANs) and Smart Buildings

In a traditional wireless sensor network, there are typically no actuators available to initiate and fulfill physical actions. In fog computing, WSANs are used[25] where actuators (fog devices) are envisaged to control measurement processes as well as to ensure the stability and oscillatory behaviors of small systems through the use of closed loop systems.

For example, such systems can be used for self-driven trains, where heat sensors on train wheels and brakes communicate threatening heat levels to rail side units that forward this information to the next train station in order to automatically stop the train, and additionally inform the train operator to take immediate actions as needed. Similar approaches can be applied to airplanes and air traffic, as well as lifesaving air ventilation systems in mines and subways etc.

Another example is decentralized, smart building controls, where wireless sensors measure temperature, humidity and the levels of various gases in the building atmosphere. These sensors exchange information, possibly on a floor by floor basis, and their readings can be combined to form reliable measurements. Sensors can then use distributed decision making processes and activate other fog devices (actuators) to react to this data. The overall system can lower the temperature and inject fresh air or increase humidity. These fog devices can be assigned to

---

[23]utc.ices.cmu.edu, Reis Andre, et al., Deploying Road Side Units in Sparse Vehicular Networks: What Really Works and What Does Not, http://www.utc.ices.cmu.edu/utc/Andre_Ozan_IEEE_TVT-2013_final_Nov25.pdf, 2013, retrieved 2017-01-12.

[24]Business Insider Deutschland, Automotive Industry Trends: IoT Connected Smart Cars & Vehicles, http://www.businessinsider.de/internet-of-things-connected-smart-cars-2016-10?r=US&IR=T, 2016, retrieved 2017-01-11.

[25]icact, Lee Wangbong, et al., A gateway based Fog Computing Architecture for Wireless Sensors and Actuator Networks, http://icact.org/upload/2016/0258/20160258_finalpaper.pdf, 2016, retrieved 2017-01-12.

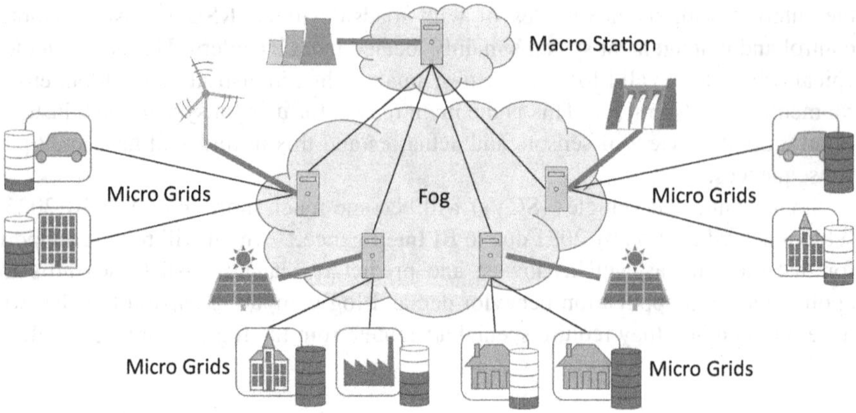

Fig. 8 Fog computing and smart grid

each floor and collaborate on higher levels of actuation. This enables smart buildings to maintain external and internal environments in order to conserve energy as well as other resources like water.

Smart Grid

To allow for better utilization and energy load balancing in electric grids, we can envisage new network edge devices such as smart meters and micro grids. These devices will be able to automatically switch to alternative energies like wind and solar, based on energy demand, energy availability and price levels. This is achieved by fog collectors at the edge that process the data generated by grid sensors and devices and send control commands to the appropriate actuators to change the power grid. In addition, all data that needs to be consumed locally is filtered from the next tier and only the rest is sent to a higher level for visualization, real-time reports and other necessary analytics. Fog computing also allows for time limited storage of data at the lowest tier, whereas data will be stored in a semi-permanent way at the highest tier. Finally, global coverage can be provided by cloud computing with the required business intelligence and analytics (see Fig. 8).

References

Bonomi, F., Milito, R., Natarajan, P., & Zhu, J. (2014). In N. Bessis & C. Dobre (Eds.), *Big data and internet of things: A roadmap for smart environments*. Berlin: Springer.

Nik, B. (2014). *Big data and internet of things: A roadmap for smart environments*. Cham: Springer.

Verdone, R., Dardari, D., Mazzini, G., & Conti, A. (2008). *Wireless sensor and actuator networks*. Amsterdam: Elsevier.

Big Data Analytics

If you can't measure it, you can't improve it

Lord Kelvin

Big Data Analytics only recently became a buzzword, promising many disruptive capabilities in combination with analytics by enabling the processing of massive volumes of structured and unstructured data. Other technology trends like virtualization, cloud computing, cloud services and finally the IoT and Fog Computing have contributed to the sudden rise of popularity for Big Data Analytics.

Big data needs to be collected and organized appropriately so that the analysis of this massive amount of data allows us to act upon the information in a timely manner thereby creating competitive advantages in the market. This allows for the better management of communities, natural resources and promotes the efficient delivery of social, healthcare and educational services. The five Vs describing volume, variety, velocity, value and veracity are key parameters defining the effectiveness and usability of the collected data.

Many drivers exist for Big Data Analytics and even a decision management framework evolved, as the different requirements for data in motion and data at rest have to be covered while also choosing the right database technology based on SQL or NoSQL databases.

Hadoop and its relation to SQL or NoSQL databases are important aspects to consider, but also what to do with unstructured data, what business view to use as well as the presentation and consumption of big data. Today numerous big data analytics cases exist, which is reflected by the exciting growth rates today and over the next decades.

M. Oppitz, P. Tomsu, *Inventing the Cloud Century*,
DOI 10.1007/978-3-319-61161-7_18

Big Data Analytics Defined

In the early 2010s Eric Schmidt, Executive Chairman of Google, highlighted the reason why Big Data is gaining importance by pointing out: *"From the dawn of civilization until 2003, humankind generated five exabytes of data. Now we produce five exabytes EVERY TWO DAYS… and the pace is accelerating."*

The term "Big Data" was coined only a few years ago and has quickly become very popular due to the many disruptive capabilities it promises in combination with analytics. Big data is used to describe a massive volume of both structured and unstructured data that is so large it is difficult to process using traditional database and software techniques (Smolan and Erwitt 2012).

Companies like eBay, Facebook, LinkedIn, Google or Amazon have built themselves around big data from the start. For these companies, big data did not have to be integrated with traditional data sources from the last century because they never had these traditional data forms. This also means that they never merged the new big data technologies with traditional IT infrastructures as these infrastructures did not exist, being built from scratch and optimized for big data and analytics.

For well-established businesses: small, medium or large ones this was and is still completely different. Usually big data in such companies started as a separate data pool and needed to be integrated into all the traditional data sources available and used in those companies, as analytics on legacy types of data have to now coexist with analytics on the new, big data sources.

Big Data and Analytics
Big data itself is a pretty useless thing as long as there are no mechanisms available to extract useful information from it, these mechanisms we refer to as analytics. The same is true for analytics, which by itself would just describe mechanisms for amassing data, but without data analytics also makes little sense. Throughout the industry and in literature both acronyms are often used alone as synonyms meaning both—big data and analytics, thus in this book we will always refer to both as Big Data Analytics.[1] In the following chapter, our major challenge is to understand what is really behind big data analytics and what are the promised disruptions.

Extension of Established IT Architectures
Big data analytics stands for a number of information management and analysis technologies that has become increasingly important in many modern organizations. They describe only a subset of the full business intelligence requirements. There is a need for coexistence between relational and non-relational big data analytics technologies and the combination of all these technologies opens different and new ways of conducting business, managing resources, caring for patients, protecting and serving citizens or teaching students.

---

[1]McKinsey&Company, How companies are using big data and analytics, http://www.mckinsey.com/business-functions/mckinsey-analytics/our-insights/how-companies-are-using-big-data-and-analytics, 2016, retrieved 2017-01-13.

One of the most disrupting requirements for the effective use of big data analytics is the extension of existing and well established IT architectures based on relational technologies to non-relational technologies.[2] It requires the combination of database technologies with modern analytics in order to allow organizations to analyze large and diverse data sets and extract the desired fact based decisions to find the right and consistent answers to business events.

Ever Growing Amount of Data

This massive amount of data needs to be collected and organized appropriately to allow for efficient analysis. Analytics allows for the discovery of patterns and any kind of useful information contained in this data, and helps to identify data most important to specific future businesses and business decisions. It is the ability to analyze and act upon information in a timely manner that creates a competitive advantage in the market, enables better management of communities and natural resources and promotes the efficient delivery of social, healthcare and educational services. All these benefits and opportunities coming from big data and analytics also cause IT decision makers significant pressure to make the appropriate decisions among all the available technologies (Foreman 2013).

The amount of data in businesses and organizations that we are typically dealing with today can easily be in the range of petabytes, exabytes or even zetabytes. To further illustrate the size of these numbers:

- 1 Terabyte could cover all X-ray films in a large hospital
- 1 Petabyte (1000 × 1 Terabyte) is 5 years' worth of data from earth observation (NASA)
- 5 Exabyte (5000 × 1 Petabyte) comprises all words ever spoken by humans
- 1.3 Zettabyte (1000 × 1 Petabyte) is equal to all Internet traffic in 2016

Due to V&C,[3] there were 2.5 quintillion bytes of data created per day in 2015.[4] To illustrate this amount, consider that a quintillion is used to refer to the mass of earth in tons or the number of molecules in the human brain. Similarly, the number of data created per time unit is growing dramatically, from 100 GB/day in 1992 to 50,000 GB/s in 2018 (same source as above) (see Fig. 1).

Volume is only one aspect of big data and can easily be covered, as we have discussed with cheap storage and advanced computing solutions. The real challenge comes from the velocity and variety of data, which, in combination with the sheer volume, exceeds an organization's storage and necessary computing capacity for timely decision making, be it business decisions or otherwise.

[2]Serra James, James Serra's Blog, Big Data and Data Warehousing, Relational databases vs Non-relational databases, http://www.jamesserra.com/archive/2015/08/relational-databases-vs-non-relational-databases/, 2015, retrieved 2017-01-13.

[3]V&C Solutions, http://www.vncsolutions.com/techtips/making-big-sense-from-big-data, 2017, retrieved 2017-02-24.

[4]vcoudnews, Walker Ben, Every Day Big Data Statistics, http://www.vcloudnews.com/every-day-big-data-statistics-2-5-quintillion-bytes-of-data-created-daily/, 2015, retrieved 2017-01-13.

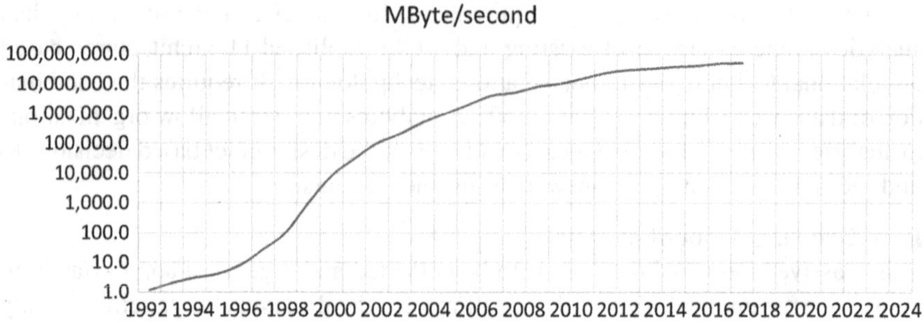

Fig. 1 Volume of data created by time unit (logarithmic plotting)

A growing number of data is held in transactional data stores, where this data describes an event and is usually the result of the increase in online activities. Data is also a part of machine-to-machine interactions like metering, environmental sensors, and RFID systems. With all of this data exploding, many originate from social media and all of it generates large amounts of unstructured or semi-structured data.

Technology Advancements as Enablers

Throughout recent years, we were faced with a continuous decline of storage costs, combined with massive advancements in computing. While initially big data was seen mainly as a data management challenge, soon this perspective became obsolete and today big data exploits internal and external information flows for the improvement of organizational performance. When used by vendors, the term "big data" refers to technologies required to handle large amounts of data and storage, including tools and processes as well.

Relevant Data

Big data might consist of billions or even trillions of records, all from different sources like the Web, customers, social media, sales, mobile data, etc. and this data is loosely structured, many times incomplete and not always easily accessible. Usually organizations have a number of difficulties creating and managing big data and, on top of that, standardized tools and procedures are not designed to search and analyze massive datasets. All of this data in big data generates an information overload. Only a portion of this total amount can be used by analytics in order to derive useful decisions, no matter for what use. As the general volume of data increases the likelihood of an information overload increases as well (see Fig. 2).

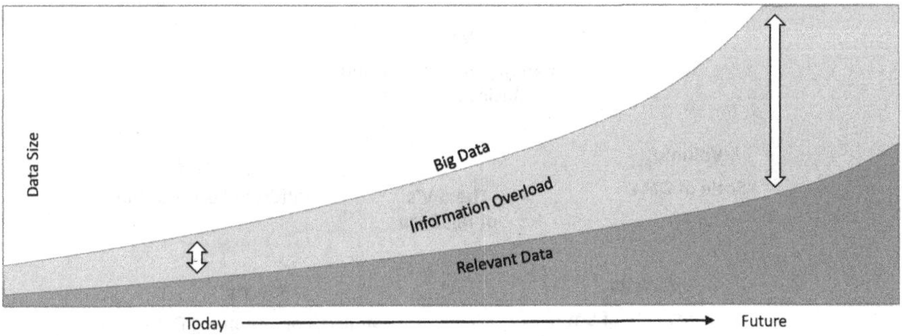

Fig. 2 Relevant data versus massive amount of data

The 5 V's of Big Data

As we have seen, Big data is a constantly increasing number and it is not only defined by volume but also other characteristics like variety, velocity, value and veracity. Big data technologies can be defined as a new generation of technologies and architectures that allow for the extraction of value from large data volumes. They use a wide variety of data economically through high velocity capture, discovery and analysis,[5] and encompass hardware, software and other services that integrate, organize, manage, analyze and present big data (see Fig. 3).[6]

Volume: There Is more than Size

The term "big" has different meanings in big data, as some organizations "only" deal with gigabytes to terabytes of data, compared to organizations, like social networks, that handle petabytes to exabytes of data.[7] No matter the amount of data, the successful use of big data requires complex and intense processing and analysis.

For example, there may be millions or billions of records handled by certain big data applications, where each record may only be a few bytes long such as stock ticker information in financial services. Contact archives including emails could contain several petabytes of data including customer suggestions, complaints, project records, contracts and proposals. In manufacturing or product design, huge numbers of prototypes might be evaluated, or in scientific experiments like

[5]IDC, Olofson Carl, et al., White Paper, Big Data: Trends, Strategies, and SAP Technology, http://www.itexpocenter.nl/iec/sap/BigDataTrendsStrategiesandSAPTechnology.pdf, 2012, retrieved 2017-01-13.

[6]Elsevier, Gandomi Amir, et al., International Journal of Information Management, Vol. 35, Issue 2, April 2015, pp. 137–144, Beyond the hype: Big data concepts, methods, and analytics, http://www.sciencedirect.com/science/article/pii/S0268401214001066, 2015, retrieved 2017-01-16.

[7]wersm, we are social media, How Much Data Is Generated Every Minute On Social Media?, http://wersm.com/how-much-data-is-generated-every-minute-on-social-media/, 2017, retrieved 2017-01-16.

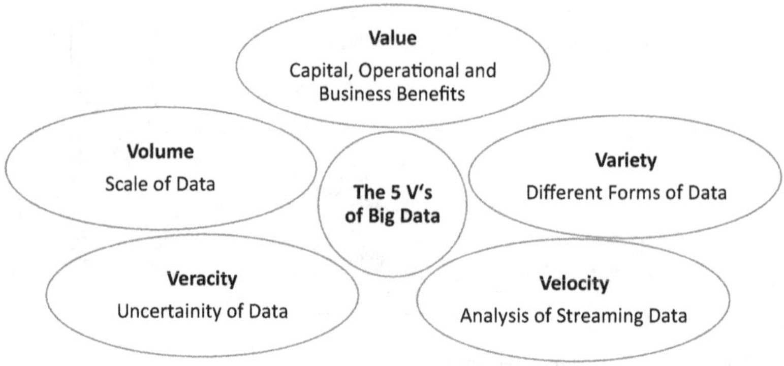

Fig. 3 The 5 V's of big data

the Large Hadron Collider[8] that can easily generate petabytes of data to be used for further input into simulation models.

Variety: Important Combination of Data Sources and Formats
The majority of an organization's data, usually up to 85%, is unstructured, but it still needs to be available for quantitative analysis to form decisions derived from this data. It is the variety of data sources and the variety of the used formats like text, video, audio and other unstructured data that needs to be handled and digested appropriately by analytics applications and requires different architectures and technologies for analysis.[9]

In addition, this data can be both internal and external to an organization, making the combination of all these different information types a complex task, like in determining the relative importance of a customer record versus a tweet, or the correlation of huge numbers of ever changing patient records with published medical research and genomic data to find the optimum treatment. Other examples are weather and climate modeling, which uses decades of weather and climate data with new, physical models of ocean water characteristics and CO_2 level changes in combination with real-time satellite data to create real-time simulations.

Velocity: Appropriate Analyzing and Delivery of Information
When talking about velocity we need to consider the variability of data, which can be daily, seasonal, or event triggered in terms of its peak loads—a huge challenge to manage. We need to distinguish between real time and near real time processing of data. Real time processing needs continuous input with constant processing and output of data, something like this is needed in radar systems, customer services etc.

[8]Cern, Large Hardon Colider, http://home.cern/topics/large-hadron-collider, 2017, retrieved 2017-01-16.

[9]MITSloan Management Review, Bean Randy, Variety, Not Volume, Is Driving Big Data Initiatives, http://sloanreview.mit.edu/article/variety-not-volume-is-driving-big-data-initiatives/, 2016, retrieved 2017-01.16.

In near real time processing, speed is important, but it is ok to have processing times in minutes or a few seconds.[10]

The use of RFID tags and smart metering requires the handling of near real time data, which, in combination with the requirements for highest agility and quick insights, results in the need for new capable infrastructures. This, combined with a necessary skill base, is definitely not an easy task.

Velocity deals with the handling of data arriving at predetermined intervals as found in traditional data warehousing and is today mainly processed using Hadoop,[11] to real-time streaming data requiring Complex Event Processing (CEP),[12,13] (Luckham, 2012), rules engines, text analytics, machine learning and event-based architectures.

In order to understand velocity requirements, it is imperative to understand the application as well as the user's business and organizational requirements. These can range from milliseconds used in trading applications, real-time face recognition or screening for airport travelers to the non-real time analysis of billions of data sets via Hadoop. Obviously, the technology infrastructures for the different cases of use can differ significantly, requiring specialized hardware to meet high performance demands. Often it will be possible to solve these issues with high availability clusters, scale out file systems or multicore processors using the commercial, off-the-shelf components we find in cloud computing.

Value: Capital, Operational, and Business Benefits

Value derived from data depends on the organization's understanding of the relationships, complex hierarchies and data linkages among all data in order to link, match and transform data across businesses.[14] But value also refers to the cost of technology since cost is a key factor that has changed dramatically over the past few years. Through the introduction and availability of new technologies like cloud computing, decreases in hardware costs such as storage and computing and more commonly available software, meant that systems, previously only affordable to governments and large organizations, became available to almost every interest party.

---

[10]syncscort, Wilson Christy, The Difference Between Ral Time, Near Real Time and Batch Processing in Big Data, http://blog.syncsort.com/2015/11/big-data/the-difference-between-real-time-near-real-time-and-batch-processing-in-big-data/, 2015, retrieved 2017-01-16.

[11]Hadoop, Welcome to Apache Hadoop, http://hadoop.apache.org, 2017, retrieved 2017-01-16.

[12]IBM, Big Data Analytics Hub, Madia Kimberly, The evolution of complex event processing, http://www.ibmbigdatahub.com/blog/evolution-complex-event-processing, 2014, retrieved 2017-01-17.

[13]cloudera, Build a Complex Event Processing App on Apache Spark and Drools, http://blog.cloudera.com/blog/2015/11/how-to-build-a-complex-event-processing-app-on-apache-spark-and-drools/, 2015, retrieved 2017-01-16.

[14]IBM, Big Data & Analytics Hub, Kobielus James, Measuring the Business Value of Big Data, http://www.ibmbigdatahub.com/blog/measuring-business-value-big-data, 2013, retrieved 2017-01-16.

Big data projects are beneficial in several ways, such as capital cost reduction through infrastructure, hardware and software cost reduction, operational efficiency through more efficient methods for data management, analysis and delivery and finally business process enhancements increasing revenue or profit resulting from the benefits of big data and analytics.

Veracity: Uncertainty of Data
Veracity describes how complete or correct the available data is, the biases, noise and abnormality of data and if it makes sense to use the data being stored to meaningfully analyze a problem. In fact, among the five Vs, veracity is easily the biggest challenge to big data analytics, as it is often overlooked yet is at least as important as the other Vs.

If incorrect data is used this can easily cause more problems to organizations and consumers than if no data was used. Not only does data need to be correct, but also the analyses performed on the data needs to be correct. This becomes especially important in automated, decision-making processes, as there are no longer humans involved correcting things.

Common Big Data Analytics Misconceptions

The market for big data in the mid-2010s is still a pretty nascent one and many vendors are trying to get a piece of the pie, so there is definitely a lot of confusion and misconception about what big data is and what big data and analytics technologies can achieve. We will briefly discuss the most common misconceptions to avoid misunderstanding and unproductive argumentation and generalization about big data and analytics. We will also suggest throughout this section that multiple big data technologies need to coexist and be applied in specific cases and for the workloads they were designed.

Big Data Analytics Is Nothing New
Big data and analytics is absolutely nothing new as big data concepts have been around for many years. What has changed are the economics for deploying big data concepts to a wide variety of the capabilities now offered by computer-aided discovery of relationships between extremely large datasets that contain a wide variety of different sources of data. What has also changed is the awareness of creating a competitive advantage that analytics allows by analyzing the right information and offering the outcome to the right decision makers in time.

Big Data Is All About Hadoop and Other Database Misconceptions
There are many different big data technologies and it is important to choose the right one for a specific job, as there is not an individual big data technology that can solve every big data requirement. One very popular big data technology today is

Hadoop[15] and while it is very efficient in many cases, it cannot handle all big data requirements (deRoos, Zikopoulos, Melnyk, Brown, & Coss, 2014; White, 2012). Although Hadoop gained a lot of acceptance in the market, it is far from being the only possibility for data management.

Another misconception is that relational databases cannot scale to very large volumes and hence cannot be considered big data technologies. It is true that NoSQL databases are gaining more popularity in big data, but relational databases still play an important role.

Big Data Is Only About Large Data Volumes

While large data sets are a key part of big data and analytics, there are other characteristics such as streaming or real-time data represented in different types or formats. There are big data deployments addressing one of these three characteristics, but there are also other deployments that might address two or all three characteristics.

Big Data as New Buzzword for Data Mining

In analytics, data mining is defined as a set of techniques used to analyze large data sets (Linoff & Berry, 2011). There are many of these techniques, which were used many decades ago while other techniques became known and popular more recently. In general, big data analytics covers much more than just data mining, as it also covers data collection, data management, data organization, data analysis, information access and operational workloads together with cases utilizing some of the same new, established big data and analytics technologies.

Big Data, Big Challenge

One can be sure that while big data and analytics offers many new business opportunities, deploying big data technology alone will not solve any business problems, nor will it help to decrease costs and increase revenue or attract any new customers. It also does not matter for the final success of an organization. It may not matter if data is stored in Hadoop clusters or relational databases and analyzing data, even if done in the most elegant and most sophisticated way, will not lead to an overall business success by itself.

As long as the most brilliant analysis is ignored or not adopted properly by the relevant decision makers, or the analysis misses its goal because it does not take into account the behavioral variables of human interaction, all big data analytics will miss the point. So, if an organization identifies new customer segments and requirements in the most accurate and meaningful way, but maybe misses a successful marketing opportunity or does not address the privacy concerns of an audience, it will fail. This means that even having the most sophisticated tools and techniques available for determining new businesses, one still requires sensibly concerted market strategies to achieve an overall success.

[15]Hadoop, Apache Hadoop, http://hadoop.apache.org, 2017, retrieved 2017-01-16.

Big Data Analytics Requirements

In 2015, most organizations and enterprises were still not up to the task of efficiently analyzing big data. This is mainly due to the volume and the different formats of all the data collected across the organization. To make it even more difficult this data was collected in different ways and was typically structured as well as unstructured, making it extremely cumbersome to analyze and find patterns that could lead to useful business decisions. One of the biggest challenges is breaking down data silos to be able to access all the necessary data form organizations that are usually stored in different places to make it even harder in different systems. On top of that, both unstructured as well as structured data needs to be handled and since the data volume is massive, traditional database and software methods will barely work as desired if at all.

Big data analytics is the method used to analyze such volumes of big data and it incorporates new specialized software tools and applications for data mining, text mining, predictive analytics, forecasting, and data optimization. All these processes are separate but, within high performance analytics, they need to be highly integrated functions to enable organizations to process their extremely large data volumes and determine the relevant data to analyze these for better business decisions in the future (Marr, 2015).

Drivers of Big Data

It is with no doubt that data generated by social media sites is naturally of much interest to big data analytics, but it is actually the analysis of operational data generated by the commercial and public sector, which is the biggest driver for big data analytics solutions. Big data, analytics and business intelligence cover a broad domain of business processes, technologies and types of expertise, which hold sheer unlimited opportunities while at the same time also ignite ambiguity and confusion just because of this broad scope. An online survey in April 2012 by the International Journal of Information Management of 154 global executives asking for the definition of big data shows the top areas for Big Data Analytics (see Fig. 4).[16]

The real goal of big data analytics solutions is to improve decision making, while allowing for advanced, fast insights for decision makers in organizations. Big data solutions describe not only the task of analyzing large data amounts by data scientists, but more importantly making use of other existing opportunities across the spectrum of various decision types as well as for different decision makers.

---

[16]Elsevier, International Journal of Information Management, Gandomi Amir, et al., Beyond the hype: Big data concepts, methods and analytics, http://www.sciencedirect.com/science/article/pii/S0268401214001066, 2015, retrieved 2017-01-16.

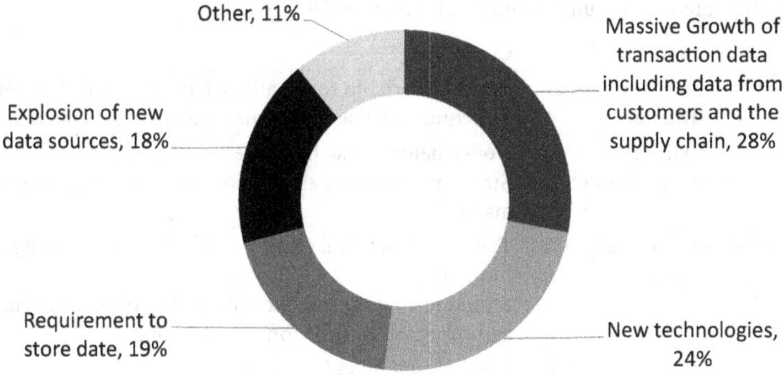

Fig. 4 Top drivers for big data analytics

Big Data Analytics Technology Landscape

One of the best methods is to classify big data technologies into either big data in motion or big data at rest (Zikopoulos, Eaton, deRoos, Deutsch, & Lapis, 2012).

Big Data in Motion

Big Data in motion is real-time, high-volume data constantly streaming and requiring immediate action upon arrival. For example, such data could be stock transaction data, RFID (Radio Frequency Identification Data) in real time inventory control systems, or smart meter data to name just a few.. In any case, data is usually received, filtered and regularized which means it is put into a consistent or readable format. This method may involve Complex Event Processing (CEP) (Luckham, 2012), which applies the newly arriving data to retained data already cached from the same stream as well as data stored in a very fast database to determine if a defined event has taken place or not. In the case that the event has occurred, the CEP will trigger actions as applications to respond to this event. It is easy to understand that big data in motion requires fast and efficient technology for receiving, formatting and responding to the data, which requires high speed data movement and transformation, in-memory handling and CEP technology.

Big Data at Rest

Contrary to big data in motion there is the huge field of big data at rest, requiring specific technologies to collect data as fast as possible upon delivery, then transform this data and analyze it before putting it in a state for meaningful search, discovery, mining, query and reporting. Big data at rest can be structured as well as

Table 1 Structured versus unstructured data (Source IDC, 2012)

	Type	Container
Structured	Extrinsic explicit structure	Fixed or variable schematic database (RDBMS, OODBMS) Structure is declared and managed by a schema manager
	Extrinsic implicit structure	Non-schematic database (Key-Value Store) Structure known to application program but not explicitly managed
Unstructured	Intrinsic explicit structure	Tagged format data (XML, CDF, standard exchange format files, etc. Structure is present in the data in the form of headings, tags and other established forms of labeling
	Intrinsic implicit structure	Content in human language Structure not explicitly declared but can be inferred from data itself, like the human language where the semantics are found in syntax and grammar Text Image Video Audio

unstructured data, where unstructured data can even have some structure, which is simply not determined by schemas or program code (see Table 1).

NoSQL Versus SQL Databases

In big data analytics both NoSQL (Non-Relational Structured Query Language) as well as SQL (Structured Query Language) databases play important roles. NoSQL databases have their strengths in supporting many different dimensions of big data, like accepting data from multiple sources in many different formats, allowing analytics programs to filter and organize this data, as in many Hadoop applications. On the other hand, SQL databases are optimally used for handling large data volumes with consistent, known structures and allow for the periodic reporting, mining and repeated analysis tasks on this data.

RDBMSs (Relational Data Base Management Systems) usually offer dynamic scalability and allow for the rapid handling of very large databases able to process requests quickly having evolved from older forms to acquire these characteristics. NoSQL is usually applied to a wide range of DBMS types. Each DBMS type serves a particular purpose and it is important to understand that in many cases multiple databases of different types can be present together in one system, forming a sensible flow of big data operations (see Table 2).

Table 2 NoSQL DBMS types (Source IDC, 2012)

Paradigm	Usage
Key value	Caching of shared data and creating ad hoc data spaces
Grid based key value	Very large and ad hoc scalable data collections for analysis
List processing	Collecting large amounts of unpredictably structured data for analysis
Graph	Unlimited levels of data entry relationships, offering rapid search capabilities
"NewSQL"	Support for queries without predefined schemes (being able to apply structures to data or infer structures from data)

Big Data Analytics Framework

Big data analytics has the potential to dramatically change how legacy technologies at many companies—small or big—are used or will be used in the near future. We see many companies replacing legacy technologies from the past with open source solutions, one of the most prominent being Hadoop.[17] We also see the increasing trend of replacing proprietary hardware with white boxes as well as custom applications, which we discussed in detail in the section on virtualization. Furthermore, completely new data visualization tools allow new, accurate and timely insights into whatever data is of interest to modern business processes, enabling new business innovations among them being faster production time as well as custom packaged products (see Fig. 5).

Big data analytics provides more advanced decision and prediction making compared to legacy tools (used until a decade ago) by introducing highly specialized features that are best described in a Big Data Analytics Framework that we will discuss in this section.

Data Source: Capture, Integration and Movement

The lowest technology layer consists of data capture, integration and movement tools. These tools provide batch data extraction, transformation, loading, or enabling of data to be streamed into target data stores. Scalability is a huge concern for the tool on this platform layer, as it covers actions like movement of multi-terabyte data sets or processing millions of streaming events.

The big data analytics platform must also be capable of providing highly reliable performance and ensure high levels of data quality. All this is necessary in addition to the complex multi-structured data transformation processes. The same platform needs to be highly scalable in a dynamic way in order to support unexpected

[17]Apache Hadoop, Opensource Project, http://hadoop.apache.org, 2016, retrieved 2016-12-29.

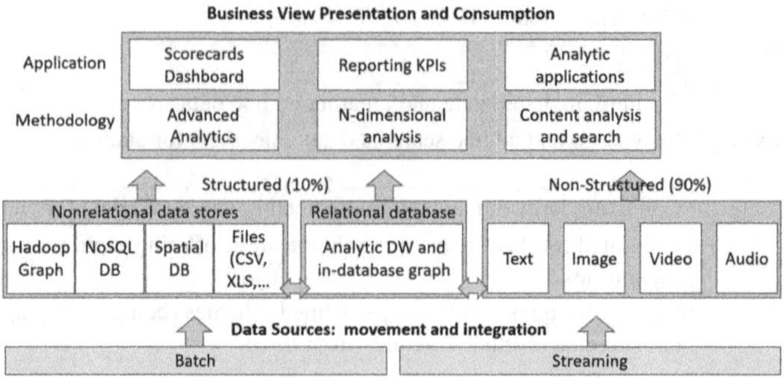

Fig. 5 Big data analytics framework

requirements, like data scientists experimenting with new data types and sources or new combinations of existing data types and sources.

Data needs to move between various data repositories, very often into Hadoop clusters for processing and after that a subset of the data has to be moved into a data warehouse. There may be a need to monitor streaming data from sensors and capture and move event patterns into a data warehouse for analysis.

Hadoop, Relational (SQL) and Non-Relational (NoSQL) Databases

Either relational or non-relational databases can be used as data stores. In a normal relational database, data is found and analyzed using queries based on SQL (Structured Query Language), which is an industry standard. Non-relational databases also use queries, but they are not constrained to SQL and can use other query languages in order to pull information out of data stores, this is why they are called NoSQL.

Hadoop, which is often used in big data analytics, cannot be seen as a real database, as it stores data and allows data to be pulled out of it, but pulling out data does not use any queries, hence also no SQL. Instead, Hadoop can be seen as a data warehouse system that uses MapReduce to actually process data. A Hadoop project called "Hive[18]" enables raw data to be re-structured into relational tables that can be accessed via SQL and incumbent SQL-based toolsets, capitalizing on the tools that a company may already be using.

[18] Apache Hadoop, Hive Project, http://hive.apache.org, 2016, retrieved 2016-12-29.

Hadoop in Detail

Hadoop is an open source software project enabling the distributed storing of enormous data sets across distributed clusters of servers and then running distributed analysis applications on all of these clusters, thus processing large, distributed data sets. Hadoop[19] consists of two parts: a data processing framework and a distributed file system for data storage. Hadoop is designed to be highly scalable, from a single server to thousands of machines and offers a high degree of fault tolerance. Instead of relying on high-end hardware, the resiliency of Hadoop clusters results from the ability to detect and handle failures at the application layer.

Hadoop Data Processing Framework and MapReduce

The processing framework used by Hadoop is a Java,[20,21] based system called MapReduce,[22] which is a YARN[23,24] based system for parallel processing of large data sets. It not only distributes data across disks, but also at the same time applies complex computational instructions to the data. Processing of instructions is done in parallel across various nodes on the big data analytics platform before new data structures are provided as answer sets.

Hadoop Distributed File System (HDFS)

The primary distributed file system used by Hadoop is called Hadoop Distributed File System (HDFS[25]) and is the component that holds the actual data, but it is worth noting that other file systems can be used as well. Basically, one can see Hadoop like a big store for data, where data is kept safe until someone wants to do something with it, like running analysis on it or even exporting data sets to other tools in order to perform analysis.

Non-Structured Data

Unstructured or non-structured data makes up by far the largest percentage of big data, as it comprises around 90% of data compared to structured data at around 10%. Typical non-structured data is text, images, video or audio.

---

[19] Apache Hadoop, Opensource Project, http://hadoop.apache.org, 2016, retrieved 2016-12-29.

[20] Oracle, Java Software, https://www.oracle.com/java/index.html, 2016-12-29.

[21] Oracle, Java+You, https://www.java.com/en/, 2016, retrieved 2016-12-29.

[22] Apache Hadoop, MapReduce, https://hadoop.apache.org/docs/r1.2.1/mapred_tutorial.html, 2016, retrieved 2016-12-29.

[23] Apache Hadoop, YARN, https://hadoop.apache.org/docs/r2.7.2/hadoop-yarn/hadoop-yarn-site/YARN.html, 2016, retrieved 2016-12-29.

[24] Apache Hadoop, YARN aka MapReduce NextGen, https://hadoop.apache.org/docs/r2.7.2/hadoop-yarn/hadoop-yarn-site/index.html, 2016, retrieved 2016-12-29.

[25] Apache Hadoop, Hadoop Distributed File System (HDFS), https://hadoop.apache.org/docs/r1.2.1/hdfs_design.html, 2016. retrieved 2016-12-29.

Data Stores: Big Data Management and Processing

Data management and processing is done in the middle layer of the big data analytics framework. As already discussed, two categories of technology are used on this layer, relational and non-relational databases, with the recent focus on Hadoop and NoSQL databases. This does not mean, however, that relational databases do not have a future in big data analytics.

While relational and non-relational technologies address different sweet spots for different workloads, it is necessary that they operate together in more scenarios than we can think of. Many organizations use a relational data warehouse as a trusted and secure source of information for performance management and structured analysis, while trusting on Hadoop or NoSQL databases if they need free format discovery, depending on the type of data and analysis required.

More and more real world deployments require movement, processing and analysis of data from both relational and non-relational technologies. Multichannel retail companies may use non-relational data management for experimenting on data sets coming from clickstream data, but also combine it with customer reference data coming from a relational data warehouse. Online media companies may use Hadoop clusters to store and process web clickstream data, before moving subsets of this data to a relational database in order to make it accessible to business analysts.

Applications Functions and Services

At the applications, functions and services layer of the big data analytics stack, we find analytics and business intelligence tools together with prepackaged analytics applications. These tools need to be able to access both relational as well as non-relational databases, support consumer centric data visualization and interaction and allow for efficient use of analytic and business intelligence tools.

Deriving intelligence from unstructured text like customer comments, social media interactions as well as emails and documents all need to be supported. In combination with transactional and operational data analysis, this enables warranty management, predictive maintenance, fraud detection and prevention as well as customer voice analysis, to mention only a few.

Business View, Presentation and Consumption

Processing via MapReduce or other custom Java code can create intermediate data structures like flat files, relational tables or statistical models. These structures can be used for additional analysis or accessed via traditional SQL query tools. In general, business views should make big data more consumable either through legacy tools or human knowledge.

Data visualization tools enable average human beings to view and understand derived information in a graphical and intuitive way. This visualization can be sent virtually anywhere, either to the mobile devices of average people or to the PCs and laptops of high level decision makers, all in a fraction of the time that was needed to find, access, load and consolidate data from multiple different billing and customer systems.

Finally, the primary output of big data analytics can be used for making automated decisions without the need for any visualization.

The Big Data Analytics Use Cases

In this section, we will discuss some of the most prominent cases of big data analytics in use, but there definitely exist more today and many more expected in the future than we can highlight in this book.

Fraud Prevention
Whenever companies want to detect fraud in real time they use big data analytics to look for patterns of fraudulent behavior in enormous amounts of either structured or unstructured data.[26] Insurance companies especially have a great interest in stopping fraud by using predictive models that analyze historical and real time data including medical claims, demographics, call center notes, voice recordings or attorney costs to spot suspected fraud claims at early stages. Return on investment for these companies are usually enormous, as they are able to analyze complex scenarios in minutes rather than in days or months.

Revenue Assurance
This is an activity often performed by telecommunications service providers and describes the process of analyzing bills against services in order to make sure there is no underperformance happening. While this is very important for telecoms, the same can be applied to many other businesses like healthcare, Internet services, streaming services etc., actually it applies anytime bills are issued against service plans.[27]

[26]Forensic Risk Alliance, Mason Greg, Big Data Analytics & Fraud Prevention, http://www.internationaltrade.co.uk/articles.php?CID=1&SCID=&AID=1719&PGID=1, 2015, retrieved 2017-01-16.

[27]tmforum, Revenue assurance tools – From case oriented black boxest o agile hypermarket instruments, https://inform.tmforum.org/features-and-analysis/2015/05/revenue-assurance-tools-from-case-oriented-black-boxes-to-agile-data-hypermarket-instruments/, 2015, retrieved 2017-01-16.

Churn Analysis

Customer churn happens when subscribers no longer do business with a specific company. Customer churn is a specifically critical metric as trying to keep existing customers is usually much less expensive compared to acquiring new customers, especially since the trust and loyalty of existing customers has already been earned. Thus, it is essential that companies use effective methods for calculating customer churn in given time periods in order to enable them to determine their customer retention success rates while finding improvement strategies.

Smart Meter Monitoring

Smart meters[28] are a perfect tool offering households and businesses ways to understand and reduce their energy usage, as they allow far more detailed metering compared to meter readings taken quarterly, for example. Using and analyzing these big data amounts helps utility companies improve accuracy in billing, helps them to better forecast and react to immediate energy usage and match supply and demand more closely and significantly cuts visits to properties to read meters.

Equipment Monitoring

This technology allows organizations that rely on large pieces of equipment to predict and prevent equipment downtime by using big data analytics. It identifies irregularities in machines by collecting and analyzing the necessary data in order to help operators optimize the service lives of machines.[29]

Pricing Optimization

The key to better pricing is to fully understand the data being at a company's disposal for optimizing prizes of certain products. It is not only that sales determines the pricing and the volume, but in addition can utilize extremely granular data, from each invoice, by product, by customer or by packaging. This enables price guidance at the individual deal level and, incentivizes performance scoring with small and relevant deal samples being essential prerequisites.[30]

Traffic Flow Optimization

Real time traffic flow prediction and optimization[31] is at the heart of intelligent and effective Intelligent Transportation Systems (ITSs) (Chowdhury, Apon, & Dey,

---

[28]DIGITALEUROPE, Energy Big Data Analytics, Unlocking the benefits of Smart Metering and Smart Grid Technologies, http://www.digitaleurope.org/DesktopModules/Bring2mind/DMX/Download.aspx?Command=Core_Download&EntryId=940&language=en-US&PortalId=0&TabId=353, 2015, retrieved 2017-01-16.

[29]Infosys, Infosys Remote Equipment Monitoring Solution, https://www.infosys.com/data-analytics/insights/Documents/remote-equipment-monitoring-solution.pdf, 2016, retrieved 2017-01-16.

[30]McKinsey&Company, Baker Walter, et al., Using big data to make better pricing decisions, http://www.mckinsey.com/business-functions/marketing-and-sales/our-insights/using-big-data-to-make-better-pricing-decisions, 2014, retrieved 2017-01-16.

[31]Hindawi Publishing Corporation, Lu Hua-pu, et al., Big Data Driven Based Real Time Traffic Flow State Identification and Prediction, Discrete Dynamics in Nature and Society, Vol. 2015, Article ID 284906, https://www.hindawi.com/journals/ddns/2015/284906/, 2015, retrieved 2017-01-16.

2017). This requires identifying and predicting traffic flow quickly, and precisely in real time, but it also requires storing the massive volume of image and video data over decent time periods like 12 or 24 months, while enabling accurate search and identification for plate numbers or skid marks in order to identify traffic violations.

Social Network Analysis
Social network analysis has its roots in the work of early sociologists like Georg Simmel or Emile Durkheim, who had already discovered the importance of studying patterns of relationships connecting social actors. Social network concepts have been used since the early twentieth century in order to identify complex sets of relationships between different members of social systems. This analysis allows us to view social relationships such as Facebook friendships, email correspondences, hyperlinks or responses and draw whatever conclusions requested of it.[32]

Life Sciences Research
Big data analytics has significant potential to improve healthcare, lower costs and save lives by taking advantage of the sheer data explosion while removing old restrictions on analytical capabilities.[33] This helps organizations to make better, informed decisions and leverage new opportunities by synthesizing and analyzing big data for healthcare providers, drug manufacturers and all companies doing business in the life sciences and healthcare industries, leading to overall better outcomes in many scenarios.

Advertising Analysis
Data driven advertising is one of the key application areas of big data analytics and had already become a powerful advertising strategy some time ago.[34] Today many enterprises increasingly rely on big data from various sources in order to produce targeted promotional content having realized the importance of big data for the effective marketing of their products. As such many new channels and platforms are in use today to launch and carry data driven advertising, the most prominent channels being social, search, email and analytics.[35]

Warranty Management
One of the major promises of big data analytics in conjunction with other modern disruptive technologies like the IoT is to help understand why things happen and

---

[32]Oracle, Rittmann Mark, Technology: Business Analytics, Social Network Analysis, http://www.oracle.com/technetwork/issue-archive/2016/16-sep/o56ba-3211403.html, 2016, retrieved 2017-01-16.

[33]EY Building a better working world, Life sciences: preparing for big data and analytics, http://www.ey.com/gl/en/services/advisory/ey-life-sciences-preparing-for-big-data-and-analytics, 2017, retrieved 2017-01-16.

[34]KDnuggets, Pal Kaushik, Big Data Influence on Data Driven Advertising, http://www.kdnuggets.com/2015/08/big-data-influencing-data-driven-advertising.html, 2015, retrieved 2017-01-16.

[35]Harvard Business Review, Nichols Wes, Advertising Analytics 2.0, https://hbr.org/2013/03/advertising-analytics-20, 2013, retrieved 2017-01-16.

how they happen. For warranty management, it is crucial to predict patterns and circumstances around product failures, while preventing them from happening again and thus big data analytics became one of the key pillars for successful warranty management (Kurvinen, Töyrylä, & Murthy, 2016).

Health Care Outcomes Analysis
Health data volume has grown significantly over the past years and is expected to grow even more dramatically in the years ahead. At the same time, healthcare reimbursement models are changing significantly as meaningful pay for performance is becoming a critical factor. Digitizing, combining and effectively using big data allows all kinds of healthcare organizations to benefit from detecting diseases at early stages, managing specific population and individual health as well as detecting health care fraud more effectively (Ghavami, 2014).

Natural Resource Exploration
Big data analytics is desperately needed today for effective ecological science in order to advance and inform resource management through data intensive approaches that find ways for readily handling these massive volumes of already existing and future data with new tools and practices.[36]

Weather Forecasting
This is one of the most obvious cases of big data analytics, especially when considering modern weather forecasting methods. These methods use a lot of actual sensor data that is assimilated and processed in modern models. It is also combined with satellite images and data taking into account temperature, pressure, humidity, satellite images etc. In the end, the more data is used for prediction the more exact the forecasts will be, which shows a clear path towards heavy usage of the IoT as well.

Customer Behavior Analysis
Data mining tools and techniques are helping modern enterprises to make better decisions. One of the most important aspects here is detecting customer behavior patterns and formulating marketing, sales as well as consumer support strategies. Typical techniques used today are cluster detection, memory-based reasoning, market basket analysis, genetic algorithms, link analysis, decision trees, and neural nets in order to find the gold mines of business solutions lying hidden in information systems of modern companies and enterprises.[37]

---

[36]Analytics Magazine, Farris Adam, How big data is changing the oil & gas industry, http://analytics-magazine.org/how-big-data-is-changing-the-oil-a-gas-industry/, 2012, retrieved 2017-01-16.

[37]Elsevier, Khade Anindita, Performing Customer Behavior Analysis using Big Data Analytics, http://www.sciencedirect.com/science/article/pii/S1877050916002568, 2016, retrieved 2017-01-16.

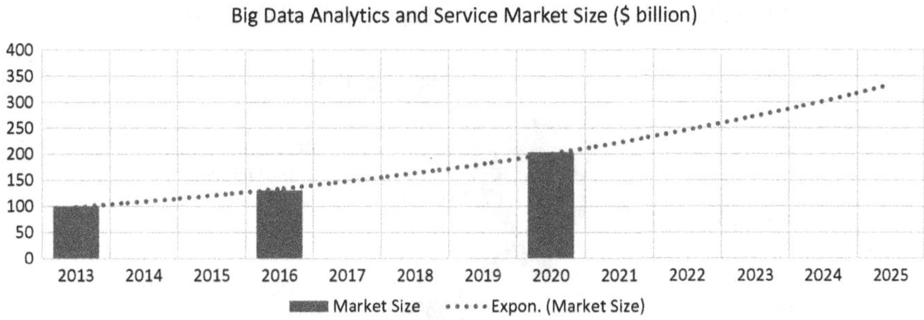

Fig. 6 Big data and analytics market size

Big Data Analytics Market

In 2013 the big data, analytics and services market broke through the $100 billion border for the first time and it is expected to reach $203 billion by 2020 due to IDC (see Fig. 6).[38]

According to IDC the biggest market share comes from banking, followed by Discrete Manufacturing, Process Manufacturing, Federal/Central Government and Professional Services (see Fig. 7).

The big data analytics market is based on a multitude of ingredients, ranging from underlying databases to storage and infrastructure platforms up to the applications layer where the goal is making sense of all the accumulated data, which makes understanding this market very confusing. We also have to acknowledge that this market is still in its infancy in 2017, where the usually big and static incumbents like Oracle, IBM, Software AG, SAP, Microsoft, EMC and HP compete with flexible startups like Datameer,[39] Alpine Data,[40] SiSense,[41] Cloudmeter, Palantir[42] and others (see Fig. 8).

The incumbents have spent billions in order to acquire software firms in the data management and analytics space like Jacada,[43] More IT Resources, Vertica[44] or Vivisimo (now IBM)[45] just to name some; startups are impressively funded by venture capital. Many big data analytics startups offer products in specific niches, like social marketing (DataSift[46]), programmatic advertising buying (Rocket

[38]IDC, Big Data Analytics Market Forecast for 2020, http://www.idc.com/getdoc.jsp?containerId=prUS41826116, 2016, retrieved 2017-01-13.

[39]Datameer, https://www.datameer.com, 2017, retrieved 2017-02-20.

[40]Alpine Data, http://alpinedata.com, 2017, retrieved 2017-02-20.

[41]SiSense, https://www.sisense.com, 2017, retrieved 2017-02-20.

[42]Palantir, https://www.palantir.com, 2017, retrieved 2017-02-20.

[43]Jacada, https://www.jacada.com, 2017, retrieved 2017-02-20.

[44]Vertica, https://www.vertica.com, 2017, retrieved 2017-02-20.

[45]IBM, Vivisimo, https://www-01.ibm.com/software/data/information-optimization/, 2012, retrieved 2017-02-20.

[46]Datasift, http://datasift.com, 2017, retrieved 2017-02-20.

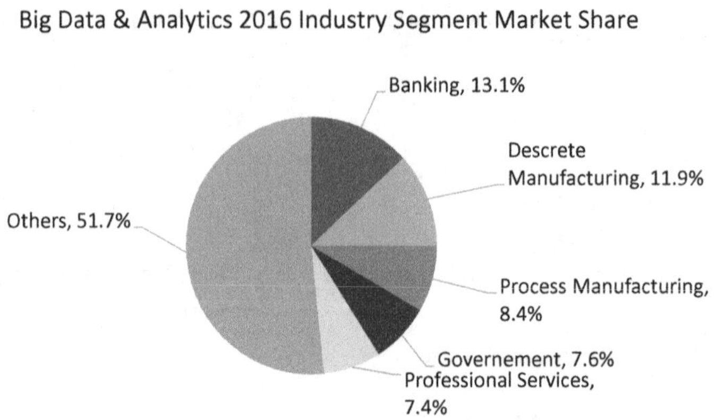

Fig. 7 Analytics 2016 industry segment market share

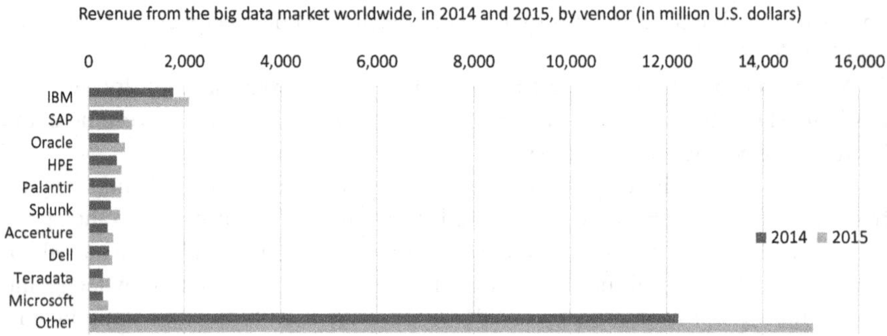

Fig. 8 Revenue from big data market worldwide in 2014 and 2015

Fuel[47]), application performance (Cloudmeter acquired by Splunk[48]) or job searches and recruiting (Bright.com acquired by Linkedin[49]). To make this market even more complex, there are also companies in this game who are just a bit too old to be called startups, including Pentaho,[50] Splunk and Jaspersoft,[51] having their foot in the door of successful big data analytics products.

[47]Rocketfuel, https://rocketfuel.com, 2017, retrieved 2017-02-20.

[48]Splunk, http://www.splunk.com/view/splunk-announces-acquisition-of-cloudmeter/SP-CAAAJD2, 2013, retrieved 2017-02-20.

[49]Bright.com, https://www.linkedin.com/company/bright.com, 2017, retrieved 2017-02-20.

[50]Pentaho, http://www.pentaho.com, 2017, retrieved 2017-02-20.

[51]Jaspersoft, http://www.jaspersoft.com/regForms/introducing-jaspersoft-6-de/?mkwid=sM8S43BtK&pdv=c&pcrid=177332879028&pmt=e&pkw=jaspersoft&campaign=ggl_s_dach_dach_jsp_brand_alpha&group=&bt=177332879028&_bk=jaspersoft&_bm=e&_bn=g&gclid=CjwKEAiAxKrFBRDm25f60OegtwwSJABgEC-Z6i63Gz82UtJkRj_Q2_1njF7P7VXKpq6FJ7YO7lr1NRoC0b7w_wcB, 2017, retrieved 2017-02-20.

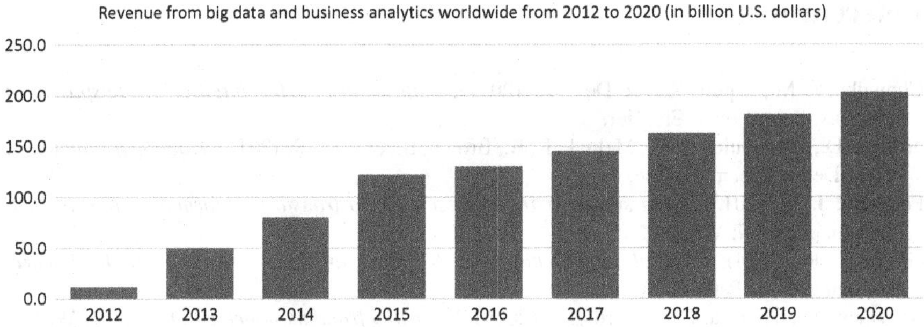

Fig. 9 Big data analytics market from 2012 to 2020

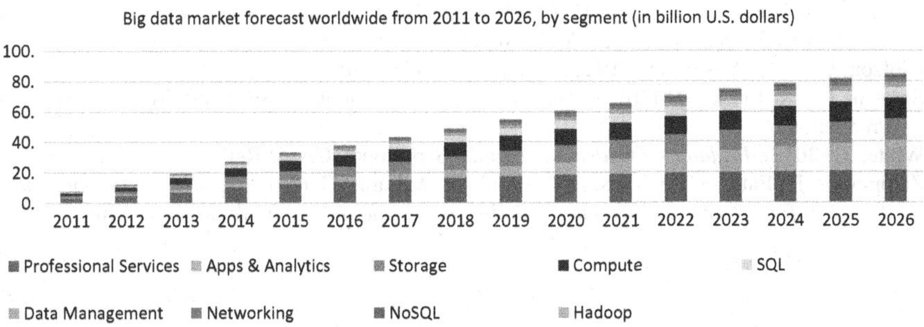

Fig. 10 Big Data market forecast worldwide from 2011 to 2026, by segment

The big data analytics market was valued at $11.4 billion in 2012, reached $122 billion in 2015 and is predicted by IDC to grow to $203 billion by 2020, which is a more than a 50% increase between 2015 and 2019. Since this market is still in a gold rush phase there is more than enough room for a number of vendors, but we also can expect some serious consolidation until the beginning of the 2020s (see Fig. 9).

It is expected that the biggest customers of big data analytics applications, tools and services will be the biggest organizations, generated by companies with more than 500 employees. Utilities, healthcare, resource industries and banking will be the fastest growing areas for this market. It is expected that the US will be the biggest market for big data analytics, followed by Western Europe, Asia Pacific and Latin America (see Fig. 10).

Big Data Analytics Trends and Future

Big data analytics has gained impressive momentum since 2010 and is poised to grow even more impressively over the next decades. As such big data analytics, which is per se nothing new, can easily become one of the next big things influencing the way we do business and maybe the only solution for handling and keeping new trends like social media or the IoT etc. under control and so turn them into big revenue sources and success factors.

References

Chowdhury, M., Apon, A., & Dey, K. (2017). *Data analytics for intelligent transportation systems*. Saint Louis: Elsevier.

deRoos, D., Zikopoulos, P. C., Melnyk, R. B., Brown, B., & Coss, R. (2014). *Hadoop for dummies*. New Delhi: Dreamtech Press.

Foreman, J. W. (2013). *Data smart: Using data science to transform information into insight*. Indianapolis, IN: Wiley.

Ghavami, P. (2014). *Clinical intelligence: The big data analytics revolution in healthcare*. Kirkland, WA: Peter Ghavami.

Kurvinen, M., Töyrylä, I., & Murthy, P. (2016). *Warranty fraud management*. Hoboken: Wiley.

Linoff, G., & Berry, M. (2011). *Data mining techniques* (3rd ed.). New York: Wiley.

Luckham, D. (2012). *Event processing for business: Organizing the real time enterprise*. Hoboken: Wiley.

Marr, B. (2015). *Big data: Using SMART big data, analytics and metrics to make better decisions and improve Performance*. New York: Wiley.

Olofson, C. W., & Vesset, D. (2012, August). www.idc.com

Smolan, R., & Erwitt, J. (2012). *The hunan face of big data*. Sausalito: Against All Odd Productions.

White, T. (2012). *Hadoop – The definitive guide*. Sebastopol, CA: O'Reilly.

Zikopoulos, P., Eaton, C., deRoos, D., Deutsch, T., & Lapis, G. (2012). *Understanding big data: Analytics for enterprise class hadoop and streaming data*. New York: McGraw-Hill.

Future Technologies of the Cloud Century

Prediction is very difficult, especially about the future.
Niels Bohr or Mark Twain or Karl Valentin

There could be a prize-winning answer to the question: "What is the next big thing?" In this chapter, we will try to read the crystal ball to determine, for the technologies we have discussed throughout this book, the likelihood that we will see some form of evolution in the coming decades and which new and disruptive technologies could still revolutionize the cloud century.

All the technology areas we have discussed throughout this book like computing, networking, the Internet, virtualization, the IoT and big data analytics will see significant changes but also significant improvements, which will form the basis of how our lives might look some decades from now. Among all these evolutions, we will definitely find some of the next big things that will surely revolutionize our future.

The Ever-Increasing Computing Power

The future of computing brings exciting new possibilities. Considering how computing has evolved over the past 50–60 years, including the revolutions of personal computers, laptops and continuing with the processing power available today on mobile devices like tablets or smart phones, we might have only seen the tip of the iceberg in terms of what is feasible in computing.

For example, the Apple iPhone 4,[1] introduced in 2010, offered the same processing power as the Cray-2[2] Supercomputer from 1985, which is roughly 1.6 GFLOPS (Giga Floating Point Operations per Second). Considering this evolution,

[1]Apple, iPhone 4 Technical Specifications, https://support.apple.com/kb/sp587?locale=en_US, 2014, retrieved 2017-01-17.

[2]Cray, Cray-2, http://www.craysupercomputers.com/cray2.htm, 2017, retrieved 2017-01-17.

M. Oppitz, P. Tomsu, *Inventing the Cloud Century*,
DOI 10.1007/978-3-319-61161-7_19

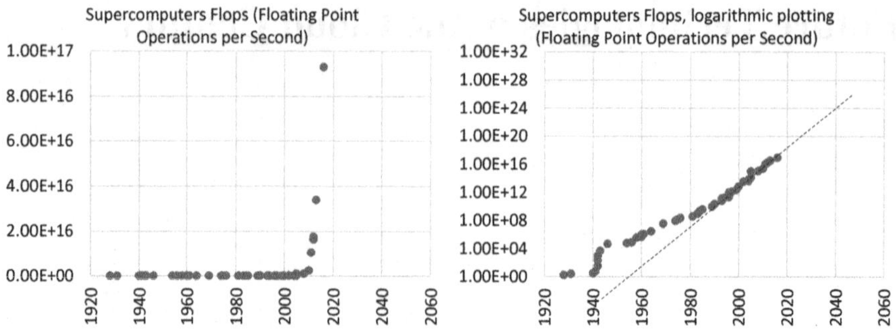

Fig. 1 Increase of supercomputing power over time

it is pretty obvious that these numbers will increase even faster over the next decades, while miniaturization will allow for the design of even smaller devices (see Fig. 1).

If we perform the comparison in the opposite direction of the Tianhe-2[3] Supercomputer (2013) versus the PlayStation 4s[4] (2016), we see that with the supercomputer it is possible to achieve 33.86 PFLOPS (Peta Floating Point Operations per Second) versus an impressive 1.84 TFLOPS (Tera Floating Point Operations per Second) using the PlayStation 4s; a clear sign of how fast the performance limits are being pushed today.

1 Trillion-Fold Increase in Computing Power from 1956 to 2015

This example also shows that computing power has been steadily increasing and, even more compelling, the processing power of today's supercomputers will be reached by stand-alone computers and mobile devices in a few years. We can expect enormous processing power to be available in mobile devices like smartphones or tablets in the next 10 years, which will easily equal the power of large stationary supercomputers operating today. This trend has brought us a 1 trillion-fold increase in performance from 1956 to 2015 and, since technology innovation has accelerated significantly over time, the expected capabilities of future computing devices will also grow dramatically (see Fig. 2).

As a result, the possibilities for future devices that we will use in our everyday lives are endless, ranging from smart mobile devices or any kind of computer used in cars, buildings, households, hospitals, etc.[5]

The conclusion from this trend is that computing power is constantly increasing and will keep doing so over the next decades, even if we only continue to use

[3] Top 500, Tihane-2 aka Milkyway-2, https://www.top500.org/featured/systems/tianhe-2/, 2013, retrieved 2017-01-17.

[4] Playstaion, Playstation 4, https://www.playstation.com/en-us/explore/ps4/, 2016, retrieved 2017-01-17.

[5] pages.experts-exchange, Processing Power Compared, http://pages.experts-exchange.com/processing-power-compared/, 2015, retrieved 2017-01-17.

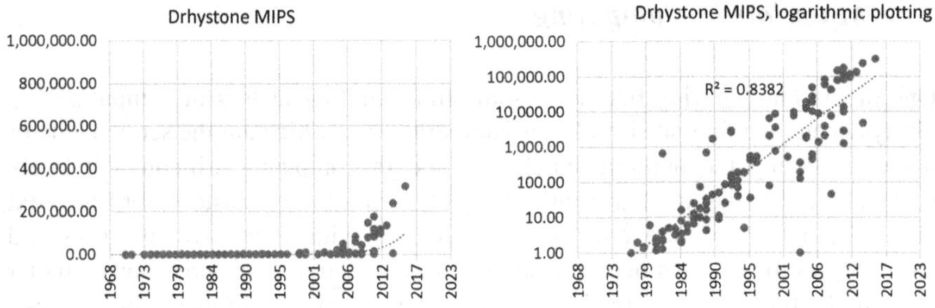

Fig. 2 Dhrystone MIPS of processors

semiconductor-based technologies. As soon as newer technologies like grapheme based semiconductors or quantum computing become commercially available, this trend will be even more heavily disrupted.

Parallel CPUs

Parallel computation is nothing new (Ananth Grama, 2003) and future computer architectures (Novoselvo, 2015) will probably expand parallel computation even more using microprocessors with a large number of cores.[6] Several CPUs will then work together on the same chip thereby increasing the number of tasks simultaneously running and, in doing so, increase the overall system performance.

At the same time, more specialized computers will start to appear because the costs will be less prohibitive than in the past, a trend we can already see today in the early stages of the Internet of Things.

Parallel computation will definitely become a very effective way of increasing computing power, but how parallel the computations can be and how many computers will be linked to a large network or to the cloud, is a different question. With advances in telecommunications and ever increasing communication speeds, it becomes much easier to link many computers to a large network. This is already happening today, as a lot of our data is stored not on desktops but in the cloud. Cloud computing will become more and more popular, while remaining based on microprocessors and electronics and current computing architectures.

As soon as new materials like grapheme become available on a manufacturing scale and replace silicon as the basic technology of computers, new functions will also be designed and new architectures optimized for these new functions, opening up new, exciting paradigm changes.

[6]Intel, Intel Many Integrated Core Architecture, http://www.intel.com/content/www/us/en/architecture-and-technology/many-integrated-core/intel-many-integrated-core-architecture.html, 2017, retrieved 2017-01-17.

New Materials in Computing

One of the most driving but at the same time limiting factors in computing has always been the minimal size of transistor structures, which are the semiconductor devices forming the building blocks of modern computers. Silicon will likely continue to be the predominantly used material for transistors for the next 10 years, but people are already experimenting with alternative materials and technologies to replace silicon, given its failure to deliver for increasingly smaller and smaller transistors. A grapheme-based transistor[7] is one of these alternatives.

Graphene (Skakalova, 2014), along with other materials, allows for the building of one-atom-thick 2-D materials and hetero-structures based on those 2-D crystals.[8] They could potentially provide an alternative to silicon technologies, but we are talking about completely new architectures here, rather than just introducing a new material into an existing system. It's hard to predict how it will develop because when one new material is introduced into a process, it's already quite a complicated step. Needless to say, changing the whole architecture would require years of research but these changes need to come and the signs of change look good.

The Networking Revolution

The need for new networking architectures comes from the sheer growing number of mobile devices, new advancements in content, virtualization and the fact that state of the art cloud services have become common today. As already discussed, conventional hierarchical but static network architectures become more and more obsolete in order to fulfill the requirements of modern networks, especially the dynamically changing requirements imposed by new types of applications as seen today. These requirements include supporting quickly changing traffic patterns, enabling easier use of mobile devices in private as well as corporate environments, cloud services and big data analytics, all while providing high level security.

---

[7]Phys Org, New type of graphene-based transistor will increase the clock speed of processors, https://phys.org/news/2016-05-graphene-based-transistor-clock-processors.html, 2016, retrieved 2017-01-17.

[8]Horizon, The EU Research & Innovation Magazine, Making materials just one atom thick, https://horizon-magazine.eu/article/making-materials-just-one-atom-thick_en.html, 2016, retrieved 2017-01-17.

Constantly Expanding Infrastructure

One of the most driving factors of networking and the Internet was the ever-increasing transmission speeds from access via distribution to the core.

Fixed Line Bandwidth Increase

If we consider 20 years ago in 1996, the speed of analog dial up modems (last mile) was already called blazingly fast with 9.6 kbps, followed pretty soon by 64 kbps shortly thereafter. In the mid-2000s access speeds were in the Mbps range. This is a 10- to 20-fold increase in speed and it was accelerated by the introduction of ADSL2[9] offering 24 Mbps or VDSL2[10] offering 200 Mbps, all still running the last mile i.e. century old telephone lines.

With the introduction of fiber to the home (FTTH[11]), speeds could easily increase to the 10 Gbps range and if this evolution continues, fixed line bandwidth will be in the Petabit per second range 20 years from now (see Fig. 3).

Mobile Bandwidth Increase

Similar to the fixed line wireless access technologies also evolved over the past 15 years and, not long ago, we were happy to have 56 kbps bandwidth offered via GPRS,[12] which quickly grew to several Mbps with the introduction of HSPA.[13] Today these speeds already reach the range of 150 Mbps using LTE,[14] and will easily offer several Gbps wireless access speeds as soon as 5G[15] technologies become available.

Over the past 20 years, these mobile access speed improvements have paved the way for the very advanced Internet experience we know today. We can expect further growth in this area by mobile devices as well as IoT and M2M communication.

---

[9]Increase Broadband Speed, ADSL2 and ADSL2+, http://www.increasebroadbandspeed.co.uk/adsl2, 2017, retrieved 2017-01-18.

[10]Ericsson, Eriksson Per-Erik, et al., VDSL2: Next important broadband technology, https://www.ericsson.com/ericsson/corpinfo/publications/review/2006_01/files/vdsl2.pdf, 2006, retrieved 2017-01-18.

[11]FTTH Council, What is FTTH, http://www.ftthcouncil.org/p/cm/ld/fid=25, 2017, retrieved 2017-01-18.

[12]GSMA, GPRS, http://www.gsma.com/aboutus/gsm-technology/gprs, 2017, retrieved 2017-01-18.

[13]3GPP, Wannstrom Jeanette, HSPA, http://www.3gpp.org/technologies/keywords-acronyms/99-hspa, 2017, 2017-01-18.

[14]3GPP, Long Term Evolution (LTE), http://www.3gpp.org/technologies/keywords-acronyms/98-lte, 2017- retrieved 2017-01-18.

[15]GSMA, GSMA Intelligence, Understanding 5G: Perspectives on future technological advancements in mobile, https://www.gsmaintelligence.com/research/?file=141208-5g.pdf&download, 2014, retrieved 2017-01-18.

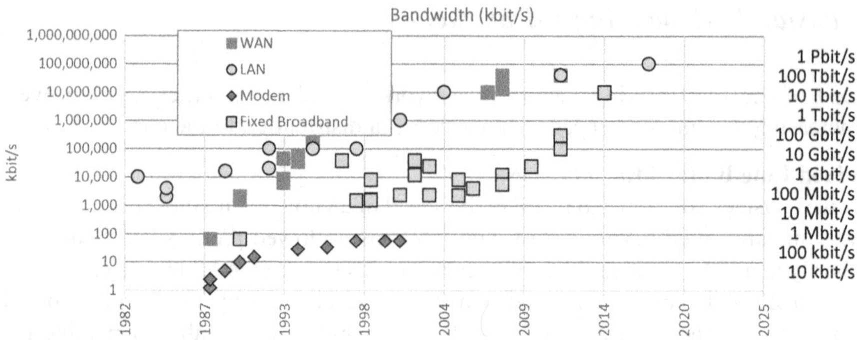

Fig. 3 Fixed line bandwidth increase, logarithmic plotting

Core Bandwidth Increase

Of course, distribution and also core networking speeds had to increase to handle the new traffic volume coming from users, devices and sensors or actors (aka IoT and Fog). While 40 Gbps was the realistic maximum speed per Internet backbone channel for a long time, this speed has already increased today using lasers that operate close to a single frequency and thus are able to carry a new amount of data through fiber optic cables.

This finally results in a 160 Gbps per Internet backbone channel and it can be assumed that 400 Gbps or even 1600 Gbps will be achievable with some further tweaking—so we can rest assured that the backbone speeds required by the cloud age are guaranteed.

Traffic Increase

According to Cisco's VNI (Visual Networking Index) in 2016,[16] the overall IP traffic will be around 194 Exabytes per month in 2020 which is triple the total amount of 2015. This equates to 511 Tbps (terabit per second) being sent or equal to 142 million people streaming high definition Internet videos simultaneously (see Fig. 4).

It is even more interesting to look at the device types filling these impressive numbers with a CAGR of 10% from 2015 to 2020. M2M communication has a strong lead with 30% by 2020 and is the fastest growing segment, a clear sign of the disruptive nature of the IoT as well as fog computing. M2M is followed by smartphones and video, while classic PCs and tablets belong to the minority of devices. This, of course, shows another important future trend in that how we use

[16]Cisco, Visual Networking Index (VNI) 2016, http://www.cisco.com/c/en/us/solutions/collateral/service-provider/visual-networking-index-vni/vni-hyperconnectivity-wp.pdf, 2016, retrieved 2017-01-17.

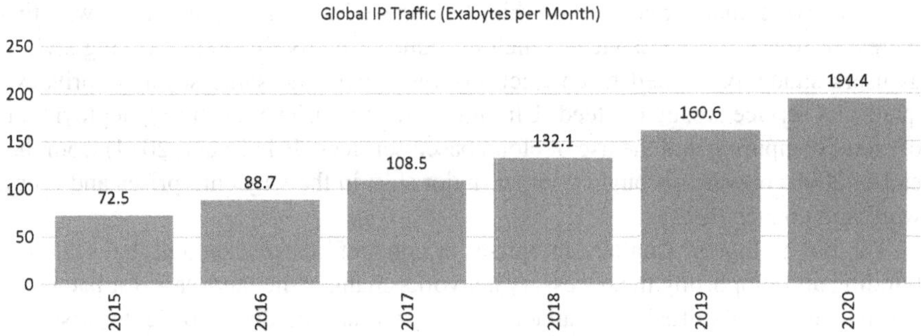

Fig. 4 Global IP traffic per month 2015–2020

and access the Internet constantly changes and will continue to be dominated by mobility.

Changing Applications Means Changing Traffic Patterns

We already discussed why traffic patterns change frequently with today's applications, which access different databases and servers. This results in east-west traffic flows before the results can be returned in the classic north-south traffic patterns to the end users. The new traffic from corporate content applications and mobile devices is added on top of other traffic patterns, which per definition implies very dynamic behavior. In addition, the enterprise cloud is usually split into private, public and hybrid clouds, creating new traffic types across wide area networks as well.

In previous times, the access to corporate networks was mainly via quasi static personal computers and later moved towards mobile laptops, today we see an increasing use of all kinds of mobile devices, including smartphones, tablets, and still notebooks. Access and security from all these personal devices needs to comply with corporate rules and fully protect corporate data in this highly dynamic and constantly changing environment (Zhang, Hu, & Fujise, 2006).

Challenges for Legacy IP Networks

The evolution towards IP only based devices was originally a great idea and made multiprotocol networking a thing of the past. Over the past decade, it quickly became difficult to scale, manage, secure and adapt IP based networks to meet new challenges from the usage and business side and this trend is poised to increase in the years to come.

There were many reasons for this, but one of the most ubiquitous was the increased use of mobile devices which introduced totally different working styles, as users suddenly wanted to connect anywhere with the same security, privacy, quality of service etc. guaranteed (Ulrich Dolata, 2013). The upcoming deployment of cloud computing and the use of cloud based services (Eric Bauer, 2014) from the end of 2000s onwards brought another major shift in the way enterprises and users work and connect today.

The possibility for smaller enterprises to conduct business on a global scale by building and supporting the necessary networks changed the landscape for bad guys and hackers who shifted their attacks from large and medium size enterprises to smaller businesses who lacked the right staff to thwart them.

Finally, the increasing complexity of today's networks with all the fancy but necessary features like more security, application intelligence and mobile orientation in addition to old IP-based networks became a constant challenge to address appropriately and in a timely manner.

SDN and Cloud Based Networking Services

Cloud computing, network virtualization and new approaches for managing these future networks radically altered the economics of business networking. A number of companies offer SDN solutions today, all with the goal of changing the face and future of networking[17] and, in so doing, reshuffle the incumbent vendor deck. Network-as-a-Service (NaaS) and Infrastructure-as-a-Service (IaaS) have already become very prominent cloud services, allowing users to connect anywhere to the IT resources they need and whenever they need them. There are still significant differences in the various products offered which, if implemented in an effective way, allows for not only the management of the data-center, servers, storage and LAN but also the WAN network, also known as the Network Function Virtualization (NFV).[18]

Cloud services became an integral and constantly growing part of today's businesses via private, public and hybrid clouds (Bloomberg, 2013). It is imperative that these services are easily accessible over the new, corporate infrastructures for applications, ensuring the additional security, compliance and auditing requirements. It is also important that they support frequent business reorganizations and mergers, which have become a constant part of modern corporate life. Thus, elasticity of all parts of the cloud infrastructure like computing, storage and network

[17]Internet Society, Kamite Yujef, et al., Software Defined Networking: The Current Picture and Future Expectations, http://www.internetsociety.org/articles/software-defined-networking-current-picture-and-future-expectations, 2013, retrieved 2017-01-17.

[18]ETSI, portal.etsi.org, Introductory White Paper, Network Function Virtualization, https://portal.etsi.org/NFV/NFV_White_Paper.pdf, 2012, retrieved 2017-01-17.

resources has become mandatory and, in these modern environments, will only be achieved through the consequent use of SDN and common management tools playing seamlessly together. This is one of the main reasons why SDN will become an even greater base technology for cloud based networking services.

Good implementation of SDN based cloud-networking services[19] needs to offer WAN overlay architectures that provide an end-to-end abstraction of the underlying carrier IP infrastructure. Data forwarding and control planes need to be nicely separated in order to create dynamic and resilient network topologies and be able to run on standard virtual machines (merchant VMs) within major, geographically dispersed cloud datacenters. This in turn will further help to afford an efficient overlay network infrastructure that can adapt in real-time as needed, reflect the quickly changing demands of end users, route around underlying infrastructure failures and be able to withstand future, threatening storms.

Ideally most or all of the topological and policy based complexity of traditional networks such as IP addressing, network address assignment and translation, access control, security certificates, authentication, DNS, etc. will reside in a cloud based control plane.

The result will be a radically simplified network building experience, which is one of the most important requirements for future networks, and network service deployments. It should be seamless as to enable dynamic and user-friendly configuration of cloud networks on demand, including users and resources and allowing for additional, future network services without any issues. The deployment of these technologies and methods marks the era of future programmable networking, which will finally fully arrive over the next decade.

Big Data Analytics Networking Requirements

The huge increase in data used by many modern applications generally referred to as "big data" requires not only new computing architectures including multi-core and parallel processing of up to several thousands of servers, but also additional, highly dynamic networking capacities[20] that must be based on SDN[21] for flexibility

[19]Research Gate, Azodolmolky Siamak, et al., SDN-based Cloud Computing Networking, https://www.researchgate.net/publication/249657889_SDN-based_Cloud_Computing_Networking_Invited, 2013, retrieved 2017-01-17.

[20]Brocade, IDC, Villars Richard, et al., White Paper, Big Data and the Network, https://www.brocade.com/content/dam/common/documents/content-types/whitepaper/idc-big-data-network.pdf, 2011, retrieved 2017-01-17.

[21]IEEE Network, Cui Laizhong, When big data meets software-defined networking: SDN for big data and big data for SDN, https://www.researchgate.net/publication/291955655_When_big_data_meets_software-defined_networking_SDN_for_big_data_and_big_data_for_SDN, 2016, retrieved 2017-01-17.

reasons. As a result, we need hyper-scale data center networks,[22] supporting not only new magnitudes of scalability, but also assuring any-to-any connectivity on demand without any failures (Xu, Lin, Misic, & Shen, 2015).

Wireless Future

WiFi has been around for almost two decades, but so far it has been optimized for wireless communication over medium to longer distances. It seems that WiGig,[23] a new standard by the Wireless Gigabit Alliance[24] involving wireless gigabit data links, is currently being pushed by HP and Dell to enable the connection of monitors, hard drives and other PC peripherals without the use of any cable and available for deployment very soon. This will allow a computer-near network similar to Bluetooth but capable of much higher speeds of up to 7 Gbps within a range up to 10 m, using beam-forming antenna technology (Poole, 2015). In addition, Intel is pioneering a wireless charging method for laptops, smartphones and other devices (Lowe, 2014a). According to a Cisco white paper,[25] the Internet traffic will increase to more than 180 PByte per month by 2020, with fixed Internet still maintaining a share of more than 60%. Mobile traffic is predicted to be around 15% in 2020 but with a yearly increase of more than 50% (see Fig. 5).

Low Power Wide Area (LPWA) networks are specifically optimized for the further development of the IoT and will be one of the key components in the success of this fast-growing field. Today three standards for different requirements in the IoT networking developed by GSMA[26] exist. These are Extended Coverage GSM for IoT (EC-GSM-IoT[27]), Long Term Evolution for Machines (LTE-M[28]) and Narrow-Band Internet of Things (NB-IoT[29]). It is very likely that new features and releases will supplement these wireless, low power wide area standards over the next years, especially as there will be a big demand for the use of the 5G spectrum for enhanced communication features.

---

[22]BigSwitch Networks, Hyperscale Networking For All, http://go.bigswitch.com/rs/974-WXR-561/images/BigSwitch_HNA_WP_FINAL.pdf, 2014, retrieved 2017-01-17.

[23]wi-fi.org, Wi-Fi Certified WiGig, http://www.wi-fi.org/discover-wi-fi/wi-fi-certified-wigig, 2017, retrieved 2017-01-17.

[24]wi-fi.org, Wi-Fi Alliance and Wireless Gigabit Alliance to unify, http://www.wi-fi.org/news-events/newsroom/wi-fi-alliance-and-wireless-gigabit-alliance-to-unify, 2013, retrieved 2017-01-17.

[25]Cisco, http://www.cisco.com/c/en/us/solutions/collateral/service-provider/visual-networking-index-vni/complete-white-paper-c11-481360.html retrieved 2017-02-10.

[26]GSMA, http://www.gsma.com, 2017, retrieved 2017-01-18.

[27]GSMA, EC-GSM-IoT, Extended Coverage – GSM IoT, http://www.gsma.com/connectedliving/extended-coverage-gsm-internet-of-things-ec-gsm-iot/, 2017, retrieved 2017-01-10.

[28]GSMA, LTE-M, Long Term Evolution for Machines, http://www.gsma.com/connectedliving/long-term-evolution-machine-type-communication-lte-mtc-cat-m1/, 2017, retrieved 2017-01-10.

[29]GSMA, NB-IoT, Narrow Band IoT, http://www.gsma.com/connectedliving/narrow-band-internet-of-things-nb-iot/, 2017, retrieved 2017-01-10.

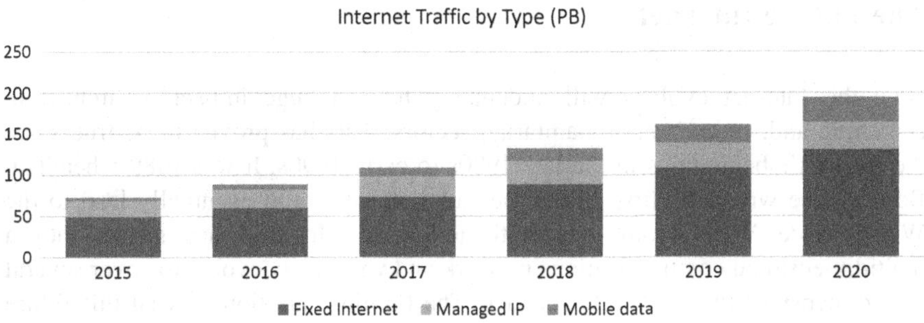

Fig. 5 Internet traffic projection by type

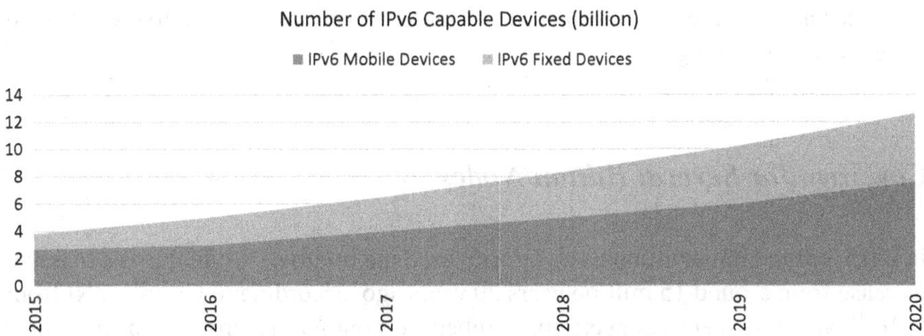

Fig. 6 Number of IPv6 capable devices and connections 2015–2020

IPv6

IPv6 is a prerequisite for the adoption of the IoT. Up to today, IPv6 has already made substantial progress, mainly because the latest devices fully support this standard as well as operators who continuously upgrade their networks to IPv6. We have already seen the exhaustion of IPv4 addresses in many regions worldwide.

Due to Cisco VNI 2016, there will be around 13 billion IPv6 devices by 2020 including both mobile as well as fixed devices. This is more than three times increase from the four billion devices in 2015. A total of 90% of all smart phones and tablets will be IPv6 capable by 2020. Extrapolating these numbers into the next decades proves how right and necessary the decision for IPv6 was more than two decades ago, as it has enabled many options from the future we will be living in. We are talking about roughly 120 billion devices by 2030 and it is anyone's guess as to where these numbers will head (see Fig. 6).

The Future Internet

How the Internet evolves will undeniably have a huge impact on industrial, economic and, of course, humanitarian sectors. This has proven to be true since the Internet's beginnings in the late 1980s to early 1990s. It was 1989 when Tim Berners Lee wrote the first online message, which would eventually lead to the World Wide Web. From then until today, the Internet has grown into a multidimensional communications network and will surely evolve to cover several new dimensions over the next decades. The burning question is what this future Internet might look like.

We will try to answer this question by looking into requirements for future coverage, speed, capacity, new and advanced security standards, next generation user interfaces, resilience as well as new dimensions imposed by the IoT, embedded systems and swarming.

Coverage for Several Billion Nodes

In 2015, around three billion users already used the Internet. While this is a massive increase from around 15 million users 20 years ago, according to Cisco's VNI from 2016[30] we can expect connectivity numbers to continue to grow even more dramatically, fueled by the growth in mobile data traffic followed by video traffic and the growing availability of cheap and easy to use smart devices. A number of companies like Apple, Google, Facebook and others are accelerating this trend through the proposed use of the Internet by satellites, smart vehicles, drones, balloon systems and, in a wider context, the Internet of Things. This will result in a projected seven billion nodes connecting to the Internet, requiring a solid infrastructure to handle all this data collection and necessary communication by 2020 (see Fig. 7).

Speed as Never Seen Before

We all know Moore's Law for computer speed, which claims a roughly 60% increase in performance per year. There is also a law for the Internet Bandwidth defined by Nielsen,[31] which says that the typical user bandwidth will grow by

---

[30]Cisco, Visual Networking Index (VNI) 2016, http://www.cisco.com/c/en/us/solutions/collateral/service-provider/visual-networking-index-vni/vni-hyperconnectivity-wp.pdf, 2016, retrieved 2017-01-17.

[31]Nielsen Norman Group, Nielsen's Law of Internet Bandwidth, https://www.nngroup.com/articles/law-of-bandwidth/, 1998, 2016, retrieved 2017-01-18.

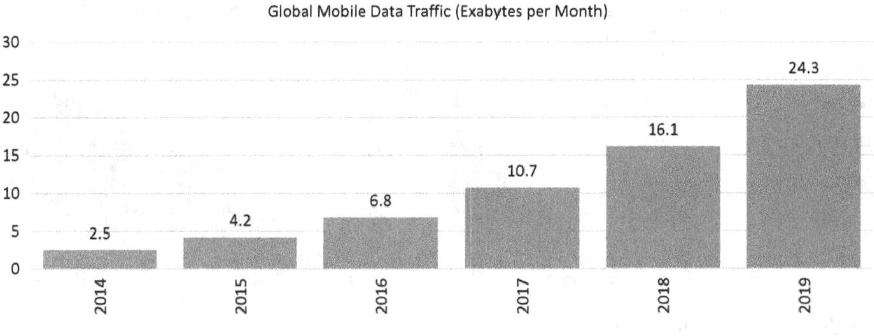

Fig. 7 Global mobile data traffic growth 2014–2019 (Source CISCO)

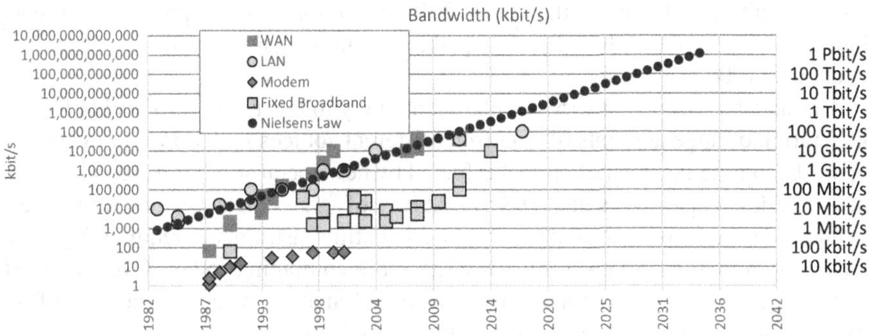

Fig. 8 Law of internet bandwidth, logarithmic plotting

around 50% per year, roughly 10% less than growth of computer speed. These numbers have proven to be correct from 1983 to 2016 and are shown below in the Figure below, which charts the respective bandwidth offered by slow modems in 1983 to typical access speeds in 2016. Keep in mind that this diagram shows an exponential growth curve in a logarithmic scale, so if we simply extrapolate the data to 2040 we get well above 1 Tbps for access speed (Frey, 2011)—some might find this shocking (see Fig. 8).

Zettabyte Capacity

Cisco projects in the Cisco VNI from 2016 that global IP traffic will almost triple from 2015 to 2020 (from 72.5 exabyte per month to 194 exabyte per month). This means a total of 2.3 zettabyte by 2020. To visualize the size of a zettabyte, if each terabyte in a zetabyte were a kilometer, it would equal 1300 round trip journeys to the moon and back, or if every gigabyte in a zettabyte were a brick, this would be enough bricks for 258 Great Walls of China (see Fig. 9).

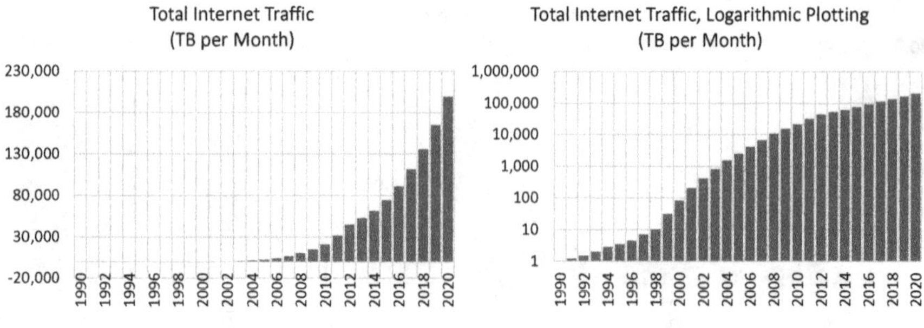

Fig. 9 Increase of internet traffic

Overall, it shows that a yearly growth rate of 20–25% is safe to assume which would result in a total of 7 zettabytes in 2025 and so forth. Just to put these numbers into context once more, 1 exabyte is equal to 1 billion gigabytes and 1 zettabyte[32] is equal to a 1000 exabytes.

Big data has become a very popular technology today, being responsible for the generation of huge amounts of data from different sources with different meanings and is becoming increasingly popular fueled by new social structures, social media, new business experiences etc. The availability of large data sets from web pages, genome datasets, as well as the output of scientific instruments requires big data analytics to take on a new form for large scale computing today. Over the next decade, these systems will have to manage and analyze data sets the size of an exabyte, sometimes even bordering on the region of zettabytes.

In order to be able to handle these amounts of data, new supercomputers are needed as well. This is also the reason why, in 2015, President Obama signed an order to fund the world's first exascale computer,[33] which should be operational by 2025 thus overtaking China who is currently leading in this field. This is the reason why a new National Strategic Computing Initiative (NSCI)[34] was established in order to develop this exascale computer, which can handle at least one exaflops calculations per second, the equivalent of a billion times a billion calculations, meaning a 1000-fold increase since the petascale computer in 2008. Even if an exascale computer is an impressive machine today, it will be outdated as soon as it is operational in 2025.

[32]SImetric, http://www.simetric.co.uk/siprefix.htm, 2016, retrieved 2017-01-18.

[33]Science, sciencemag.org, Service Robert, Obama orders effort to build first exascale computer, http://www.sciencemag.org/news/2015/07/obama-orders-effort-build-first-exascale-computer, 2015, retrieved 2017-01-18.

[34]NITRD, The Networking and Information Technology Research and Development Program, The National Strategic Computing Initiative (NSCI), https://www.nitrd.gov/nsci/, 2015, retrieved 2017-01-18.

From the above we see that requirements are not only increasing for transmission speeds, but also for storage and computing, fueled by new ways of using the Internet like the IoT and many others.

Based on the exponential growth stated in Moore's law, by 2030 a micro SD card will have the storage capacity equivalent to 20,000 human brains and, by 2043, the storage capacity of more than 500 billion gigabytes. This is equal to the entire content of the Internet in 2009. By 2050 this storage capacity will be three times the brain capacity of the entire human race as projected by FutureTimeline.net.[35]

Balance Between Privacy and Security

In today's networked and connected world, security has become a major concern, and the number of bad actors intending to harm major businesses or even entire countries is increasing exponentially. Of course, there are surveillance technologies to ensure secure operation but it is a constant race between evil minded and good minded actors. On the other hand, everybody wants to have privacy, which is in contradiction to security, as more convenience means less security.

While, in theory, a radically transparent world would ensure a safer environment, it also means that every detail about every person would be known, like bank accounts, credit card numbers, passwords etc., which is hard to envisage based on the principles on which we have built our existing world.

The future Internet will need to strike a balance between privacy, security, trust and ethics. President Obama for a Privacy Bill of Rights reflects this in the 2015 proposal,[36] but this is just a US initiative and actually we would require a global initiative and solution to address these issues. This needs to be pushed by the UN and should reflect a kind of Geneva Convention for Privacy as the basis for global standards and practical implementation guidelines, as well as legal definitions, monitoring tools and tools to handle cases of abuse. Without this and without more philosophical work and preparation, the future Internet will suffer and will unlikely be able to reach its full potential.

Next Generation User Interfaces

User interfaces have come a long way since the use of keyboards given that the Internet was mainly text based in its beginning stages. The mouse and the invention

[35]FutureTimeline.net, http://futuretimeline.net, 2017, retrieved 2017-01-18.

[36]Whitehouse, whitehouse.gov, Administration Discussion Draft: Consumer Privacy Bill of Rights Act 2015, https://www.whitehouse.gov/sites/default/files/omb/legislative/letters/cpbr-act-of-2015-discussion-draft.pdf, 2015, retrieved 2017-01-18.

of hyperlinks can be seen as the start of the surfing experience, but graphics were still pretty poor due to the small bandwidth allowed by connection links in those days. As bandwidth increased, animated graphics and videos became very popular and can no longer be excluded from a modern Internet experience. Speech recognition allows for further improvements in user communication with input devices, but there are still devices operating between what humans think and what devices can fulfill, as we still have to articulate our thoughts before a machine can understand what we want.

Before we can take the next, big and revolutionary step by finally allowing a brain interface, we will see many minor, but important steps in that direction, like optical interface devices for augmented reality and virtual reality such as Google Glass,[37] Facebook's Oculus VR,[38] Microsoft's HoloLens,[39,40] or Samsung's Gear VR.[41]

Resilience and Survivability

For a network to become as popular, important and vitally necessary as the Internet, which is just at the beginning of its evolution if we take into account the future projections of features like the IoT, Fog Computing or Big Data Analytics, the network needs to be extremely resilient to any kind of attacks. It needs to withstand not only hacker attacks, but also survive military strikes and potential downturns in the global economy.

Even more serious than these external threats is the constant aging of the systems of which the Internet is composed. It needs things like distributed intelligence between its core root servers. In mid-2015, ten root servers were located in the US, two in Europe and one in Japan[42] and all of these root servers operate in multiple geographical locations via anycast addressing, using redundant equipment to provide uninterrupted service in case of hardware or software failures. To address the growing demands of the future Internet, this root server architecture will no longer be sufficient and enhancing it needs technical understanding as well as the ability to envisage economic, military and environmental disasters.

---

[37]techradar, Swider Matt, Google Glass review, http://www.techradar.com/reviews/gadgets/google-glass-1152283/review, 2016, retrieved 2017-01-18.

[38]Facebook, Oculus, https://www.facebook.com/oculusvr/, 2017, retrieved 2017-01-18.

[39]Ted Talks, Kipman Alex, Microsoft deoms the Hololens, https://www.youtube.com/watch?v=kjAHwLaLjUw, 2016, retrieved 2017-01-26.

[40]Microsoft, Microsoft HoloLens, https://www.microsoft.com/microsoft-hololens/en-us, 2017, retrieved 2017-01-18.

[41]Samsung, Gear VR, http://www.samsung.com/global/galaxy/gear-vr/, 2017, retrieved 2017-01-18.

[42]lifewire, Mitchell Bradley, Why There Are only 13 DNS Root Name Servers, https://www.lifewire.com/dns-root-name-servers-3971336, 2016, retrieved 2017-01-18.

In the years to come, a special emphasis will have to be put on the durability and long term survivability of the future Internet in order to ensure its stable and secure operation given the requirements of the coming decades. Especially when we are looking at security vulnerabilities in the area of the IoT, it becomes pretty obvious that there will be a huge need for better security protection.[43]

New Dimensions Through IoT and Embedded Systems

The original Internet mainly linked simple computers with one another. Smaller and more capable laptops, smartphones, smart watches, tablets and a number of other devices allowing humans new ways of accessing the Internet have replaced personal computers. In addition, the IoT with all its sensors, actuators and embedded devices continues to constantly change the dimensions that the Internet must support.

Janus Bryzek, a Fairchild executive, organized the Trillion Sensor Summit[44] in 2013 with 200 executives from around the world in attendance. The numbers coming out of this event are breathtaking. Sensors will exceed the trillion devices threshold by 2020 and could reach 100 trillion sensors by 2030,[45] adding many new data streams from virtually anywhere. Not only will these sensors be embedded in almost every article we can imagine like cars, houses or clothing etc., but the computational power of these things will increase immensely as seen in skin sensors and body sensors that will be able to constantly monitor a person's condition or health.

The advancements in 3D printing will soon allow us to embed things consisting of sensors, microchips and transmitters into printed objects.[46] We already have solar cells that are printed on roof foil, connected vehicles will use sensors in the pavement and at crossroads and, for travelers, identity devices will allow automatic security and customs checks. Wireless energy will provide cordless charging,[47]

---

[43]CSA, Cloud Security Alliance, Future-proofing the Connected World: 13 Steps to Developing Secure IoT Products, https://downloads.cloudsecurityalliance.org/assets/research/internet-of-things/future-proofing-the-connected-world.pdf, 2016, retrieved 2017-01-19.

[44]TSensors, http://www.tsensorssummit.org, 2017, retrieved 2017-01-19.

[45]Motherboard, Merchant Bryan, With a Trillion Sensors, the Internet of Things Would Be the "Biggest Business in the History of Electronics", http://motherboard.vice.com/blog/the-internet-of-things-could-be-the-biggest-business-in-the-history-of-electronics, 2013, retrieved 2017-01-19.

[46]Advanced Materials Technologies, Ota Hiroki, et al., Application of 3D Printing for Smart Objects with Embedded Electronic Sensors and Systems, http://nano.eecs.berkeley.edu/publications/Advanced_Materials_Technologies_2016_3D%20printed%20sensors.pdf, 2016, retrieved 2017-01-19.

[47]PowerbyProxi, Wireless Charging, https://powerbyproxi.com/wireless-charging/, 2017, retrieved 2017-01-19.

which we have seen through the introduction of wireless networking during the 2010s.

Swarming and Collaboration

Today the advancements in the Internet have enabled us to collaborate much more efficiently than in the 1990s. Collaboration tools allow efficient pairing of human intelligence to specific projects and, in this way, groups of people can easily become dynamic unified systems, solving problems unsolvable before. Nature demonstrates repeatedly that creatures functioning together in systems can easily outperform individual creatures in problem solving and decision-making tasks. Internet technologies allow humans to build these groups, often called swarms, as seen in studies at the California State University.[48]

Swarming will evolve to encompass powerful tools in the next decades, enabling groups to unleash intelligence in many fields and applications. Furthermore, the future Internet will help us to understand intelligence in order to make better use of it.[49]

Storage Virtualization Future

Software Defined Data Center

To consider the future of storage virtualization, we need to first understand one of the latest evolutions in data centers, the so-called Software Defined Data Center (SDDC).[50] SDDC (Darin, 2014), which is also referred to as Virtual Data Center (VDC) is an IT infrastructure vision extending virtualization concepts like abstraction, pooling and automation to all data center resources and services in order to finally achieve something like IT as a Service (ITaaS).

In an SDDC, all infrastructure elements such as networking, CPU, storage and security are virtualized and can be delivered on demand as a service.[51] The

---

[48]Sacramento State, Swarm Day, http://www.csus.edu/soal/campus%20events/swarm%20day.html, 2016, retrieved 2017-02-09.

[49]European Commission, Cocoro: robot swarms use collective cognition to perform tasks, https://ec.europa.eu/digital-single-market/en/news/cocoro-robot-swarms-use-collective-cognition-perform-tasks, 2015, retrieved 2017-01-19.

[50]VMware, Software Defined Data Center (SDDC), http://www.vmware.com/at/solutions/software-defined-datacenter/in-depth.html, 2017, retrieved 2017-01-19.

[51]VMware, Technical White Paper, Software Defined Data Center (SDDC), http://www.vmware.com/content/dam/digitalmarketing/vmware/en/pdf/techpaper/technical-whitepaper-sddc-capabilities-itoutcomes-white-paper.pdf, 2015, retrieved 2017-01-20.

deployment, provisioning, configuration, operation, monitoring and automation of the infrastructure is abstracted from hardware and implemented in software. This means that the entire infrastructure is virtualized and delivered as a service—hence Infrastructure as a Service (IaaS). SDDC seeks integrators and datacenter builders and not as tenants, as data center infrastructure software awareness should not be visible to tenants.

SDDC is not very well accepted, as many critics see this as a marketing tool more than a new vision with many future implementation scenarios. On the other side of the camp, there are individuals who are sure that software will define future data centers and see this trend as a work in progress.[52] The growth potential for SDDC looks pretty promising anyway, according to many analysts who expect that some components of SDDC will see strong growth soon. The Software Defined Storage (SDS) market alone should grow to $22.56 billion by 2021,[53] which is due to exponential growth of data volume across enterprises and a general rise in software defined concepts.

Storage Virtualization as Last Missing Link in SDDC
Meanwhile the believers in SDDC have started to see storage as the final missing link of the complete virtualization and SDN story. What is needed now is a clear separation of storage control planes, where software controls and manages data, as well as a storage data plane for storing, copying and retrieving data, both working seamlessly with storage infrastructures.

One of the most important reasons to separate the control and data planes is to free the storage control software from the hardware. Software defined storage (Lowe, 2014b) enables offloading of the computationally heavy parts of storage management functions as seen in RDMA (remote direct memory access) protocol handling, data lifecycle management compression and caching. This computational power can come from vast amounts of CPU power available in private and public clouds and opens up unknown possibilities for both network and storage management, options which were not feasible before.

Revolutionized Non-volatile Memory Design

Advancements in non-volatile memory (NVM) technology make solid state memory, especially flash memory more affordable, while there are also numerous new possibilities promised by next generation storage technologies like Phase Change

[52]Gartner, Gartner Says the Future of the Data Center in Software Defined, http://www.gartner.com/newsroom/id/3136417, 2015, retrieved 2017-01-19.

[53]MarketsandMarkets, Software Defined Storage Market by 2021, http://www.marketsandmarkets.com/Market-Reports/software-defined-storage-sds-networking-sdsn-market-1067.html?gclid=Cj0KEQiAh4fEBRCZhriIjLfArrQBEiQArzzDAdnIKrtMrJxj-uC2haL6i9TzCFRjGVmptdWGA1BvQ30aAsCA8P8HAQ, 2016, retrieved 2017-01-20.

Memory (PCM)[54] (Moinuddin K. Quershi, 2011) and Spin Transfer Torque Random Access Memory (STT-RAM)[55] (Xiaobin Wang, 2010). Both PCM and STT-RAM offer access speeds as well as byte addressable characteristics of Dynamic Random Access Memory (DRAM), which is mainly used in servers today, but they both have the same advantage of solid-state persistence, like flash memory.

As soon as these two prototype technologies become cheaper than flash memory, one or both of them will revolutionize memory design and within a few years most server storage will be based on solid-state cache within the server itself, which will have huge implications on storage design in combination with evolving network technologies and software supporting distributed architectures. If every server had terabytes of super-fast solid-state memory connected via ultra-fast and low latency networking, this would make today's implementation of shared storage for critical applications a thing of the past, an evolution which is definitely necessary.

Optimized Capacity Large Disk Drives

Today we are increasingly faced with the demand for massive amounts of data storage and processing, mainly fueled by big data analytics, and coinciding with a dramatic cost reduction for data storage. Meanwhile, drive capacity has started to exceed 10 TB per disk using conventional magnetic storage technologies. When using new storage techniques, for example, those based on quantum physics[56] (Vathsan, 2015), future storage characteristics will increase significantly compared to those used today.

Large cloud providers have totally new needs in terms of scalability of computing power in combination with close data proximity, which means the value proposition for storing an organization's cold data becomes very compelling. This opens up new requirements for securing and finding this data as well as managing the lifecycles of this huge amount of information, which requires new structures that are more capable than today's methods of files, folders and directories. Thus, new management and access technologies are required based on object-based access to data, like we have already seen in Amazon S3[57] and the open standards based Cloud Data Management Interface (CDMI)[58] (SNIA, 2015).

---

[54] Poplab, Stanford, EU, Raoux Simon et al., Phase change materials and phase change memory, http://poplab.stanford.edu/pdfs/Raoux-PCMreview-mrsbull14.pdf, 2014, retrieved 2017-01-20.

[55] AIP, Journal of Applied Physics, Ando K., et al., Spin-transfer torque magnetorestrictive random-access memory technologies for normally off computing (invited), http://aip.scitation.org/doi/pdf/10.1063/1.4869828, 2013, retrieved 2017-01-20.

[56] MPG, Max Planck Gesellschaft, Hunger David, Quantum storage system with long term memory, https://www.mpg.de/5856755/quantum_storage_memory, 2012, retrieved 2017-01-20.

[57] Amazon Web Services, Amazon Simple Storage Service (Amazon S3), http://docs.aws.amazon.com/AmazonS3/latest/dev/Welcome.html, 2017, retrieved 2017-01-20.

[58] SNIA, Cloud Data Management Interface, http://www.snia.org/cloud/cdmi, 2017, retrieved 2017-01-20.

Why Software Defined Storage (SDS) Infrastructure

Using Software Defined Storage (SDS)[59] infrastructures is the only way to ensure the effective utilization of performance as well as speed of solid-state storage and the scalability advantages of capacity optimized storage. The separation of the control plane especially enables data center designs to make effective use of these new storage trends.

Server Virtualization Evolution

Organizations can achieve big impacts and a greater level of benefits by appropriately virtualizing their server infrastructures. The more sophisticated server virtualization technologies are deployed, the greater the final value derived from these activities will be. It is safe to assume that server virtualization will continue to grow and accelerate as it has already done over the past decade, so that participating organizations can achieve greater value. We will look into the future of server virtualization in this section, which is closely related to the evolution of storage virtualization.

The Overall Virtualization Problem

Back in the days of mainframes, computing was centralized and expensive, but it was also predictable and controllable and, moreover, it was manageable. One of the major drivers for decentralized, distributed computing was the reduction of CAPEX due to the introduction of low cost commodity servers. These servers must be connected to proprietary, expensive, monolithic storage boxes, keeping the overall virtualization solution for computing and storage proprietary and expensive. Servers have meanwhile become reasonably cheap, but storage today is still complex, incompatible and highly priced.

The big question remains, how all this virtualization will evolve if we are to create data centers based on virtualized assets. In order to enable this, all virtual layers need to coexist supporting the same functionalities, while appropriately reacting to changing conditions within their areas of discipline.

Future Server Virtualization Requirements

We need elements (boxes) that are able to self-optimize and reconfigure themselves to changing workload requirements, and self-healing infrastructures that can deal with fault scenarios autonomously and rebuild themselves without affecting the applications. This involves self-scaling infrastructures that extend virtually to all requirements imposed by workloads and self-managing infrastructures that automatically adapt to changing scenarios based on policies. These are the requirements that need to be addressed in future virtualized infrastructures but, in 2017, we are

[59]SNIA, Software Defined Storage, http://www.snia.org/sds, 2017, retrieved 2017-01-20.

still far from such a solution, although there is hope when looking at some of the solutions, which are already around.

From Fiber Channel to Ethernet

While servers are connected to Ethernet today, storage is still largely connected via Fiber Channel (FC) networks,[60] which were used in the mainframe times to carrying ESCON and FICON traffic.[61] It would require further price drops in 10 Gbps Ethernet and cheaper offers of 40 Gbps and 100 Gbps Ethernet to transfer all intra data center connectivity to Ethernet.

There are, at least, no longer "religious" fights between FC and Ethernet, but we are still ages away from a complete Ethernet connectivity solution. Meanwhile FC vendors push Fiber Channel over Ethernet (FCoE),[62,63] in order to at least preserve the existing protocol over the Ethernet network layer, allowing them to run FCoE on Ethernet switches. This should not be seen as anything more than an interim step towards full Ethernet implementation, as FCoE does not reduce complexity nor challenge costs.

ISCSI defined in RFC 3720[64] and NAS already allow full Ethernet, nicely supporting a number of modern environments, while pure forms of Ethernet storage have appeared like ATA over Ethernet (AoE).[65] This enables design and support of flexible and cost efficient virtualized and cloud based architectures.

The Single Data Center Networking Solution

Ultimately, the future data centers need to converge on a single networking solution. There are examples in nature of this, like the human body's nervous system, which shares a common structure while supporting a network of different functions. The same must become true for the future data center and virtualization networking solution.

---

[60]FCIA, Fiber Channel Industry Association, http://fibrechannel.org, 2017, retrieved 2017-01-20.

[61]IBM, IBM Knowledge Center, https://www.ibm.com/support/knowledgecenter/STCMML8/com.ibm.storage.ts3500.doc/ipg_3584_ficon_escon_interface.html, 2017, retrieved 2017-01-20.

[62]Cisco, Fiber Channel over Ethernet (FCoE), http://www.cisco.com/c/en/us/solutions/data-center-virtualization/fibre-channel-over-ethernet-fcoe/index.html, 2017, retrieved 2017-01-20.

[63]EMC, Fiber Channel over Ethernet, https://www.emc.com/collateral/hardware/white-papers/h5916-intro-to-fcoe-wp.pdf, 2015, retrieved 2017-01-20.

[64]IETF, RFC 3720, Satran J., et al., Internet Small System Interface (iSCSI), https://www.ietf.org/rfc/rfc3720.txt, 2004, retrieved 2017-01-20.

[65]alyseo, Coile Brantley, The ATA over Ethernet Protocol, http://ftp.alyseo.com/pub/partners/Coraid/Docs/AoE/AoEDescription.pdf, 2017, retrieved 2017-01-20.

Ethernet Based Storage Area Networks (SANs)
One of the secrets of future data centers is the support of Scale Out Storage.[66] In an Ethernet connected network, virtual servers can be established everywhere allowing virtual workloads to run almost instantly and wherever needed. However, with the monolithic storage that we often still use today, the overall advantage of total flexibility stays pretty limited. The answer to this is Commodity Based Storage,[67] which supports full native Ethernet connectivity[68] and allows for the scaling out of capabilities either matching or exceeding the capabilities of the server layer.

Full Data Mobility for Storage and Server Virtualization
The ESG (Enterprise Strategy Group) has pushed for the similar mobility and ease of use in storage since the early 2010s, in order to make data mobility as seamless as virtual machine mobility. While server mobility is an instant task, moving virtual machines to another server and immediately increasing CPU power has not achieved enough speed or efficiency.

For this to happen, storage needs to become a fully virtualized complement to server and network layers, running on commodity hardware components, supporting full Ethernet connectivity, remaining self-managing and self-healing and, finally, able to scale to whatever demand is required. Most of the technologies to make this happen exist already and will soon be required by users. Vendors not jumping on this bandwagon will lose a significant market share and probably disappear.

Storage administration will become something fully automated, allowing performance, protection and recovery policies to be enforced by a virtualization orchestrating layer while assets are provisioned almost instantaneously in order to allow for seamless operation of server and storage virtualization.

Network Virtualization of the Next Decade

Virtualization became the foundation and enabled many new technology trends like cloud computing, IoT, fog computing, and big data analytics to mention just the most important. If we want to understand the evolution and future of virtualization,

[66]NetworkComputing, Marks Howard, Scale-Out Storage: How Does It Work?, http://www.networkcomputing.com/storage/scale-out-storage-how-does-it-work/545167674, 2016, retrieved 2017-01-20.

[67]Data Center Knowledge, Kleyman Bill, Commodity Data Center Storage: Building Your Own, http://www.datacenterknowledge.com/archives/2015/06/23/commodity-data-center-storage-building-your-own/, 2015, retrieved 2017-01-20.

[68]The Register, Mellor Chris, Which Ethernet SAN storage protocol is best? http://www.theregister.co.uk/2010/10/04/ethernet_san_storage_protocols/, 2010, retrieved 2017-01-20.

we need to understand why virtualization and cloud computing are so closely related.

Cloud has come a long way since the end of the 2010s and it is still constantly evolving, but just as cloud computing evolves virtualization must too. Most clouds still use 10-year-old virtualization technology, but new ways of virtualization are needed to build the future clouds.

Future virtualization needs more IO intensive network and storage workloads, while fully ensuring the support for open industry standards, and making these standards applicable to new hypervisor designs. Cloud computing software like OpenStack, sitting on the highest layer, will manage cloud infrastructures and whatever is needed for virtualization.

SDDC and Networking

This evolution of virtualization and cloud computing is a continuous process and finds a new settling point in the Software Defined Data Center (SDDC) that we have already discussed in Storage Virtualization Breakthroughs. Here, however, we will concentrate on the networking aspects of SDDC. Virtualization and cloud evolution are tightly intertwined and while virtualization already was used decades ago, the latest revolutions in virtualization and cloud will significantly shape the future of cloud services and offerings (see Fig. 10).

Virtualization and Cloud Computing Based on SDDC

There is a continuous evolution of virtualization happening today through the presence of cloud computing and culminating in SDDC. Cloud computing saw a new operational model for IT services based on virtualization technologies and new, IaaS approaches. SDDC will allow for the delivering of even more intelligent services in combination with advanced management solutions for cloud and standard virtualization technologies.

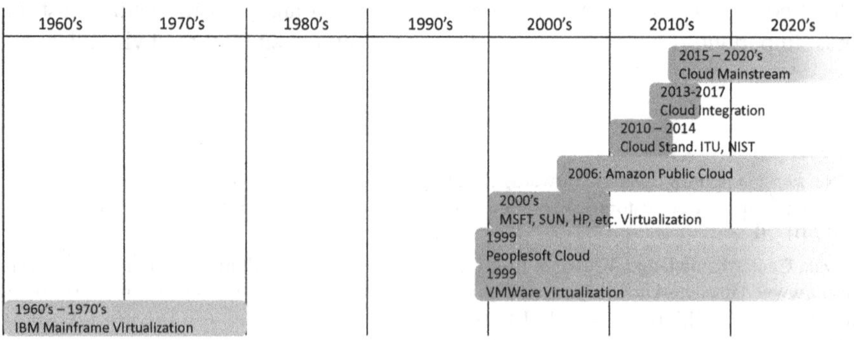

Fig. 10 Virtualization and cloud evolution

The most significant changes in cloud environments today are driven by consolidation and the need for private as well as hybrid cloud environments. One of the secrets of SDDC is to look into applications and how they run, or how better to run them, in the data center. Most applications are network distributed (aka known as multi-tier apps) and require instant distribution, which is fine as long as we are dealing with servers. For tasks like re-configuring the network elements such as firewalls, load balancers and setting up new virtual LANs (VLANs) as well as new IP addresses for all applications, network distributed applications is still very slow taking hours, days and even weeks.

Network Distributed Applications and SDDC

SDDC provides the foundation for this new, flexible data center by defining a container, a virtual data center or a virtual application for network-distributed applications. This container can be easily manipulated in almost the same way as virtual machines, but now with the capability to manipulate the complete application. This technology was first driven by VMware,[69] which used the ideas of the SDN implementation they got from the acquisition of Nicira[70] in July 2012.

With SDDC, virtualization is expanded from servers to include storage and networking while separating applications from the infrastructure, all being encapsulated in a container. As soon as applications are in containers, the lifecycle of these containers is automated and, likewise, the applications in these containers. This concept of lifecycles sounds familiar to OVF [see Open Virtualization Format (OVF)][71] and this is not a surprise, as in the cases of OVF and SDDC the same standards organization is behind and driving both. VMware, so again no surprise in this context, also heavily drives DMTF.

This container concept[72,73] becomes especially important for large enterprises running thousands of applications, which would not be possible to provision and manage individually. With the application-container concept provisioning and management of applications becomes very similar to classical virtualization operations including provisioning, moving to scale up or down, moving for availability and moving to and from the cloud.

[69]VMware, Developer Center, https://developercenter.vmware.com/sddc-getting-started, 2017, retrieved 2017-01-20.

[70]VMware, VMware and Nicira, https://www.vmware.com/at/company/acquisitions/nicira.html, 2012, retrieved 2017-01-20.

[71]DMTF, Open Virtualization Format, https://www.dmtf.org/standards/ovf, 2017, retrieved 2017-01-20.

[72]SNIA, SNIA Global Education, Vasudeva Anil, Containers: The Future of Virtualization & SDDC, http://www.snia.org/sites/default/files/AnilVasudeva_Containers_the_Future_Virtualization_SDDC.pdf, 2015, retrieved 2017-01-20.

[73]DMTF, White Paper from the OSDCC Incubator, Software Defined Data Center (SDDC) Definition, http://www.dmtf.org/sites/default/files/standards/documents/DSP-IS0501_1.0.1a.pdf, 2014, retrieved 2017-01-20.

New Demands on Hypervisors

With all this additional flexibility, it is pretty obvious that SDDC places a lot of new demands on hypervisors, as these will now have to handle IO-intensive storage and network virtual appliances plus traditional applications. This, in turn, requires increased processing and thread capabilities, resulting in increasing demand for processor cores.

There are already some hypervisor vendors entering this market like ZeroVM.[74] This hypervisor is specifically designed for the new SDDC model allowing application isolation and efficiency combined with the necessary deployment speed, and can separate every single task into its own container, only virtualizing the parts of the server that are required to do the work. This approach is more optimized compared to existing clouds where giant server farms are used wasting precious resources through the virtualization of unneeded things.

This hypervisor creates a new VM for every incoming request and its UNIX style processes communicate through pipes such as VMware, XEN[75] and KVM.[76] Multiple physical servers can be aggregated and are represented as a single virtual system as it is possible to represent a number of virtual systems backed up by any number of virtual servers. This enables dividing the hypervisor into huge number of processes thereby allowing totally new levels of virtualization. This opens up a new concept of virtualizing from the application to the process and user levels.

SDN Controller Future

Future SDN controllers can evolve from where they are today in several possible directions, by either becoming the network operating system, evolve into single function solutions, become cloud orchestration platforms or evolve into policy renderers.

When controllers become the network operating system they need to evolve into generic platforms that form the basis for network applications with support for services like file systems or memory management of traditional operating systems. They also need to offer APIs for application developers etc. This means that controllers will, in general, become bigger and much more complex.

---

[74]ZDNet, Shamah David, Rackspace picks up ZeroVM's built-for-cloud hypervisor, http://www.zdnet.com/article/rackspace-picks-up-zerovms-built-for-cloud-hypervisor/, 2013, retrieved 2017-01-20.

[75]Xenproject, Linux Foundation Collaborative Project, https://www.xenproject.org, 2017, retrieved 2017-01-20.

[76]KVM, Linux-KVM, Kernel Virtual Machine, http://www.linux-kvm.org/page/Main_Page, 2017, retrieved 2017-01-20.

As soon as controllers end up as single function solutions tied into specific applications like optimizing network virtualization applications or software defined WAN applications, the scope and size of the controller would likely be limited because functionality would be implemented within the applications and no longer as a separate entity.

If controllers become cloud orchestration platforms with OpenStack being a very prominent example, the network would become an abstracted view of the cloud infrastructures. In this example, the applications need to deal with the cloud orchestration platform, which moves control and agility to the cloud orchestration and easily can limit proactive adjustments the network could otherwise have made. These controllers would finally be parts of the orchestration platform and no longer stand-alone entities.

When controllers become pure policy renderers, this makes them policy and intent based systems that translate higher-level device policies into lower level device configurations. This, in turn, sets all the controls within the policy platform and reduces the controller to a pure translator for the network configuration.

At this time, it is not completely sure that all the described scenarios will become a reality in the future, but it is likely that they will happen to some degree and it is even more likely that we will see mixed implementations depending on their infrastructure parts. It is easily feasible that even in the same data center, different instantiations could be deployed on demand.

One fact does not change: the controller will remain the strategic control point of the SDN network and will keep the role of being the network brain. It is also unlikely that networking vendors will voluntarily hand over control of their networking equipment to others, because otherwise they would most likely be pushed out of business sooner than later.

From Closed to Open SDN Environments

Finally, we are left with two possible ways to go in this virtualized networking ecosystem: either to follow dominant networking vendors and use the controllers they provide for orchestrating their proprietary although somehow open equipment, or trust vendors using open controllers supported by many vendors, in the best case we can do both. This open approach we discussed as the OpenDaylight[77] (ODL) Project and it has gained tremendous vendor support over the recent years. The open controllers allow end users as well as cloud and service providers to operate, develop and deploy in an open environment, making it much easier and faster to add proprietary implementations and offer differentiated solutions for their specific needs.

---

[77]OpenDaylight, Linus Foundation Collaborative Project, https://www.opendaylight.org, 2017, retrieved 2017-01-20.

IoT and Fog: The Next Big Disruption?

The IoT has become very popular over the last few years, but we are still far from where this technology, that makes the connection between intelligent devices, can lead us. Fog computing, which extends cloud computing and services to the edge of the network, can be seen as an enhancement for the overall IoT movement and will hence be part of the greater IoT ecosystem that will evolve.

We have discussed many cases and solution samples and have seen that the IoT requires a vast range of new technologies and skills, which are unfortunately, still absent in many organizations and not yet mastered by many vendors in that area. Hence, how this immaturity is architected and the many risks managed will be key for the success of the IoT.

This also means that the picture we have of IoT today will change massively over the coming years, which is of course a good thing as we are dealing with a greenfield market where new players making use of new business models and solutions can easily outperform incumbents. This market is big, ranging from wearable devices, implants, connected homes, smart cities and healthcare to new businesses and enterprises, and as soon as the latter starts deploying IoT solutions the rise will be even multiplied.

Things in the IoT will become increasingly inexpensive and connectivity will become extraordinarily cheap. As this happens, more applications will exist further propelling the IoT market. This will all contribute significantly to the emergence of new ecosystems. Think about the trillions of devices that will generate tons of zettabyte data; not only big data analytics and cloud computing will see new heights, but also millions of new apps will fight with each other for success and a market share, all of this translating into enormous economic possibilities.

From Internet Age to IoT Age

Since we are able to identify things in a unique way, we are able to use Internet technology (in a simple way: the internet protocol stack) to connect them and to build new types of solutions based on the smart interaction between things and applications. In the old Internet, connections were built between components like servers, PCs, routers and switches as typical computer devices. In the IoT, we are dealing with integration and connectivity of any type of item like cars, house infrastructure, production machines or other types of things (Greengard, 2015).

The fast-growing technology of microminiaturization allows the design and production of items that are easy to integrate into things making them a potential part of the IoT. The basic functions making a "thing" a part of the IoT are simple: sensor, actor and connectivity. Sensor functions like temperature, position, movement or any other kind of valuable information are used to collect the important data and information used by applications. Actor functions are necessary to let the

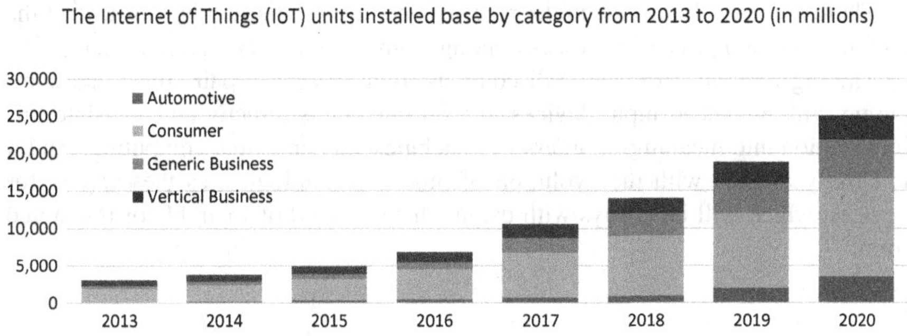

Fig. 11 IoT units installed base by category

device react to new information, by changing their status, influencing their environment or informing users. Last but not least, connectivity to the Internet is necessary to integrate things and applications. Gartner says that 8.4 billion connected things will be used in 2017, which is an impressive 31% increase compared to 2016.[78] If that will "only" stay a 30% increase per year, we end up with 18.4 billion connected things in 2020, but this yearly increase is only the conservative assumption, as estimated from Cisco and Intel for connected IoT devices range from 50 to 200 billion in 2020 (we already discussed that in chapter "IoT") (see Fig. 11).

Compared to the old Internet, these IoT functions need to be seen in a much wider context and related to the real world, as this was the case with the typical computing devices. Sensor functions will react on different parameters in the real world. Actor functions will control not only digital data but analog and technical parameters like power level, water drainage, position and movement of mechanical structures and so on. Connectivity will include not only classical Internet access but also all types of wireless as well as low bandwidth connections.

Self-Driving and Flying Cars

Hand in hand with these developments, devices will gain immense levels of intelligence and make many decisions for us. Examples are driverless cars as we have seen in many trials since the mid-2010s by Google,[79] BMW,[80] Apple,[81] etc., or smart homes.

[78]Gartner, 2017: http://www.gartner.com/newsroom/id/3598917 retrieved 2017-02-15.

[79]Google, Waymo, https://waymo.com, 2017, retrieved 2017-01-18.

[80]BMW, BMW Blog, BMW autonomous car, http://www.bmwblog.com/tag/bmw-autonomous-car/, 2017, retrieved 2017-01-18.

[81]skyNEWS, Apple reveals it is investing "heavily" in driverless cars, http://news.sky.com/story/apple-reveals-it-is-investing-heavily-in-driverless-cars-10683224, 2016, retrieved 2017-01-18.

There are even flying cars in the pipeline and due to be tested by Airbus[82] at the end of 2017, and passenger drones coming from a Chinese company E-Hang.[83]

The big revolutionary step will come as soon as we can pair contextual computing with advanced input devices to scan our brains directly and translate this information into meaningful orders for machines via contextual computing. And it is closely aligned with the evolution of mobile technologies, as these allow for devices, which will be always with us, in whatever kind of wearable or implanted form.[84]

Enormous Economic Benefits Through IoT

The IoT has a big potential to be the next big thing and initialize an unprecedented disruption that might be even bigger and more impactful than the invention of the Internet and the Internet browser put together. The estimated number of objects that the IoT will consist of in 2020 is 50 billion! McKinsey Global Institute sees the potential for the IoT to create an economic impact of $11.1 trillion per year in 2025 for IoT applications.[85,86] These numbers alone clarify the hype about the IoT as the next big revolution, which might happen some decades after the Internet and the Internet Browser (see Fig. 12).

---

[82]The Telegraph, Technology, Flying cars to be tested by end of 2017, says Airbus, http://www.telegraph.co.uk/technology/2017/01/17/flying-cars-end-2017-says-airbus/, 2017, retrieved 2017-01-18.

[83]The Telegraph, Technology, Flying robot taxi to start trials in Las Vegas, http://www.telegraph.co.uk/technology/2016/06/08/flying-robot-taxi-to-start-trials-in-las-vegas/, 2016, retrieved 2017-01-18.

[84]IEEE, Stachel Joshua, et al., The impact of the Internet of Things on implanted medical devices including pacemakers and ICDs, https://www.researchgate.net/publication/261147816_The_impact_of_the_internet_of_Things_on_implanted_medical_devices_including_pacemakers_and_ICDs, 2013, retrieved 2017-01-18.

[85]McKinsey&Company, Manyika James, et al., The Internet of Things: Unlocking the potential of the Internet of Things, http://www.mckinsey.com/business-functions/digital-mckinsey/our-insights/the-internet-of-things-the-value-of-digitizing-the-physical-world, 2015, retrieved 2017-01-20.

[86]McKinsey&Company, Manyika James, et al., The Internet of Things: Mapping The Value Beyond The Hype, https://www.google.at/search?client=safari&rls=en&q=mckinsey+the+inernet+of+things:+mapping+the+value+beyond+the+haype&ie=UTF-8&oe=UTF-8&gfe_rd=cr&ei=uxuCWNqwKsXV8geB65fYCw, 2015, retrieved 2017-01-20.

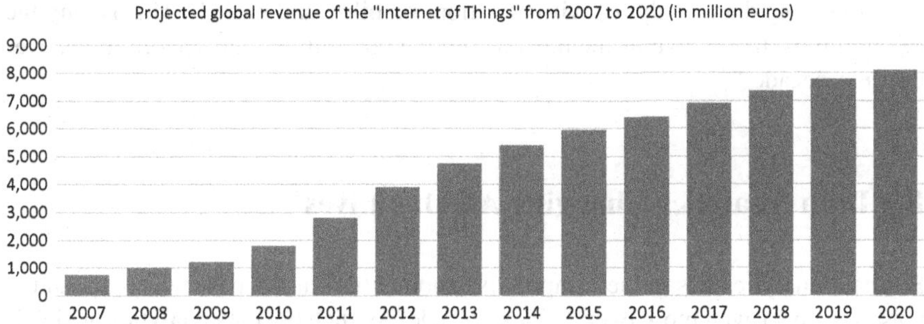

Fig. 12 Projected global revenue of IoT 2007–2020

Next Big Disruption Through IoT

It is obvious that the real value created by the IoT results from the intersection of collecting data by sensors and using this data in a meaningful way by machines to influence our environment.[87] If we only collect information using all the sensors we have available at a given point in time, all this data would be pretty meaningless without the right, real time infrastructure to analyze it. The only way we know how to achieve this today is with cloud-based applications, which are necessary to interpret all the data coming from the sensors and in turn start machines (we call actuators) to take action as needed like control temperature or pressure, light, etc. with whatever necessary electromechanical possibility.

As a result, the IoT enables not only the collection and analysis of data in order to allow better control of business processes for certain enterprises or user groups, it also offers fully automated processes to control a smarter future world as we discussed in IoT solution samples like Smart Connected Vehicles, Precision Agriculture, etc. This means enabling smart cities, smart ports, smart cars, smart roads, etc., with sensors monitoring and tracking all sorts of data, cloud based applications that translate this data into useful intelligence and communicate with actuators (machines) enabling mobile real-time responses. It is not just about business optimization and money savings, it is a much larger and fundamental shift introduced and enabled by the IoT, as making things intelligent is the major engine for creating new products, services and possibilities.

This is why the IoT is probably the biggest and most disruptive technology trend right now and will be for the years to come. It is fueled by a lot of supporting technologies and new methods that were enabled by recent technology evolutions

[87]Brookings, West Darrell, et al., How robots, artificial intelligence, and machine learning will affect employment and public policy, https://www.brookings.edu/blog/techtank/2015/10/26/how-robots-artificial-intelligence-and-machine-learning-will-affect-employment-and-public-policy/, 2015, retrieved 2017-01-20.

like cloud, big data, analytics, virtualization, mobility and more.[88] This is why the IoT will most likely give us the most disruptive possibilities and opportunities over the next decade![89]

Big Data Analytics Changing All Our Lives

Big data analytics has gained impressive momentum in the past few years and is poised to do so even more over the next decade. As such big data analytics, which is per se nothing new, can easily become another one of the next big things influencing the way society[90] evolves, we do business and maybe the only solution for handling and keeping new trends like social media, tens of millions of connected people, many billion sensors, trillions of transactions etc. under control[91] and, in so doing, turn them into big revenue sources and successes.

Triumph of Open Source Tools

While gaining momentum, today there is naturally a big emphasis on open source tools to break down and analyze data. Clearly Hadoop[92] and NoSQL[93] databases seem to be the winners in this game and proprietary technologies are determined to disappear quickly. One of the main goals is to unlock data from proprietary data silos and keep the big data as open and accessible as possible.[94]

---

[88]Gartner, Panetta Kasey, Gartner's Top 10 Strategic Technology Trends for 2017, Artificial intelligence, machine learning, and smart things promise an intelligent future, http://www.gartner.com/smarterwithgartner/gartners-top-10-technology-trends-2017/, 2016, retrieved 2017-01-20.

[89]Wired, Amyx Scott, Amyx-McKinsey, Why The Internet of Things Will Disrupt Everything, https://www.wired.com/insights/2014/07/internet-things-will-disrupt-everything/, 2017, retrieved 2017-01-20.

[90]Journal of Biometrics & Biostatistics, Roy Ajit, Impact of Big Data Analytics on Healthcare and Society, https://www.omicsonline.org/open-access/impact-of-big-data-analytics-on-healthcare-and-society-2155-6180-1000300.php?aid=75499, 2016, retrieved 2017-01-23.

[91]PewResearchCenter, Anderson Janna, et al., Big Data: Experts say new forms of information analysis will help people be morenimble and adaptive, . . ., http://www.pewinternet.org/files/old-media//Files/Reports/2012/PIP_Future_of_Internet_2012_Big_Data.pdf, 2012, retrieved 2017-01-23.

[92]datanami, Muglia Bob, Hadoop Past, Present, and Future, https://www.datanami.com/2016/05/17/hadoop-past-present-future/, 2016, retrieved 2017-01-23.

[93]NetworkWorld, Dix John, et al., What's better for your big data application, SQL or NoSQL? http://www.networkworld.com/article/2226514/tech-debates/what-s-better-for-your-big-data-application--sql-or-nosql-.html, 2014, retrieved 2017-01-23.

[94]InfoWorld, Asay Matt, Beyond Hadoop: The streaming future of big data, http://www.infoworld.com/article/2900504/big-data/beyond-hadoop-streaming-future-of-big-data.html, 2015, retrieved 2017-01-23.

Big Data Analytics and New Market Segments

Meanwhile a large number of big data analytics platforms have already hit the market and this is only the beginning, which is impressively shown by all the new companies emerging to focus and cover specific niches of the big data analytics market (see also big data analytics use cases and market).[95] Currently, though not many vertical specific applications are available on top of the general analytics platforms, the market is still not mature enough and it could still be a bit unsafe to bet on Hadoop or NoSQL as the general-purpose platforms for underlying databases.

Thus, we can expect more vertical tools to emerge, targeting specific analytic challenges, which are common to business sectors like marketing, online shopping, shipping, social media and more. Small-scale analytic engines are being built into software suites, such as social media management tools like Hootsuite[96] and Nimble,[97] which include data analysis as key feature and are also interesting for future market segmentation.

Predictive Analytics

It has always been the big desire of mankind to be able to predict future events and behaviors and while some things are easy to predict (like bad weather suppressing voter turnout), other predictions are much harder like swing voters that are alienated rather than influenced by push polls. This is also the reason why machine learning, modeling, statistical analysis and big data are often mixed together, hoping to better predict future events and behaviors.[98]

Of course, we have the ability to run large scale experiments on our accumulated data in a continuous way, as when online retailers do a redesign of shopping carts in order to find out which specific design yields the most sales, or doctors who are able to predict future disease risks based on data about family history, diet or the amount of exercise one gets every day.

Many of these predictions date back to the beginnings of human history, but instead of just basing predictions on gut feelings or incomplete data sets, predictive analytics has made a lot of progress in many areas like fraud detection, risk

[95]Forbes, Columbis Louis, Roundup Of Analytics, Big Data & BI Forecasts And Market Estimates, 2016, http://www.forbes.com/sites/louiscolumbus/2016/08/20/roundup-of-analytics-big-data-bi-forecasts-and-market-estimates-2016/#37ef54fb49c5, 2016, retrieved 2017-01-23.

[96]hootsuite.com, https://hootsuite.com, 2017, retrieved 2017-02-15.

[97]nimble, http://www.nimble.com, 2017, retrieved 2017-02-15.

[98]PredictiveAnalyticsWorld, IBM, Diegel Eric, Seven Reasons You Need Predictive Analytics Today, http://www.predictiveanalyticsworld.com/patimes/wp-content/uploads/2015/11/7-Reasons-Predictive-Analytics.pdf, 2015, retrieved 2017-01-23.

management for insurance companies and customer retention, just to name a few. We will see a lot of new possibilities in the field of predictive analytics in the years to come, as the tools become better and more stable in collecting, storing and analyzing big data.

Is Human Decision Making Still Necessary?

In times with extensive machine-to-machine communication and constantly improving machine learning, the human factor seems to become less important at first glance. This is only natural because eliminating human error has always been a very critical factor. Consider the simple mistakes humans usually make in the area of security by using weak passwords or getting caught up in phishing attacks or even worse clicking links they should not have. There is much hope that once machines can take over critical actions, a lot of these weaknesses imposed by human beings might go away.

On the downside, this is only half of the truth, as machines will only be able to do what human beings have initially taught them.[99] Relate this to big data and it becomes instantly clear that there are limits to what we can learn from machines and how much we can rely on these conclusions and, at the end of the day, the human element will always remain important.

Nate Silver, who is considered to be one of the big data pioneers, (Silver, 2012) outlines that what matters most for predictions is not the machinery which is used to collect data and run the initial analysis, but the human being and human intelligence necessary to find out what all the results from big data analytics and even more so from predictive analytics means. In the example of Silver, he analyzes reams of data, looks at historical results, calculates in factors that could influence margins of error and finally emerges with very accurate predictions.

As soon as big data analytics becomes a state of the art technology, it will be seen as just another tool we can use to help human beings derive the best possible decisions for whatever business, research etc. is needed. It is important to understand that what one does with the results of big data analytics is what really matters and the success of this task will remain with human beings for a long time.[100]

[99]Accenture, Harris Jeanne, Does Intuition Matter in a Big Data World, 2017, https://www.accenture.com/us-en/insight-does-institution-matter-in-the-big-data-world, retrieved 2017-01-23.

[100]Harvard Business Review, McAfee Andrew, et al., Big Data: The Management Revolution, https://hbr.org/2012/10/big-data-the-management-revolution, 2012, retrieved 2017-01-23.

References

Ananth Grama, A. G. (2003). *Introduction to parallel computing* (2nd ed.). New Delhi/New York: Pearson/Addison Wesley.

Bloomberg, J. (2013). *The agile architecture revolution: How cloud computing, REST-Based SOA, and mobile computing are changing enterprise IT*. Hoboken, NJ: Wiley.

Darin, A. (2014). *Software defined data center for dummies*. El Segundo, CA: Nexenta.

Eric Bauer, R. A. (2014). *Service quality of cloud based applications*. Hoboken, NJ: Wiley-IEEE Press.

Frey, T. (2011). *Communicating with the future*. Aurora, Co: CGX Publishing.

Greengard, S. (2015). *The internet of things*. Cambridge, MA: MIT Press Essential Knowledge Series.

Lowe, M. (2014a). *Intel envisions a wireless future: WiGig and wireless charging to eliminate all cables by end of 2015*. Retrieved from Pocket-lint: pocket-lint.com

Lowe, S. D. (2014b). *Software defined storage for dummies*. Hoboken: Wiley.

Moinuddin K. Quershi, S. G. (2011). *Phase change memeory: From devices to systems* (1st ed.). San Rafael, CA: Morgan & Claypool Publishers.

Novoselvo, K. (2015, January 23). A graphene discoverer speculates on the future of computing. *Scientific American*.

Poole, I. (2015). *Radio-electronics.com/info/wireless/wi-fi/ieee-802-11ad-microwave.php*. Retrieved from Radio-Electronics.com: radio-electronics.com

Silver, N. (2012). *The signal and the noise. Why so many predictions fail – and others not*. New York: Penguin Press.

Skakalova, K. (2014). *Graphene* (1st ed.). Amsterdam: Elsevier.

SNIA. (2015). *Cloud data management interface (CDMI) v1.1.1*. Colorado Springs, CO: SNIA.

Ulrich Dolata, J.-F. S. (2013). *Internet, Mobile Devices und die Transformation der Medien*. Berlin: Deutsche Nationalbibliothek.

Vathsan, R. (2015). *Introduction to quantum physics and information processing* (1st ed.). Boca Raton: CRC Press.

Xiaobin Wang, Y. C. (2010). *Magnetization switching in spin torque random access memory: Challenegs and opportunities*. Berlin: Springer.

Xu, S., Lin, X., Misic, J., & Shen, X. (2015). *Networking for big data*. Boca Raton: CRC Press/Taylor & Francis Group.

Zhang, Y., Hu, H., & Fujise, M. (2006). *Resource, mobility and security management in wireless networks and mobile communications*. Hoboken: CRC Press/Taylor & Francis Group.

References

[illegible]

New Paradigms and Big Disruptive Things

Today the Internet, the cloud computing technologies and the cloud services business have reached a state where a broad foundation of technology is supporting a dynamic and rapidly growing industry. As we have seen the status of cloud services is the result of 40 years of evolution and it seems clear that we have not yet reached the end-point. Predicting future aspects, evolutions and—maybe—revolutions is difficult and in most cases, nearly impossible.

Since the introduction of digital computers and networks in the 1950s and the creation of the Internet, the World Wide Web and cloud services, technology, economy and society have gone through several paradigm changes. We already accept that:

- Digital computers are generally usable machines
- Software is reusable
- An open society needs open systems
- The network is a common resource
- We live in a global village
- Access to information is free
- Data is value and currency
- Production costs are no longer the driver
- Sharing is better than owning
- The default status of users is mobile
- Services are delivered from a virtual space to a global space

In extension to those paradigm shifts, there is an open number of paradigms or theses that are still discussed today and will shape the future economy and society and the way we use technology and move our daily life into cloud and network structures. Three of those theses are:

- Peer-to-peer can replace centralized services
- Machines (can be) intelligent

M. Oppitz, P. Tomsu, *Inventing the Cloud Century*,
DOI 10.1007/978-3-319-61161-7_20

- New computing technologies will guarantee the continuation of Moore's law after the silicon age is over

Being the candidates for disruption in economy, technology and society, we will drill down into these three themes in more detail.

Decentralize: Peer-to-Peer

If the blockchain has not shocked you yet, I guarantee it will shake you soon
William Mougayar, 2016 in "The Business Blockchain"

On October 31st, 2008, an unknown programmer, who used the pseudonym Satoshi Nakamoto, published a post introducing "*...a new electronic cash system that's fully peer-to-peer, with no trusted third party*" on the "The Cryptography Mailinglist."[1] In this post[2] "Satoshi Nakamoto" describes the new approach:

- *Double-spending is prevented with a peer-to-peer network.*
- *No mint or other trusted, third parties.*
- *Participants can be anonymous.*
- *New coins are made from a Hashcash style proof-of-work system.*
- *The proof-of-work for new coin generation also powers the network and prevents double-spending.*

A complete description was published on the same day under the title "Bitcoin: A Peer-to-Peer Electronic Cash System"[3] baptizing the new technology "Blockchain" and the new application "Bitcoin." It was nothing less than the birth of a completely new type of currency—later called "Cyber Currency" or "Crypto Currency" because of its focus on encrypted peer-to-peer exchanges of currency between individuals. Cryptocurrencies are completely different from currencies like the Dollar, Euro or Yen, which are controlled by governments and national or global financial organizations. There is, however, an exchange rate between a cryptocurrency and national currencies. The idea of creating a peer-to-peer transaction network using existing internet infrastructure has its origins in the open society movement that began to discuss privacy of communication and anonymity on the net as principles and basic rights for users. Starting as mailing lists like the Cypherpunk-mailing[4] list in the 1990s, different groups of activists experimented with technologies enabling communication without using centralized services, organizations and bypassing federal control. Projects like Phil

[1]The Cryptography Mailing List: http://www.metzdowd.com/mailman/listinfo/cryptography

[2]Satoshi Nakamoto: "Bitcoin P2P e-cash paper", http://www.mail-archive.com/cryptography@metzdowd.com/msg09959.html, retrieved 2016-11-16.

[3]Satoshi Nakamoto: "Bitcoin: A Peer-to-Peer Electronic Cash System" http://www.bitcoin.org/bitcoin.pdf, retrieved 2016-11-16.

[4]The Cypherpunk Mailing List: https://www.cypherpunks.to/list/, retrieved 2016-11-16.

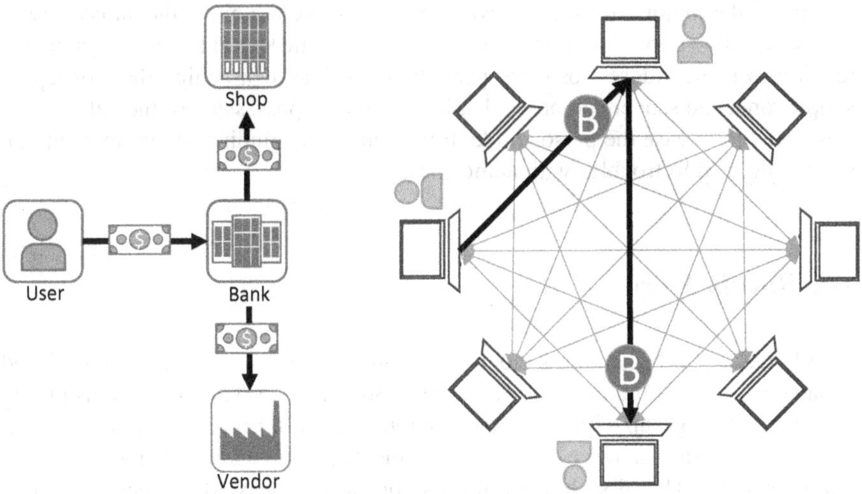

Fig. 1 Centralized money transfer versus distributed ledger money transfer

Zimmermann's PGP (Pretty Good Privacy) and TOR (The Onion Router) have their origins in the Cypherpunk movement (see Fig. 1).

Connected to the concept of "Bitcoin" is the blockchain technology that allows for the secure peer-to-peer exchange of cryptocurrency values between individuals without using centralized banks or third party money accounts. The theory or philosophy of bitcoins is simple: why use a centralized financial corporation to transfer money? This can be done much quicker and with fewer costs in a direct way between the sender and the receiver. Bitcoin was introduced as a new currency, easy to use and based on secure peer-to-peer transactions.

Creating a new currency also had to manage the volume of money in circulation. Satoshi Nakamoto found a clear solution for that: at the start of bitcoin the volume was zero. New bitcoins can be "mined" or "minted" by bitcoin-users or "nodes" by taking part in the approval process for new transactions as so-called miners. The miner who successfully approves a new transaction adds a new block to the bitcoin blockchain and receives not only a transaction fee but also a mining reward payed in the form of newly minted bitcoins. At the beginning of bitcoin in 2009, this bounty was 50 bitcoins, and as an additional rule the reward is halved every 210,000 blocks. This already happened two times: in November 2011 and in July 2016, so the current award for mining a new block is 12.5 bitcoins, which is equal to approximately $8800 (November 2016) making bitcoin mining an interesting business field.[5] As an additional limitation, the maximum volume of bitcoins in circulation is capped at 21 million. Currently (November 2016), the bitcoin volume in circulation is around 16 million, and according to estimations the last bitcoin will be minted around the year 2140.

[5]Bitcoin Controlled Supply: https://en.bitcoin.it/wiki/Controlled_supply, retrieved 2016-11-16.

One of the major attributes of cybercurrencies like bitcoin is the anonymity of the users. Value transactions are signed using the public key of the user without any relation to name, address or other personal data. The idea behind that concept is simple: once the sender can prove that he or she is in possession of the value, there is no need to release the personal identity. From the beginning, it was evident that this would lead to troubles with authorities.

The Bitcoin Story

Based on Sakoshi Nakamoto's blockchain implementation, bitcoin was introduced in January 2009 as the first cryptocurrency. Soon after Nakamoto's publishing in October 2008, a group of programmers (among them Hal Finney, Nick Szabo, Wei Dei and Gavin Andresen) downloaded the software and started to work on improvements. The first bitcoin transaction was processed in January 2009: Nakamoto had created the first 50 bitcoins and transferred them to Hal Finney. After its start in 2009, bitcoin went through many ups and downs. It soon turned out that cryptocurrencies like bitcoin also needed interfaces to the world of traditional currencies to provide exchange platforms between bitcoin and those currencies. One of the first exchange platforms was introduced in Japan under the name Mt.Cox in 2010 followed by other exchange platforms in the US and later in Europe. In the beginning, Mt.Cox was extremely successful in terms of exchange volume but soon went through severe technical and security issues leading to its insolvency and shutdown in February 2014. In parallel it soon turned out that cryptocurrencies like bitcoin are the perfect tool for illegal trade and financing of criminal and terrorist activities.

The availability of a private and anonymous currency soon led to the introduction of black markets for drugs, like "Silk Road" launched in February 2011 on the TOR darknet. All these activities drew the attention of federal organizations to bitcoin and especially to the bitcoin exchange platforms. Some of the early supporters of bitcoin faced severe accusations including money laundering and supporting drug trade. Between 2012 and 2015, it seemed on several occasions that the blockchain and bitcoin idea was dead and would never evolve to become a stable alternative to traditional currencies. The exchange rate diagram of bitcoin to USD is a perfect track record of its roller coaster-like history (see Fig. 2).

In September 2012, Gavin Andresen and others founded the Bitcoin Foundation[6] as a stable platform for the blockchain and bitcoin community to support the idea and organize the further development of the blockchain software. It soon appeared that the weak points of bitcoin were not the blockchain architecture but the integration of traditional currencies and the forming of a proper legal status.

---

[6]Bitcoin Foundation: http://bitcoinfoundation.org/, retrieved 2016-11-16.

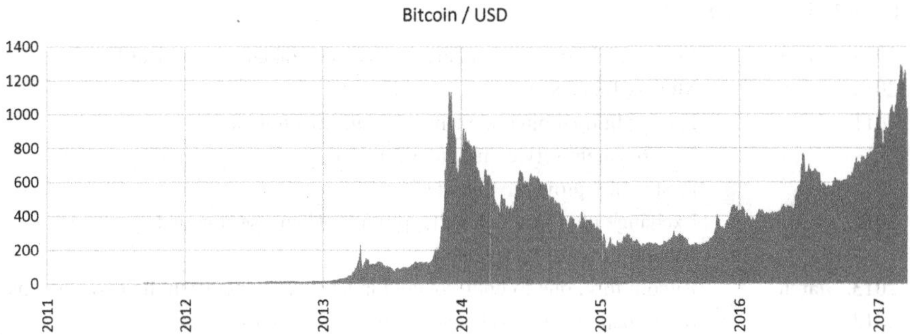

Fig. 2 Bitcoin price

After the collapse of the early exchange services like Mt.Cox, a growing number of bitcoin exchange platforms entered the market. Today there are more than 15 exchange platforms worldwide. The leading platforms, in volume of bitcoins and transactions, are itBit (USA), Bitstamp (Slovenia and Luxembourg), BTC-e (Russia), OKCoin (China). In parallel payment service providers started to provide exchanges of bitcoin payments into traditional currencies thus offering a service for businesses like retailers, restaurants or hotels that wanted to accept bitcoins as payment, but needed a way to transfer the money to their bank accounts. By 2011, there were currently half a dozen payment providers in the US and Europe offering those services.

The anonymity of the bitcoin users led to another challenge. Owning bitcoins simply means that there are transactions in the bitcoin blockchain signed with the key of the user. To spend this money or transfer it to another user, the owner must have the private transaction key to prove that he or she is the owner of the coins. Losing the key means losing the money. So, users end up storing their private keys on paper or on a storage device locked in a safe-deposit box. Beginning in 2011, wallet providers offered so-called online wallets to store the keys of the user in the cloud, which is, of course, based on the complete trust in the wallet provider. To complete the service infrastructure, ATMs (Automated Teller Machines) were introduced to offer exchange services in public places like shops, train stations or airports. Currently the number of bitcoin ATMs worldwide is over 800, two-thirds of them are in North America and approximately 25% in Europe.

As a public cybercurrency, the bitcoin transactions are public although the users are anonymous. Several platforms provide bitcoin exploration services and allow for the tracking of transactions on bitcoin blockchains.[7]

Today, the total number of bitcoins in circulation is 16 million, with a total market cap of $10 billion. In February 2015, the number of merchants accepting bitcoins as currency for services or goods exceeded 100,000 including PayPal, Dell, Microsoft and Time Inc.

[7]https://bitcoinchain.com/, https://www.biteasy.com/, http://blockr.io/, https://bchain.info/BTC/

Table 1 Bitcoin timetable

2009, January	First Bitcoin transaction between Satoshi Nakamoto and Hal Finney
2010	Mt.Cox founded
2011	Five additional bitcoin exchange platforms founded
	First payment service provider founded
	First wallet providers founded
2012	Exchange platform and wallet provider Coinbase founded
	Bitcoin Foundation founded
2013, March	Bitcoin fork, due to undiscovered inconsistency between the two versions
2013	More than 10 new exchange platforms founded
	Four new wallet providers founded
2014, February	Mt.Cox shutdown, price is falling from $1000 to a level around $500
2014, March	Chinese government discusses legal status of bitcoin, price is falling
2015	Price is recovering reaching a maximum of $550
2016	Price reaching a maximum $750
2016, December	Price is peaking at $978
2017, March	All time high of $1280

Even today there is still doubt about the real identity of Satoshi Nakamoto. A number of attempts including text and language analysis of Satoshi Nakamoto's posts and papers have been undertaken leading to a number of assumptions but no final proof of his or her real identity. Being the first miner of bitcoins in 2009, Nakamota owns around 1 million bitcoins today (see Table 1).

The Technology Behind Bitcoin: How Does Blockchain Work?

Blockchain is the technology behind cryptocurrencies like bitcoin and others and can be used not only for cryptocurrencies but for a much wider field of applications. To understand the advantages and disadvantages of blockchains we must drill down to the paradigm of peer-to-peer methodologies. The basic idea of blockchain is to use the Internet and Web for organizing safe and secure transactions between individuals but also to remove the third-party component.

Before blockchain, the centralized third party service (like a bank managing accounts using a central database and service) was responsible for guaranteeing the proper execution of a transaction. With blockchains, this role is taken over by the blockchain participants. In a centralized architecture, the bank holds the account's status in its database and collects and processes all transactions on the accounts in a centralized ledger. Blockchain is based on the idea that the ledger is held by all participants of the blockchain as a so-called distributed ledger. As well as in the centralized ledger architecture, a request for a new transaction must be created by the requesting user. To prove his or her identity, the user must sign the transactions with a private key (see Fig. 3).

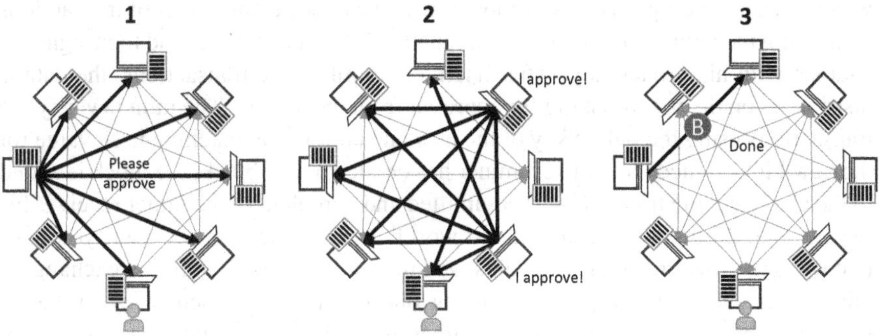

Fig. 3 Blockchain and distributed ledger

Instead of passing the request to a central authority, it is distributed to the blockchain network by asking for a commit or as it is called "asking for distributed consensus." Commitment of newly requested transactions is approved only if a new block is generated successfully, added to the blockchain and distributed to all participants. At that point, a specific group of blockchain participants take over: the so-called miners. Miners have the job of verifying the proposed transactions and creating or "mining" a new block in the chain containing those transactions. To create a new block successfully, miners must calculate cryptographic hash functions using the new transaction data, the hash value of the previous block and a random number: the "nonce". As additional rule, the newly generated hash value must have a certain form with a certain number of leading zeros. It may take several attempts for a miner to successfully create a new hash number and it also may require a lot of computing power.

Miners work in competition to each other. The motivation for a miner is the reward he or she receives for successfully mining. In a cryptocurrency environment, the miner receives newly minted cryptocurrency coins in his or her digital wallet. After the successful creation of the new block hash value, the block is added to the blockchain and distributed as a newly approved block entry to all ledgers. That means that at any point in time all ledgers consist of a seamless sequence of all blocks and transactions. Each block in the distributed ledger or blockchain contains the ID and hash value of the previous block thus securing the consistent sequence and integrity of the data.

Cryptocurrencies Replacing Banks

Cryptocurrencies like bitcoin were the first applications of blockchain. In a cryptocurrency community, each user has his or her own wallet holding the current amount of crypto-coins. The wallet itself is created or updated according to the ledger containing all crypto-coins received or payed by the user. Whenever a user wants to transfer crypto values to another user, he or she creates a set of transactions containing the value in number of coins and the ID of the receiver and then signs the data set with his private key. After having committed the transactions, the data is distributed on the net by asking for approval from the miners. The miners verify the transactions using the public key of the user and compute a matching hash value for the new block. Once the quickest miner has created a new block, the block is added to the chain and the transaction is committed and completed. The miner receives his reward, e.g. 25 bitcoins in the bitcoin world. As a cryptocurrency user, you have two ways to fill your wallet with coins: you may simply "buy" coins via exchanging other currencies with crypto-coins using one of many cryptocurrency exchange services. The other way is to start a career as a miner and collecting rewards for mining. Professional miners use mining IT infrastructure optimized for calculating hash values in a short time. In that case, the miner is simply acting.

The basic motivation behind cryptocurrencies was to offer a payment method that is independent from central banks and not regulated by federal laws. Soon this led to the discussion on the legal status of cryptocurrencies in comparison to traditional currencies. A currency like the dollar or euro is always backed by the national bank of the state either by insuring the currency with gold (gold standard) or simply accepting tax payments in the national currency. This is not the case with cryptocurrencies. Cryptocurrencies define their value through the acceptance of the digital coins as payment by providers of goods and services and—in parallel—through buyers of cybercurrency coins who use the currency as an investment vehicle. In the case of bitcoin, the value (as in exchange rate to currencies like USD or EUR) was determined in the beginning by traders buying bitcoins as investments. This was also motivated by estimations of the future target value of bitcoins reaching more than $1000.

Paying with cryptocurrencies offers many advantages. Cryptocurrency transactions are cheaper and quicker than sending money via banks or credit card companies. This is efficient when the transaction value is small like in micropayments for services. Paying for services like rooms, shared cars, crowd services or private energy trading could be organized much easier using blockchain and cryptocurrencies. The major question is if a small number of public cryptocurrencies will achieve global acceptance or if there will be a monopoly of a single cryptocurrency like bitcoin. Since 2015 the concept of distributed ledgers including strengths, weaknesses, threats and opportunities has been widely discussed (Table 2).

Bypassing banks as the centralized intermediary was the major motivation for the growing community of bitcoin users. The introduction of bitcoins in 2009 was partly a consequence of the financial crisis in 2008. Consumers had lost trust in centralized banks and the effectiveness of national rules and regulations. Bypassing banks and moving to a system which guarantees a higher level of trust and anonymity for financial transactions was a major motivation. From the current perspective, there are four groups of cybercurrency users aiming at different goals:

Table 2 Strength, weaknesses, opportunities and threats of cryptocurrencies

Strength	**Weaknesses**
Low cost of transactions Micropayments Fast execution Globally available	Technology is not yet mature Energy consumption for mining is huge Legally not covered Acceptance as payment method limited
Opportunities	**Threats**
Building a new ecosystem of public ledger based currencies Reduce process costs for financial transactions Open access to financial transactions to more people also in low developed countries Empower crowd sourcing collaboration models	Governments could block Powerful exponents of the old economy take over Misuse through criminal organizations Blockchain as job-killer

Cryptocurrencies as Investment

Investors are fascinated by the rapidly growing value of the bitcoin currency. Like every other currency investment, investment in bitcoins is risky but seems to offer huge opportunities. Having invested in a cybercurrency like bitcoin in January 2016 would have tripled your return within 1 year.

Omitting Risk with Weak Fiat Currencies

Consumers in countries with weak local fiat currencies or restrictive regulations embrace the option to save their money using a more stable currency and bypass national rules restricting the exchange of the local currency in currency markets like USD or Euro. In January 2017, the bitcoin popularity reached a new peak in Mexico due to the announcements by the Trump administration to restrict USD transfers between the USA and Mexico.[8]

Simplify Cross-Border Payments

Offering cross-border services in a crowd sourcing ecosystem is sometimes hindered by the lack of easy-to-use bank connections. Individuals or small firms in less developed countries that provide services like translation, web design, presentation design or manufacturing are handicapped by the complexity of money transfers between their home location and their customers. Using cybercurrencies enables those businesses and create new business opportunities in a global service ecosystem.

Criminal Payments

Cybercurrencies like bitcoin follow the principle of anonymity. The sender and the receiver of a transaction exchange their public keys, but no private data like name, address or mail-account. Thus, money transfers are completely anonymous, allowing payments without knowing the identity of the sender or receiver. This is

[8]Peter Chawaga, Bitcoin Magazine, 2017: http://www.nasdaq.com/article/has-trump-made-mexico-the-next-hotspot-for-bitcoin-cm750695 retrieved 2017-03-10.

the perfect setting for those who want to stay anonymous. Anonymity is the basic assumption for criminal activities like ransom or dealing with weapons or drugs.

Legal Status of Cryptocurrencies

In the United States and many other countries, the discussions about the legal status of cybercurrencies started around 2012. In November 2016, the situation is still unclear and differs from country to country—far from perfect for a global approach to a new international money transaction framework.[9] In the United States, New York is the only state with a law on bitcoin use. It was released in August 2015 and commonly referred to as a Bit License, which is not accepted by several bitcoin businesses who announced stopping business in New York. In the European Union, the Court of Justice ruled that "*The exchange of traditional currencies for units of the 'bitcoin' virtual currency is exempt from VAT*" and that "*Member States must exempt, inter alia, transactions relating to 'currency, bank notes and coins used as legal tender.*'"[10] In October 2016, the European Commission followed by the European Central Bank proposed extending the rules and regulations for money laundering to cybercurrency exchanges, which would simply require the checking and tracking of the identity of all transactions of cybercurrency. In China, banks and payment systems are prohibited from dealing in bitcoins, but individuals are free to trade them. These examples exactly show the area of conflict between the interests of centralized governmental organizations and the supporters of free and anonymous transaction processing: you simply cannot have both.

Beyond Bitcoins

After bitcoin was released in 2009, other cryptocurrency projects were started based on blockchain technology. In 2016, there are more than 600 cryptocurrencies in circulation, only four of them having a market cap of more than $100 million[11] (see Table 3).

The total market capitalization of all cryptocurrencies is slightly above $12 billion, thus giving bitcoin a market share of over 80%. The monthly trade volume of the leading cryptocurrency bitcoin is around $1.6 billion. Compared to national "fiat" currencies, the share of cryptocurrencies is still very small. The worldwide volume of money in circulation (coins and banknotes) is estimated at $3 trillion, thus the share of cryptocurrencies is around 0.2% (see Table 4).

Cryptocurrencies have become a hot topic since the introduction of bitcoin in 2009, but have also created a lot of open questions regarding security, involvement of tax or legal authorities and their relation to the "real" world represented by the global financial industry and market. At least the last question seems to be up for

[9]Szczepański, Marcin (November 2014). "*Bitcoin: Market, economics and regulation*" (PDF). European Parliamentary Research Service. Annex B: Bitcoin regulation or plans therefor in selected countries. Members' Research Service. p. 9. Retrieved 18 February 2015.

[10]Court of Justice of the European Union: "*The exchange of traditional currencies for units of the 'bitcoin' virtual currency is exempt from VAT*", http://curia.europa.eu/jcms/upload/docs/application/pdf/2015-10/cp150128en.pdf

[11]Cryptocurrencies Market Capitalization: https://coinmarketcap.com, retrieved 2016-11-16.

Table 3 Top ten cryptocurrencies by market cap in October 2016

#	Name	Volume ($)	Price ($)	Available supply
1	Bitcoin	10,478,558,933	657.50	15,936,996 BTC
2	Ethereum	1,020,051,977	11.96	85,259,399 ETH
3	Ripple	328,188,522	0.009248	35,488,165,563 XRP
4	Litecoin	188,349,150	3.91	48,110,354 LTC
5	Ethereum Classic	89,374,899	1.05	85,160,315 ETC
6	Monero	86,326,407	6.55	13,182,885 XMR
7	Dash	69,693,758	10.21	6,829,307 DASH
8	Steem	39,470,210	0.213650	184,742,380 STEEM
9	NEM	33,416,190	0.003713	8,999,999,999 XEM
10	WAVES	27,618,300	0.276183	100,000,000 WAVES

Table 4 Volume of currencies in circulation (coins and banknotes)

Name	Volume ($ trillion)
Bitcoin	0.010
All cryptocurrencies	0.012
USD	1.500
Euro	1.200
All fiat currencies	5.000

discussion and the attempts to get involved in that new type of business are made by financial service organizations around the world. The motivation is easy to understand: transfer of values between individuals in different countries using blockchain technology and a cryptocurrency like bitcoin can be executed within 10 min. Comparing that to the third-party services provided by banks using technologies like SWIFT and based on complex implementations, rules and regulations caused by international or bilateral agreements lead to a process and transfer time of days, higher costs and sometimes much effort for the sender or receiver. All of these considerations indicate a clear advantage for cryptocurrencies and peer-to-peer blockchain technology.

Back to Blockchain as a Basic Technology

Blockchain is a technology that was introduced with cryptocurrencies and bitcoin in 2008, but has a much more wider scope than cryptocurrencies alone. Blockchain as a paradigm follows the idea of moving away from centralized services to distributed ledgers in all situations where a contractual agreement should be closed and approved between individuals or organizations. Money transfer is one example; the same method can be applied to every type of value transfer or any type of agreement.

Typical patterns and candidates include the trading of stocks, company shares, real estate, cars, yachts or diamonds. In all those cases, the traders are still dependent on third parties as brokers. Using distributed ledgers could completely change the game. The same pattern could be used for services like car sharing, electric power and house or apartment sharing.

Private or Public Blockchains?

Bitcoin was introduced as a public blockchain application for digital money transactions. Its open business model is based on individual or corporate consumers using bitcoin for money transfer or as an investment vehicle. The operation is executed by miners, using bitcoin to generate revenue.

On the other hand, the distributed ledger method can also be used for closed communities to organize their internal value transfer or process integration. In that case, the blockchain is implemented as a private blockchain, and usage and membership is restricted to community members. Within the last 2 years, several projects and initiatives had been started in order to use distributed ledger and blockchains beyond the public cryptocurrency approach. The motivation behind that approach is process optimization and cost reduction in global cross-linked ecosystems.

2015 Ethereum and Smart Contracts

Ethereum was founded as a project in 2015 by Vitaly Burin and introduced the "Ether" as cryptocurrency. Ethereum is the second largest platform for blockchain technology and alternative cryptocurrencies today. More importantly, Ethereum is more than a cryptocurrency blockchain: it provides the blockchain technology as a platform on top of a public ledger and also focuses on smart contracts as the methodology to create and execute distributed programs. This enables users of the Ethereum blockchain service to implement their own applications based on distributed ledgers. The payment is still cryptocurrency based: executing Ethereum transactions or smart contracts is done using Ether, the Ethereum cryptocurrency.

2015 IBM, Cisco and Others Join in the Hyperledger Project

The IT industry has already recognized the huge potential behind distributed ledgers and it seems that they have also recognized that a parallel world of peer-to-peer networks could became a major threat for centralized services. In December 2015, the Linux Foundation announced the "Hyperledger Project"[12] as a platform for blockchain technologies for the industry. Hyperledger is an open source collaborative effort created to advance cross-industry blockchain technologies. Compared to bitcoin or Ethereum, the Hyperledger project is organized as an "open governance" project. Among the more than 100 members are CISCO, IBM, Fujitsu, Intel, VMWare and others.[13]

[12]Hyperledger project: https://www.hyperledger.org/, retrieved 2016-11-16.

[13]Hyperledger members: https://www.hyperledger.org/about/members retrieved 2017-03-20.

The Hyperledger project focus on developing blockchain technologies including smart contracts rather than introducing a new cryptocurrency.[14] Hyperledger is "*...an open source collaborative effort created to advance cross-industry blockchain technologies. It is a global collaboration, hosted by The Linux Foundation, including leaders in finance, banking, Internet of Things, supply chains, manufacturing and Technology.*"[15] Members of Hyperledger contribute to the community by creating proposals and providing resources and software to several Hyperledger initiatives and projects. The result is open source software provided under open source license agreements and can be downloaded from github.[16]

2016 Microsoft, Google and Amazon
By 2016, Microsoft had partnered with Ethereum and started a new project called "Bletchely"[17] using the Azure cloud as its foundation. In March 2016, AWS (Amazon Web Services) announced the offer of a blockchain "laboratory" environment for enterprises to test their ideas. Google supports the blockchain startup Ripple, which has raised up to $93 million by 2016.

First Applications

Since 2015, the blockchain wave has gained speed and the first real-life projects have been started in a variety of industry segments using blockchain as the distributed and private or community ledger between a closed group of organizations or corporations. From a current point of view, there are several industries that could become the first movers and shakers in this field. At the same time, many public blockchain projects have already started to evolve, exploring blockchain and distributed ledger technology for the transfer of different types of values like energy, luxury goods or medical data. Following the ongoing discussion and tracking activities, it seems that the top three areas of application are:

- Financial transactions
- Supply chain processes
- Energy trading

Financial Transaction
The financial sector is a major candidate for this type of approach. In September 2015, the R3 consortium consisting of nine financial companies including Barclays,

[14]Brian Behlendorf, executive director of the Hyperledger Project, October 2016: http://www.eweek.com/cloud/hyperledger-blockchain-project-is-not-about-bitcoin retrieved 2017-01-10.

[15]Hyperledger: https://www.hyperledger.org/about retrieved 2017-02-20.

[16]Hyperledger on github: https://github.com/hyperledger retrieved 2017-01-20.

[17]Microsoft Azure: „Project Bletchley – Blockchain infrastructure made easy" https://azure.microsoft.com/en-us/blog/project-bletchley-blockchain-infrastructure-made-easy/, retrieved 2016-11-16.

Credit Suisse, Goldman Sachs, J.P. Morgan, Royal Bank of Scotland, and UBS started to work on future applications of blockchain technology for the global financial industry. UBS also started to work on a prototype virtual currency to be used by banks and financial institutions as a way to handle mainstream financial market transactions. By October 2016, the R3 initiative included 22 of the world's largest banks. Visa has announced a new payment platform based on blockchain technology, ICICI Bank has started cooperation with Emirates NBD to execute cross border blockchain transactions, and plans to build a network consortium for further participation by interested member banks. A group of major banks including UBS, Deutsche Bank, Santander and BNY Mellon have announced a cooperation to develop a new form of digital cash that will help to set an industry standard to clear and settle financial trades over a distributed ledger. The London based company Everledger[18] is using blockchains for diamond certification and transaction history.

In October 2016, the five largest European insurance companies (Aegon, Allianz, Munich Re, Swiss Re and Zurich) launched the Blockchain Insurance Industry Initiative B3i "*...aiming to explore the potential of distributed ledger technologies to better serve clients through faster, more convenient and secure services.*"[19]

The World Economic Forum estimates that over $400 million will be invested by banks in digital technologies such as blockchain by 2019. It seems that the pendulum that was originally pulled by Sakoshi Nakamoto in the direction of peer-to-peer without third parties was caught by those third parties who have started to pull in their direction by using blockchain as an efficiency tool.

Supply Chain

The rapidly growing world trade ecosystem leads to an extension of supply chains between different types of producers, logistic partners, distributers and retailers and involves a growing number of administrational entities like customs authorities, insurances or quality assurance organizations. Organizing the digital data flow between those entities and synchronizing it with the material flow of goods or services in a more efficient way to reduce costs, shorten delivery times and provide a more effective quality tracking.

Using blockchain as the basic technology for tracking data and goods is recognized as a promising option. The Hyperledger partner IBM claims to work with more than 600 customers on blockchain projects, many of them are looking at blockchain as productivity tools for their supply chains like the retail company Walmart[20] or container shipping company Maersk.[21]

---

[18]Everledger: http://www.everledger.io/, retrieved 2016-11-16.

[19]https://www.allianz.com/en/press/news/commitment/sponsorship/161018-insurers-and-reinsurers-launch-blockchain-initiative-b3i/, retrieved 2016-1-16.

[20]IBM blockchain and Walmart: https://www.ibm.com/blogs/blockchain/2016/11/leveraging-blockchain-improve-food-supply-chain-traceability/ retrieved 2017-03-10.

[21]IBM blockchain and Maersk: https://www-03.ibm.com/press/us/en/pressrelease/51712.wss retrieved 2017-03-20.

Energy Trading and Distribution

The energy trading and distribution ecosystem is undergoing major transitions triggered by the growing share of renewable energy and the liberation of markets. The price level of electric energy is low, distributors and network providers are challenged by changing consumer behavior which is triggered by the availability of alternative energy production using photo voltaic technologies and changing consumption and storage patterns using electric vehicles and batteries.

Thus, the ecosystem of energy producers, providers and distributors is going through a major transition. Trading with electrical energy in the macro-grid as well as in so-called microgrids formed by private producers and consumers could reduce the pressure on the electrical grid by balancing production, storage and consumption of energy. Blockchain is a key technology for organizing this new ecosystem. Several initiatives have been started to provide blockchain based energy trading on the level of micro-grids and macro-grids.

The Brooklyn Micro Grid[22] uses blockchain to administer electric energy transactions between consumers and providers within a community in Brooklyn, NY. The Australian company, Power Ledger,[23] is currently applying its blockchain-based software to open peer-to-peer energy trading behind the meter and across the network. At the same time, energy producers and macro-grid operators are looking at options to improve their trading processes and reduce costs by following the changing production and demand patterns.

Distributed Ledger as Disruptive Business Model

The possible applications of distributed ledgers and blockchains are nearly endless. Peer-to-peer computing and distributed ledgers could be one of the major future trends, not only shaping the way we design IT systems but also influencing the role and importance of central organizations like banks, insurance companies, federal administrations or trading platforms for goods and services.

In mid-2016, the total venture investment in blockchain startups crossed the $1 billion mark and is projected to reach more than $1500 million by the end of 2016. Blockchain seems to be a candidate for the next big innovation (see Fig. 4).

In December 2015, Deloitte published a report on the disruptive nature of blockchains for different industries[24] focusing not only on financial services but also on technology, media, telecommunications, consumers, industrial products, life science, health care, the public sector and energy. Within each of these

[22]Brookly Micro Grid: http://brooklynmicrogrid.com/, retrieved 2016-11-16.

[23]Power Ledger: http://powerledger.io/, retrieved 2016-11-16.

[24]David Schatsky, Craig Muraskin (Deloitte): „Beyond bitcoin / Blockchain is coming to disrupt your industry", December 2015, https://dupress.deloitte.com/dup-us-en/focus/signals-for-strategists/trends-blockchain-bitcoin-security-transparency.html, retrieved 2016-11-16.

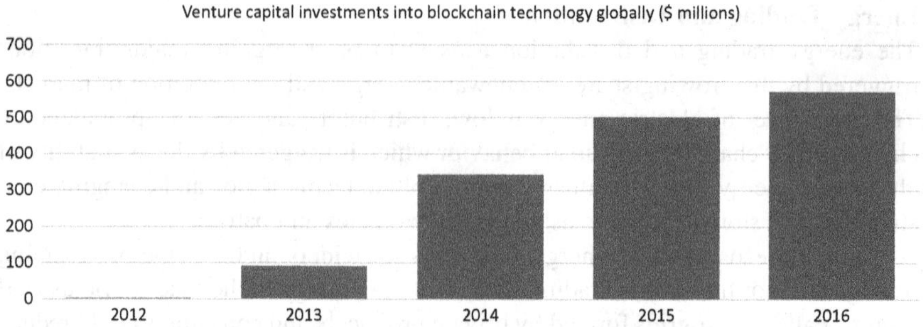

Fig. 4 VC investments in blockchain

segments, blockchain can be applied to reduce costs, simplify processes and gain efficiency. In a nutshell, blockchain can be applied to every type of process where values or data is transferred between people or organizations. More than that, blockchain could be a major trigger for disruptive business models in many industries.

Cognitive Computing and Machine Intelligence

Every aspect of learning or any other feature of intelligence can in principle be so precisely described that a machine can be made to simulate it.
John McCarthy, 1956

One of the most promising new areas in computing that we are entering is cognitive computing—often called artificial intelligence, which offers fundamental differences in how systems are built and how they interact with humans. Cognitive computing systems can build on existing knowledge and learn new things, understand natural language and reason and interact more naturally with human beings than traditional systems. This opens totally new and so far, not thought of possibilities of using computers in everyday life.

We are already surrounded by smart services delivered from different types of clouds. This is an example of a quite normal day in the life of a "digital" person, let's call him Dave (as a reference to Arthur C. Clark's Space Odyssey):

"On a fine and sunny day Dave enters his car to drive to work in the morning. Immediately his navigation agent informs him that the usual road is blocked and proposes a detour which will bring him to his target only 5 min later. The agent is not only a navigation system receiving traffic information, but also relates to other driver's agents by exchanging real time data about their progress and optimizes the suggestions to clients. The target is to balance the traffic load between all possible bypass routes to prevent traffic jams. Instead of getting stuck on the detour route, the service offers another option to reach the office in time. On the way to the office, Dave receives a message from his bank informing him that there were some weird

transactions on his credit card during the night and made in another country. The bank concluded that this was not Dave, rejected the transactions and is now asking for confirmation. The system behind that process is an automatic fraud detection system that uses cognitive software and tracks all money transactions to detect misuse. Dave is happy that his money is safe and confirms. Arriving at the office, Dave starts his PC and his favorite search engine and is informed that the designer office chair he has been looking at for weeks is now being offered at a special price. This information was triggered by the search engine's knowledge base containing all his search activities on the Internet, matching the words used in offers from product or service providers and combines that data with typical patterns of behavior from other users. Dave is happy, orders the chair and immediately receives confirmation of delivery within 1 week. At home again, Dave learns that his smart power agent has decided to sell the electric power stored in the wall battery to the market at a good price this morning and has already loaded the battery again using the solar panel on the roof of Dave's home. In the morning, Dave's smart car had automatically switched on the heating at 7 a.m. knowing that Dave will use it at 7:30."

Well, sounds great, doesn't it? The good thing is all those cloud-powered services are already there. Smart navigation agents like the one described are on the market, banks and credit card companies use automatic fraud detection systems, and search engines and social networks make up to 90% of their revenue through focused advertising. Finally, the smart power management is on its way to the market. Like every new technology, these services have their dark side and risks as well.

Dave's story could be told another way: The day before, a security breach caused the loss of all user data of the navigation service including bank account numbers and mail addresses. The mail was not sent from the bank, but a phishing mail sent from a bot-net and Dave was so careless to confirm the message with his password, that his money has been withdrawn within minutes. The great offer for the stylish office chair came from a company that spends its money on advertising and not on producing chairs, and the chair will never be delivered. At night, Dave will find out that somebody hacked into his smart meter and interrupted the power supply, and now his lighting and heating are gone. Visiting his garage in the morning, he learns that his car is gone, the keyless entry systems was hacked by well-equipped thieves using a sophisticated software that attacks a zero-day exploit in the car manufacturer's system. Not such a good day!

It is easy to say that every new technology may have its risks and sometimes dangerous impacts, but the threats we perceive may be more important if the new technologies move in a direction where the distinction between human behavior and machine functionality becomes difficult. Smart services like the real examples described above seem to have a certain level of intelligence without being human. Are we moving towards an intelligent behavior of cloud services and where is the boundary between machine intelligence and human intelligence?

Today we are confronted with more sophisticated technologies in interactions with machines, software and services enabled by providers of smart devices, cloud services and global big data sources. The question of how far all these things are from intelligence is more valid than 70 years ago. More than that, the question is not only if there is something like machine intelligence but whether there could be something like a super-intelligence someday that overrules human intelligence and thus takes over our lives. To get a better understanding of what is going on today and where the journey may lead us, we must drill down into intelligence and what is called machine intelligence or artificial intelligence today.

Looking Back to Cybernetics and Artificial Intelligence

Since machines have become a more important part of daily life, the question of intelligent machines taking over tasks from humans has occupied our imagination. Machine intelligence is a paradigm that can be traced back to the first mechanical devices of the eighteenth century.

One of the most popular examples is the chess playing Turk introduced by Baron Wolfgang von Kempelen in Vienna around 1780 and later all over Europe. Although the machine was a deception, operated by a dwarf human chess player hidden in the machines case, many people—including the Austrian empress Maria Theresia—were convinced of the machine's ability to play chess on the same level as a chess grandmaster.

With the rapid development of mechanical and later electric and electronic devices that adopted mental tasks like counting and calculation from human workers, the idea of an intelligent machine propelled the imagination of engineers, scientists, artists and the public audiences. The first electronic computers were called "electronic brains," science fiction authors like Isaac Asimov, Stanislav Lem, Arthur C. Clark and Douglas Adams described worlds where intelligent computers and robots worked side-by-side with humans and started to discuss the social and ethical impact of such ecosystems.

1948 Cybernetics

In the 1940s, machine intelligence was connected to the work of Norbert Wiener and his theory of "Kybernetik" or "Cybernetics." Wiener worked on the automatic aiming and firing of anti-aircraft guns during World War II. This caused him to create a theory on information and control of systems based on the concept of feedback-loops. Wiener defined cybernetics in 1948 as "the scientific study of control and communication in the animal and the machine."[25] The word "Kybernetik" ("Cybernetics" in English) is taken from the ancient Greek "kybernetes" which was the designation for the helmsman of a ship. Wiener recognized that complex systems in nature or technology are controlled via feedback-loops. A feed-back loop receives information, processes the information

[25]Norbert Wiener: "Cybernetics or Control and Communication in the Animal and the Machine", 1961.

and feeds back new output to control the system or interact with its environment. There are numerous examples of feedback loops: the helmsman of a boat or the driver of a car is acting along a feedback-loop when steering his boat or car, safety valves in steam engines are feedback-loops and even a human conversation can be seen as two people using feedback-loops to interact. Today we find the concept of feed-back loops in many technical systems like navigation systems, manufacturing control systems or communication systems. From a more distant point of view, feedback-loops are a basic element of every complex system. The science of "Kybernetik" has since lost its fascination and was replaced by more extensive views on the mystery of intelligent machines. The word "cybernetic" later became a famous and frequently used term for visionary concepts of new machine influenced worlds and imaginations like cyberspace, cyborg or cyber physical system.

1950 Turing's Test

One of the major questions discussed was how to distinguish an intelligent machine from a human being. Around 1950, Alan Turing tried to find an answer to that question by introducing a scenario, which later became known as the "Turing Test." In this thought experiment a human player, the "interrogator," is confronted with two other players, A and B, both hidden behind a curtain or wall, and one of them is a computer, the other one a human. The interrogator can communicate to both of them using a typewriter and receive answers on a printer or screen. If the interrogator is not able to tell the machine from the human, the machine has passed the test. For a time, this scenario became the main definition of machine intelligence.

In the mid-1960s, MIT professor Joseph Weizenbaum tried to show that the Turing test is too narrow avdefinition of intelligent behavior. He wrote a computer program named ELIZA after the character in George Bernhard Shaw's play "Pygmalion." ELIZA could simulate the conversation with a psychotherapist following the strategy of a person-centered therapy, where the therapist tries to answer statements by the client with questions referring to key phrases and words. The story goes that Weizenbaum's assistant claimed that everybody must leave the room when she was working with ELIZA because of the intimate and personal character of the conversation. Weizenbaum wanted to show that a comparably simple computer program like Eliza could pass the Turing test although was far from intelligent.

1956 Artificial Intelligence

Back in 1956 a young assistant professor of mathematics at Dartmouth College, John McCarthy, organized a conference about the future aspects of computers. In the conference invitation McCarthy proposed that *"every aspect of learning or any other feature of intelligence can in principle be so precisely described that a machine can be made to simulate it."*[26] Some of the attendees of that conference like John McCarthy, Marvin Minsky, Allen Newell, Arthur Samuel and Herbert

[26]http://www-formal.stanford.edu/jmc/history/dartmouth/dartmouth.html, retrieved 2016-11-14.

Simon became the leaders of the new discipline, which was named "Artificial Intelligence."

The young community of AI (Artificial Intelligence) researchers around Minsky started to develop a large number of activities in the directions of semantic networks, speech recognition, text understanding, reasoning systems and learning theories. The major obstacle was that the technology was far behind the theory. Computers were simply too slow, networks not yet available and image processing devices clumsy and expensive. Artificial intelligence remained a theoretical field until the 1970s.

1980 Rise and Fall of Artificial Intelligence

With the availability of faster and less expensive equipment, the community started a new approach in the 1980s with the intention that new programming languages like LISP and PROLOG were developed to break the border of "old fashioned" procedural programming. Several software companies started to offer application frameworks for artificial intelligence. The goal was to introduce so-called "expert systems" as applications which contained the knowledge of experts and could be used to multiply, distribute and preserve their know-how. It soon turned out that the available hardware and the options to collect and organize knowledge were still far from efficient. Around 1990, the interest in AI seemed to disappear.

1996 IBM Takes the Challenge

It was IBM that started a new initiative and interestingly enough returned to Wolfgang Kempelen's idea of a chess playing machine, this time without a dwarf. The first project, named "Deep Thought," was started at Carnegie Mellon University and then continued by IBM. The Carnegie Mellon "Deep Thought" team of Feng-Hsiung Hsu, Thomas Anantharaman, and Murray Campbell were hired by IBM and started the "Deep Blue" project: a chess-playing super computer that could beat chess legend Garry Kasparov in 1996 in one game and in 1997 by winning a six game match against Kasparov.

Instead of a human dwarf, the IBM team used a parallel RS 6000 architecture running under AIX and powering a C-program to analyze many possible options before taking the next move on the board. With its 11.38 GFLOPS, Deep Blue was among the top 25 supercomputers of its time. Besides the fact that the Deep-Blue project was a tremendous and extremely popular engineering achievement, it was a "brute-force" approach and far from machine intelligence. Intelligence is more than being able to evaluate 200 million positions per second. In its aftermath, discussions and critiques were heard on the level of creativity in Deep Blues strategy. IBM also refused to disclose the machine's log of evaluations although it was later published on the Internet. The computer itself was dismantled, one of the racks is on display in the National Museum of American History.

2011 Jeopardy!

Five years later, IBM started a new approach with the IBM Watson research initiative. Again, the idea was to compete against human brain power in a field which was popular and to prove that computers can perform at the same level as humans. In 2011, the IBM Watson computer, named after IBM's first CEO Thomas J. Watson, was able to beat human competitors in "Jeopardy!", an American television game show.

IBM's Watson project was intended to show that machines can understand and analyze text, transfer the results in a base of deep and profound knowledge and finally use that stored knowledge for complex tasks like medical prognosis, financial analysis or crime detection. Selecting Jeopardy as the first test field was an excellent idea. Jeopardy is a game show where the candidates are confronted with so-called clues, that describe a piece of common knowledge within a certain category. The candidates must respond, but formulate their response in the form of a question. An example clue is: "*Embracing the future & new technology in 1962, Purdue established the 1st college dept. in the US for this 2-word discipline*," correct answer: "*What is Computer Science?*"

To build a machine that can compete in that type of knowledge processing, IBM had to solve several problems. First, the input and output data is in natural language, the machine had to understand text and convert it to a data structure that contains not only the blank words, but also their meanings and relations. In our example, the machine must analyze the text and understand that "1962" is the date of an event that happened in a place called "Purdue," also understanding that this is a university. "College department" and "discipline" would create the relation to the category of a scientific discipline and "new technology" would help to reduce the possible answers. Sounds easy for a well-educated human, but difficult for a machine. The other challenge was to build a knowledge base containing all relevant data to search for finding a match.

The IBM team, led by David Ferrucci, took those challenges and used nearly every type of software or platform that was available: "*The system we have built and are continuing to develop, called DeepQA, is a massively parallel probabilistic evidence-based architecture. For the Jeopardy Challenge, we use more than 100 different techniques for analyzing natural language, identifying sources, finding and generating hypotheses, finding and scoring evidence, and merging and ranking hypotheses. What is far more important than any particular technique we use is how we combine them in DeepQA such that overlapping approaches can bring their strengths to bear and contribute to improvements in accuracy, confidence, or speed.*"[27]

The IBM's team approach was to slice the complete task into a large number of single steps starting with a "linguistic preprocessor" to parse the structure of the natural language text finding nouns, verbs, pronouns, the grammatical structure of the sentence and also identifying possible classes of answers (e.g. Country, Epoch, Artist, etc.). This results in a tree-shaped data structure formulated in a PROLOG-code. This code is handed over to the candidate generation. Candidate generation uses different types of search engines like Lemur INDRI, Apache Lucene and

[27]David Ferrucci et al.: "Building Watson: An Overview of the DeepQA Project", AI Magazine Fall, 2010 http://www.aaai.org/Magazine/Watson/watson.php, retrieved 2016-11-14.

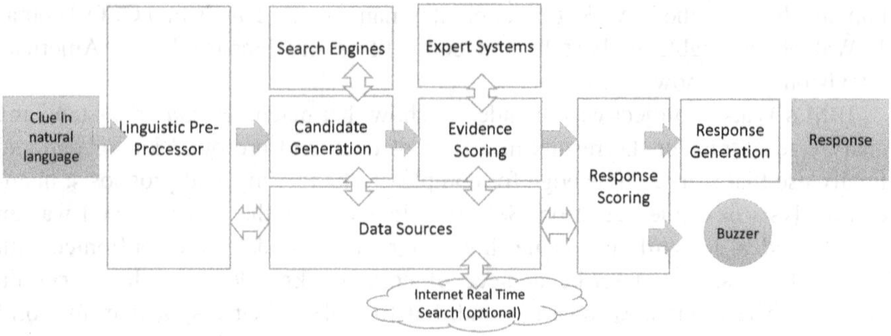

Fig. 5 Watson architecture for Jeopardy!

SPARQL to search for possible matches in many of data sources. In the Jeopardy configuration, Watson uses lexical databases to find synonymous words, text documents, and structured data bases. The data comes from data sources like DBpedia, Wordnet, Yago, Cyc, Freebase, Wikipedia, IMDb, World Book Encyclopedia, the Bible and other sorts of taxonomies and ontologies. The text data base also contained articles from Newswire and the New York Times. In the Jeopardy configuration, there was no real-time access to Internet sources like Wikipedia (see Fig. 5).

The result of this step was a list of 100–250 candidates as hypothesis for the correct answer. This list is moved to evidence scoring, which checks the evidence of each of those candidates using a large number of up to 10,000 parallel agents to try to find proofs for the correctness of the hypothesis by revisiting and searching the data sources again. Each agent comes back with a score from the response. Once a hypothesis with a high score is found, the buzzer is activated and the response is formulated. The whole process must be performed within seconds, in Jeopardy the fastest response wins.

DeepQA

DeepQA uses Apache UIMA (Unstructured Information Management Applications) as the framework implementation of the unstructured information. UIMA can be used to convert unstructured data to relational tables.

UIMA was developed by IBM and was later made available on the website of Apache Software Foundation. Apache Hadoop was used to organize the distributed processing of data. To achieve that high-performance level, IBM applied the top hardware products available at that time. Watson's DeepQA software was powered by a cluster of 90 IBM Power 750 servers, each of them using a 3.5 GHz Power7 eight core processors. In total the system had 2880 Power7 processors threads and 16 TB of RAM.

Watson

IBM's Watson was one of the first working cognitive based systems according to (John E. Kelly III, 2013), who identifies three basic types of capabilities for cognitive systems in the areas of engagement, decision and discovery (Bellisimo, 2015).

Engagement Systems fundamentally change the way humans and systems interact and significantly extend the capabilities of humans by taking advantage of people's ability to provide expert assistance and understanding. This is achieved by developing deep domain insights and presenting information in a timely, natural and useable way. Cognitive systems play the role of an assistant, by consuming vast amount of structured and unstructured information, and can reconcile ambiguous and even self-contradictory data plus they can learn.

Decision Systems offer decision-making capabilities, where decisions made by cognitive systems are evidence based and continually evolve based on new information, outcomes and actions. Currently cognitive computing systems perform more of an advisory role by suggesting sets of options to human users, who ultimately make the final decisions. This is achieved by confidence scores, which are based on quantitative values, representing the merit of a decision after evaluating multiple options, and help users to make the best possible choice.

Discovery Systems help to discover insights that perhaps could not be discovered even by most intelligent human beings, which involves finding insights and connections and understand the vast amounts of information available around the world. The constantly increasing volume of data makes it necessary to have systems that help exploit information more effectively than humans could do on their own. These systems are still in early stages, already some discovery capabilities have emerged and the value proposition for future applications is compelling, especially in the areas of medical research, where a robust amount of information already exists.

Since Watson's 2010 Jeopardy contest win, IBM has expanded the Watson architecture to other fields like health and fraud detection. Also, Watson is no longer called "Artificial Intelligence" but "Cognitive Computing" and it seems that this approach is leading to more relevant applications and business cases.

The Watson project showed that cognitive computing can be used to perform tasks which were restricted to human thinking until now. Cognitive computing is not artificial intelligence but it could be tremendously useful. It can be used to not only compete in television game shows but in many other applications. Today cognitive computing is already used in health care, consumer advice, banking, fraud detection and legal advice.

Cognitive Computing Elements

Today the four elements of understanding, reasoning, learning and interaction and their reference to machines and computing are described as "Cognitive Computing." In comparison to the classic term "Artificial Intelligence" (AI), the cognitive computing definition is structured and has clear references to methods and tools.

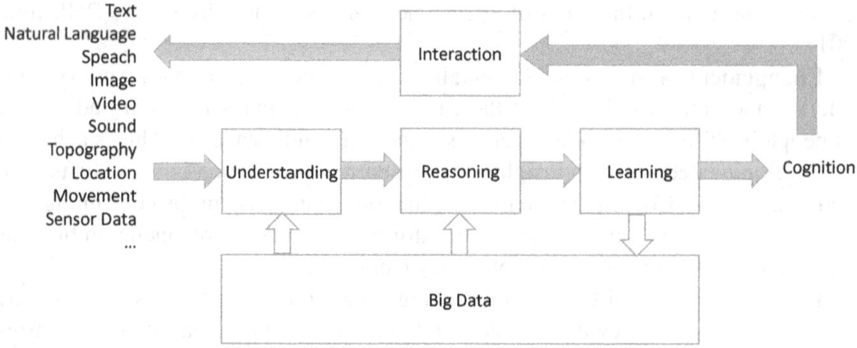

Fig. 6 Cognitive computing

Together those elements could lead us to a target, which is more closed to intelligence than conservative calculating machines (see Fig. 6).

One of our expectations from intelligent systems is to get results which are predictive, accurate and profound. All these attributes require not only understanding of different communication methods, reasoning software and learning methods, but also need a large amount of data. This is where network, cloud and big data comes in. The efficiency and the effectiveness of cognitive computing is dependent on the data delivered, distributed and stored in cloud environments and global network structures.

Understanding

One of the fundamentals of intelligent behavior is the natural language. Humans have developed a large set of skills and talents to communicate. This includes speaking and understanding, but also creating and understanding communication using our face, body, conduct or also how we dress. These communication channels use our body as the sensor, actor and generator of noise and movement in many different and parallel ways. It is no question that current technology can provide an impressive and rapidly growing set of communication methodologies and some of them could overtake our human capabilities in performance or range.

It is also true that there is currently no single device that could beat a human being in variety and richness of communication and sensing its environment for general life situations. Smart cars may react faster to traffic situations than a human driver, but there are not able to train your dog or take the right decision in a business negotiation.

Understanding what we see, hear, smell or feel, simply means translating signals from our environment into signals that can be processed by our brain. This seems to be simple for humans but it is still one of the major challenges for machines.

For a long time, machines where restricted to a very specialized input stream based on a limited set of characters that had to follow a limited set of syntactical rules. Designing more comprehensive cognitive computing systems soon created the demand for a wider and deeper ability in understanding text, speech, and

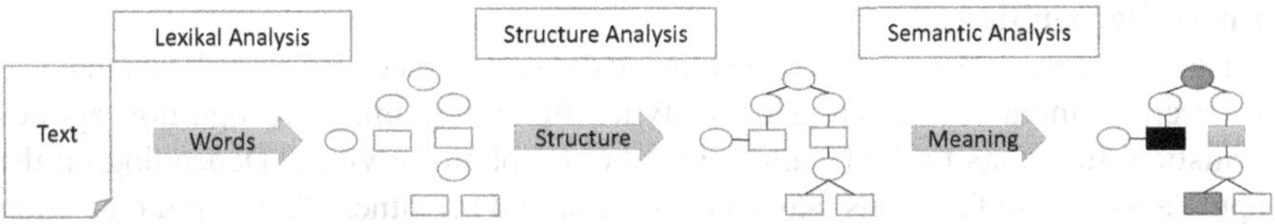

Fig. 7 Understanding text

images. The basic task in all those cases is to translate different types of unstructured data into structured data that can be processed by reasoning or machine learning algorithms. The most important instance of use in the field of understanding is text or speech understanding followed by image recognition leading to perceptual computing.

Text Understanding or Text Mining

Text understanding systems can analyze natural language text and translate that text into a data structure depicting not only the syntax of the text (nouns, verbs, adverbs. etc.), but also identifying the structure of a sentence and the semantic relations to categories or synonyms. Thus, natural text understanding systems generate a data structure forming a tree or network representing the meaning of the text or sentence which can then be used by reasoning or learning systems (see Fig. 7).

Taxonomies and Ontologies

Taxonomies and ontologies are collections of domain specific terms together with their meaning in the specific domain and relations to other terms in the domain. They are the foundation of text understanding and may consist of hundreds of thousands of words, categorizations and relations. Building taxonomies is sometimes extremely time consuming especially for complex domains like medical language, juridical language, technical or scientific domains.

Natural Language Translation

Translation of a given text into another language is one of the major challenges in cognitive computing. There are different competing approaches. Today most of the machine translation programs use the interlingua method or the example based method. In the first case, the text is analyzed and translated into an abstract language-independent representation and in a second step the target text is generated from the abstract representation. The example based method uses text samples in both languages to find a matching translation.

Speech Recognition

Speech recognition uses voice as the input and creates text that can then be analyzed by text recognition programs. Those programs use statistical methods to match the input stream of frequency and amplitude data to existing patterns. Speech recognition systems may be designed as "speaker-dependent" including training and learning functions for a specific person or "speaker independent" applications. The major speech industry players include Google, Microsoft, IBM, Baidu, Apple, and Amazon integrating those services in their products and platforms.

Image Recognition

Image or video recognition analyzes image data and creates structured data that can be used as input to reasoning or analytics functions. Image recognition applies statistical functions to the bit-mapped data of a photo or video. Depending on the application, these functions may analyze color and brightness, but also the position of edges, significant points and compile typical forms included in the image. The result is a set of structured data that can be compared to reference data to create information like: "this a person, female, aged between 25 and 30, smiling and sitting in a sunny garden". Additional analysis is used to compare the result with existing images to recognize specific persons or images of specific locations: "this is Albert Einstein at the age of 60 in front of his house in Mercer Street, Princeton".

Image recognition is also the kernel technology for biometric access controls using fingerprints or iris scans and became one of the kernel technologies for autonomous driving cars.

3D Space Recognition

Creating a digital model of the physical environment of a person, sensor or machine is the target of three-dimensional recognition including the geographical position, the movement, and also the physical elements like their shape surrounding the person or sensor. Methods used for 3D recognition are combinations of GPS, movement sensors, signal triangulation, radar, infrared or ultrasonic. 3D imaging is already used in various application including robotics, aviation and autonomous driving.

Reasoning

The missing link here is not only understanding the environment and communicating decisions but also the reasoning behind a decision. Reasoning is the second important ingredient in intelligence. Intelligent beings can create an internal model of their situation and deduce the best reaction. Reasoning is based on knowledge, rules and patterns that are part of this knowledge. Human beings are perfect in reasoning, which does not mean that the conclusion to act, to react or to do nothing is always the best. Reasoning is also one of the skills our human brain provides, but is also one of the methods we still don't understand. Computers are also able to reason and reasoning systems have been a major part of machine intelligence research for many decades. Computer based reasoning systems use data structures like semantic networks and reasoning algorithms to create possible actions or answers out of new input delivered by their input channels. Those reasoning methods have become more and more efficient over the last few years. Powered by the growing performance of computers and their ability to access large amounts of data, computer reasoning systems outperform humans in many specific disciplines. It is also true that those "artificial" reasoning systems have nothing to do with the inner-working of the human brain and, more importantly, each computer reasoning system is dedicated to a certain discipline, it is not generally usable, like human reasoning.

Reasoning is the process of creating a conclusion from certain premises using a given methodology, e.g. logic. It is part of human intelligence and one of the major

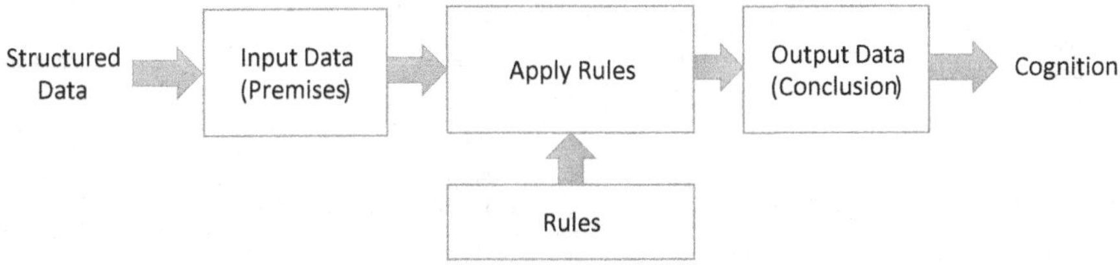

Fig. 8 Reasoning

strengths of our brain called logical deduction. A simple sample for logical reasoning is the following (see Fig. 8):

Premises

1. Thunderstorms are preceded by low air pressure and high temperature.
2. At the moment, it is hot and air pressure is falling.

Conclusion:

- A thunderstorm is to be expected.

Within cognitive computing, reasoning systems are software algorithms that are able to generate conclusions out of given premises as structured input data using logic deduction methods. Today reasoning systems are used in a wide variety of fields. They help find conclusions and causes of illness in medical systems, they create alerts for frauds attempts detected in financial transactions and they are used to create targeted advertisements for users of social networks and search engines. The efficiency of reasoning systems depends on a rich and complete set of rules. Creating those rules is the major mission in machine learning and analysis.

Learning

The process of reasoning is much more effective if the volume of knowledge is not static but growing. Learning as enrichment of knowledge is the third important discipline we need for intelligence. Humans are perfect at learning, they start learning on the first day of their life and never stop. Learning is a basic element of intelligence. Humans use a simple learning method: they create experiences from real day situations. Machine learning as technology has also been a major research field for many decades. Today many different machine learning methods are used for specific fields of knowledge. All those methods are based on the idea that learning means the generation of new rules out of a given input and a possible output value. Machine learning systems are able to make data-driven predictions or decisions, rather than following strictly static program instructions.

The target of learning is to collect rules that can be applied to a set of input data and immediately create output data as a result, prediction or call to action without using trial-and-error strategies or wasteful calculations. A simple daily life example is your personal weather forecast: if it is a hot day, the air pressure declines at high speed and there are dark clouds on the horizon then the probability of a thunderstorm is high. This is a simple rule and as a human being you will learn that rule

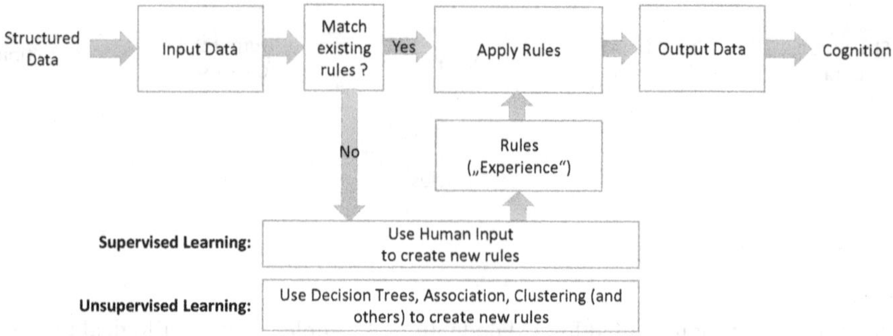

Fig. 9 Learning

either via the experience of getting wet several times or via your friends telling you to be cautious. In the first case, it is unsupervised learning using association, in the second case it is supervised learning thanks to your friends or parents (see Fig. 9).

Machine learning follows the same strategy. It tries to collect many applicable rules to match new input data to already existing output data. A chess playing computer is more effective if it has already stored successful moves based on past games, a fraud detection system for insurance companies is more successful if frequently used patterns of fraud are already stored as rules and a medical consultant application can find possible causes quicker if the number of data connecting symptoms with possible causes is extensive.

Creating new rules can follow two strategies. Supervised learning follows the idea that human experts complete or add the solution or output data to a give set of input data. Much more interesting is unsupervised or automatic learning. In that case, a specialized algorithm is applied to find new rules to match given input to possible output using statistical or logical methods. A typical sample is clustering data. Clustering is a statistical method to find patterns in a large data set. If you have a lot of data about fraud attempts, a cluster analysis algorithm will find correlations between the single data points and form groups of data that share the same attributes. Thus, new rules can be created leading to a much more efficient detection of new fraud attempts.

Interaction

Finally, the result of cognitive processes must be fed back to the individual or machine to complete the feedback circle. Given the target for the feedback is a machine, the result can be delivered in structured data creating input for an actor or machine function. In this case, the target for feedback is a human and the result should be delivered in easily understandable data: pictures or speech.

Speech Synthesis

Speech synthesis takes text as input and delivers speech as output. This is much easier than speech analysis. The text input must be stripped down to words consisting of phonemes. Given the structure of the sentence, the accentuation can be added as additional information. The result is then delivered to an audio processor which generates the audio output. Today many speech synthesizers are built using application specific processors.

2D and 3D-Image Generation
The loop of cognitive systems interacting with humans is closed if results can be presented as images integrated into a virtual or augmented reality. Image creation is a field of its own and was powered by the gaming industry long ago, developing graphical representations of artificial worlds using static or animated 2D and 3D image generation software. In an augmented environment, the merge of real world and digital world is conducted using overlapping rendering, or holographic projections like Microsoft's HoloLens.

Convergence of Technologies

The number of cognitive application is growing rapidly in many different areas. All of them are based on elements of cognitive computing and combine some of those elements in a new and innovative way to create new applications and sometimes new business models. Technologies like text recognition, reasoning or speech synthesis are converged upon by innovative corporations or startups to build new applications like chatbots, smart homes or self-driving cars. We will drill down into some of these applications in then following sections.

Chatbots
Chatbots became popular with the availability of text recognition programs. The name was coined by the combination of "chat" and "bots" derived from "robots." Chatbots are programs that can perform a meaningful conversation with a human in a specific domain. They are mainly used by service desk and customer support departments of airlines, retail companies or travel agencies. Chatbots may be integrated into the web-pages of companies or integrated into live chat, email, SMS and messaging services like Facebook messenger or Snapchat. By November 2016, more than 34,000 chatbots had been integrated into Facebook messenger. Tools for building are easily available at low cost from various open source platforms or on the PaaS platforms provided by Microsoft or IBM.

Smart Homes
Smart homes combine the technology of IOT with cognitive computing including the technical infrastructure for lighting, heating, security with smart assistance like Amazon Alexa or Google Home. The global market for smart home infrastructure and services will more than double by 2020 and spending for smart home is expected to reach $120 billion in 2020.[28]

[28]Strategic Analytics, 2016: https://www.gearbrain.com/strategic-analytics-reports-smart-home-market-62-billion-service-provi-2064971380.html retrieved 2017-01-20.

Smart Cars

From a consumer perspective, smart cars and autonomous driving is the holy grail for machine intelligence. In fact, the field of self-driving cars combines most of the cognitive computing elements. It needs environment, speech and text recognition, as well as image and 3D recognition, reasoning, learning and has some vital elements of interaction with the human driver like augmented reality and speech synthesis. Offered for the first time by the car manufacturer Tesla for Tesla Model S in 2014, within a short time the major car manufacturers followed the trend and started their own autonomous driving projects. In the meantime, many new models are already equipped with cognitive elements like image recognition of traffic signs, navigation systems, augmented reality such as head-up-displays for different types of distance and speed controls and automated parking assistance systems.

Perceptual Computing

Perceptual user interfaces combine sensor systems to interact with a machine using natural senses instead of traditional input/output devices like keyboards or mouse (Jerry M. Mendel, 2010). Hand gestures can be used to control the computer, or the machine might even track your eyes to know the action to perform. The technology is already widely used in cars and access control systems.

Virtual Reality

Virtual reality is a computer technology that tries to replicate either real or imagined environments while simulating the physical user presence and environment and allowing for user interaction (Jason Jerald, 2016; Rheingold, 1992). It creates virtual experiences in the areas of sight, touch, hearing and even smell.[29]

In the past, the big problem with VR was simply how small or integrated with human physics the devices could get. Glasses that are the size of half a football will never allow one to really feel at home in a virtualized environment but as soon as such glasses become smaller or finally can be replaced by implants that allow easy entrance to a virtual world just by closing the eyes and turning on the implants, this would guarantee an almost perfect virtual illusion. The same of course is true for sound or taste and other virtualized, sensual experiences that all could culminate in cyborgs, which are beings with both organic and biomechatronic body parts, a term that was coined in 1960 by Manfred Clynes and Nathan S. Kline.[30]

With today's first generations and incarnations of VR, there is already a big and constantly growing market for VR applications mostly in the gaming sector as can be seen in the Virtual Reality Society.[31] But there are more applications for VR than just gaming like architecture, sports, medicine, arts or entertainment.

---

[29] IEEE Spectrum, Nordum Amy, The Fuzzy Future of Virtual Reality and Augmented Reality, http://spectrum.ieee.org/tech-talk/consumer-electronics/gadgets/can-you-see-it-the-future-of-virtual-and-augmented-reality, 2016, retrieved 2017-01-23.

[30] MIT, Clynes Manfred, et al., Cyborgs and Space, http://web.mit.edu/digitalapollo/Documents/Chapter1/cyborgs.pdf, 1960, retrieved 2017-01-23.

[31] VRS, Virtual Reality Society, http://www.vrs.org.uk/category/virtual-reality-games/, 2017, retrieved 2017-01-23.

Augmented Reality

Another revolution was started with augmented reality, which is very close to VR but creates a digital overlay on top of the physical world. Examples are Microsoft's Holo Lens[32] that may allow medical students to view a three-dimensional model of a heart in the middle of a classroom or help technicians to successfully repair broken things.[33]

As such, augmented reality is either a direct or indirect live view of the physical world and is called augmented because the real-world elements are enhanced (augmented) by computer-generated simulations like sound, video, graphics or GPS data.

Augmented reality was invented or projected as a concept in the 1960s by the subculture of cyberpunks who focused on a high tech future full of cyborgs, androids and virtual reality all based on ubiquitous computer performance and network connection. The cyberpunk community has created a kind of parallel society to the normal world that includes hackers, rebels, transhumanists and other outsiders.[34] It is also a fact that the cyberpunk culture of the 1960s and 1970 made major contributions to the areas of personal computing, global networks and human-machine interaction.

Contextual Computing

Contextual Computing (Porzel, 2011) is based on the idea that machines understand you and everything you care about, anticipate your behavior and emotions, absorb your social graphs, interpret your intentions and, in so doing, hopefully make life easier. Imagine you are involved in an unexpected situation, and your brain is trying to figure out what to do. You suddenly become very sensitive to all kind of signs, like looking for threats around you. Your brain then cycles through many scenarios seeking the answers.

This is called situational awareness. The way we respond to the world we live in is so seamless that it is almost unconscious. We sense a multitude of information, put it in context to past experiences and out of that we develop sets of options for our reactions, and finally choose a preferred path on which to act. This basic ability to interpret and act on all kinds of different situations is something, which has evolved since early mankind. But the big problem is that our experience has evolved over millions of years and suddenly we find ourselves in a modern world, where much of the data we need for good decisions is either unreliable or nonexistent.

---

[32]Microsoft, Microsoft HoloLens, https://www.microsoft.com/microsoft-hololens/en-us, 2017, retrieved 2017-01-23.

[33]Ted Talks, Kipman Alex, Microsoft deoms the Hololens, https://www.youtube.com/watch?v=kjAHwLaLjUw, 2016, retrieved 2017-01-26.

[34]MUO, makeuseof, Albright Dann, What Is Cyberpunk? An Introduction to the Sci-Fi Genre, http://www.makeuseof.com/tag/cyberpunk-introduction-sci-fi-genre/, 2016, retrieved 2017-01-23.

Contextual computing makes use of always present computers, which can sense the objective and subjective aspects of given situations by augmenting our ability to perceive and act in the moment based on where we are, who we are with and our past experiences, which will become our sixth, seventh and eighth senses.

We experience applications of this technology, as mobile devices deliver location-based services based on GPS, which serves as the baseline for many ways to make life easier. Amazon or Netflix recommend a book to you or a video based on your behavior and ratings. Or valuations from Facebook or Twitter promising to leverage your acquaintances and interests to create relevant content and market to you in the most effective ways.

Data Graphs

Contextual computing needs four data graphs to make it work: social, interest, behavioral and personal. Future dominating players of the Web will need to master all four of these graphs. Sure enough, there are several ethical concerns mainly based on privacy policies that we are currently concerned with. How much information is ok to disclose on your social graph? Do you fear being swamped with marketing information because of information shared in your interest graph? Do you fear any harm based on the information of your behavioral graph and finally, for the personal graph, are you allowing an outside entity to read your mind? But even with all these legitimate worries today, companies are already constructing these graphs.

- The social graph is all about connections, how you connect with other people and how they are connected to one another.
- The personal graph contains all your beliefs, core values and personality. This data set is not very developed today and it is difficult to design. As psychology is still struggling to explain how personal identities function, documentation of this information cannot be expected to emerge quickly.
- The interest graph is all about what you like, your tastes and preferences and describes the overlaps in tastes between individuals. Several companies are already active in this area like Twitter by trying to fully chart how the individual subject connects to all others. Applications based on this can analyze and predicting our interests, for example what books you might buy based on the ones you recently bought. This is a typical example of what Amazon uses to make purchasing suggestions to individuals.
- The behavior graph is one of the easiest to be built, as it is easy for data to depict what you do instead of what you claim to do. In this example, sensors are employed to detect what you do. This data may be in contrast to your interest graph, which makes it easy for contextual computing to know how likely you are to do some action. The behavior graph provides input to Google Search, Netflix, Amazon Recommendations, iTunes Genius and many other applications today.

The future of contextual computing is that these four graphs are filled with data and connect and resonate between each other in a meaningful way. Early examples like Google's Now and Glass projects, Highlight, and Siri are just beginning to experiment with these technologies.

True, contextual computing is a little further away than the most optimistic pundits would have you believe. That should not be mistaken for the claim that it's unlikely to fully arrive. As Bill Gates astutely pointed out, "*There's a tendency to overestimate how much things will change in 2 years and underestimate how much change will occur over 10 years.*"

Within a decade, contextual computing could very well be the dominant paradigm in technology. Even office productivity will move to such a model. By combining a task with broad and relevant sets of data about us and the context in which we live, contextual computing will generate relevant options for us. Then we will have truly wearable intelligence and not just wearable computing as we see today.

Cognitive Tools in the Market Today

We are already surrounded by systems that use cognitive computing methods. In using your smartphone, you may experience speech recognition introduced by Apple's Siri in its iPhone 4s in October 2011. By typing text in various text processing tools you may be confronted with helpful text recognition and correction features, which also use simple learning algorithms. The car you bought is already equipped with smart features supporting save driving and parking. Navigation systems like "Waze" help us to find the shortest way by combining information from various sources and using reasoning software to calculate an optimized solution. Fraud detection systems applied by credit card and insurance companies help to detect fraud attempts as early as possible to prevent financial damage. Service desks use chatbots to provide efficient information to calling customers at low costs and with high availability. The search engine or e-commerce platform you use is constantly collecting data reflecting your behavior and resulting in smart and personalized information about new offers for you. It is based on sophisticated analytics and learning algorithms. Cognitive computing is less of a revolution, rather it is an evolution that is gaining speed and turning into a major commodity for nearly every type of business or service.

The market for artificial intelligence is growing rapidly and spreading out to many different industry segments. A research study from Transparency Market Research estimates that the growth in the AI market will move from $600 million (2016) to more than $35 billion by 2025.[35] IDC reports the results of its research with larger numbers forecasting a revenue growth to reach $47 billion in 2020 and also claims: "*...According to a new Worldwide Semiannual Cognitive/Artificial*

[35]Transparency Market Research, 2016: http://www.transparencymarketresearch.com/pressrelease/artificial-intelligence-market.htm retrieved 2017-01-20.

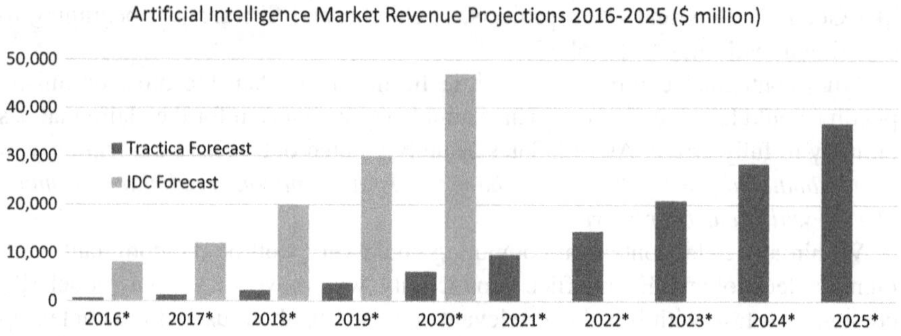

Fig. 10 Artificial intelligence market revenue projections

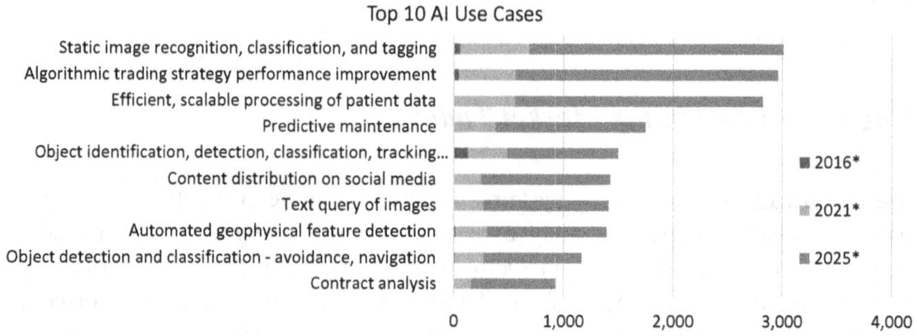

Fig. 11 Top ten artificial intelligence use cases 2016–2025

Intelligence Systems Spending Guide…the market for cognitive/AI solutions will experience a compound annual growth rate (CAGR) of 55.1% over the 2016–2020 forecast period." (see Fig. 10).[36,37]

The research company, Tractica, has collected data about the focus segments up to 2025 and reports contract image recognition, trading strategy, processing of patient data and predictive maintenance as the top 4 (see Fig. 11)[38]:

Players in the Cognitive Computing Market

The traditional IT companies as well as the Internet companies have already started to follow this new path and work hard to gain the lead.

Companies providing cloud platform-as-service like Amazon, Apple, IBM, Google, and Microsoft have been investing in research and development to power the capabilities of machine intelligence agents based on existing technologies like IBM's Watson, Apple's Siri, Amazon's Alexa or Microsoft's Cortana.

---

[36]IDC, AI Revenue Forecast, 2016: http://www.idc.com/getdoc.jsp?containerId=prUS41878616 retreived 2017-01-20.

[37]IDC, Worldwide Semiannual AI Systems Spending Guide, 2016: http://www.idc.com/getdoc.jsp?containerId=IDC_P33198 retrieved 2017-01-20.

[38]Tractica, Artificial Intelligence Market Forecast, 2016: https://www.tractica.com/research/artificial-intelligence-market-forecasts/ retrieved 2017-01-10.

IBM

IBM has started the new party by introducing its DeepQA in 2010, later renamed Watson and provided this in two product families known as Watson Analytics and Watson Cognitive. Together, with cloud computing, IBM has defined these segments as "strategic imperatives" for the company to underline IBM's transition to a high-level service company based on cutting-edge technology. At the World of Watson conference in October 2016, IBM CEO Gini Rometty said: *"In 5 years, there is no doubt in my mind that cognitive computing will impact every decision. Bringing cognitive capabilities to digital business will change the way we work and help solve the world's biggest problems."*[39] IBM also reported that Watson is already used by more than 700 companies worldwide including medical service organizations, financial and retail companies. In the 2016 annual report, IBM reports revenue from cognitive computing at $18.2 billion which is more than 22% of $79.9 billion, which also includes revenue from Watson analytics.

Today IBM offers cognitive computing elements in its Watson developers cloud, which is based on the IBM bluemix Platform as-a-Service including Apps for language identification, machine translation, sentiment analysis, question and answer, speech to text, text to speech, visual recognition, visualization rendering and many others.[40]

Microsoft

Microsoft began to provide machine intelligence products and solutions with its Azure Machine Learning Studio in 2015. With the release of Microsoft Cognitive Services in 2016, the Azure platform was expanded into a set of APIs providing cognitive basic elements like face recognition. The services are offered with a transaction based pricing model.

In October 2016, Microsoft announced global expansion of their HoloLens product which is the logical extension of Windows Mixed Reality, a platform based on Windows 10. HoloLens is a head-mounted display and enables the use of augmented reality applications with holographic projections for presentations, collaborations or integration of real world images with digital data. The platform works by enabling applications in which the live presentation of physical, real-world elements is incorporated with that of virtual elements such that they are perceived as existing together in a shared environment.

In January 2017, Microsoft also announced the Microsoft Connected Vehicle Platform for car manufacturers and named Renault-Nissan as the first customer.[41]

---

[39]IBM, World of Watson, 2016: https://www.ibm.com/blogs/internet-of-things/watson/ retrieved 2017-01-20.

[40]IBM Watson developers cloud: https://www.ibm.com/watson/developercloud/ retrieved 2017-03-20.

[41]Microsoft Blog, 2017: https://blogs.microsoft.com/blog/2017/01/05/microsoft-connected-vehicle-platform-helps-automakers-transform-cars/#sm.000doamg214eccrhr341frs6wxdfw retrieved 2017-01-20.

Google

Google follows the strategy of founding specific research departments for innovative new technologies. In 2011, Google X was introduced as the research arm of Goggle (now Alphabat). Google Brain is a deep learning project within Alphabete's Google X. Other Google X projects include the self-driving car, now outsourced to a new company called Waymo, and the Quantum Artificial Intelligence Lab (QuAIL) as a joint initiative of NASA, Universities Space Research Association, and Google. The goal of QuAIL is to pioneer research on how quantum computing might help with machine learning.

Google Home is a smart speaker developed by Google as part of its "Made by Google" product line. The product is a rival to Amazon's Echo in the smart speaker industry. In November 2016, Google released its Google Home appliance to rival the Echo. Google Home is a smart speaker that uses speech recognition, text recognition and speech generation to implement a smart personal assistant providing smart home automation functions and smart access to media streaming services like Google Play Music, YouTube Music, Spotify and others.

Amazon

In 2014, Amazon announced its Amazon Echo smart speaker and Amazon Alexa as an integrated personal home assistant. Both products became available in 2016. As an extension to other smart home assistants, Amazon included an ordering service for the Amazon platform.

Amazon launched its new AI platform, Amazon AI, in November 2016.[42] The new platform-as-service contains image recognition, text-to-speech conversion in 47 voices and 24 languages and Amazon Lex to build conversational applications like chatbots.

AI Startups and Investments

Artificial Intelligence and cognitive computing is an intense field of activity in the startup business world as well as for classical IT or Internet companies. Analyzing the investment in AI startups, the focus is on the following segments:

1. Machine learning (applications)
2. Smart robots
3. Gesture control
4. Computer vision (general)
5. Machine learning (general)
6. Natural language processing
7. Computer vision (applications)
8. Video content recognition
9. Virtual personal assistants
10. Speech recognition
11. Speech to speech translation

[42] Techcrunch, 2016: https://techcrunch.com/2016/11/30/amazon-launches-amazon-ai-to-bring-its-machine-learning-smarts-to-developers/ retrieved 2017-01-20.

Table 5 Top cognitive computing startups by size of funding

Company	Country	Founded		Funding ($ million)
Sentient Technologies	USA	2007	Distributed AI platform	144.00
Ayasdi	USA	2008	Machine intelligence platform for big data	106.00
Vicarious Systems	USA	2010	Visual perception based on the recursive cortical network	67.00
Context Relevant	USA	2012	Near real-time analytics for human behaviour, sales intelligence, pricing strategy	44.00
Cortica	USA	2007	Image and video recognition	39.00
WorkFusion	USA	2010	AI powered cognitive automation of enterprise business processes	71.00
RapidMiner	USA	2007	Predictive analytics platform	36.00
Digital Reasoning Systems	USA	2000	Machine learning, natural language processing, big data analytics	73.00
H2O.ai	USA	2011	Open source machine learning platform	33.00
Viv Labs	USA	2012	Open intelligent conversational interface by the makers of siri	30.00

12. Recommendation engines
13. Context aware computing

A listing of the top ten startups ranked by funding demonstrates that innovations in cognitive computing are currently accelerating at a high speed (see Table 5).

Cognitive computing is not a revolution, it is an evolution from traditional data processing to smart and cognitive services. It was started in 1956 by John McCarthy, went through the "winter of AI" in the 1980s and was restarted by IBM's Watson only 7 years ago. Cognitive computing is a wave and many individual users, companies and technology providers have started to ride it and it seems to be accelerating. The interesting question is: will that wave peak at some point when we can refer to something as an intelligent machine and will our cloud ecosystem turn into a nexus of intelligent clouds?

Truly Intelligent Clouds

"O Deep Thought computer," he said, "the task we have designed you to perform is this. We want you to tell us. . ." he paused, "The Answer."

Douglas Adams, The Hitchhiker's Guide to the Galaxy

Will Clouds Be Intelligent?

The field of machine intelligence went through a few dramatic stages starting with Norbert Wiener's "Kybernetik" in 1948, moving to "Artificial Intelligence" in the

1950s and now evolving into "Cognitive Computing" in the last decade. Although the term AI or "Artificial Intelligence" or "Machine Intelligence" is used, there is still the open question of whether the applications and services summarized today under "cognitive" are also "intelligent." It is no question that cloud services like Google, Amazon or Facebook use cognitive methods to be successful in their business. Does that mean that Google, Amazon and Facebook are "intelligent?"

At this point we will drill down into the terms "intelligence" and "machine intelligence." The basic question is "what is intelligence?" Frankly said: we don't know. The best definition that can be found is that intelligence is what humans are. There are numerous definitions of intelligence, most of them refer to human thinking, reasoning, behavior and learning. It also seems clear that human intelligence has its fundaments in the human brain and this leads to a new challenge. If the human brain is an "Intelligent Machine" then how does it work?

Neurologists and psychologists tell us that they are working on that question but are still far from a complete answer. We know a lot about the human brain, but in the end, we can only describe some topography and the basic elements like the functionality of synapsis but not much beyond that. The hope in building an intelligent machine by copying the design of the human brain is futile, we simply don't have that complete blueprint.

To achieve more intelligent, better and smarter machines must follow another strategy. The good news is that we understand which elements are required and useful. Here comes the bad news: we have no comprehensive understanding of what is missing and what is the complete picture. We will see that this could be one of the major topics.

Some of the ingredients for intelligence are at least partly understood like language, learning or reasoning. Some of them seem to be an important part of what we call intelligence but are still far from being fully comprehended.

The target of what is called artificial intelligence is to build machines that can cover each of the aspects of human intelligence. As it is easy to see that there are still many gaps and attributes which are not covered by machine intelligence like awareness, emotion or creativity. What is really missing is a "Great Unified Theory" of intelligent behavior. Otherwise, we have already moved along this road for a while and made some progress. Some of the parts on the map above have their counterpart in machine behavior. We are surrounded by systems and machines that can understand language, reasoning, learning and create cognition, and they are useful cognitive applications, but they are not intelligent.

Super Intelligent Systems or Strong AI

Superintelligence or strong AI is defined as a type of machine intelligence that not only matches human intelligence but goes beyond it. There are several open questions in this scenario. What we have built today as machine intelligence or cognitive systems is still based on "silicon processing" and binary, logical computing. The human brain works differently. The brain's synapsis and neurons are very slow compared to computer circuits but they work in a massively parallel way. The architecture of the brain is a combination of analog and digital processing. The topology is digital, synapsis are connected or not, their functions are analog and sometimes based on random. The human brain is also based on a combination of electrical and chemical phenomena, it is deeply and redundantly connected and rewires itself constantly. And we are still far from a complete understanding of the detailed functionality of our brain.

Transhumanism

The Swedish philosopher Nick Bostrom is an active researcher in the field of transhumanism and superintelligence. In his book "Superintelligence: Paths, Dangers, Strategies" he concentrates not on the technical feasibility but on the ethical and social impacts and effects. He describes several scenarios of how superintelligence would influence today's political and social systems. A part of his analysis concentrates on the question of how long it would take before a super machine intelligence would take over and if it could still be controlled by humans. In one of his scenarios, the time between the first implementation of a super intelligent machine and its assuming absolute power is a question of minutes (Bostrom, 2014).

In his book "The Singularity Is Near" (Kurzweil, 2005) the U.S. computer scientist and futurist Ray Kurzweil describes his theory about how mankind is moving to a future where biological intelligence is melted together with machine intelligence. Based on a rich collection of data showing the rapid increase of computing power, Kurzweil wants to show that by the middle of the twenty-first century, computing power supported by new processor technologies like 3-D molecular computing, nanotubes or quantum computing will provide a level of processing capacity beyond the computational capacity of the human brain.

In parallel Kurzweil mentions that neurological science and research is also making rapid progress in scanning the human brain and analyzing its architecture and functionality in greater detail. The key technology behind this could be nanotechnology and nanobots operating within the human body in the bloodstream.

Kurzweil's assumption is that both parallel paths could lead to a situation where a human brain can be scanned for every functional detail using nanotechnology and the result of the scan can be copied to a computing machine. This machine could provide a powerful matrix in terms of computing power for the scanned and copied functional image of the brain. The result would be super intelligence. Kurzweil's prediction is that this point could be reached around 2050, and he calls it "singularity." If one follows that assumption, the question would not be "will clouds be intelligent?" but "how does it feel to move yourself into a cloud?

Following this approach would consequently mean that mankind will transform itself into another type of existence where the biological existence, the body, will be stepwise transformed into a machine-based existence. Supporters of this theory on future evolution propagate "Transhumanism," which aims at the transformation of the today human biological status into advanced sophisticated technology-based machines. They claim that we already experience a slow but accelerating trend in exchanging parts of the human body for machines and that we already use computing and cognitive systems in important parts of our daily life.

Disruptive Future Computing Technologies

In 1965 Moore described what is generally known as “Moore’s Law,” which is the prediction that the number of components per integrated circuit would double every year. This law was revised in 1975 by claiming that chip performance would double every 18 months with the combination of more and faster transistors. The numbers of components today per chip are in the region of many billions. There have already been multiple predictions on the death of Moore’s Law, but, in 2017, we see the progress from 14 nm technologies to 7 nm and, furthermore, the death of this law is still some years, perhaps decades away.

Traditional computing, with its continually more microscopic circuitry etched in silicon, can take us only so far. Moore’s law, which dictates that the amount of computing power you can squeeze into the same space will double every 18 months, is about to run into a silicon wall.

The alternatives to silicon are the use of newer disruptive computer technologies that will continue Moore’s curve even after the end of silicon technology. These technologies can bring about significant miniaturization, much higher density and significantly reduced power consumption, all very much needed to keep the progress of increased computing performance up and running. Cramming more circuits onto chips is one way to solve the problem that will be possible with 3-D molecular computing, while DNA computing offers an enormous increase in the density of information packaging. Using light, spin or quantum mechanics for computing could further multiply the capacity and performance of digital computers by powers of ten (see Fig. 12).

Even if we are sure that computing based on silicon and the ever-growing amount of processor cores per chip will keep evolving over the next decade or two before finally hitting the Moore’s law wall, there are already many technologies around today that are poised to take over conventional silicon based computing in the years to come. In the following sections, we will describe some of the most promising technologies and paradigms likely to mature for mass production and use in daily life over the next decades.

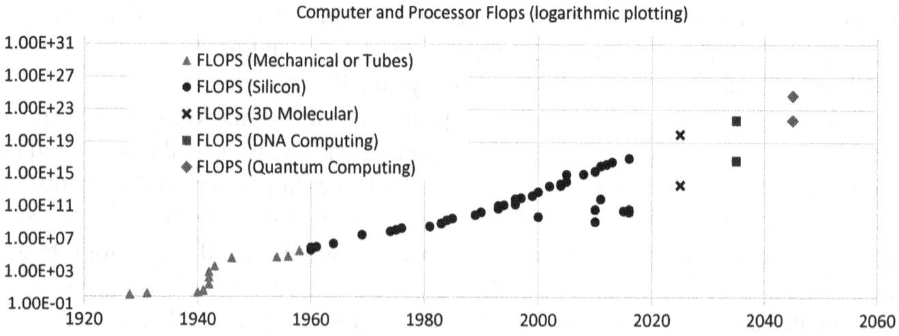

Fig. 12 Silicon, 3-D molecular computing, DNA and quantum computing (logarithmic plotting)

3-D Molecular Computing and Nanotubes

The first approach that comes to mind for raising computing power is to build three-dimensional circuits based on conventional silicon. One of the first companies to do so was Matrix Semiconductor[43] founded in 1999 and acquired by SanDisk in 2005.[44] This allowed SanDisk to benefit from the huge 3-D chip know how that is based on multiple layers of antifuse-based one-time programmable ROM.

Some other approaches in this area are cylinder-like memory designs where working prototypes have been shown at the MIT Media Lab, etc., and also Nippon Telegraph and Telephone Corporation (NTT) has shown 3-D technology based chips.

Based on 3-D silicon technology, it is also possible to build 3-D computer chips[45] that promise to be up to 1000 times faster than existing silicon chips. This is made possible by using carbon nanotubes that allow stacking memory and processor layers in three dimensions enabling a much higher density and resulting in increased speed.

Molecular Computing

The next step beyond nanotubes and 3-D structures for computing could be molecular computers. Arieh Aviram suggested this idea very early in his PhD dissertation more than 40 years ago.[46] His thesis was that it should be possible to miniaturize electronic components down to a molecular size based on silicon transistors and diodes.

It would take until 2002 before scientists at the University of Basel and the University of Wisconsin could create an atomic memory drive using atoms emulating the hard drive. Today we are still in the phase of understanding how to fabricate arrays of molecular architectures that can communicate with each other in the macro world. This needs a self-assembly of all molecular components to direct them to their right location and orientation to generate integrated nanosystems, which is usually done using lithography and by adjusting the patterned structures with dimensions as small as 45 nm.

---

[43]Bloomberg, Matrix Semiconductor, https://www.bloomberg.com/profiles/companies/639832Z:US-matrix-semiconductor-inc, 2017, retrieved 2017-03-23.

[44]eetimes, SanDisk buys Matrix Semi for $238 million, http://www.eetimes.com/document.asp?doc_id=1157072, 2005, retrieved 2017-03-23.

[45]LIVESCIENCE, 3D Computer Chips Could Be 1000 Times Faster Than Existing Ones, 2015, retrieved 2017-03-23.

[46]IEEE Spectrum, Kelly K., et al., What Happened to the Molecular Computer, http://spectrum.ieee.org/biomedical/devices/whatever-happened-to-the-molecular-computer, 2015, retrieved 2017-03-23.

For the success of molecular computing, it is key to understand how to assemble functional structures with the ability to compress huge numbers of computing components into given areas with the potential to revolutionize the nano-macro world and finally allow for the design of more powerful computing devices.

DNA Computing

In DNA computing, computations are performed using biological molecules instead of silicon chips. Using molecules for computing dates back to 1959 when the American physicist Richard Feynman[47] had the idea of using nanotechnology for DNA computing. Computations are always executions of algorithms, usually defined as step-by–step lists of instructions.

Even today, DNA computers exist mainly in test tubes or as DNA-based logic switches and may even be fully programmable. The main difference between conventional and DNA computing methods is the way instructions are executed. In conventional computers, data is processed sequentially, which causes an increase in the time required to solve a problem, while DNA computers process data in a parallel manner that dramatically speeds up the process.

DNA provides enormous amounts of information storage, as 1 g of DNA can hold around 1×10^{14} MB of data. Storing this amount of data on CDs requires 145 trillion CDs, which would circle the earth 375 times if the CDs were lined up edge-to-edge.

DNA computing can use the information present by the four-character genetic alphabet A (Adenine), G (Guanine), C (Cytosine) and T (Thymine) instead of the binary alphabet used by traditional computers. This allows for representing algorithm inputs by DNA molecules with specific sequences and sorting the molecules according to length, while the result can be a property of the final set of molecules.

Today many scientists believe that DNA based computers could bring a number of advantages, as they would allow for the use of the inherent parallelism of biology and biological operations acting on all DNA strands at the same time. This would allow DNA computers to solve problems far beyond the ones solvable by conventional silicon based computers lacking this massive parallelism. Anyway, we should be aware that it would still take years if not decades to develop a practical and workable DNA computer.

There is another fundamental change coming with DNA computers and many other new disruptive technologies, which is the need for new architectures offering Single Instruction Multiple Data (SIMD) capability. For example, in DNA computing each of the many trillion computers needs to perform the same operation at the same time and on different data, thereby making this impossible to use for

[47]Encyclopaedia Britannica, Richard Feynman, https://www.britannica.com/biography/Richard-Feynman, 2017, retrieved 2017-03-24.

general purpose algorithms where each computer needs to execute the specific operation to complete its particular mission.

Controlling Spin

Research has already made progress and single electrons have already been made to adjust their spin, while subatomic circuitry is already within our reach. One of the first commercial applications is MRAM (magnetic random-access memory), which guarantees enormous speeds. MRAM gets its speed from something called the Giant Magneto Resistive (GMR).

GMR has to do with the fact that if you place layers of ultrathin magnetic film on top of one another and alternate their polarity, you get resistance. That is, the electrons can be spun in one direction or the other. Electrons spin like a top or a billiard ball in some direction relative to a magnetic field. Flip the direction of the field, and the electron flips the direction of its spin. This very basic quantum effect can be used like a binary bit, its direction labeled "0" or "1" and employed to store digital information (Almadena Chtchelkanova, 2003).

Conventional computing creates these zeroes and ones by switching an electric current on and off. Spins are less affected by the environment than electric charges and take longer to decay. Also, keeping an electric charge in position requires continuous power, which means when computers lose power, the charge goes away. With a magnetic device, the memory stays put when the power shuts off. On top of that, if you take electricity out of the equation, you get rid of the overheating problem that undercuts Moore's law. This memory breakthrough was possible in a large part by the research done by DARPA, the Defense Advanced Research Projects Agency—the same Pentagon gang that gave us the Internet and it was developed by a physicist named Stuart Wolf. While MRAM is just about memory, the ability to control spin in a computational device is the next step and was called "Spintronics" by Wolf (Weisheng Zhao, 2015; Wolf, 2001).

Using Light for Computing

While light has already been used for many decades to transfer data between different sites, which is known as optical transmission or communication and has reached a pretty stable and well-known status, the use of light for computing is still lacking today. Computing with light requires a chip that can integrate electrical and optical devices on the same piece of silicon. There is development under way by IBM[48] to solve this problem as part of their Exascale computing program. IBM

[48]MIT Technology Review, Zax D., Computing with Light, https://www.technologyreview.com/s/508606/computing-with-light/, 2012, retrieved 2017-03-24.

wants to use nanophotonics to compute light, which would help overcome the constraints like heating and processing power of current supercomputers working as electrical computers based on silicon.

Light computing also marks another approach to solve the SIMD computing problem as it can use multiple laser beams with encoded information where optical components can perform arithmetic and logical operations on these encoded streams like the micro lens arrays built by Thorlabs.[49] Another example of light computing is done with replacing all metal wiring between components inside devices with fiber-optic links.[50] This is effectively achieved by having lasers built into the transmitter chips that are capable of supporting wavelength multiplexing and allow for transmission speeds of up to 1000 Gbps compared to the traditional 10 Gaps speeds available for longer distances today.

Quantum Computing

Thanks to quantum physics there is hope that even after that phase of the next generation computing technologies we discussed, there can be growth in computing power and speed. The time until quantum computing would really become commercial available is estimated at 40–60 years from today. When we consider just some of the possibilities which quantum computing promises, the age of computing will only start then. If we are successful in the major challenge of approaching the fundamental laws of physics, then what we have today are only tiny toys.

The Quantum Computer: Using Spin and Superposition for Computing
Already in 1959 Richard Feynman noted the effects of quantum mechanics when components begin to reach microscopic scales. A quantum computer is a device that makes use of properties described by quantum mechanics and Feynman suggested this would lead to significantly more powerful computers in 1982 (Milburn, 1999).

A fundamental prerequisite for successful quantum computing is to make spin actually work for computation. A team at the University of California at Santa Barbara, led by David Awschalom, has made big progress in this direction by controlling electron spins in semiconductors and other materials a few nanometers in size (Awschalom, 2004). This could mean not just an end to overheating worries but the possibility of moving computer technology into the molecular realm. With molecular-level chips, a laptop could have more computing power than trillions of

[49]THORLABS, Microlens Array, https://www.thorlabs.com/newgrouppage9.cfm?objectgroup_id=2861, 2017, retrieved 2017-03-24.

[50]MIT Technology Review, Simonite T., Computing at the Speed of Light, https://www.technologyreview.com/s/420082/computing-at-the-speed-of-light/, 2010, retrieved 2017-03-24.

today's supercomputers. But even molecular-level computers could soon be old-fashioned, as Dan Rugar of IBM performed in 2004 what the American Institute of Physics dubbed the most important experiment of the year by using a magnet to control the spin of a single electron. In theory, that means we could have subatomic-scale circuitry. At that level, the behavior of particles is more complicated and could—again, in theory—do even more powerful things.

In the subatomic world, the same magnetic spin can be up and down and everything in between—all at the same time. This is known in quantum mechanics as superposition,[51] made possible because electrons sometimes behave more like waves than particles. If you picture a piece of string, fixed at both ends and vibrating, if you get the vibration right, the string will be moving up at one end and down at the other. And as a wave, it will have every value in-between. In the binary math of computers as we know them today, each bit represents either a zero or a one but if each electron in a row of atoms can be in two or more places at once, and all these positions could be used for computing, this would mean a real revolution.

Qubits

Consider a quantum bit, or qubit, that can represent two values simultaneously. Two qubits linked together can represent four values at once, three can represent eight, and so on. Twenty qubits already can represent almost a million numbers (2 to the power of 20) simultaneously.[52]

With the power of this exponential growth, one can tackle any problem that gets exponentially larger, and there are lots of important ones that fall in this category. We cannot reliably predict weather or traffic nor the mutation of viruses today because the massive number of variables and possible interactions is too extensive for current computers, but qubits could change that. Unlike ordinary bits, the power of qubits grows exponentially with their number.

A single, 30-qubit quantum computer would be comparable to a digital computer capable of performing 10×10^{12} or 10 trillion floating-point operations per second (TFLOPS)—comparable to the speed of the fastest supercomputers existing today. For many tasks, a comparatively small quantum computer made up of only 100 qubits could outperform the world's best future supercomputers and deliver new levels of computing power to humanity.

Quantum Computing for Dummies

What does that all mean in reality? As soon as a practical quantum computer can be built, it will be able to break current encryption schemes that are usually based on multiplying two large primes. This may sound scary, but we also need to consider

---

[51]physics.org, What is superposition?, http://www.physics.org/article-questions.asp?id=124, 2017, retrieved 2017-03-24.

[52]CB Insights, WTF Is Quantum Computing? A 5-Minute Primer, https://www.cbinsights.com/blog/quantum-computing-explainer/?utm_source=CB+Insights+Newsletter&utm_campaign=d9dea193e6-ThursNL_10_20_2016&utm_medium=email&utm_term=0_9dc0513989-d9dea193e6-87759901, 2016, retrieved 2017-01-23.

that quantum mechanical effects will offer new methods of secure communication known as quantum encryption, so as soon as we can all afford a quantum computer this encryption problem will be solved again.

On one side, the potential possibilities of a quantum computer are enormous, but their requirements are equally stringent. Most importantly, a quantum computer must keep coherence between the qubits it uses, which is also known as quantum entanglement, at least for the time it performs a certain algorithm. There are always inevitable interactions with the environment known as decoherence and hence practical methods of error detection and correction are mandatory. On top of that, whenever a quantum system is measured its state is disturbed, which means reliable methods for extracting information and results are a must.

Quantum Computing Evolution

This new technology of quantum computing can be used in many different areas as we will outline throughout the rest of this section and these areas do not only cover general computing. There is a huge potential in quantum computing through its encryption possibilities based on quantum mechanical properties, where actually quantum key-distribution is the best-known example today.

Once quantum computers become available there could be solutions for cracking not only normal security keys we use and know today but even quantum encryption. Specific quantum processors have already been realized up to 50 qubits and these capabilities will increase over the next decades.

The general quantum computer or quantum processor that could be used as we know from silicon computers today for general applications is very likely to be further in the future, estimates range from 40 to 70 years (see Fig. 13).

Drivers and Evolution of Quantum Computing

Quantum computing looks like a very promising solution to keep computing progress and evolution on track after hitting the silicon miniaturization wall (Peter Schwartz, 2006) and once all other possibilities for developing future computers have been exploited. There are several quantum computer demonstrations that show the fundamental principles, all still in an experimental stage.

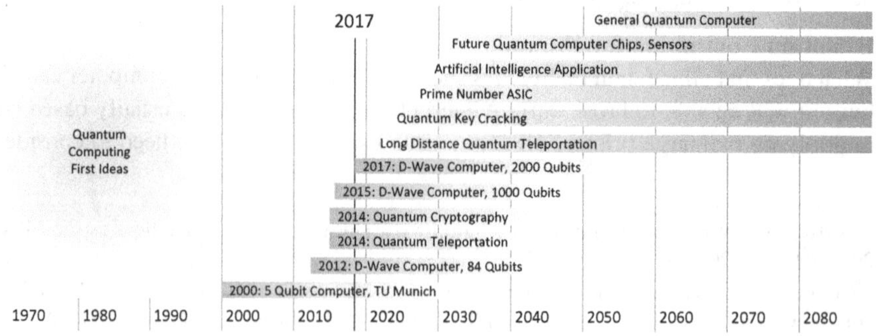

Fig. 13 Evolution of quantum computing

Table 6 Quantum computing use cases and companies

Company	Use case
Airbus	Aerospace enhancements
Alibaba	Security for e-commerce sites and data centers
Booz Allen Hamilton	Offer for government and business clients
British Telecommunications (BT)	Protect transfer of sensitive information on their network
Google	Artificial intelligence and complex optimization problems
Hewlett Packard (HP)	Small-scale quantum computers and simulators for advanced real-world use cases
IBM	Superconducting circuits coupled with error correction
Intel	Improve advanced manufacturing and better systems for architectural design
KPN	Cryptographic algorithms for secure telecommunications
Lockheed Martin	State of the art software verification and validation
Microsoft	Develop software and hardware for quantum computers
Mitsubishi Electric	Quantum cryptography equipment for mobile communications
NEC & Fujitsu	Quantum encrypted communications over long haul distances
Nokia	Development of quantum computing algorithms
NTT	Quantum-based computer chips relying on photons for information processing
Raytheon	Future sensors, computers, data security, imaging technology
SK Telecom	Quantum secure cryptographic communications in South Korea
Toshiba	Quantum key distribution and secure communication

Today quantum computing faces increasing investments and many companies are involved in quantum computing projects[53] like Airbus, Alibaba, Booz Allen Hamilton, British Telecommunications, Google, Hewlett Packard, IBM, Intel, KPN, Lockheed Martin, Microsoft, Mitsubishi, NEC, Nokia, NTT, Raytheon, SK Telecom and Toshiba (see Table 6).

On top of that, the European Union will launch a 1 billion euros quantum technologies flagship initiative,[54] and the US government, NASA[55] with NASA Quantum Artificial Intelligence Laboratory (QuAIL), NSA[56] and Los Alamos

[53]CB Insights, 18 Corporations Working On Quantum Computing, https://www.cbinsights.com/blog/quantum-computing-corporations-list/?utm_source=CB+Insights+Newsletter&utm_campaign=e2f6321a32-TuesNL_9_13_2016&utm_medium=email&utm_term=0_9dc0513989-e2f6321a32-87759901, 2016, retrieved 2017-01-23.

[54]EC, European Commission, European Commission will launch €1 billion quantum technologies flagship, https://ec.europa.eu/digital-single-market/en/news/european-commission-will-launch-eu1-billion-quantum-technologies-flagship, 2016, retrieved 2017-01-23.

[55]NASA, Nasa Quantum Artificial Intelligence Laboratory (QuAIL), https://ti.arc.nasa.gov/tech/dash/physics/quail/, 2017, retrieved 2017-01-23.

[56]CNET, Farber Dan, NSA working on quantum computer to break any encryption, https://www.cnet.com/news/nsa-working-on-quantum-computer-to-break-any-encryption/, 2014, retrieved 2017-01-23.

Laboratory[57] as well as the Chinese government are all working on building commercial quantum computers. In addition, there are three main private companies investing in quantum computing, D-Wave,[58] Cambridge Quantum Computing,[59] and Quantum Biosystems.[60]

Quantum Internet: An Early Practical Implementation

China launched their first quantum satellite in August 2016.[61] This satellite will provide ultra-secure quantum communications by sending uncrackable keys from space to the ground. Quantum communications gets extremely secure, as any tinkering with a communication would immediately be detected. The two communicating parties share an encryption key that is encoded in a polarization of a string of photons. The Chinese Academy of Sciences and the Austrian Academy of Sciences collaborate in this mission.

One of the key goals of that mission is to perform scientific measurements to prove that entanglement can exist between particles that are separated a distance of 120,000 km, which should actually work as entanglement should be possible at any distances, but this test—the so-called Bell test would prove it.

In total China plans to launch over 20 satellites if the tests go well. This satellite network should be enough to enable secure communications throughout the world. The Austrian researcher Anton Zeilinger, who is a physicist at the Austrian Academy of Sciences in Vienna, is a key person behind this quantum internet that will involve a combination of satellite- and ground-based links. There are of course more possibilities opening up with that technology than secure encryption, for example quantum teleportation in space would allow combining photons from different satellites in order to build a distributed telescope that would be similar in size to the earth. This would allow enormous resolution like being able to read books located on one of Jupiter's moons.

When Will Quantum Computing Be Available?

The secret of building a functioning quantum computer is how to increase the number of qubits that can be linked together. It has been the fragile nature of many existing qubit approaches that restricts successfully networking them so far. But there is a new approach called Topological qubits, which still relies on the extraordinary rare quantum state. Once formed, they behave like sturdy knots and are

[57]Los Alamos National Laboratory, Science of a New Century, Quantum information science and technology research, http://quantum.lanl.gov, 2017, retrieved 2017-01-23.

[58]DWave, https://www.dwavesys.com, 2017, retrieved 2017-03-27.

[59]Cambridge Quantum Computing, http://www.cambridgequantum.com, 2017, retrieved 2017-03-27.

[60]CB Insights, 6 Charts Breaking Down The Nascent Quantum Computing Startup Ecosystem, https://www.cbinsights.com/blog/quantum-computing-startup-ecosystem/, 2016, retrieved 2017-01-23.

[61]QUARTZ, https://qz.com/760804/chinas-new-quantum-satellite-will-try-to-teleport-data-outside-the-bounds-of-space-and-time-and-create-an-unbreakable-code/, 2016, retrieved 2017-03-27.

resistant to disturbances, that would kill the properties of other known kinds of qubits.

While quantum computing (Gribbin, 2013) has so far still not made its breakthrough, it is a tantalizing possibility. Computers could be everywhere, painted on walls, in furniture, in our bodies, allowing constant communication with one another, while requiring no more power than that which they can extract from radio frequencies in the air. Then we would no longer need conventional laptops or cellphones but we would wear them as headbands. Or instead of using computer screens we could directly couple the visual information into the right side of our brain via ultrasonic technology, Sony already holds a patent on this technology. In this new world humans could be made to see, hear, taste, touch or smell anything and get instructions back from the brain via mind reading computers allowing for new communication devices for the disabled.

That this future can become a reality is not a question of if, but rather when it will happen. As we have seen there is already worldwide research in these areas,[62] not only in the US, but also Europe, Japan, India and China. Today's main limitations are the imaginations of software engineers, who need to take computing and networking made available by this new technology and create interfaces simple enough for all human beings to understand. Therefore, it seems reasonable to predict some 40–60 years before quantum computers will reach wide use by everyone.

References

Almadena Chtchelkanova, S. W. (2003). *Magnetic interactions and spin transport*. Arlington: Springer-Science-Business Media, LLC.

Awschalom, D. (2004). *Spin electronics*. Berlin: Springer.

Bellisimo, J. (2015, February 23). What's the future of cognitive computing? IBM Watson. *Forbes/Leadership*.

Bostrom, N. (2014). *Superintelligence. Paths, dangers, strategies*. Oxford: Oxford University Press.

Gribbin, J. (2013). *Computing with quantum cats*. London: Bantam Press.

Jason Jerald, P. (2016). *The VR book*. New York: ACM – Associates for Computing Machinery.

Jerry M. Mendel, D. W. (2010). In D. B. Fogel (Ed.), *Perceptual computing*. Hoboken, Piscataway, NJ: Wiley, IEEE Press Series on Computational Intekkigence.

John E. Kelly III, S. H. (2013). *Smart machines*. New York: Columbia University Press.

Kurzweil, R. (2005). *The singularity is near*. New York: Penguin.

Milburn, G. (1999). *The Feynman processor: Quantum entaglement and the computing revolution*. Reading, MA: Basic Books.

[62]CB Insights, Leaders in quantum computing discuss the challenges and potential for this technology across finance, AI, and many other fields, https://www.cbinsights.com/blog/quantum-computing-investor-commentary/?utm_source=CB+Insights+Newsletter&utm_campaign=c3495e7f52-TuesNL_10_25_2016&utm_medium=email&utm_term=0_9dc0513989-c3495e7f52-87759901, 2016, retrieved 2017-01-23.

Peter Schwartz, C. T. (2006, August 2). Quamntum leap. *Fortune Magazine*.
Porzel, R. (2011). *Contextual computing*. Heidelberg: Springer.
Rheingold, H. (1992). *Virtual reality*. New York: Touchstone.
Weisheng Zhao, G. P. (2015). *Spintronics based computing*. Berlin: Springer.
Wolf, S. A. (2001). Spintronics: A spin-based electronics vision for the future. *Science*, *294*, 1488–1495.

Arrival in the Cloud Century

We become what we behold. We shape our tools and then our tools shape us.

Marshall McLuhan

Since the beginning of industrial revolutions in the eighteenth century, the creation of network-based ecosystems has had a huge impact on economic structures and social developments. Early forerunners like water pipelines or transportation networks proved that networks had leveraged the efficiency of businesses and technical applications. It was up to the scientific revolution as part of the intellectual movement of enlightenment to prepare the ground for disruptive innovations like steam machines, railways, electricity, electronics and wireless communication. Based on these technologies the second and third industrial revolutions were empowered by network-based ecosystems like railroad networks, worldwide telegraph connections, telephone, radio and TV broadcasting. These revolutions created not only new industries but also influenced social changes and supported the formation of new political systems in the twentieth century. Likewise, the nineteenth century was influenced by steam technology and electricity, the twentieth century became the electronic century, generating applications and businesses based on the electronic tube, the transistor and the integrated circuit technology. The success of those technologies was again based on the rollout of network based ecosystems delivering communication and transportation networks and resulting in completely new ecosystems powered by electronic devices, new production methods and new forms of media.

Electronics was also the key technology for the move to digital computers as general usable machines in the 1950s. While steam or electricity amplified the physical strength of humans, digital computers provided the option to scale the mental performance needed for the processing and storage of data. Within the second half of the twentieth century, digital computing became a major driver behind economic growth. Combining digital computing with networks and network base ecosystems was a logical step and led to the first concepts of global computer networks in the 1960s. The basic idea was similar to other types of network-based ecosystems. By distributing and connecting digital processing machine capacity via

M. Oppitz, P. Tomsu, *Inventing the Cloud Century*,
DOI 10.1007/978-3-319-61161-7_21

a network the benefits and usefulness of data processing and data storage would increase.

Though research groups in the USA and Europe soon recognized packet switching as the optimal technical solution, the discussion about a globally accepted standard lasted for more than a decade. Finally, a research group at Stanford University, with its roots in the US Arpanet project, introduced the Internet Protocol (IP) as the widely-accepted standard thus building the technical foundation of the Internet. The first specification was released in 1974 and within one decade the Internet Protocol became the primary standard for global networks. One of the success factors of the IP design was the open collaboration approach within the Internet Engineering Task Force (IETF) based on discussion and consensus within the group also cultivating openness to other concepts. Born as the standard for research computer networks and military use, the Internet Protocol was released as open standard leading to a rapidly growing acceptance by the IT industry and to the creation of the commercial Internet connecting more than 300,000 host computers in 1990. At the beginning of the last decade of the electronic century, the foundation of the Internet as global computer network was vested but still in its infancy.

Parallel to the creation of the Internet, several other revolutionary developments changed the computer industry and led to major paradigm changes between 1970 and 1990. With the introduction of personal computers, a class of low-cost devices became available for a large community, software development based on standardized programming languages became a rapidly growing business of its own and the move to open source software facilitates innovation and development of new products.

The final step on the way to a worldwide network based ecosystem for digital services was made with the introduction of the World Wide Web at the end of 1990 by Tim Berners-Lee at the CERN. Within 10 years, the Web would generate 1 billion users growing to 3.5 billion in 2015 and reaching 50% of the world population. Applications like email and social networks developed into a new form of communication generating competitive offers to traditional media and creating new forms of personal contacts.

The last decade of the twentieth century was characterized by the rollout of the Web to a quickly growing community of more than 400 million Internet users in 2000 having access to more than 17 million websites. This rapid evolution was accelerated by the introduction of web browsers for all different types of computers and the extension of faster network infrastructure and led to a fast-growing economy of new businesses focusing on the provision of services for the new market. At the beginning of the twenty-first century, the mood was heated up to a point where believe in the limitless success of the new economy was stronger than rational analysis. The result was the burst of the dotcom bubble in the first quarter of the new millennium causing a dramatic loss of investments and demanding a restart of the network based new businesses.

The economic shock was heavy but also created a rampant learning curve for the implementation of new business models based on the Web as a network based service ecosystem. It soon became evident that the market was still sound and within the first decade of the twenty-first century many new services were created focusing on search, social networks, media and e-commerce. Disruptive payment models were introduced, many of them based on advertising revenue and powered by a rapidly growing numbers of platform users. The introduction of smart mobile devices, guided by Apples iPhone in 2007, added a new dimension to the network ecosystem and introduced mobility as new paradigm for the default status of users.

The growing acceptance of the Internet and the Web by the consumer market was accompanied by a developing ecosystem focusing on network-based services for enterprises. The new paradigm introduced the idea of sharing resources and was coined "cloud service" being powered by cloud computing infrastructure. Delivering a centralized infrastructure, software or platforms as service became a growing market segment addressing companies as well as private users. This evolution was enabled by the constantly expanding technologies for the virtualization of computing resources. Virtualization of storage, servers, networks or complete data centers became the basic technology for providing cloud services to consumers. The increasing capacity of network infrastructure followed the raising demand for more agility and larger data volumes including graphics, audio and video distribution and storage.

At the end of the first decade of the new millennium, it became evident that the Internet and the Web has started to influence all industry segments and a rapidly growing part of private consumers. The dimension of digitalization is huge: data and information stored and communicated was moved from analog media to digital media reaching more than 80% today, the worldwide capacity of computing and storage resources increased by the factor of 1000 within 10 years, the total Internet traffic increased by the factor of 1000 between 2000 and 2016.

The impact of digitalization on industries is massive and influences internal production processes as well as customer and provider relations. More than that, digitalization is the breeding ground for new disruptive business models replacing traditional businesses. The computer industry itself went through major transitions since the last decade of the twentieth century. New companies born on the web overhauled traditional enterprises in revenue, market cap and growth in a sometimes-dramatic speed. Traditional companies had to undergo transitions from hardware, software or service providers to cloud service companies introducing and following new business models. In parallel new businesses companies started to attack traditional business in their own ecosystem. Today industries like commerce, travel, transportation, energy or health are facing new challenges based on the vast opportunities offered by the Internet and the Web as network-based ecosystem.

Innovation and the quest for new paradigms contribute largely to this evolution and became an industry of its own. Change, transition and innovation gained enormous speed and turned out to be a question of survival for many enterprises. This rapidly accelerating train has passed several gates during the last couple of years: The Internet of Things and Fog Computing extended the range of applications to new areas in industry and private life, big data and analytics technologies expanded the usefulness of information and raised the value of data.

Furthermore, we are still at the beginning of technical, economic and social changes caused by the Internet, the Web and cloud services. Computing and networking capacities are still growing and today's silicon based technologies could be replaced by other technologies like molecular computing, DNA computing or computing and communication technologies based on light or quantum technologies. Faster computers and networks will also form the base for smarter machines. Cognitive computing methods like understanding, reasoning and learning has already found its place in a vast variety of applications influencing many industries and the private life of consumers. With the introduction of cryptocurrencies, the new paradigm of decentralization started to evolve. Applied to other areas, decentralization and the blockchain technology could lead to a digress from centralized cloud services to the distribution of data, processing and trust to the users and consumers causing dramatic changes in the role of intermediaries like financial services, legal administration and authorities.

Hand in hand with economic and technical revolutions the growth of cloud based services influenced social behavior and created impacts on politics. The number of social media users increased to three billion in 2016. Being on the Web became a significant part of daily life for much of individuals. This trend seems to continue and Internet, Web and cloud services are on their way to become a phenomenon covering the whole planets population within the next two or three decades.

Taking the first year of the new millennium as the checkpoint, the economic, social and political environment went through vast transitions since the burst of the dotcom bubble. Based on technical innovations in the last three decades of the twentieth century, Internet, Web and cloud services transferred the electronic century into the cloud century. The voyage started with a big bang and gained speed since then. We are still at the beginning of this journey accelerated by ongoing technical innovation, rising social acceptance, the hope for an increasing quality of life but also challenged by security fears and slow developing alignments of legal rules and regulations. It is evident that cloud services already made huge contributions to quality of life like providing borderless communication and personal networking, offering easy access to products and services, making admission available to Terabytes of information and knowledge and providing availability of storage or computing resources at low costs. Social networks and knowledge platforms support the exchange of new ideas and help to form and organize new communities. Using cloud services together with other innovative technologies is already improving health services, lifesaving emergency procedures and is helping to understand and monitor nature and environment.

Introducing and using this nexus of communication and interaction is also accompanied by the rise of new forms of crime, cyberwarfare and misuse of social media platforms for offenses, fake news, mobbing or organizing terroristic activities. Trust, ethical correctness and the value of privacy became a new extended meaning for individuals, organizations and enterprises when resigning data or personal information to third parties or cloud service providers. Living in a virtual world demands new rules of interaction based on guaranteed trust. The increasing number of privacy violations and damages caused by cybercrime is the major obstacle for a relaxed occupation of this new world.

The question comes up what can be done to enjoy the benefits of technology and avoid the bad sides that could happen. Do we really need to accept every technology or can we avoid some that would provide special risk for our human society and the way we evolved over the past few thousand years? This requires broad measures, one of the most important being to guarantee that our social systems are kept alive healthily. It will become important to judge the introduction of new technologies by independent experts already in advance of using these technologies. International organizations like United Nations or European Union will face the challenge to work together with standard organizations and governments in creating legal systems and the supervision of compliance rules balancing freedom of communication with security and privacy.

Internet, Web and cloud follow the path of other network-based ecosystems which proved to be a successful approach in the past. Compared to railway, telegraph, telephone or broadcasting the impact of a global cloud service based ecosystem is dramatically bigger, it will change our life with so far unknown accelerating speed. Although innovation is the trigger for improvement, we still carry the responsibility to make the right decisions as individuals as well as a society.

Persons

M. Oppitz, P. Tomsu, *Inventing the Cloud Century*,
DOI 10.1007/978-3-319-61161-7

Printed in the United States
By Bookmasters